Results and Problems in Cell Differentiation

Volume 58

More information about this series at http://www.springer.com/series/400

Rafal P. Piprek
Editor

Molecular Mechanisms of Cell Differentiation in Gonad Development

Springer

Editor
Rafal P. Piprek
Institute of Zoology
Jagiellonian University
Krakow
Poland

ISSN 0080-1844 ISSN 1861-0412 (electronic)
Results and Problems in Cell Differentiation
ISBN 978-3-319-31971-1 ISBN 978-3-319-31973-5 (eBook)
DOI 10.1007/978-3-319-31973-5

Library of Congress Control Number: 2016943438

Printed on acid-free paper

This Springer imprint is published by Springer Nature
The registered company is Springer International Publishing AG Switzerland

Preface

The Gonad: A Complex Organ Containing Cells of Diverse Origin

The sex is a set of structural and functional features that classify an individual as a male or female. In males, the testes produce sperm and androgens, which orchestrate the creation of male sex traits. In females, the ovaries produce eggs and estrogens, which drive the female sex features. In the majority of vertebrates, anatomically recognized gonads form late during embryogenesis; thus, the sex of an early embryo is defined exclusively by the presence of sex chromosomes. Interestingly, although the testes and ovaries have different structure and perform diverse functions, they develop from a common sexually undifferentiated bipotential anlage. In addition, the gonads consist of several cell types, which originate and migrate from different embryonic germ layers.

Studies of gonadal development are fascinating because they reveal how the distinct cell lines mix and organize into a unique male or female structure. In this book, we describe how various cell types create the gonad and what is the molecular and cellular machinery driving these processes. Figure 1 summarizes key processes in mouse gonadal development.

Summary of Gonadal Development in Mammals

At the earliest stages of gonadogenesis, the gonadal primordia, termed the genital ridges, are sexually undifferentiated. In all vertebrates, the genital ridges form at the ventral surface of mesonephroi and the first cells in the gonadal primordium originate from the coelomic epithelium (Chap. 1). Mutational analyses revealed a series of genes (e.g., *Gata4*, *Wt1*, *Sf1*, and *Lhx9*) key for the formation of genital ridges in mice. The first morphological sign of gonadogenesis is the thickening

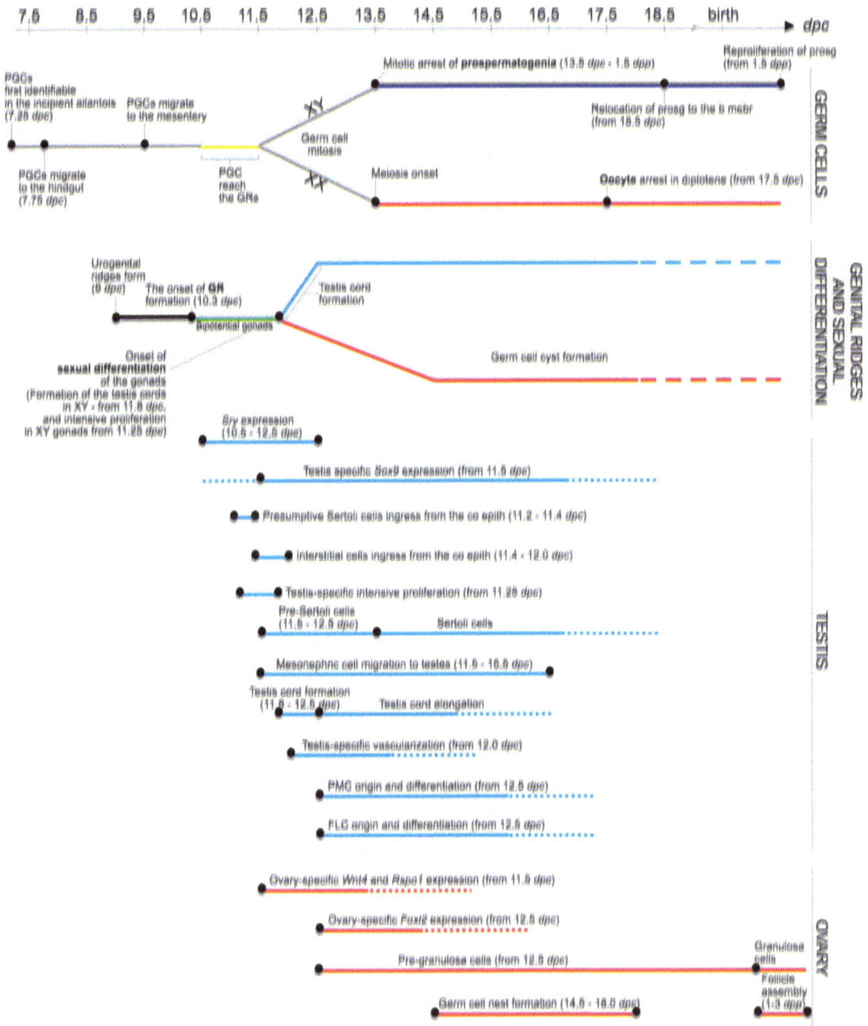

Fig. 1 Mouse timeline for major processes of male and female gonadal development. This diagram summarizes various processes occurring in germ cells and gonadal soma during the fetal development in mice starting from PGC specification to follicle assembly and spermatogonia reproliferation just after birth. Abbreviations: *b member* basement membrane, *co. epith.* coelomic epithelium, *dpc* days *post coitum*, *dpp* days *postpartum*, *FLC* fetal Leydig cells, *GR* genital ridge, *PMC* peritubular myoid cells, *prosg* prospermatogonia

(proliferation) of coelomic epithelium, accompanied by disintegration of its basement membrane. These processes transform the monolayer epithelium into a cluster of coelomic epithelium-derived cells, called the genital ridge. These first gonadal cells, called the gonadal precursor cells or GREL cells, express SF1 protein and during further development of the gonad will differentiate into supporting cells, i.e., the Sertoli cells in the testes and follicular cells (granulosa) in the ovaries that assist the germ cells in the process of spermato- or oogenesis, respectively.

The second cell type present in developing gonad are the primordial germ cells (PGCs) that immigrate to the genital ridges (Chap. 2) from the distant embryonic locations. In mice, the germ line is established by epigenetic mechanisms in extraembryonic tissues before gonads develop and acquires totipotency through genetic and epigenetic regulation of genome function. We discuss the molecular and cellular mechanisms underlying the formation of PGCs in extraembryonic tissues and their migration toward the genital ridges.

After reaching the genital ridges, the PGCs settle and become enclosed by coelomic epithelium-derived gonadal precursor cells. Battle between male sex-determining genes (*Sry*, *Sox9*, and *Fgf9*) and female sex-determining genes (*Rspo1*, *Wnt4*, and *Foxl2*) decides on the differentiation of these somatic cells into Sertoli or follicular cells, i.e., the male or female supporting cells (Chap. 3). Subsequently, in the testes, the Sertoli cells gather into groups enclosing the germ cells; these groups are organized into the testis cords that give rise to the sperm producing seminiferous tubules. In the ovaries, gonadal precursor cells enclose the oocytes and differentiate into granulosa.

The third cell type present in developing gonad are cells immigrating from the embryonic kidney (mesonephros) (Chap. 4). The migration of mesonephros-derived cells seems critical for the formation of testes. In mice, the majority of cells immigrating from the mesonephros give rise to the endothelial cells contributing to the formation of blood vessels that are key for the establishment of testis cord structure and number. The VEGF, PDGF, and neurotrophins seem to be signaling factors crucial for the migration of mesonephros-derived cells to developing gonads. It has been suggested that mesonephros or gonad–mesonephros border region is also a source of the peritubular myoid cells and steroidogenic cells of the gonad.

The next step in gonadogenesis is the differentiation of the steroidogenic cells (Chap. 5). This cell line differentiates into the interstitium of the gonad, i.e., the region located outside of the testis cords/seminiferous tubules. While steroidogenic cells of the male gonad, termed Leydig cells, originate during fetal development, steroidogenic cells of the ovary, termed theca cells, differentiate around ovarian follicles at perinatal stages at the beginning of follicle formation. The origin of steroidogenic cells is still obscure. The potential sources of steroidogenic cells include the coelomic epithelium, mesonephros, adrenogonadal primordium, neural crest cells, gonad–mesonephros border region, and perivascular cells. Several factors (such as DHH and PDGF) are believed to play a role in steroidogenic cell differentiation.

After the onset of sexual differentiation of somatic cells in the gonads, the germ cells commit to the male or female pathways of gametogenesis (Chap. 6). In the developing female gonads, the germ cells enter meiosis just after the beginning of ovarian differentiation; thus, the oocytes form before birth. In the fetal testes, the germ cells are mitotically arrested in the spermatogonial stage, and in the male gonad, the meiosis is triggered during puberty. It seems that retinoic acid is a meiosis-inducing substance in both male and female gonads, but the difference in the pattern of expression of enzymes synthesizing and inactivating retinoic acid in the developing testes and ovaries causes sex-specific differences in meiotic entry.

The final process of ovarian cells' assembly occurs during folliculogenesis (Chaps. 7 and 8). The oocytes in the fetal ovaries gather to form germ cell clusters. Around birth, the clusters break down to form primordial follicles, each containing a single oocyte enclosed by flat follicular (granulosa) cells. Most of the primordial follicles remain quiescent; however, a selected subpopulation develops into primary follicles containing oocytes enclosed by cuboidal granulosa cells. The transition from primary to secondary follicles occurs when a single-layer granulosa proliferates to form a multilayer granulosa and the theca cells differentiate around the secondary follicle. Subsequently, the antrum forms within the secondary follicle. Signaling between oocyte and granulosa regulates oocyte growth and maturation. After ovulation, the remaining granulosa cells and theca cells undergo terminal differentiation into corpus luteum.

The final assembly of cells in developing testes is the formation of elongated tubular structures termed the testis cords containing germ cells enclosed by pre-Sertoli cells (derived from the aggregate of SF1-positive cells). The pre-Sertoli cells commit to epithelial differentiation and polarize and gather into groups surrounding the germ cells. A basement membrane is deposited outside of the cords. Peritubular myoid cells differentiate outside the basement membrane of the testis cords. Soon after birth, the cords transform into seminiferous tubules when a lumen appears inside. Sertoli cells assist in spermato- and spermiogenesis which transform spermatogonial stem cells (SSCs) into spermatozoa (Chaps. 9 and 10). Specific cell junctions of Sertoli cells form a testis–blood barrier that is critical for the process of spermatogenesis.

The continuity of function of the testes and ovaries is ensured by stem cells which, owing to their abilities of self-renewal, constitute a source of somatic and germ cells in developing and adult gonads (Chap. 11). The continuity of spermatogenesis is ensured by the aforementioned spermatogonial stem cells (SSCs). Interestingly, despite the general belief that the number of oocytes is determined during the perinatal period, there is evidence that ovaries may have regenerative capability in adult females, ensured by the presence of female germline stem cells (FGSCs). Among somatic stem cells in the adult testes, stem Leydig cells (SLCs) have been described; however, so far, there is no evidence for the presence of stem cells for Sertoli or peritubular myoid cells. In adult ovaries, granulosa stem cells and thecal stem cells have been found.

In this book, we also present information on the complex molecular machinery regulating gonad development. One of the important regulators are microRNAs (miRNAs), i.e., small noncoding RNAs with a major role in posttranscriptional regulation of gene expression. miRNAs are differentially expressed in the developing male and female gonads, and thus, these molecules may be responsible for sexual differentiation (Chap. 12). miRNAs have crucial roles in gonad development by either directly silencing the expression of proteins in somatic or germ cells or indirectly acting at the hypothalamus–pituitary level. The fundamental role of miRNAs in follicle assembly, growth, differentiation, and ovulation is especially well documented.

The molecular mechanisms driving sex determination and sexual differentiation, i.e., creation of the testis vs. ovarian structure and function, constitute a cascade of factors, and the action of genes involved in these processes is interconnected in a complex network. Any disturbance in this complex cascade can lead to far-reaching consequences such as gonadal dysgenesis or sex reversal. Although a number of human sexual developmental disorders have been traced to various known mutations, the genetic causes for many disorders still remain unknown. The effects of genes involved in sex determination are discussed in Chap. 13.

Much is known about the mechanisms of gonadogenesis; however, the further one delves into this area, the more complicated it becomes. The first descriptions of gonad development made in the nineteenth century were based on a simple histological staining and light microscopy observations. The development of molecular techniques in the second half of the twentieth century greatly contributed to the study of the genetic control of gonadogenesis. In Chap. 14, the history of gonad development studies is presented and techniques that launched work on the origin of cell lineages and the roles of molecular and cellular mechanisms driving the creation of testis vs. ovarian structure are described.

Krakow, Poland Rafal P. Piprek

Contents

Chapter 1
Early Development of the Gonads: Origin and Differentiation of the Somatic Cells of the Genital Ridges

Rafal P. Piprek, Malgorzata Kloc, and Jacek Z. Kubiak

Abstract The earliest manifestation of gonadogenesis in vertebrates is the formation of the genital ridges. The genital ridges form through the transformation of monolayer coelomic epithelium into a cluster of somatic cells. This process depends on increased proliferation of coelomic epithelium and disintegration of its basement membrane, which is foreshadowed by the expression of series of regulatory genes. The earliest expressed gene is *Gata4*, followed by *Sf1*, *Lhx9*, *Emx2*, and *Cbx2*. The early genital ridge is a mass of somatic SF1-positive cells (gonadal precursor cells) that derive from proliferating coelomic epithelium. Primordial germ cells (PGCs) immigrate to the coelomic epithelium even in the absence of genital ridges, e.g., in mouse null mutants for *Gata4*. And conversely, the PGCs are not required for the formation of the genital ridges. After reaching genital ridges, the PGCs become enclosed by somatic cells derived from coelomic epithelium. Subsequently, the expression of sex-determining genes begins and the bipotential gonads differentiate into either testes or ovaries. Gonadal precursor cells, derived from coelomic epithelium, give rise to the somatic supporting cells such as Sertoli cells, follicular cells, and probably also peritubular myoid and steroidogenic cells.

R.P. Piprek (✉)
Department of Comparative Anatomy, Institute of Zoology, Jagiellonian University, Gronostajowa 9, 30-387 Krakow, Poland
e-mail: rafal.piprek@uj.edu.pl

M. Kloc
Department of Surgery, The Houston Methodist Hospital, Houston, TX, USA

The Houston Methodist Research Institute, Houston, TX, USA

J.Z. Kubiak
CNRS, UMR 6290, Institute of Genetics and Development of Rennes, Cell Cycle Group, 35043 Rennes, France

Université Rennes 1, UEB, UMS Biosit, Faculty of Medicine, 35043 Rennes, France

R.P. Piprek (ed.), *Molecular Mechanisms of Cell Differentiation in Gonad Development*, Results and Problems in Cell Differentiation 58,
DOI 10.1007/978-3-319-31973-5_1

1.1 The Earliest Events of Gonadogenesis

The gonads initially form as sexually bipotential structures termed the genital ridges (gonadal ridges). Such precursors of testes and ovaries arise at the ventromedial surface of the embryonic kidneys (mesonephroi) as twofolds running along both sides of the dorsal mesentery (Brambell 1927; Gropp and Ohno 1966; Pelliniemi 1975; Wartenberg et al. 1991). The formation of the primordial gonads starts with the increasing proliferation of the coelomic epithelial cells within a strictly defined site. This leads to the transformation of a monolayer epithelium into a dense and pseudostratified layer underlined by the basement membrane that subsequently disintegrates (Fig. 1.1a, b) (Hu et al. 2013; Karl and Capel 1998; Kusaka et al. 2010; Paranko 1987). Due to disintegration of the basement membrane, the epithelial cells are able to migrate inward and form multilayered structure, whose cells undergo epithelial–mesenchymal transition (EMT). The quintessence of the genital ridges formation is the differentiation of the monolayer coelomic epithelium into multilayered thickening to which the primordial germ cells immigrate. The molecular mechanisms driving coelomic epithelium transformation into the genital ridge remain unclear.

In mice, the first morphological sign of the genital ridge formation—appearance of the coelomic epithelium thickening and disintegration of the basement membrane underneath—is noticeable at 10.3–10.4 *dpc* (days *post coitum*), i.e., stage 5–6 ts (tail somite stages) (Hu et al. 2013). Kusaka et al. (2010) showed that as early as at 10.25 *dpc*, the coelomic epithelial cells are already able to migrate and begin formation of thickened gonadal anlage. In the literature there is a discrepancy in the timing of genital ridge formation onset, which may reflect diversity in the rate of mouse development. Some studies indicate that the genital ridges begin development at 9.0, 9.5, 10.0, or 10.5 *dpc* (Chen et al. 2012; Hacker et al. 1995; Karl and Capel 1995; Nef and Parada 2000; Tanaka and Nishinakamura 2014). Thus, the onset of gonadal development occurs soon after the intermediate mesoderm starts to differentiate into the pronephros and mesonephros, which takes place at 9.5 *dpc*. Just after 10 *dpc*, the genital ridges start to develop at the surface of the anterior part of the mesonephroi. It has been postulated that the gonads share common primordium with adrenal glands (AGP, adrenogonadal primordium) (Fig. 1.2). In mouse, AGP forms as a thickening of coelomic epithelium as early as 9.5 *dpc*, and then by 10.5 *dpc*, the AGP splits into adrenal cortex primordium (anterior) and gonadal primordium (posterior) (Bandiera et al. 2013; Hatano et al. 1994; Ikeda et al. 1994).

The thickening of the coelomic epithelium begins at the anterior part of a mesonephros at 10.3 *dpc*. Then, the process of the genital ridge formation precedes toward the posterior half of the mesonephros. At 10.4 *dpc*, the coelomic epithelium is multilayered in the anterior part; however, the middle and posterior part of a future genital ridge is still single layered (Hu et al. 2013). Meantime, the primordial germ cells (PGCs) migrate to the genital ridges from the basis of allantois via the hindgut and mesentery. They settle down in the genital ridges between 10.0 and 11.5 *dpc* (Fig. 1.1c) (Gomperts et al. 1994; Molyneaux et al. 2001). From 11.5 *dpc*

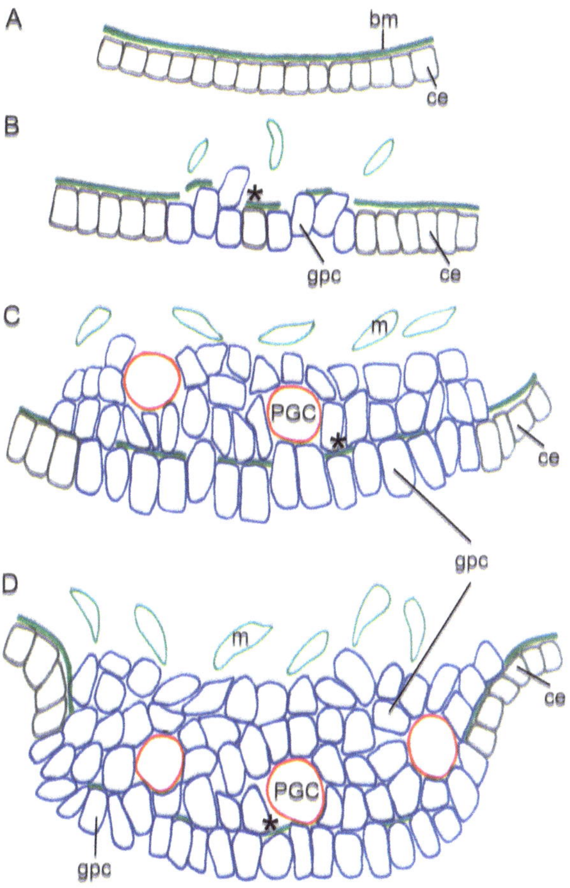

Fig. 1.1 Mouse genital ridge development. (**a**) Coelomic epithelium (ce) lined by a basement membrane (bm). (**b**) Some coelomic cells lose epithelial features, transform into SF1-positive gonadal precursor cells (gpc), and ingress through a disintegrating basement membrane. (**c**) The early genital ridge forms a cluster of coelomic epithelium-derived cells; PGCs settle among SF1-positive gonadal precursor cells; fragments of the basement membrane (*asterisk*) present near the surface of the genital ridge. (**d**) The genital ridge grows and is not covered by a true epithelial layer; mesonephros-derived cells (m) immigrate to the genital ridges

onward, the sex-specific features appear in the gonads signaling the beginning of the sexual differentiation of the gonads. In human fetus, the genital ridges form between 4.5th and fifth week of gestation and remain sexually undifferentiated by the seventh week (Francavilla et al. 1990). At the end of the fifth and during the sixth gestational week, human primitive gonads are colonized by PGCs.

1.2 Genetic Mechanisms Initiating Genital Ridge Development

The molecular control of genital ridge formation is not well understood. However, studies on mutant mice have provided key information on the genes regulating initiation of gonadogenesis. Genes participating in the regulation of the genital ridge formation were summarized in Table 1.1. It is clear that the first molecular

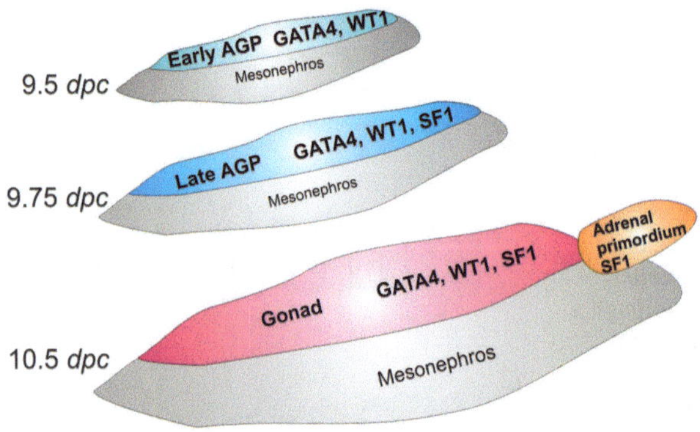

Fig. 1.2 Mouse adrenogonadal primordium (AGP) development. In early AGP at 9.5 *dpc*, the *Gata4* and *Wt1* are expressed. In late AGP at 9.75 *dpc*, *the Sf1* expression begins. At 10.5 *dpc* the gonad and adrenal primordium split; expression of *Gata4*, *Wt1*, and *Sf1* continues in the developing gonads, whereas *Sf1* is expressed in the adrenal primordium (modified from Bandiera et al. 2013)

signs of the genital ridge formation precede the earliest morphological manifestation of the genital ridge. One of the first expressed genes is *Wt1* (Wilms' tumor suppressor 1), which regulates formation of the genital ridge and kidney. Starting from 9.5 *dpc*, *Wt1* is expressed throughout the whole urogenital ridges, i.e., in the future gonads and developing kidneys. *Wt1* gene encodes 24 isoforms of zinc-finger transcription factor, among which a short isoform WT1-KTS is a key for the development of the genital ridges. Mice lacking functional WT1-KTS have impaired genital ridges due to increased cell death (Hammes et al. 2001; Kreidberg et al. 1993). *Gata4* (GATA-binding factor 4) is the earliest expressed gene specific for the genital ridge development. In mice, the expression of this zinc-finger transcription factor begins at 10.0 *dpc* (26–27 total somite stage) in the anterior half of the nascent genital ridge and then gradually extends posteriorly within the coelomic epithelium (Hu et al. 2013). In the posterior half, the expression of *Gata4* starts at 10.2 *dpc* (2 ts). The expression of *Gata4* is followed by the thickening of the coelomic epithelium in anteroposterior direction and leads to the formation of genital ridges. In mutant mouse embryos, the loss of *Gata4* expression results in single-layered and unthicken coelomic epithelium and thus in the total lack of genital ridges (Hu et al. 2013). The loss of the *Gata4* expression leads to decreased BrdU incorporation, which suggests that lack of *Gata4* abrogates proliferation of coelomic epithelium. Additionally, in *Gata4*-deficient mouse embryos, the basement membrane underlying coelomic epithelium does not disintegrate (Hu et al. 2013). In wild-type mice, the *Gata4* expression is followed by the expression of other genital ridge markers such as *Sf1* (SF1, steroidogenic factor 1, also known as Nr5a1, nuclear receptor subfamily 5, group A, member 1 *s.* Ad4BP—adrenal 4-binding protein), *Lhx9* (LIM homeobox 9), and *Emx2* (empty spiracles homeobox 2) (Birk et al. 2000; Kusaka et al. 2010; Luo et al. 1994; Miyamoto et al. 1997, 2008). These genes differ slightly in the site of their expression. *Gata4* is expressed not only in the nascent genital ridges but also in their vicinity, i.e., in the adjacent mesenchyme, mesentery, and gut endoderm (Hu et al. 2013). *Emx2* is expressed in the coelomic epithelium at the site of genital ridge formation, adjacent mesenchyme, nephric duct, and mesonephric tubules (Kusaka et al. 2010). *Wt1* is expressed in the whole urogenital ridges (gonads and mesonephroi) from 9.5 *dpc*. *Lhx9* is expressed in the genital ridges and partially in the mesentery (Hu et al. 2013). In contrast, the expression of the orphan nuclear receptor *Sf1* is restricted to the genital ridges. Thus, the SF1 marks the identity of true gonadal somatic precursor cells. The expression of *Sf1* begins at 10.2 *dpc* (2 ts) along the anterior half of the mesonephros and then proceeds posteriorly (Hu et al. 2013). There is also a striking pattern of gene expression in the AGP, i.e., the common primordium of the gonads and adrenal gland (Fig. 1.2). In the early AGP (9.5 *dpc*), the *Wt1* and *Gata4* are expressed. In the late AGP (9.75 *dpc*), *Sf1* is expressed along with *Wt1* and *Gata4*. After AGP splits into gonadal primordium and adrenal primordium (10.5 *dpc*), the *Sf1*, *Wt1*, and *Gata4* remain expressed in the genital ridge, and only *Sf1* is expressed in the adrenal primordium (Bandiera et al. 2013). Although the timing of gene expression pattern described by Bandiera et al. (2013) differs slightly from the timing described by Hu et al. (2013), undoubtedly the differential expression of these genes is a key for development of gonads and adrenal glands from common primordium. Later in development, in sexually differentiating gonads, *Sf1* expression ceases in the coelomic epithelium covering the genital ridge but remains in the pre-Sertoli cells and fetal Leydig cells at 12.5 *dpc* and later persists only in the fetal Leydig cells (Ikeda 1996). In the developing ovaries, *Sf1* expression ends by 13.0 *dpc*.

In contrast to *Gata4* null mutants, in which the genital ridges do not form at all, in the *Wt1*, *Sf1*, *Lhx9*, and *Emx2* null mutant mice, the genital ridges start to form but degenerate about 12.5 *dpc*. The *Sf1*, *Wt1*, and *Emx2* mutants show increased cell death in genital ridges, while *Lhx9* mutants show disrupted cell proliferation (Table 1.1; Birk et al. 2000; Bland et al. 2004; Hammes et al. 2001; Kusaka et al. 2010; Luo et al. 1994; Mazaud et al. 2002). WT1-KTS (splice form lacking KTS tripeptide) is required for genital ridge formation through regulation of *Sf1* promoter in cooperation with LHX9. The expression of *Sf1* is also regulated by *Pod1* (podocyte expressed-1, basic helix-loop-helix transcription factor). POD1 represses *Sf1* expression (Cui et al. 2004; Tamura et al. 2001). *Pod1* null mutation leads to the ectopic expression of *Sf1* in gonads and mesonephroi and to the formation of hypoplastic gonads lacking the testis cords or ovarian follicles (Cui et al. 2004). Other factors involved in regulation of the genital ridge formation belong to insulin/insulin-like growth factors (IGFs). In mice lacking insulin receptor (*Insr*) and insulin-like growth factor 1 receptor (*Igf1r*), development of the gonads is impaired due to decreased *Sf1* expression and reduced proliferation of somatic cells in the genital ridges (Pitetti et al. 2013). *Emx2* null mutants also show reduced expression of *Sf1* (Kusaka et al. 2010). In *Emx2* mutants, cell migration from coelomic epithelium through the basement membrane and thus thickening of the coelomic epithelium and the formation of multilayer epithelium in the genital ridges are impaired (Kusaka et al. 2010). All these data indicate that *Gata4* is

Table 1.1 Genes participating in regulation of the early gonadogenesis and genital ridge formation

Gene symbol	Gene name	Function/role	Site and time of expression in the gonads	Loss-of-function mutation effects in mouse gonads	Loss-of-function mutation effects in human gonads	Reference
Cbx2 (*M33*)	Chromobox homolog 2	Component of the polycomb multiprotein complex	Genital ridges	Hypoplastic gonads	Gonadal dysgenesis	Katoh-Fukui et al. (2012)
Emx2	Empty spiracle homeobox 2	Homeodomain transcription factor/gonadal development	Coelomic epithelium in the site of genital ridge development, adjacent mesenchyme, nephric duct, and mesonephric tubules	Genital ridge degeneration due to the increase in cell death	Gonadal dysgenesis	Miyamoto et al. (1997); Kusaka et al. (2010)
Gata4	GATA-binding factor 4	Zinc-finger transcription factor/ gonadal development	Coelomic epithelium in the nascent genital ridges from 10.0 *dpc*, adjacent mesenchyme, mesentery, gut	No genital ridges due to lack of coelomic epithelium proliferation and thickening	Normal	Hu et al. (2013); Miyamoto et al. (2008)
Insr, *Igf1r*	Insulin receptor, insulin-like growth factor 1 receptor	Receptors of insulin/IGF signaling	Coelomic epithelium descendants and germ cells	Genital ridge impaired due to decreased cell proliferation	NR	Pitetti et al. (2013)
Lhx9	LIM homeobox 9	Homeodomain transcription factor/gonadal development	Coelomic epithelium in the genital ridges and partially the mesentery	Genital ridge degeneration due to the impaired cell proliferation	NR	Birk et al. (2000); Mazaud et al. (2002)
Pod1	Podocyte expressed-1	Helix-loop-helix transcription factor	Coelomic epithelium of the genital ridge and the boundary between the gonad and mesonephros	Ectopic expression of *Sf1* and hypoplastic gonads	NR	Tamura et al. (2001); Cui et al. (2004)

Sf1 (*Nr5a1*)	Steroidogenic factor 1, nuclear receptor 5a1	Nuclear receptor/ gonad and steroidogenic gland development	Coelomic epithelium from 10.2 *dpc* and descendants	Genital ridge degeneration due to the increase in cell death	Dysgenetic or absent gonads	Luo et al. (1994); Bland et al. (2004)
Six1, *Six4*	Sine oculis homeobox homologs 1 and 4	Homeobox transcription factors	Coelomic epithelium in the nascent genital ridges from 9.5 *dpc*	Disrupted coelomic epithelial cell ingression, decreased SF1-positive gonadal progenitor cells, and decreased gonad size	NR	Fujimoto et al. (2013)
Wt1 (-KTS isoform)	Wilms' tumor factor 1	Zinc-finger transcription factor/ gonad and kidney development	Urogenital ridges from 9.5 *dpc* in mice	Genital ridge degeneration due to the increase in cell death	Dysgenetic gonads	Kreidberg et al. (1993); Hammes et al. (2001)

NR not reported

required for the initiation of genital ridge formation, whereas *Sf1*, *Wt1*, *Lhx9*, *Emx2*, and insulin/IGF signaling are necessary for the maintenance of genital ridges and their further development (Fig. 1.3a). The expression of all these genes is interdependent. For example, the expression of *Sf1* is significantly downregulated in *Gata4*-, *Wt1*-, *Lhx9*-, or *Emx2*-deficient genital ridges (Hu et al. 2013; Kusaka et al. 2010; Wilhelm and Englert 2002). In *Gata4* null mutants, *Wt1* and *Emx2* but not *Sf1* or *Lhx9* are expressed in the genital ridges (Hu et al. 2013). *Lhx9* expression depends on the *Gata4* but is not altered by the lack of *Sf1*, *Wt1*, and *Emx2* (Fig. 1.3a). The lack of *Wt1* and *Emx2* expression does not impair *Gata4* expression or the thickening of coelomic epithelium. Functional genetic experiments have revealed a complex network of interactions between the genes expressed in the primitive gonads (Fig. 1.3a). Interestingly, in mice, the set of these genes involved in the formation of the genital ridges is also responsible for the regulation of male sex-determining genes, such as *Sry* and *Sox9* (Piprek 2009a). The described above genes seem to be evolutionarily conserved among vertebrates; they are also expressed in developing gonads of zebra fish, tilapia, *Xenopus laevis*, *Trachemys scripta*, alligator, and chicken (Barske and Capel 2010; Kawano et al. 2001; Kent et al. 1995; Li et al. 2012; Oreal et al. 2002; Smith et al. 1999).

Some of the homeodomain proteins may also be engaged in the early development of gonads. These genes are key regulators of the body plan and thus may be important for spatiotemporal patterning of gonadal development. *Hoxa10*, *Hoxa9*, *Hoxa11*, and *Hoxa13* expressions have been detected in the urogenital system (Taylor et al. 1997) and thus presumably may be responsible for precise determination of the site of genital ridge formation. Mutation of *Hoxa11* affects gonadal development in mice, leading to both male and female sterility (Hsieh-Li et al. 1995). Other homeodomain proteins SIX1 and SIX4 (sine oculis homeobox homologs 1 and 4) are expressed from the onset of the genital ridge formation. Double null mutation of these genes leads to reduction in *Sf1* expression, delayed and decreased epithelial–mesenchymal transition (EMT), and disrupted coelomic epithelial cell ingression during the formation of the genital ridges, decreased SF1-positive gonadal precursor cells, and decreased gonad size (Fujimoto et al. 2013). It has been shown that SIX1 and SIX4 transactivate *Sf1* promoter (Fujimoto et al. 2013). Similarly, *Pbx1* (pre-B-cell leukemia homeobox 1) contributes to regulation of genital ridge development via upregulation of proliferation of SF1-positive cells (Schnabel et al. 2003). Mice with null mutation in *Pbx1* display decreased cell proliferation in genital ridges and highly reduced expression of *Sf1* resulting in limited expansion of SF1-positive cells and decreased gonadal growth (Schnabel et al. 2003). In addition, a polycomb protein CBX2 (chromobox homolog 2, *M33*) may play a role in regulation of HOX gene expression in the developing urogenital system. Mice lacking functional *Cbx2* show defects in gonadal development; decreased *Lhx9*, *Sf1*, and *Gata4* expression; male-to-female sex reversal; and hypoplastic gonads of both sexes (Katoh-Fukui et al. 2012). Interestingly, *Gata4*, which is the first gene expressed specifically in the genital ridges, is expressed in the coelomic epithelium located only along the mesonephros, and its expression does not extend caudally on the surface of the metanephros. It remains to be seen how the

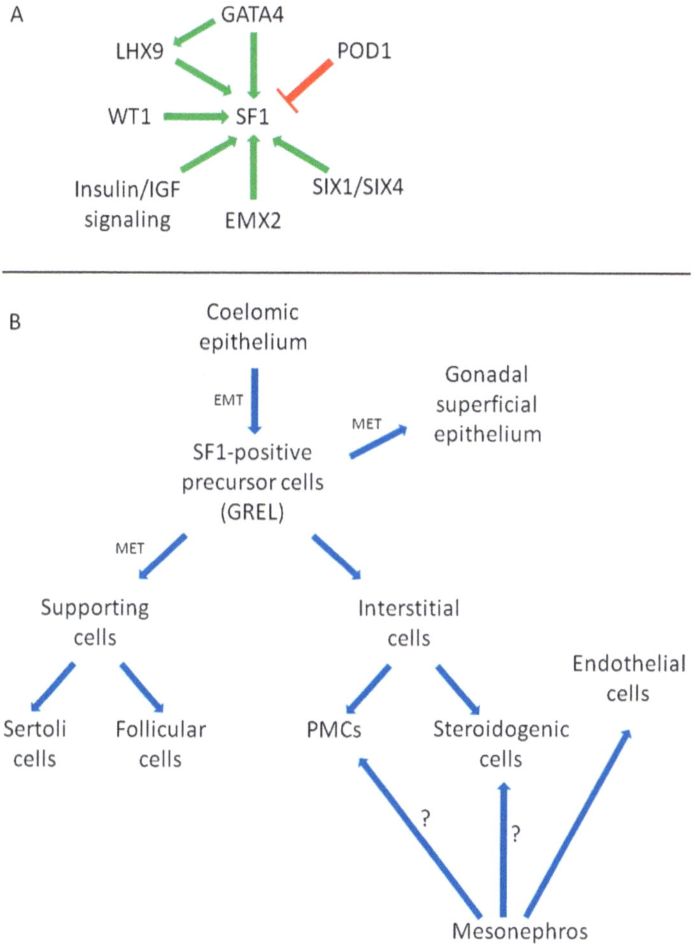

Fig. 1.3 (**a**) Interactions between genes regulating genital ridge development. *Green arrows* indicate upregulation and *red arrow* indicates downregulation. (**b**) Diagram of gonadal cell lineage origin, fate, and differentiation

genital ridge formation becomes restricted only to the site running along the mesonephros and if the mesonephros induces gonadal development.

The mentioned above genes differ not only in the spatiotemporal pattern of expression but also in their role in the genital ridge formation. It has been suggested that (1) *Gata4*, *Six1*, and *Six4* regulate formation of SF1-positive gonadal precursor cells in the coelomic epithelium; (2) *Lhx9*, *Wt1*-KTS, and IGF influence *Sf1* expression and promote genital ridge formation; (3) *Emx2* and possibly *Six1* and *Six4* contribute to EMT and cell ingression in forming genital ridges; and (4) after formation of the coelomic epithelium thickening, the *Cbx2* and *Pod1* modulate *Sf1* expression regulating cell proliferation and differentiation of the genital ridges (Tanaka and Nishinakamura 2014).

Interestingly, expression of these genes is not necessary for PGC migration and colonization. In mutants lacking *Lhx9*, *Sf1*, *Wt1*, or *Emx2* expression, the PGCs are able to colonize genital ridges. In the *Gata4* null mutants, the genital ridges do not form; nevertheless, PGCs immigrate to the monolayer coelomic epithelium at the surface of the mesonephroi (Hu et al. 2013). This shows that the presence of genital ridges is not required for the PGC migration, guidance, or settlement. It has been postulated that developing gonads secrete SDF1 factor that attracts PGCs (Doitsidou et al. 2002; Molyneaux et al. 2003). However, it must be emphasized that in *Gata4* null embryos, the SDF1 continues to be secreted by the mesentery and mesonephros starting from 9.0 *dpc*, and probably because of this, the PGCs can still migrate to the site of presumed genital ridges. On the other hand, the PGCs are not required for the initiation or development of the genital ridges (McLaren 1991).

1.3 Cell Proliferation in Early Gonadal Development

The gonads' development depends not only on cell proliferation but also on cell migration from the adjacent mesonephric mesenchyme. The primordial gonads initially derive from, at least, three cell lines: coelomic epithelium-derived cells, mesonephros-derived cells, and germ cells. These three cell lines give rise to a multitude of cell types in adult testes and ovaries. In early mouse gonads, the most intensive proliferation has been described for coelomic epithelium-derived cells, which cover genital ridges (Schmahl et al. 2000; Schmahl and Capel 2003). In addition, the most robust cell divisions have been reported for the XY gonads (genetic males) indicating the importance of cell proliferation for early stages of testis development. There are two phases of male-specific proliferation in developing gonads. The earliest phase of proliferation affects SF1-positive cells in coelomic epithelium that will give rise to the precursors of Sertoli cells (pre-Sertoli cells) and possibly also other cell lineages (Fig. 1.3b). This phase takes place between 11.0 and 11.5 *dpc* and is promoted by fibroblast growth factor 9 (FGF9) acting via FGFR2 (Karl and Capel 1998; Piprek 2010; Schmahl et al. 2000). *Fgfr2* null mutants show decreased proliferation of somatic cells in XY gonads and impaired differentiation of Sertoli cells (Kim et al. 2007; Piprek 2010). It is not clear if Sertoli cells originate from SF1-positive gonadal precursor cells preexisting in the genital ridges or from cells that ingress inward the genital ridges at later stages. Lineage tracing indicated that Sertoli cells in mice originate from cells that enter the interior of the genital ridges from the superficial layer within 2-h window of time between 11.2 and 11.4 *dpc* (Karl and Capel 1998; Schmahl et al. 2000). After *Sf1* expression in the coelomic epithelium ceases, the precursors of Sertoli cells stop dividing, but SF1-negative cells of coelomic epithelium continue to proliferate. Cells originating in this later phase of coelomic epithelium proliferation (11.4–12.5 *dpc*) give rise to interstitial cells only. This later phase of male-specific proliferation is promoted by PDGF rather than by FGF9, since the loss of *Pdgfrα* receptor disrupts only this latter phase of proliferation and does not influence the early proliferation of

SF1-positive cells (Brennan et al. 2003; Piprek 2010). As a result of enhanced proliferation of somatic cells in the male gonads, the differentiating testis becomes almost twice the size of the ovary. Proliferation in developing ovaries is not as robust as in the male gonads, and additionally the follicular cells probably originate from coelomic epithelium-derived cells ingressing at the earliest stages of gonadogenesis, which give rise to FOXL2-positive granulosa of medullary follicles, but also during later development (around birth) from LGR5-positive cells giving rise to the granulosa of cortical follicles (Mork et al. 2012).

1.4 Cellular Events in Genital Ridge Formation in Mice and Other Vertebrates

The ventral surface of the urogenital ridges is covered by coelomic epithelium (monolayer of cuboidal cells in the mouse or flat cells in the frog *X. laevis*) and underlined by a thick basement membrane separating the epithelium from the mesonephric mesenchyme and nephrons (Fig. 1.1). As it was mentioned above, the first cellular sign of genital ridge formation is disintegration of the basement membrane under coelomic epithelium at the site of gonadal development. In mouse, the fragments of disintegrating the basement membrane are visible beneath the epithelium-like layer at the surface of genital ridges (Karl and Capel 1998). Disintegration of the basement membrane under the epithelium covering the genital ridges was also observed in human, prosimian *Galago*, swine, and bovine (Hummitzsch et al. 2013; Pelliniemi et al. 1998; Pereda et al. 2001; Satoh 1991; Yoshinaga et al. 1988). The disintegration of the basement membrane occurs through the action of extracellular matrix (ECM) digesting enzymes such as metalloproteinases. It would be interesting to study the ECM-remodeling enzymes during the formation of the genital ridge and to test how these enzymes influence early gonadogenesis. Disintegration of the basement membrane allows the proliferating cells of the coelomic epithelium to ingress and form a cluster of SF1-positive gonadal somatic precursor cells constituting the genital ridge (Fig. 1.1). In bovine embryos these coelomic-derived gonadal somatic precursor cells have been named the GREL cells (genital ridge epithelial-like cells) (Hummitzsch et al. 2013). In summary, during the primordial gonad formation, the coelomic epithelial cells of the genital ridges proliferate, partially lose their epithelial features, and transform into gonadal precursor cells (GREL cells) and thus undergo EMT transition and move inward (Hummitzsch et al. 2013; Karl and Capel 1998; Schmahl and Capel 2003). It is still unknown if these superficial cells actively migrate or passively ingress due to the high rate of proliferation. The enhanced divisions of superficial cells (indicated by high incorporation of BrdU) of the genital ridges lead to the growth of gonadal primordium (Schmahl and Capel 2003). Superficial cell labeling of the genital ridges with fluorescent lipophilic dye DiL in mice or with a mixture of MitoTracker and rhodamine derivative in a red-eared slider turtle *Trachemys scripta* showed the ingression of coelomic epithelium-derived cells inward into the gonadal primordium (Karl and Capel 1998; Yao et al. 2004). This proved the origin of the first gonadal cells from the coelomic epithelium. The onset of gonadogenesis, which is marked by the formation of coelomic epithelium thickening, is similar in mouse, bovine, chicken, and *T. scripta* and is probably characteristic for all amniotes. However, *T. scripta* and chicken early gonadogenesis is an exception due to the lack of basement membrane disintegration under proliferating epithelium. Thus, in these species the increase in cell number does not lead to the formation of an unorganized cell mass but to the formation of the fingerlike sex cords covered by the basement membrane (Fig. 1.4). The genital ridge formation is slightly different in lower vertebrates such as *X. laevis*. Here the first sign of gonadal development occurs when a monolayer coelomic epithelium composed of flat cells bulges into the coelomic cavity forming mounds containing PGCs (Wylie and Heasman 1976; Fig. 1.5). Interestingly, in amphibians the coelomic epithelium does not form thickenings at the onset of genital ridge development, which is characteristic for amniotes.

Studies of the early gonadal development have been facilitated by identification of molecular markers. Several molecular markers have been identified following the microarray analysis of gene expression in the gonadal precursor cells (GREL cells) of bovine embryos (Hummitzsch et al. 2013). GREL cells differentiating from typical epithelial cells of gonadal surface still express some epithelial-specific markers such as cytokeratins 18 and 19, plakophilin 2, and desmoglein 2, but also express StAR, which marks the steroidogenic cells. However, GREL cells do not express genes specific for gonadal surface cells such as adipophilin, fibulin 2, merocin, or mucin 1. This implies that GREL cells are intermediate cell type between the coelomic epithelium cells and the mesenchymal cells.

Early studies indicated that the developing gonad is covered by a superficial epithelium, which proliferates and gives rise to the somatic cells of the gonad (Karl and Capel 1998), but later studies showed that developing gonad is not covered by a true epithelium. A definitive epithelium is always composed of a layer of polarized cells lying on a basement membrane; however, the genital ridge is a mass of cells and is not organized into the cell layer (Figs. 1.1 and 1.6). This was clearly showed in the studies on early gonadogenesis in bovine embryo (Hummitzsch et al. 2013). The gonadal hilum (basal region) is the only region of the primitive gonad covered by a true epithelium and is continuous with the mesonephric superficial epithelium. This region of gonadal epithelium is a source of the LGR5-positive stem cells that are also present in adult ovaries and play an important role in ovarian surface regeneration after ovulation (Flesken-Nikitin et al. 2013).

The early genital ridge is built of a cluster of somatic cells among which the primordial germ cells settle (Fig. 1.1). Later, the stromal cells migrate from the mesonephric mesenchyme into the developing genital ridge (Figs. 1.1 and 1.6) (Capel et al. 1999). Consequently, the primordial gonads are composed of cells of various origins. The stromal (mesenchymal-like) cells derived from mesonephroi invade the genital ridges, which in mice occurs at 11.5 *dpc* and in bovine between 70 and 130 days of gestation (Hummitzsch et al. 2013; Tilmann and Capel 1999).

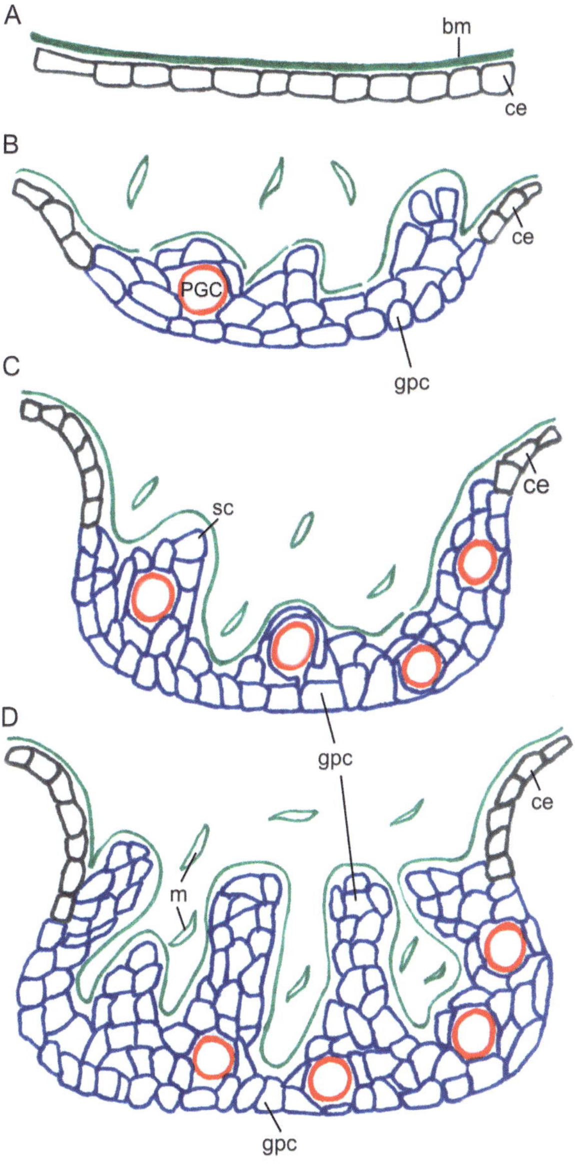

Fig. 1.4 Diagram of genital ridge development in the turtle *Trachemys scripta* and chicken. (**a**) Coelomic epithelium (ce) lined by the basement membrane (bm). (**b**) Coelomic epithelial cells transform into SF1-positive gonadal precursor cells (gpc or GREL cells) and ingress forming fingerlike cords covered by continuous basement membrane. (**c**) Proliferation of the gonadal precursor cells leads to the growth of primitive sex cords (sc) that contain primordial germ cells (PGCs). (**d**) Sex cords (sc) grow; in the undifferentiated gonads, the germ cells are located in the peripheral, cortical region. Mesonephros-derived cells (m) invade the genital ridges and locate between the sex cords

The mesonephric cell migration was also observed in amphibians but not in the turtle *T. scripta* (Piprek et al. 2010; Yao et al. 2004). In bovine embryo the stromal cells migrating from the mesonephroi have a mesenchymal character and are surrounded by extracellular matrix containing collagen type I, fibrillin 1, fibronectin, and decorin. In contrast, the extracellular matrix of GREL cells consists of collagen types IV and XVIII, laminin, perlecan, and nidogen (Hummitzsch et al. 2013). Both the coelomic epithelium-derived and mesonephros-derived

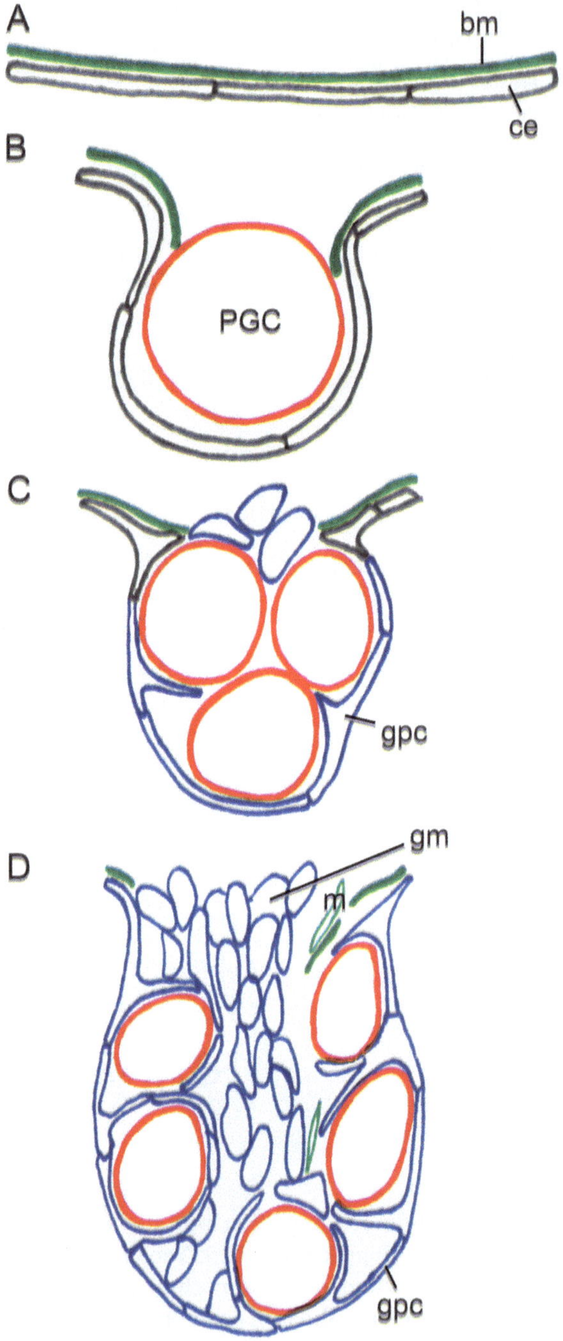

Fig. 1.5 Diagram of genital ridge development in the African clawed frog *Xenopus laevis*. (**a**) Coelomic epithelium (ce) lined by the basement membrane (bm). (**b**) PGC settlement results in formation of the genital ridges that bulge to the coelomic cavity. (**c**) Coelomic epithelial cells transform into gonadal precursor cells (gpc) that proliferate, ingress, and enclose the PGCs. (**d**) Coelomic epithelium-derived cells locate in the center of the genital ridges forming the gonadal medulla (gm). Mesonephros-derived stromal cells (m) ingress between the medulla and superficial region (cortex)

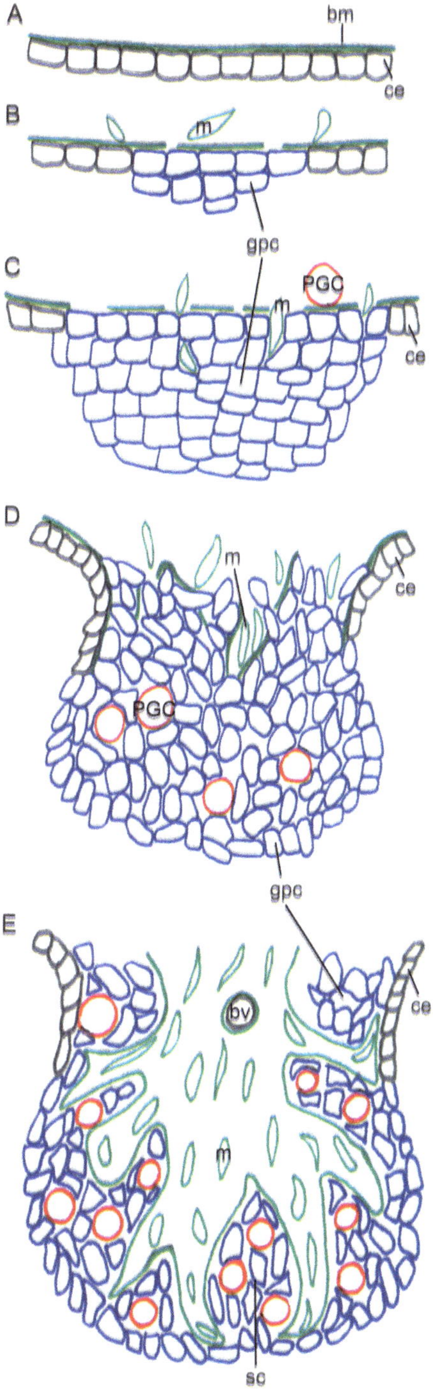

Fig. 1.6 Diagram of genital ridge development in the bovine embryo. (**a**) Coelomic epithelium (ce) lined by the basement membrane (bm). (**b**) Coelomic epithelial cells transform into SF1-positive gonadal precursor cells (gpc) also termed GREL cells that proliferate and form a cluster; the basement membrane disintegrates. (**c**, **d**) PGCs and mesonephros-derived cells (m) invade the genital ridges. (**e**) The basement membrane forms at the interface between the mesonephros-derived and GREL cells, and the sex cords (sc) develop; blood vessels (bv) form in the stromal region (modified from Hummitzsch et al. 2013)

stromal cells produce ECM in the primitive gonads. The basement membrane forms at the interface between the GREL and stromal cells, which contributes to the establishment of gonadal structure. The stromal cells penetrate the gonads toward their surface separating the sex cords. Migrating stromal cells reach the space under the surface of the gonads and induce differentiation of the superficial cells into the true epithelium covering the gonads.

The formation of sex cords is a first sign of compartmentalization in developing gonads. Within the sex cords, the coelomic epithelium-derived gonadal precursor cells (GREL cells) enclose germ cells, and afterward they differentiate into Sertoli cells in the developing testes or follicular (granulosa) cells in the developing ovaries (Fig. 1.6). In mice, the sex cords do not form in the undifferentiated gonads; they form only in the differentiating testes. The formation of testis cords depends on the cell migration from the mesonephros (Combes et al. 2009). In bovine, the sex cords (testis and ovigerous cords) develop in XX and XY gonads when the mesonephros-derived cells invade the gonads. Subsequently, the basement membranes form at the interface of GREL and stromal cells. This indicates that the mesonephros-derived cells and timing of their migration are the keys for the organization of gonadal structure. As it was mentioned before, in *T. scripta* and chicken, the sex cords appear earlier in development when the cells of thickened coelomic epithelium proliferate, which leads to the growth of the fingerlike primitive sex cords protruding inward the genital ridges from the thickened coelomic epithelium (Smith and Sinclair 2004; Yao et al. 2004) (Fig. 1.4). In these species the sex cords are covered by the basement membrane that is continuous with the basal membrane of the coelomic epithelium.

In amphibians, the structure of the primitive gonads is quite unique; the sex cords are not present, the germ cells remain in the peripheral (cortical) position in connection with the superficial cells of the gonads, and the proliferating somatic cells gather in the center of the primitive gonad forming a sterile medulla (Fig. 1.5). The medullary cells originate from the coelomic epithelium (Falconi et al. 2004; Iwasawa and Yamaguchi 1984; Merchant-Larios 1979; Merchant-Larios and Villalpando 1981; Piprek et al. 2010; Tanimura and Iwasawa 1988, 1989). Later in development, mesenchymal cells immigrate from the mesonephroi into the gonads and locate between the gonadal cortex and medulla. Eventually the basement membrane forms at the interface of the coelomic epithelium-derived cells and the stromal cells (Fig. 1.5).

The fate of the bipotential gonads is established in the sex determination process that occurs in sexually undifferentiated gonads. The molecular mechanism determining sexual differentiation of the gonads has been described in Chap. 3 of this volume. In mice, in XY gonads the SF1-positive gonadal precursor cells derived from the coelomic epithelium start expressing *Sry* gene followed by the expression of *Sox9* gene (Piprek 2009a). These two genes determine the male sex in mammals and trigger the differentiation of gonadal precursor cells into Sertoli cells that enclose the germ cells and gather into the testis cords, which later in development give rise to the seminiferous tubules (Combes et al. 2009). In XX murine, gonads *Rspo1* and *Wnt4* are upregulated; here, the SF1-positive gonadal precursor cells gather into the ovarian follicles in which they enclose (as a follicular or granulosa cells) a single oocyte (Piprek 2009b; Yao et al. 2004). Thus, Sertoli and follicular cells share common origin (Albrecht and Eicher 2001). It seems that the ingressing coelomic epithelial cells first undergo EMT (they lose epithelial features to form a cluster of somatic cells inside the genital ridges) and then some of these cells commit to MET (mesenchymal–epithelial transition) when they regain epithelial features differentiating into supporting cells. Nevertheless, gonadal precursor cells do not have typical mesenchymal character; and they are rather an intermediate stage between epithelial and mesenchymal cells. In bovine, unlike in mice, both in XX and XY gonads, the somatic gonadal precursor (GREL) cells initially gather into sex cords (Hummitzsch et al. 2013). In XY bovine gonads, GREL cells in the cords differentiate into the Sertoli cells, and the cords develop into seminiferous tubules; however, in XX gonads, GREL cells give rise to follicular cells and the sex cords are fragmented into ovarian follicles.

Interestingly, the supporting cells are not the only cells derived from coelomic epithelium in the primitive gonads. Cell-tracking experiment revealed that the SF1-positive gonadal precursor cells are also present between the sex cords and probably give rise to the interstitial lineages (Fig. 1.3b) (Karl and Capel 1998). Figure 1.3b depicts origin and differentiation of gonadal cell lineages. Originally, it has been postulated that the interstitial cells (including steroidogenic cells) originate from the stromal cells migrating from the mesonephros (Buehr et al. 1993; Martineau et al. 1997; Merchant-Larios et al. 1993). Studies on mouse testis development showed that the mesonephros-derived cells give rise mainly to the endothelial cells and thus the gonadal vasculature; however, the interstitial cells probably originate from the coelomic epithelium (Combes et al. 2009; Cool et al. 2008). It has been shown that two interstitial cell lineages, the peritubular myoid cells (PMCs) and the fetal Leydig cells (FLCs), and possibly also the adult Leydig cells (ALCs) originate from the SF1-positive coelomic epithelium-derived cells (Barsoum et al. 2013; Brennan et al. 2003; Combes et al. 2009; Cool et al. 2008). Importantly, Liu et al. (2015) showed that steroidogenic cells in developing mouse ovaries have a dual origin: coelomic epithelial and mesonephric. It seems that some of the theca cells originate from WT1-positive cells migrating from the coelomic epithelium, and others differentiate from Gli1-positive mesenchymal cells migrating from the mesonephros (Liu et al. 2015). This may indicate that the coelomic epithelium-derived and mesonephros-derived cells form a pool of mixed multipotent precursor cells that randomly commit to different cell fates in the genital ridges. However, it is more probable that the genital ridge is not an unorganized mass of multipotent cells that will differentiate into various gonadal cell lines but that the fate of gonadal cells is determined early: for example, cells originating from the first wave of coelomic epithelium proliferation give rise to the supporting cells, whereas later wave of coelomic epithelium proliferation gives rise to steroidogenic cells. Thus, timing of cell origin may determine the fate of the cells. The origin of gonadal cell fates and the mechanisms of their differentiation still require further studies.

1.5 Conclusion

The analysis of early gonadogenesis reveals a sequence of processes leading to the formation of the bipotential gonad anlage, which differentiates into the ovaries or testes. The first step of gonadal development determines the site of genital ridge formation. Although the molecular mechanism of this process is still obscure, studies of mutant mice have pointed to a group of genes involved in this earliest step of gonadal development. At the site of gonadogenesis, the cells of the coelomic epithelium of the ventromedial surface of mesonephroi undergo transformation; they proliferate, the underlying basement membrane disintegrates, and these result in the formation of genital ridges. The basement membrane disintegration depends on activation of ECM-remodeling enzymes such as metalloproteinases. The proliferating cells leave the coelomic epithelium, lose their epithelial features, undergo EMT, and invade the interior of the primitive gonad either individually (mice, frogs) or in the form of cords (reptiles, birds). The cells in the site of gonadogenesis secrete chemoattractants, which guide migrating PGCs toward the genital ridges. PGCs settle and associate with the cells of the thickened coelomic epithelium. The genital ridges are not required for PGC immigration to the site of gonadogenesis, and PGCs are not needed for genital ridge formation. In the genital ridges, the cells of the coelomic epithelium transform into a cluster of SF1-positive cells (termed GREL cells in bovine embryos), and thus the genital ridges are not covered by true epithelium. In all studied vertebrates, the cells derived from the coelomic epithelium give rise to the supporting cells and presumably also to other cell types, such as Leydig and theca cells. The cells of the developing gonads segregate and the gonad compartmentalizes. In amphibians, distinct cortex and medulla appear in the developing gonad. In reptiles and birds, the thickened coelomic epithelium form the cortex from which the sex cords protrudes constituting the medullary region of the gonad in both sexes. In mice, the testis cords and ovarian follicles emerge from the cell cluster just after the sex determination period. Importantly, the gonads originate not only from the cells deriving from the coelomic epithelium but also from the mesonephric mesenchyme, and a specific sequence of changes is cell proliferation, cell adhesion, cell movement, aggregation, deposition of the basement membrane, and cell differentiation that lead to the formation of the testes or ovaries.

Grant Support RPP was supported by the project financed by the National Science Centre assigned on the basis of the decision number DEC-2013/11/D/NZ3/00184.

References

Albrecht KH, Eicher EM (2001) Evidence that *Sry* is expressed in pre-Sertoli cells and Sertoli and granulosa cells have a common precursor. Dev Biol 240:92–107

Bandiera R, Vidal VPI, Motamedi FJ (2013) WT1 maintains adrenal-gonadal primordium identity and marks a population of AGP-like progenitors within the adrenal gland. Dev Cell 27:5–18

Barske LA, Capel B (2010) Estrogen represses SOX9 during sex determination in the red-eared slider turtle *Trachemys scripta*. Dev Biol 341:305–314
Barsoum IB, Kaur J, Ge RS, Cooke PS, Yao HH (2013) Dynamic changes in fetal Leydig cell populations influence adult Leydig cell populations in mice. FASEB J 27:2657–2666
Birk OS, Casiano DE, Wassif CA, Cogliati T, Zhao L, Grinberg A, Huang S, Kreidberg JA, Parker KL, Porter FD, Westphal H (2000) The LIM homeobox gene Lhx9 is essential for mouse gonad formation. Nature 403:909–913
Bland ML, Fowkes RC, Ingraham HA (2004) Differential requirement for steroidogenic factor-1 gene dosage in adrenal development versus endocrine function. Mol Endocrinol 18:941–952
Brambell FWR (1927) The development and morphology of the gonads of the mouse—Part I. The morphogenesis of the indifferent gonad and of the ovary. Proc R Soc Lond B Biol Sci 101:391–409
Brennan J, Tilmann C, Capel B (2003) *Pdgfr-α* mediates testis cord organization and fetal Leydig cell development in the XY gonad. Genes Dev 17:800–810
Buehr M, Gu S, McLaren A (1993) Mesonephric contribution to testis differentiation in the fetal mouse. Development 117:273–281
Capel B, Albrecht KH, Washburn LL, Eicher EM (1999) Migration of mesonephric cells into the mammalian gonad depends on *Sry*. Mech Dev 84:127–131
Chen H, Palmer JS, Thiagarajan RD, Dinger ME, Lesieur E, Chiu H, Schulz A, Spiller C, Grimmond SM, Little MH, Koopman P, Wilhelm D (2012) Identification of novel markers of mouse fetal ovary development. PLoS One 7(7):e41683
Combes AN, Wilhelm D, Davidson T, Dejana E, Harley V, Sinclair A, Koopman P (2009) Endothelial cell migration directs testis cord formation. Dev Biol 326:112–120
Cool J, Carmona FD, Szucsik JC, Capel B (2008) Peritubular myoid cells are not the migrating population required for testis cord formation in the XY gonad. Sex Dev 2:128–133
Cui S, Ross A, Stallings N, Parker KL, Capel B, Quaggin SE (2004) Disrupted gonadogenesis and male-to-female sex reversal in Pod1 knockout mice. Development 131:4095–4105
Doitsidou M, Reichman-Fried M, Stebler J, Koprunner M, Dorries J, Meyer D, Esguerra CV, Leung T, Raz E (2002) Guidance of primordial germ cell migration by the chemokine SDF-1. Cell 111:647–659
Falconi R, Dalpiaz D, Zaccanti F (2004) Ultrastructural aspects of gonadal morphogenesis in *Bufo bufo* (Amphibia Anura) 1. Sex differentiation. J Exp Zool A 301:378–388
Flesken-Nikitin A, Hwang CI, Cheng CY, Michurina TV, Enikolopov G, Nikitin AY (2013) Ovarian surface epithelium at the junction area contains a cancer-prone stem cell niche. Nature 495:241–245
Francavilla S, Cordeschi G, Properzi G, Concordia N, Cappa F, Pozzi V (1990) Ultrastructure of fetal human gonad before sexual differentiation and during early testicular and ovarian development. J Submicrosc Cytol Pathol 22:389–400
Fujimoto Y, Tanaka SS, Yamaguchi YL, Kobayashi H, Kuroki S, Tachibana M, Shinomura M, Kanai Y, Morohashi K, Kawakami K, Nishinakamura R (2013) Homeoproteins Six1 and Six4 regulate male sex determination and mouse gonadal development. Dev Cell 26:416–430
Gomperts M, Wylie C, Heasman J (1994) Primordial germ cell migration. Ciba Found Symp 182:121–134
Gropp A, Ohno S (1966) The presence of a common embryonic blastema for ovarian and testicular parenchymal (follicular, interstitial and tubular) cells in cattle Bos taurus. Z Zellforsch Mikrosk Anat 74:505–528
Hacker A, Capel B, Goodfellow P, Lovell-Badge R (1995) Expression of *Sry*, the mouse sex determining gene. Development 121:1603–1614
Hammes A, Guo JK, Lutsch G, Leheste JR, Landrock D, Ziegler U, Gubler MC, Schedl A (2001) Two splice variants of the Wilms' tumor 1 gene have distinct functions during sex determination and nephron formation. Cell 106:319–329
Hatano O, Takakusu A, Nomura M, Morohashi K (1994) Identical origin of adrenal cortex and gonad revealed by expression profiles of Ad4BP/SF-1. Genes Cells 1:663–671

Hsieh-Li HM, Witte DP, Weinstein M, Branford W, Li H, Small K, Potter SS (1995) Hoxa 11 structure, extensive antisense transcription, and function in male and female fertility. Development 121:1373–1385

Hu YC, Okumura LM, Page DC (2013) Gata4 is required for formation of the genital ridge in mice. PLoS Genet 9(7):e1003629

Hummitzsch K, Irving-Rodgers HF, Hatzirodos N et al (2013) A new model of development of the mammalian ovary and follicles. PLoS ONE 8(2), e55578

Ikeda Y (1996) SF-1: a key regulator of development and function in the mammalian reproductive system. Acta Paediatr Jpn 38:412–419

Ikeda Y, Shen WH, Ingraham HA, Parker KL (1994) Developmental expression of mouse steroidogenic factor-1, an essential regulator of the steroid hydroxylases. Mol Endocrinol 8:654–662

Iwasawa H, Yamaguchi K (1984) Ultrastructural study of gonadal development in *Xenopus laevis*. Zool Sci 1:591–600

Karl J, Capel B (1995) Three-dimensional structure of the developing mouse genital ridge. Philos Trans R Soc Lond B Biol Sci 350:235–242

Karl J, Capel B (1998) Sertoli cells of the mouse testis originate from the coelomic epithelium. Dev Biol 203:323–333

Katoh-Fukui Y, Miyabayashi K, Komatsu T, Owaki A, Baba T, Shima Y, Kidokoro T, Kanai Y, Schedl A, Wilhelm D, Koopman P, Okuno Y, Morohashi K (2012) Cbx2, a polycomb group gene, is required for Sry gene expression in mice. Endocrinology 153:913–924

Kawano K, Furusawa S, Matsuda H, Takase M, Nakamura M (2001) Expression of steroidogenic factor-1 in frog embryo and developing gonad. Gen Comp Endocrinol 123:13–22

Kent J, Coriat AM, Sharpe PT, Hastie ND, van Heyningen V (1995) The evolution of WT1 sequence and expression pattern in the vertebrates. Oncogene 11:1781–1792

Kim Y, Bingham N, Sekido R, Parker KL, Lovell-Badge R, Capel B (2007) Fibroblast growth factor receptor 2 regulates proliferation and Sertoli differentiation during male sex determination. Proc Natl Acad Sci U S A 104:16558–16563

Kreidberg JA, Sariola H, Loring JM, Maeda M, Pelletier J, Housman D, Jaenisch R (1993) WT-1 is required for early kidney development. Cell 74:679–691

Kusaka M, Katoh-Fukui Y, Ogawa H, Miyabayashi K, Baba T, Shima Y, Sugiyama N, Sugimoto Y, Okuno Y, Kodama R, Iizuka-Kogo A, Senda T, Sasaoka T, Kitamura K, Aizawa S, Morohashi K (2010) Abnormal epithelial cell polarity and ectopic epidermal growth factor receptor (EGFR) expression induced in Emx2 KO embryonic gonads. Endocrinology 151:5893–5904

Li J, Chen W, Wang D, Zhou L, Sakai F et al (2012) GATA4 is involved in the gonadal development and maturation of the teleost fish tilapia, *Oreochromis niloticus*. J Reprod Dev 58:237–242

Liu C, Peng J, Matzuk MM, Yao HH (2015) Lineage specification of ovarian theca cells requires multicellular interactions via oocyte and granulosa cells. Nat Commun 6:6934

Luo X, Ikeda Y, Parker KL (1994) A cell-specific nuclear receptor is essential for adrenal and gonadal development and sexual differentiation. Cell 77:481–490

Martineau J, Nordqvist K, Tilmann C, Lovell-Badge R, Capel B (1997) Male-specific cell migration into the developing gonad. Curr Biol 7:958–968

Mazaud S, Oreal E, Guigon CJ, Carre-Eusebe D, Magre S (2002) *Lhx9* expression during gonadal morphogenesis as related to the state of cell differentiation. Gene Expr Patterns 2:373–377

McLaren A (1991) Development of the mammalian gonad: the fate of the supporting cell lineage. Bioessays 13:151–156

Merchant-Larios H (1979) Origin of the somatic cells in the rat gonad: an autoradiographic approach. Ann Biol Anim Biochim Biophys 19:1219–1229

Merchant-Larios H, Villalpando I (1981) Ultrastructural events during early gonadal development in *Rana pipiens* and *Xenopus laevis*. Anat Rec 199:349–360

Merchant-Larios H, Moreno-Mendoza N, Buehr M (1993) The role of the mesonephros in cell differentiation and morphogenesis of the mouse fetal testis. Int J Dev Biol 37:407–415
Miyamoto N, Yoshida M, Kuratani S, Matsuo I, Aizawa S (1997) Defects of urogenital development in mice lacking Emx2. Development 124:1653–1664
Miyamoto Y, Taniguchi H, Hamel F, Silversides DW, Viger RS (2008) A GATA4/WT1 cooperation regulates transcription of genes required for mammalian sex determination and differentiation. BMC Mol Biol 9:44
Molyneaux KA, Stallock J, Schaible K, Wylie C (2001) Time-lapse analysis of living mouse germ cell migration. Dev Biol 240:488–498
Molyneaux KA, Zinszner H, Kunwar PS, Schaible K, Stebler J, Sunshine MJ, O'Brien W, Raz E, Littman D, Wylie C, Lehmann R (2003) The chemokine SDF1/CXCL12 and its receptor CXCR4 regulate mouse germ cell migration and survival. Development 130:4279–4286
Mork L, Maatouk DM, McMahon JA, Guo JJ, Zhang P, McMahon AP, Capel B (2012) Temporal differences in granulosa cell specification in the ovary reflect distinct follicle fates in mice. Biol Reprod 86:37
Nef S, Parada LF (2000) Hormones in male sexual development. Genes Dev 14:3075–3086
Oreal E, Mazaud S, Picard JY, Magre S, Carre-Eusebe D (2002) Different patterns of anti-Mullerian hormone expression, as related to DMRT1, SF-1, WT1, GATA-4, Wnt-4, and Lhx9 expression, in the chick differentiating gonads. Dev Dyn 225:221–232
Paranko J (1987) Expression of type I and III collagen during morphogenesis of fetal rat testis and ovary. Anat Rec 219:91–101
Pelliniemi LJ (1975) Ultrastructure of gonadal ridge in male and female pig embryos. Anat Embryol (Berl) 147:20–34
Pelliniemi LJ, Frojdman K, Sundstrom J, Pollanen P, Kuopio T (1998) Cellular and molecular changes during sex differentiation of embryonic mammalian gonads. J Exp Zool 281:482–493
Pereda J, Pozo J, Motta PM (2001) Is there a mesonephric cell contribution to the gonadal primordium before sexual differentiation in humans?: an ultrastructural study. Ital J Anat Embryol 106:143–154
Piprek RP (2009a) Genetic mechanisms underlying male sex determination in mammals. J Appl Genet 50:347–360
Piprek RP (2009b) Molecular mechanisms underlying female sex determination—antagonism between female and male pathway. Folia Biol 57:105–113
Piprek RP (2010) Molecular machinery of gonadal differentiation in mammals. Int J Dev Biol 54:779–786
Piprek RP, Pecio A, Szymura JM (2010) Differentiation and development of gonads in the yellow-bellied toad, *Bombina variegata* L. 1758 (Amphibia: Anura: Bombinatoridae). Zool Sci 27:47–55
Pitetti JL, Calvel P, Romero Y, Conne B, Truong V, Papaioannou MD, Schaad O, Docquier M, Herrera PL, Wilhelm D, Nef S (2013) Insulin and IGF1 receptors are essential for XX and XY gonadal differentiation and adrenal development in mice. PLoS Genet 9:e1003160
Satoh M (1991) Histogenesis and organogenesis of the gonad in human embryos. J Anat 177:85–107
Schmahl J, Capel B (2003) Cell proliferation is necessary for the determination of male fate in the gonad. Dev Biol 258:264–276
Schmahl J, Eicher EM, Washburn LL, Capel B (2000) Sry induces cell proliferation in the mouse gonad. Development 127:65–73
Schnabel CA, Selleri L, Cleary ML (2003) Pbx1 is essential for adrenal development and urogenital differentiation. Genesis 37:123–130
Smith CA, Sinclair AH (2004) Sex determination: insights from the chicken. Bioessays 26:120–132
Smith CA, Smith MJ, Sinclair AH (1999) Expression of chicken steroidogenic factor-1 during gonadal sex differentiation. Gen Comp Endocrinol 113:187–196

Tamura M, Kanno Y, Chuma S et al (2001) Pod-1/Capsulin shows a sex- and stage-dependent expression pattern in the mouse gonad development and represses expression of Ad4BP/SF-1. Mech Dev 102:135–144

Tanaka SS, Nishinakamura R (2014) Regulation of male sex determination: genital ridge formation and Sry activation in mice. Cell Mol Life Sci 71:4781–4802

Tanimura A, Iwasawa H (1988) Ultrastructural observations on the origin and differentiation of somatic cells during gonadal development in the frog *Rana nigromaculata*. Dev Growth Differ 30:681–691

Tanimura A, Iwasawa H (1989) Origin of somatic cells and histogenesis in the primordial gonad of the Japanese tree frog *Rhacophorus arboreus*. Anat Embryol 180:165–173

Taylor H, Vanden Heuvel GB, Igarashi P (1997) A conserved Hox axis in the mouse and human female reproductive system: late establishment and persistent adult expression of the Hoxa cluster genes. Biol Reprod 57:1338–1345

Tilmann C, Capel B (1999) Mesonephric cell migration induces testis cord formation and Sertoli cell differentiation in the mammalian gonad. Development 126:2883–2890

Wartenberg H, Kinsky I, Viebahn C, Schmolke C (1991) Fine structural characteristics of testicular cord formation in the developing rabbit gonad. J Electron Microsc Tech 19:133–157

Wilhelm D, Englert C (2002) The Wilms tumor suppressor WT1 regulates early gonad development by activation of Sf1. Genes Dev 16:1839–1851

Wylie CC, Heasman J (1976) The formation of the gonadal ridge in Xenopus laevis. I. A light and transmission electron microscope study. J Embryol Exp Morphol 35:125–138

Yao HH, DiNapoli L, Capel B (2004) Cellular mechanisms of sex determination in the red-eared slider turtle, *Trachemys scripta*. Mech Dev 121:1393–1401

Yoshinaga K, Hess DL, Hendrickx AG, Zamboni L (1988) The development of the sexually indifferent gonad in the prosimian, *Galago crassicaudatus crassicaudatus*. Am J Anat 181:89–105

Chapter 2
The Formation and Migration of Primordial Germ Cells in Mouse and Man

Massimo De Felici

Abstract In most multicellular organisms, including mammals, germ cells are at the origin of new organisms and ensure the continuation of the genetic and epigenetic information across the generations.

In the mammalian germ line, the primordial germ cells (PGCs) are the precursors of the primary oocytes and prospermatogonia of fetal ovaries and testes, respectively. In mammals such as the primates, in which the formation of the primary oocytes is largely asynchronous and occurs during a relatively long period, PGCs after the arrival into the XX gonadal ridges are termed oogonia which then become primary oocytes when entering into meiotic prophase I. In the fetal testes, germ cells derived from the PGCs after gonad colonization are termed prospermatogonia or gonocytes.

One of the most fascinating aspect of the mammalian germline development is that it is probably the first cell lineage to be established in the embryo by epigenetic mechanisms and that these inductive events happen in extraembryonic tissues much earlier that gonad develop inside the embryo proper. Moreover, such events prepare the germ cells for totipotency through genetic and epigenetic regulations of their genome function. How this occurs remained a mystery until short time ago.

In this chapter, I will report and discuss the most recent advances in the cellular and molecular mechanisms underlying the formation in extraembryonic tissues and migration of PGCs toward the gonadal ridges made primarily by studies carried out in the mouse with some perspective in the human. Established concepts about these processes will be only summarized when necessary since they are widely described and discussed in many excellent reviews; most of them are cited in the text below.

M. De Felici (✉)
Dipartimento di Biomedicina e Prevenzione, Università di Roma Tor Vergata,
Via Montpellier, 1, 00133 Roma, Italy
e-mail: defelici@uniroma2.it

R.P. Piprek (ed.), *Molecular Mechanisms of Cell Differentiation in Gonad Development*, Results and Problems in Cell Differentiation 58,
DOI 10.1007/978-3-319-31973-5_2

2.1 Introduction

The formation of PGCs in mammals occurs through cell "induction" (or "epigenesis") by extracellular signaling in contrast to other vertebrate groups such as anurans, in which it occurs through "preformation" driven by the presence of germline cytoplasmic determinants in the egg. As a rule, the signaling induction resulting in germline specification occurs early in the embryo in the epiblast during the pregastrulation period. Subsequently, the PGC precursors move into extraembryonic tissues where the PGC' fate is determined; finally PGCs move to the sexually indifferent gonadal ridges (GRs) developing inside the embryo proper. It has become increasingly evident that the repression of the somatic differentiation programs is a key event in germline specification. Moreover, in mammals, a wide genome epigenetic reprogramming occurs in PGCs during the journey toward the GRs and after their arrival into the developing gonads. While the repression of the somatic programs likely establishes a latent pluripotency in PGCs preparing the germ cells for totipotency, the genome reprogramming is probably necessary to eliminate epimutations and erase the parental imprinting in the germline (reviewed in De Felici 2011).

In the last decades, studies aimed to characterize the phenotype of mouse embryos with natural or induced mutations of genes encoding certain growth factors, adhesion molecules, and transcription factors have allowed elucidating some basic mechanisms governing germline specification in such species representing the most studied model of mammalian development. Moreover, analyses of single-cell gene expression in selected embryonic populations have identified key regulators and global transcription events associated with mouse PGC specification (reviewed in Hayashi et al. 2007; Irie et al. 2015a).

Although the mouse is the key mammalian model for germline studies, some aspects of human PGC development might be distinct from that in mice. Actually, some embryonic tissues involved in PGC development in mice have no clear counterpart in the human. For example, there is no apparent structure in the human embryo equivalent to the mouse extraembryonic ectoderm, which, as reported in the next sections, plays a crucial role in the specification of the mouse PGCs via BMP4 signaling. Nevertheless, the data available to date testify important progresses made in unraveling the basic steps of PGC formation in the mouse model and the rapidly ongoing information for human germ cells. The available results indicate that mouse and human PGC specification occurs by epigenesis and depends on common and distinct players acting within diverse tissue context. In vitro models of derivation of PGC-like cells from stem cells appear promising to study human PGC specification and determination and to elucidate some of these differences. If and how such differences may have a critical effect on PGC specification mechanisms remain to be clarified.

Although it is not exactly clear when PGCs undergo irreversible determination, PGCs formed in extraembryonic tissues are subsequently incorporated into the hindgut and move inside the embryo proper (Chiquoine 1954; Ginsburg et al. 1990; Lawson and Hage 1994; Mikedis and Downs 2012). Several cell types migrate during development, the length and varied terrain that PGCs traverse is, however, unrivaled. Available literature suggests that basic principles of cell migration are valid also for PGCs. These include their dependence on receptor tyrosine kinases for motility and a combination of chemokine signaling via G-protein-coupled receptors and cell–cell/cell–extracellular matrix adhesion interactions for directionality.

Two mechanisms have been proposed to explain the movement of PGCs from the allantois/yolk sac wall to the GRs. PGCs are considered migratory cells with active motility capability or, on the contrary, are thought to become passively enclosed in the endoderm of the wall of the hindgut and then transported by morphogenic movement and proliferative expansion of this tissue and the underlying mesenchyme into or very near the GRs. Alternatively, PGC translocation might result from a combination of active movement and passive transport by tissue morphogenesis (Freeman 2003).

In the chapter, I will report and discuss the most recent advances in the cellular and molecular mechanisms underlying the crucial events of germ cell development outline above, the formation, and migration of PGCs in the mouse and human embryo.

2.2 Specification and Determination of Primordial Germ Cells

2.2.1 Origin of Mouse PGCs

In the mouse embryo, putative PGCs were first localized as a small population of 50–100 cells positive for alkaline phosphatase (AP) within the posterior region of the embryo around 8.5 days *post coitum* (*dpc*) (Chiquoine 1954). Using an enhanced histochemical method, AP-positive PGCs were identified to form a distinct cluster of about 40 cells at the base of the incipient allantois in the extraembryonic mesoderm (ExM) at around 7.25 *dpc* (Ginsburg et al. 1990; Figs. 2.1a and 2.3). Such cells were subsequently shown to originate from a few (about 6) of the most proximal epiblast (PEpi) cells, immediately adjacent to the extraembryonic ectoderm (ExE) (Lawson and Hage 1994).

Several in vivo and in vitro studies focused on local molecules able to induce PGC formation and showed that, prior to and at the beginning of gastrulation, bone morphogenic proteins (BMPs, members of the TGFβ family) secreted from the ExE (BMP4, BMP8b) and visceral endoderm (VE) (BMP2) ultimately converged on proximal epiblastic (PEpi) cells to promote PGC specification (Lawson et al. 1999; Pesce et al. 2002; Ying et al. 2000, 2001; Ying and Zhao 2001). B lymphocyte-induced maturation protein 1 (BLIMP1, also known as PRDM1), a PRDI-BF1 and RIZ (PR) domain-containing transcriptional regulator, is likely downstream BMP4

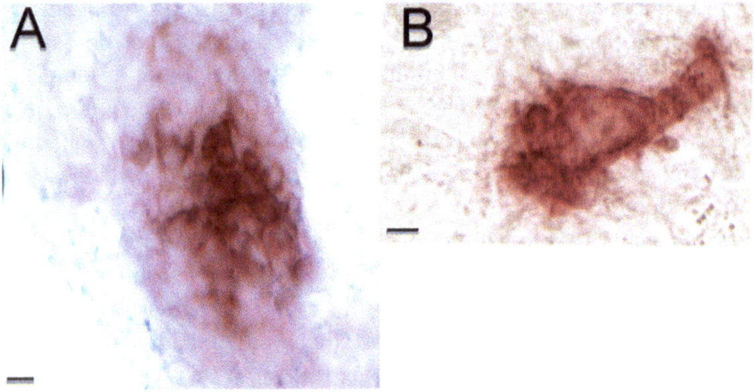

Fig. 2.1 (**a**) A cluster of 40–50 AP-positive cells in the allantois of 8.5 *dpc* mouse embryo, these cells likely constitute the initial population of fate-specified PGCs which subsequently move inside the embryo proper to colonize the gonadal ridges. *Bar* approximately 20 μm. (**b**) A single human AP-positive PGCs isolated from a GR of a human embryo at the 5th–6th week of gestation showing motile features onto a mouse fibroblast monolayer. *Bar* approximately 2.5 μm

induction and marks the origin of the PGC precursors in the epiblast around 6.25 *dpc* (Ohinata et al. 2005). Because such putative PGCs were identified as a cluster, it was hypothesized that direct cell–cell contact, perhaps mediated by E-cadherin, contributed to their specification (Okamura et al. 2003).

Other studies indicated that WNT3 (wingless-type MMTV integration site family, member 3) signals originating from the posterior visceral endoderm (PVE) and epiblast conferred a receptive environment for PEpi cells to interpret BMP4 activity (Ohinata et al. 2009). Yet, BMP signals alone could not explain why only a subset of PE cells is induced to a germ cell fate. Thus, it was proposed that both positive and negative signals directed PGC fate. An interesting model emerged that depicts a complex interaction between yet unidentified inhibitory signals originating from the anterior visceral endoderm (AVE) and positive influences from BMP4, BMP8b, and WNT3 factors on epiblast cells to ensure that only a small population of such cells progressed into the PGC fate (reviewed in Tam and Loebel 2009).

The BLIMP1-expressing cells increase in number, move in the posterior embryonic region, and develop into PGCs with AP activity, *Dppa3* (developmental pluripotency-associated 3, also known as *Stella*) expression, and *Hox* gene repression at around 7.25 *dpc*. The PGC precursors appear to be initially driven toward a somatic mesodermal fate but then regain a latent pluripotency typical of PGCs. The transcription factor SALL4 critically cooperates with BLIMP1 in suppressing the somatic cell program (Yamaguchi et al. 2015). It has recently been suggested that the RNA-binding protein LIN28 may play a role as an upstream regulator of the BLIMP1 transcript translation (West et al. 2009). PRDM14, another PR domain-containing protein, is also a critical regulator for PGC specification (Yamaji et al. 2008). The *Prdm14* gene is expressed in mouse PGC precursors at around 6.5 *dpc* and continues exclusively in PGCs until the end of gonad colonization. In the absence of *Prdm14* gene, AP-positive PGC-like cells are formed and repress *Hox* genes but fail to upregulate *Sox2* gene (SRY-related HMG-box gene 2) encoding a transcription factor crucial for mouse stem cell pluripotency. The PGC precursors of *Prdm14*$^{-/-}$ *embryos* also fail to undergo proper genome-wide epigenetic reprogramming and eventually disappear, perhaps due to the failure to proliferate. WNT3 induces the expression of many transcription factors associated with mesoderm in epiblast-like cells through β-catenin. Among these, T (brachyury), a classical and conserved mesodermal factor, is essential for robust activation of *Blimp1* and *Prdm14* genes (Aramaki et al. 2013). The transcription factor AP2C (TCFAP2C, also known as activating enhancer-binding protein 2 gamma, AP2γ) is probably downstream BLIMP1 and also plays a key role in mouse PGC specification since, in its absence, PGCs were specified but were lost around 8.0 *dpc* (Weber et al. 2010).

It is to be pointed out that the PGC precursors are not intrinsically different from other cells within the epiblast. Indeed, this region, including the posterior primitive streak and extending into the base of the allantois, has been identified as the allantoic core domain (ACD), which houses a unique pool of precursor cells co-expressing AP, BLIMP1, fragilis, POU5F1 (POU class 5 homeobox 1, also known as OCT4), and STELLA (Downs et al. 2009). It is important to note that these cells give rise to putative PGCs in addition to cells of extraembryonic lineages and all three embryonic layers, thus bringing into question the timing and definition of the true PGC identity (Mikedis and Downs 2012, 2013). In this regard, we have recently reported that the cell population usually considered PGCs for the expression of several markers moving toward the GRs includes a subset of cells co-expressing several germ cell and hematopoietic markers and possessing hematopoietic activity (Scaldaferri et al. 2015).

There are a number of molecules that are now known to play critical roles in the subsequent phases of PGC development, presumably after the PGC specification (for review, see Saitou and Yamaji 2012). For example, the hypoxia-inducible factor-2a (HIF-2a) has been shown to play a critical role for PGC specification and/or early PGC development, presumably by regulating the expression of *Pou5f1* (*Oct4*) (Covello et al. 2006). The precise roles of key pluripotency genes such as *Oct4*, *Sox2*, and *Nanog* expressed at high level by specified PGCs remains also to be clarified, although OCT4 and Nanog proteins have been shown to be critical for their survival at relatively later stages (Chambers et al. 2007; Kehler et al. 2004; Okamura et al. 2008; Yamaguchi et al. 2009). The receptor tyrosine kinase Kit and its ligand stem cell factor (SCF or Kit ligand, KL) regulate the migration, proliferation, and/or survival of PGCs after they move into the hindgut (Buehr et al. 1993; Gu et al. 2011; McCoshen and McCallion 1975; Mintz and Russell 1957; Pesce et al. 1993; Runyan et al. 2006), presumably through the Akt/m-TOR/Bax signaling pathway (De Miguel et al. 2002; Farini et al. 2007; Runyan et al. 2006; Stallock et al. 2003). Similarly, mutations of RNA-binding proteins such as Nanos3, TIAR (T-cell-restricted intracellular antigen-1-related protein), and DND1 (dead-end protein homologue 1) have been known to lead to the loss of germ cells due to the failure of survival and/or proliferation of migratory and/or gonadal PGCs (reviewed in De Felici and Farini 2012).

2.2.2 Origin of Human PGCs

Fuss (1911, 1912) and Felix (1911) were apparently the first to describe the extragonadal location of PGCs in human embryos. The place where PGCs were first identified in human embryos around the end of the 3rd week of gestation is the same as in the mouse: the wall of the yolk sac at the angle with the allantois. The putative PGCs were distinguished by their large size, spherical shape, a prominent nucleolus, and the presence of abundant glycogen granules in the cytoplasm. Subsequently, histochemical methods for periodic acid-Schiff (PAS)-positive materials and AP activity were applied successfully to identify human PGCs by McKay et al. (1953). Only recently, the first gene and protein expression data on human PGCs at different stages of development were produced. Of note, these studies show that gene expression patterns in early post-migrating human 7th–9th week PGCs are similar to mouse PGCs of comparable stages. In particular, they reported simultaneous expression of both pluripotency (*OCT4*, *NANOG*, *PRDM14*, and *LIN28*) and germline-specific (*BLIMP1*, *AP2γ*, *DAZL*, *Kit*, and *DDX4*) genes (Guo et al. 2015). Moreover, like in mouse PGCs, the surface oligosaccharide SSEA1 (CD15) and the CXCR4 receptor are expressed by human PGCs at these early stages (see, De Felici 2013 and references herein). Some important differences have been, however, described. In particular, SOX2 and SOX17 expression is completely different: while *Sox2* is expressed and essential for mouse PGC development, its human counterpart is not expressed in PGCs which instead express SOX17 (Avilion et al. 2003; Campolo et al. 2013). Moreover, human PGCs express KLF4 (Krupper-like factor 4), a gene encoding a trascription factor involved in pluripotency maintenance, whereas mouse PGC do not express it (Guo et al. 2015).

In order to compensate for the lack of information about the molecular mechanisms of PGC formation in the human embryo, in vitro culture systems able to reproduce some of these processes are now becoming available. These systems are based on the possibility to induce embryonic stem (ES) cells derived from blastocyst or induced pluripotent stem (iPS) cells produced from differentiated somatic cells to develop into specific cell lineages including the germline.

Using in vitro culture ES and iPS cells, PGC differentiation has been diagnosed primarily by the analysis of mouse germ cell gene and protein expression and, more recently, by the use of reporter constructs with the expression of green fluorescent protein (GFP) under control of germ cell specific promoters. Progressively increasing numbers of studies show that human ES and iPS cells possess a certain propensity to spontaneously differentiate into PGC-like cells, albeit at a low frequency (around 5 %) (see De Felici 2013 and references herein). Most interesting, the efficiency of spontaneous differentiation to PGCs was increased with the addition to the culture medium of BMP4, 7, and 8b, the same growth factors governing the formation of mouse PGCs (see above). Small changes in stem cell culture conditions or coculture onto human fetal gonad stromal cell, or mouse embryonic fibroblast (MEF) monolayers in the presence of basic fibroblast growth factor (bFGF), have been also reported to favor the formation of putative human PGCs in vitro. In addition, silencing the *DAZ* family and *NANOS3* genes in human ESCs resulted in a marked reduction in their capability to give rise in vitro to PGC-like cells.

While these data have provided the first evidence that BMPs, and probably bFGF, are involved in the formation of human PGCs and that DAZ and NANOS3 proteins function at the early stage of their development, a "*SOX*" role for human PGC specification has been shown to be played by *SOX17*, which was recently shown to be required for a robust induction of PGC-like cells from epiblast stem cell-like cells (EpiSCLCs), probably under the control by BLIMP1 (Irie et al. 2015b).

Recently, Sugawa et al. (2015) and Sasaki et al. (2015) described similar defined and stepwise robust differentiation system for inducing pre-migratory PGC-like cells from human iPS cells. In particular, Sugawa et al. (2015) reported that in response to cytokines (a combination of activin A, BMP4, and FGF2 under serum-free conditions for 2 days), iPS cells differentiate first into a heterogeneous mesoderm-like cell population ($OCT4^{+}/BlIMP1^{+}/T^{+}$ cells) and then (following culture for 4 days in G-MEM/20 % KSR, containing BMP4, LIF (leukemia inhibitory factor), and Y-27632) into SSEA1/TRA-1-81/Kit-positive PGC-like cells (7–20 %), which exhibit minimal PRDM14 expression and the suppression of neural induction and of de novo DNA methylation.

Most important, when epigenetic of the PGC-like cells was analyzed, it was found that these cells were characterized by erasure of global DNA methylation, ongoing removal of parental imprinting, and histone modifications similar to endogenous PGCs (Guo et al. 2015; Irie et al. 2015b; Sugawa et al. 2015).

Taken together, these data testify the important progress made in unraveling the basic steps of PGC formation in the mouse model and the rapidly ongoing information available for human germ cells (Fig. 2.2). These indicate that mouse and human PGC specification occurs by epigenesis and depends from some common (i.e., BMP4, BLIMP1) but also different players (i.e., SOX2 and SOX17, KLF4, low PRDM14 level) acting within diverse tissue context. In vitro models of derivation of PGC-like cells from stem cells appear promising to study human PGC specification and to elucidate some of these differences. If and how such differences may have a critical effect on PGC specification mechanisms remain to be clarified.

Taking into account the knowledge reported in the previous sections and the different organization of the mouse and human embryos at the pregastrulation stage (Fig. 2.3), a hypothetical model for the human PGC formation can be drawn. After implantation and before gastrulation, around the end of the 2nd week of gestation, the human embryo consists of a flat disk with two cell layers, epiblast and hypoblast. It seems plausible that at this stage, the precursors of PGCs are set aside within the epiblast following the action of BMP signals coming from the

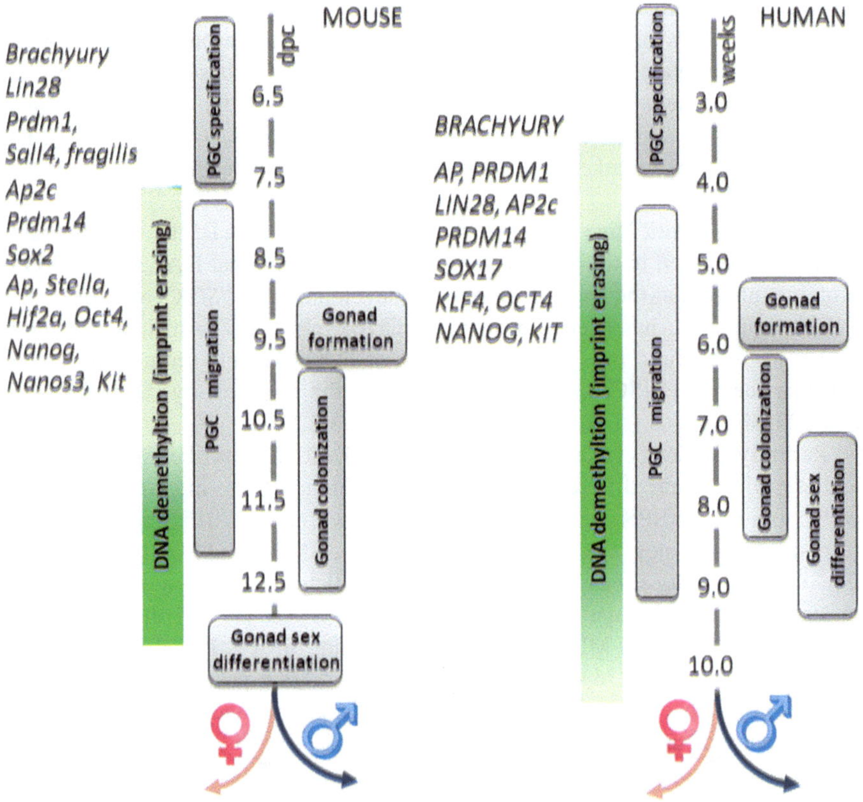

Fig. 2.2 A schematic representation of developmental timelines of mouse and human PGCs, note distinct periods of epigenetic reprogramming. Genes reported to be expressed in PGC precursors and PGC are also shown; for human, the time expression of most genes is largely hypothetical and mainly based on in vitro model of PGC-like production from stem cells

trophoblastic cells (including the developing syncytio- and cytotrophoblast) immediately surrounding the periphery of the embryonic disk (or perhaps from the amnioblasts located above the epiblast) and the hypoblast that can be regarded as rodent ExE and VE, respectively. These precursors move with the forming extra-embryonic mesoderm out the embryo proper and during the 3rd week reach the region of the wall of the definitive yolk sac where the allantois originates around day 16; here they are specified as AP-expressing PGCs.

A summary of the factors implicated in the formation of mouse and possibly of human PGCs is reported in Table 2.1.

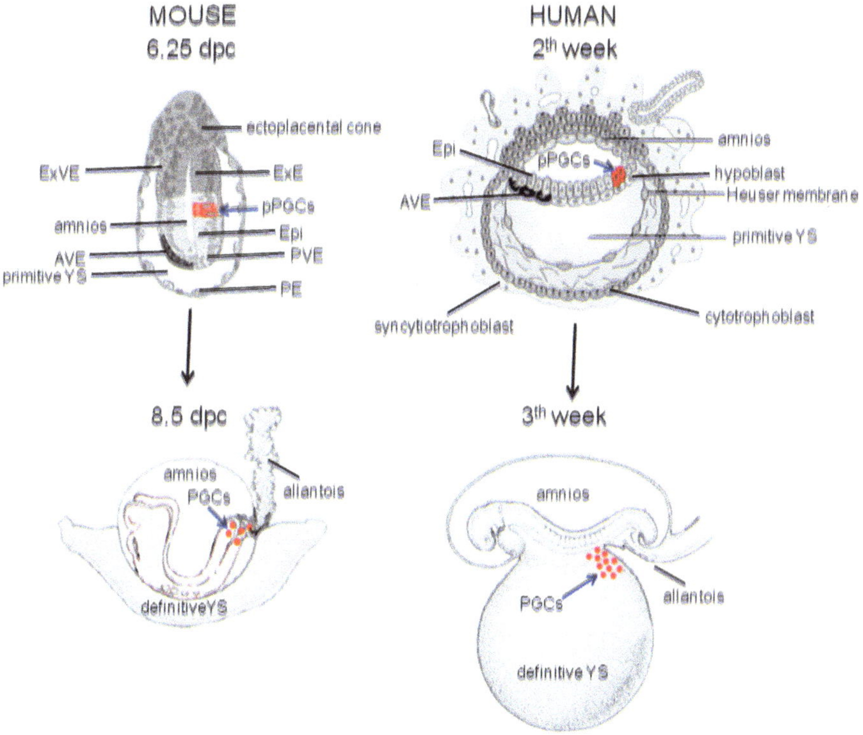

Fig. 2.3 A schematic representation of the organization of the mouse and human embryo during the period of the PGC induction in the epiblast and of their subsequent localization in the posterior region of the embryo at the angle between the allantois and the yolk sac wall. For information about the present notion on the formation of mouse PGCs and a hypothetical model of human PGC formation, see text. *ExVE* extraembryonic visceral endoderm, *ExE* extraembryonic ectoderm, *pPGCs* precursors of PGCs, *Epi* epiblast, *PE* parietal endoderm, *VE* visceral endoderm, *AVE* anterior visceral endoderm, *YS* yolk sac

2.3 PGC Migration

2.3.1 Migration of Mouse PGCs

According to Molyneaux and Wylie (2004), the journey of mouse PGCs to the GRs can be subdivided into five stages, depending on their activity: (1) putative PGCs enter the primitive streak to take residence within the allantois/posterior embryo; (2) PGCs are incorporated into the hindgut; (3) the hindgut elongates and forms a tubular structure, while PGCs are located within the gut epithelium; (4) PGCs move directionally into the dorsal mesentery between a discontinuous basal lamina and the underlying mesenchymal cells and finally into the GRs (stage 5). The putative PGCs left behind may undergo apoptotic death or may not be PGCs at all, instead they may represent cells of different lineages that mature into other somatic cell

Table 2.1 A list of the main genes involved in the formation and migration of the primordial germ cells (PGCs) in the mouse; genes in bold might be involved in such processes also in humans

	Gene symbol	Gene name	Role of Gene	Effect of mutation/inhibition	References
PCG origin	*Bmp4*	Bone morphogenic protein 4	GF acting as a major inducer of PGC formation	Lack of PGCs	Lawson et al. (1999)
	Bmp8b *Bmp2*	Bone morphogenic protein 8b and 2	Co-inducer of PGC formation	Lack or reduced number of PGCs	Ying et al. (2001)
	CdH1	E-cadherin	Adhesion molecule responsible of PGC precursor clustering	Antibodies inhibit PGC formation in vitro	Okamura et al. (2003)
	Wnt3	Wingless and int 3	GF necessary for epiblast responsiveness to BMP4 (through expression of the transcription factor T)	Lack of PGCs	Aramaki et al. (2013); Tanaka et al. (2013)
	Iftm3 (Mil/fragilis)	Interferon-induced transmembrane protein 3	Adhesion molecule; PGC marker	No effect on PGC formation or detectable consequence for germline development	Lange et al. (2008); Saitou et al. (2002)
	Dppa3 (PGC7, Stella)	Developmental pluripotency-associated 3	Protein involved in epigenetic chromatin reprogramming; may have a role in maintaining cell pluripotency	No effect on PGC formation	Bortvin et al. (2004)
	Prdm1 (Blimp-1)	PR domain-containing 1 (B lymphocyte-induced maturation protein 1)	TF necessary for PGC specification controlling epigenetic modification and gene expression	PGC precursors do not proliferate and migrate; inconsistent repression of *Hoxa1a* and *Hoxa1b* genes	Ohinata et al. (2005)
	Prdm14	PR domain-containing 14	TF necessary for PGC specification	Male and female infertility for lack of germ cells in the adult gonads	Yamaji et al. (2008)
	Lin28	Lin28 homologue A	RNA-binding protein upstream of *Prdm1* activation	No formation of PGC-like cells from ES cells in vitro (through inhibition of the microRNA *let-7)*	West et al. (2009)

	Tcfap2c (AP2g)	Transcription factor AP-2 gamma	TF putative downstream target of Prdm1	PGCs are specified but were lost around 8.0 dpc	Weber et al. (2010)
	Sall4	Spalt-like 4	TF suppresses somatic cell program by recruiting histone deacetylase repressor complex together with PRDM1	PGCs fail to translocate from mesoderm to endoderm and underwent apoptosis	Yamaguchi et al. (2015)
PGC migration	*Iftm1 and Iftm3 (Mil/fragilis)*	Interferon-induced transmembrane protein 1	Adhesion molecules controlling PGC migration from YS wall into hindgut endoderm	No detectable effect on PGC formation	Lange et al. (2008); Tanaka et al. (2005)
	Cdh1	E-cadherin	Adhesion molecule involved in PGC-PGC adhesion	Perturbation in PGC-PGC aggregation and gonad colonization by antibodies in vitro	Bendel-Stenzel et al. (2000); Di Carlo and De Felici (2000)
	Kitlg (Steel)	Kit ligand gene	Cytokine involved in stimulation of PGC motility and possible PGC chemoattractant	Reduction in PGC motility and number in the GRs; reduced directional migration in vitro	Gu et al. (2011); Farini et al. (2007); Mahakali Zama et al. (2005)
	Kit (White)	Kit	Encodes the Kitl tyrosine kinase receptor Kit	Defects in PGC migration and number in the GRs; reduced adhesiveness to cell monolayers by antibodies in vitro	Buehr et al. (1993); Pesce et al. (1997)
	Sdf1 (CXCL12)	Stromal cell-derived factor 1	Cytokine; possible PGC chemoattractant	Reduced number of PGC s in the GRs	Ara et al. (2003)
	Cxcr4	Chemokine C-X-C motif receptor 4	Encodes the SDF1 tyrosine kinase receptor CXCR4	Defects in PGC migration and reduced number in the GRs	Molyneaux et al. (2003)
	Itgb1	Integrin beta1	Integrin receptor mediating adhesion of PGC to ECM molecules	Defects in GR colonization by PGCs	Anderson et al. (1999)
	Wnt5a	Wingless and int 5a	GF with possible parallel action with Kitl/Kit and SDF1/CXCR4	Reduction in gonadal germ cell numbers (aberrant PGC migration along the hindgut endoderm)	Chawengsaksophak et al. (2012)

(continued)

Table 2.1 (continued)

	Gene symbol	Gene name	Role of Gene	Effect of mutation/inhibition	References
	Ror2	Receptor tyrosine kinase-like orphan receptor 2	Encodes the WT5a tyrosine kinase-like receptor ROR2; mediates polarity and directional migration of PGCs	Similar to Wnt5a−/−	Laird et al. (2011)
	Foxc1	Forkhead box c1	TF necessary for the correct differentiation of the gonadal somatic cells	Many PGCs do not migrate to the GRs, remaining trapped in the hindgut	Mattiske et al. (2006)
	Lhx1	LIM homeobox 1	TF influencing PGC localization by modulating *Ifitm1*-mediated repulsive activity	Ectopic localization of PGCs	Tanaka et al. (2010)
	Sox17	SRY (sex-determining region Y)-box 17	TF necessary for hindgut morphogenic movement guiding PGC movement	Scattering of PGCs in the YS endoderm	Hara et al. (2009)

GF growth factor, *TF* transcription factor, *GR* gonadal ridge, *YS* yolk sac types and become incorporated into other tissues (Molyneaux et al. 2001; Molyneaux and Wylie 2004; Mikedis and Downs 2012, 2013; Scaldaferri et al. 2015; Stallock et al. 2003).

Mouse PGCs were described to initiate active migration just after specification. Time-lapse photography documented that OCT4-GFP/AP-positive cells extended leading edges and actively migrated into the allantois/posterior embryo (Anderson et al. 2000). Before putative, PGCs begin this first step of their journey; however, expressions of factors associated with adhesion were localized to these cells, implicating that their role in determining PGC fate includes restricting their movement while they receive signals from neighboring cells. Such factors include interferon-inducing transmembrane protein (IFITM, fragilis) family members, IFITM1, IFITM2, and IFITM3 and E-cadherin. It was mentioned above that blockage of E-cadherin-mediated adhesion using specific antibodies under in vitro culture disrupted PGC specification (Okamura et al. 2003). Immotile proximal epiblast cells corresponding to putative PGCs were found to express *Ifitm* genes, and ectopic expression studies showed that IFITM proteins worked together to properly home these cells into the endoderm. Actually, it was proposed that IFITM1 activity in the ExM acts as a repulsion guidance cue for IFITM3-positive and IFITM1-negative PGCs into the IFITM-free endoderm (Lange et al. 2003; Saitou et al. 2002; Tanaka et al. 2005). However, genetic analysis by targeted disruption of the entire *Ifitm* locus clearly showed no detectable influences on migration and development of the germ cell line in mouse embryos lacking *Ifitm* genes (Lange et al. 2008), indicating that IFITM proteins play a minor role in the guidance of PGC migration from the allantois into the hindgut endoderm.

Activation of cell motility, dynamic adhesion and displacing to and from extracellular matrix components and/or surrounding cells along outlined pathways, repulsive displacement and orientated attraction, and finally homotypic adhesion have been all implicated in active PGC migratory. Actually evidence from in vivo and in vitro observations and characterization of spontaneous or induced mutations supported the importance of all such factors.

Shortly after PGCs arrive within the hindgut, they are thought to receive signals that first promote random motility by 9.0 *dpc*, followed by new signals that promote directional migration toward the developing GRs between 9.0 and 9.5 *dpc*. E-cadherin expression disappears in PGCs that have colonized the hindgut, whereas its expression in the hindgut epithelium remains high (Bendel-Stenzel et al. 2000). Together with decreased E-cadherin expression and consequential reduction in adhesion, receptor tyrosine kinase activity via Kit/KL interactions promotes PGC motility. This receptor/ligand pair has long been appreciated for their role in PGC migration, proliferation, and survival (Buehr et al. 1993; Mahakali Zama et al. 2005; McCoshen and McCallion 1975; Pesce et al. 1997; Runyan et al. 2006). Recent studies showed that PGCs harboring Kit tyrosine kinase receptors were continuously surrounded by somatic cells expressing KL throughout their migratory journey in the hindgut, which provided a niche friendly toward migratory activity (Gu et al. 2011). In collaboration with Kit/KL activity, another receptor tyrosine kinase receptor/ligand pair, receptor tyrosine kinase-like orphan receptor 2 (ROR2)/WNT5A, was found to promote polarization and elongation of migrating PGCs; disruption of either component caused PGCs to round up and cease migration (Chawengsaksophak et al. 2012; Laird et al. 2011).

The final part of the journey occurs between 10.5 and 11.5 *dpc* and requires PGCs to migrate from the hindgut through the dorsal body wall to the GRs. Chuva de Sousa Lopes et al. (2005) found that, in the absence of TGFβ signaling (ALK5 deficiency), PGC migration out of the hindgut was promoted, due to the reduced deposition of collagen type I surrounding the gut. Migratory PGCs adhere strongly to collagen; therefore, reduced collagen type I along the gut may result in reduced adhesion, facilitating migration into the dorsal mesentery and GRs. In this migratory step, chemokine signals likely provide the attractants and homing mechanisms. In particular, stromal cell-derived factor 1 (SDF1), which emanates from neighboring mesenchyme and the developing GRs, interacting with PGCs via its G-protein coupled receptor, CXCR4, (chemokine C-X-C motif receptor 4) (Ara et al. 2003; Molyneaux et al. 2003) and/or KL, stimulating its tyrosine kinase receptor Kit (Farini et al. 2007), are involved in such activities. Null mutations in any of these genes lead to a failure of PGCs to properly colonize the GRs. Null-mutant mice for either *Forkhead box c1* (*Foxc1*) or *LIM homeobox 1* (*Lhx1*) also show a similar phenotype, pointing toward a complex network of genes involved in such process (Mattiske et al 2006; Tanaka et al 2010).

Other studies reported that cell–cell and cell–extracellular matrix interactions are also critical for PGC migration and GR colonization. It was hypothesized that PGC extracellular matrix interactions were required to maintain PGCs on the narrow path toward the gonadal ridges in the context of diffuse attractant signals. Interactions between PGCs and proteoglycans or extracellular matrix glycoproteins integrin β1, laminin, fibronectin, and collagen IV were found and were dependent on the stage of PGC migration (Alvarez-Buylla and Merchant 1986; De Felici and Dolci 1989; De Felici et al. 1998; Ffrench-Constant et al. 1991; García-Castro et al. 1997; Soto-Suazo et al. 1999, 2002, 2004). Specific to PGC homing, integrin β1-deficient PGCs failed to colonize the gonads (Anderson et al. 1999). After PGCs emigrated from the hindgut, they were also found to extend long cellular processes that interacted with each other and their surroundings while finding their way to the gonad (Gomperts et al. 1994). Subsequent analysis revealed that E-cadherin expression was reestablished once PGCs left the hindgut and was concentrated with β-catenin at sites of PGC-PGC interactions (Bendel-Stenzel et al. 2000; Di Carlo and De Felici 2000). In this case, treatment with E-cadherin-blocking antibodies prevented PGCs from directional migration in both in vitro and in vivo experiments, suggesting its role in mediating cell–cell interactions and promoting migration to the gonad. Finally, E-cadherin was also found to play an important role as PGCs settle within the gonad, stop migration, and take up a rounded appearance (Di Carlo and De Felici 2000).

Gonadal ridges are probably dispensable for the directional migration of PGCs, but they are required for their precise positioning at the final step of migration (Chen et al. 2013). In fact, in *Wt1* or *Gata4*-mutated mice in which GRs do not form, most of the PGCs reached the mesenchyme under the coelomic epithelium and no ectopic PGCs were noted; however, the precise positioning of PGCs was disrupted

(Hu et al. 2013). Since somatic factors that might guide PGC migration, such as the chemokine SDF1 and the transcription factor FOXC1, are expressed independently of *Gata4* in the mesentery and mesonephros, starting as early as 9.0 *dpc* (Ara et al. 2003; Mattiske et al. 2006), it is possible that these factors are sufficient to direct PGCs to their final destination or to its proximity, even in the absence of the GRs.

Finally, accumulation of cholesterol within the GRs has been proposed to control signaling interactions required for PGCs to colonize the gonads (Ding et al. 2008), but the precise mechanisms involved remain elusive.

Actually, some authors claimed that despite such results and observations, on the basis of morphological observations, the movement of PGCs from the allantois/ yolk sac wall to the hindgut and then to the GRs can be explained by a passive translocation thought first an enclosure in the endoderm forming the wall of the hindgut and then a transport by the proliferative expansion of this tissue enclosed to the mesenchyme into or very near the GRs. In this model, only the last short track might be accomplish by active PGC motility (Freeman 2003). In supporting such a model, at least for the first part of the journey, it has been reported that defective hindgut expansion in *Sox17* mouse mutants caused the failure of further lateral PGC movement, resulting in the immobilization of PGCs in the hindgut entrance. In contrast, the majority of the remaining PGCs moves into the visceral endoderm layer but relocate outside of the embryonic gut domain. This leads to a scattering of PGCs in the extraembryonic yolk sac endoderm (Hara et al. 2009). A possible explanation of these data is that hindgut morphogenic movement is crucial for directing PGC movement toward the embryonic gut side, but not for their relocation from the ExM into the endoderm.

The two models discussed above are not mutually exclusive. In fact, PGC translocation toward the GRs might be a combination of active cell motility and tissue movements.

2.3.2 *Migration of Human PGCs*

The two models of active and passive translocation are applicable also for the journey of human PGCs toward the GRs. In the human embryo, the superior part of the yolk sac wall containing the PGCs becomes incorporated in the developing gut during the lateral folding during the 4th week of gestation, suggesting that the PGC translocation into the gut may be passive. The gonadal anlagen may be recognized by a bilateral thickening of the coelomic epithelium near the developing root of the gut mesentery in the posterior embryonic body cavity around the 5th week of gestation. PGCs present in the endodermal gut epithelium were seen from the 5th week onward (reviewed in De Felici 2013). Thereafter, they leave the epithelium and invade the surrounding mesenchyme moving through the gut mesentery toward the bipotential GRs.

Early electron microscopic studies described human PGCs in vivo as having an irregular appearance and possessing pseudopodia during their journey toward the

GRs. These features were interpreted as a manifestation of active migration (Fujimoto et al. 1977; Makabe et al. 1989). An interpretation confirmed by several in vitro observations reporting that human PGCs, as those of mouse, show several features of motile cells and are able to move actively both on cellular and extracellular matrix substrates (Kuwana and Fujimoto 1983) (Fig. 2.1b). However, Freeman (2003), reinterpreting these observations, concludes that morphogenic movements and local cell divisions rather than active migration are mainly responsible for PGC displacement in the different regions of the human embryo. Even Freeman, however, admits that PGCs might migrate actively to cover a distance of approximately 50 mm separating the pre-aortic region from the GRs. In support, the possibility of passive relocation of PGCs within the GRs, Aeckerle et al. (2014) showed that, in the marmoset monkey, PGCs are located in the hindgut endoderm in close spatial proximity to the prospective GRs (about 40 μm only), making a long-range migration of PGCs dispensable. These authors therefore favor a migratory model based primarily on passive translocation of PGCs from the mesenchyme that surrounds the gut to the prospective gonad through the expansion of mesenchymal tissue which contains the PGCs.

As in the mouse, during their journey, human PGCs appear to be surrounded by extracellular matrix components in which glycosaminoglycans have been histochemically detected (Pereda et al. 2006). PGCs seem also interacting with the mesenchymal cells through different type of junctions, such as desmosomes, gap junctions, and focal contacts (Fujimoto et al. 1977; Pereda et al. 2006). No specific attractants have been proposed for human PGCs although they possess Kit and are surrounded by cells expressing KL (Gu et al. 2011). Interestingly, mutations in *WNT5A* encoding a protein implicated as the ROR2 receptor ligand promoting polarization and elongation of migrating mouse PGCs (see above) have been recently linked to Robinow syndrome displaying a wide variety of phenotypes, including shortened limbs, craniofacial dysmorphology, and genital hypoplasia (Person et al. 2010). Actually, mouse loss-of-function phenotypes of *Wnt5a* and *Ror2* are very similar and both grossly resemble those observed in Robinow syndrome. The expression of transcripts for a member of the olfactory receptor gene family in human PGCs from 10-week-old fetuses makes this class of receptor additional candidates for PGC attractants (Goto et al. 2001). Finally, Møllgard et al. (2010) observed that human PGCs preferentially ascended from the mesentery of the hindgut to the gonadal anlage by migration along autonomic nerve fibers close to the Schwann cells and proposed that these nerve fibers and/or Schwann cells may release chemoattractants supporting PGC migration. In humans, primary sites of ectopic germ cells are within the CNS but they are also seen in the pelvic region, the mediastinum and thorax (Mamsen et al. 2012).

During migration, human PGCs contain a large PAS-positive cytoplasmic store of glycogen granules and several lipid droplets (see De Felici 2013 and references herein). Following their arrival in the GRs, the glycogen content is diminished. Round mitochondria with a pale matrix and small tubulovesicular cristae were observed near the nucleus. They significantly increase in number during PGC migration and settlement in the GRs. Migratory PGCs have less than

10 mitochondria, while 100 mitochondria are present in ovarian PGCs and 200 in oogonia (Motta et al. 2000). These observations suggest that PGCs might prevalently employ an anaerobic metabolism during migration and undergo a transition in their energy metabolism after reaching the GRs.

In the attempt to reconcile these various results and observations, it can be postulated that both mouse and human PGCs are capable of cytokine-dependent active directional migration that requires interaction with extracellular matrix molecules and cell substrates. At the same time, however, it is likely that PGCs can be passively transported by morphogenic movements. While active migration appears to be involved to cover short distance maybe within the same tissue, the second are likely responsible to cover longer distance. In this contest, it is also possible that a first period of GR colonization when such anlagen are close to the hindgut before the mesentery extends might occur almost entirely by passive translocation of PGCs located in the gut epithelium, whereas active migration is initiated only subsequently.

A summary of the factors implicated in the migration of mouse and possibly of human PGCs is reported in Table 2.1.

2.4 Epigenetic Reprogramming

Shortly after specification, throughout migration, and for a couple of days after gonad colonization, mouse PGCs of both sexes undergo extensive epigenetic reprogramming. Such process is a key feature of PGC development and allows erasure of the DNA methylation marks associated with imprinted genes, allowing establishment of sex-specific imprints during gametogenesis. There are also extensive changes in several histone marks both prior to and during imprint erasure. It is likely that these chromatin changes are necessary to facilitate DNA demethylation but may have other functions such as the erasure of "somatic" epigenetic marks established during post-implantation development, to allow X-chromosome reactivation in female germ cells or to prevent inheritance of epimutations. A detailed description and references about the epigenetic reprogramming in the mouse PGCs can be found in De Felici (2011). Here a brief account of this process is reported.

Immediately following specification, a marked reduction in heterochromatic H3K9me2 (7.5–8.25 *dpc*) followed by an increase in H3K27me3 (associated with inactive gene promoters) (8.25 and 9.5 *dpc*) occur. The decrease in H3K9me2 is most likely due to the reduction in G9a-like protein (GLP), a SET domain protein histone methyltransferase (HMT) that forms a heteromeric complex with G9a. Such reduction correlates with a period of mitotic arrest and transcriptional quiescence in PGCs. The repressive H2A/H4R3me2s is also enriched in PGCs from 8.5 *dpc* onward likely as a consequence of the PRMT5 activity, an arginine-specific HMT that mediates symmetrical dimethylation of arginine 3 on the tails of these histones. The maintenance methyltransferase DNMT1 [DNA (cytosine-5-)-methyltransferase

1] is likely active in nascent PGCs but the efficiency of its recruitment into replication foci is severely impaired after 9.5 *dpc* consistently with undetectable ubiquitin-like with PHD and ring finger domains 1(UHRF1), which targets DNMT1 to replication foci for maintenance DNA methylation (Ohno et al. 2013). A decrease in the protein level of the de novo methyltransferases DNMT3a and DNMT3b accompanies the PGC post specification and gonad colonization periods. Finally, reactivation of the inactive X chromosome in female PGCs also commences during migration to be concluded after the GR colonization.

Although DNA demethylation begins during PGC migration, the bulk of DNA demethylation occurs rapidly after gonad colonization (between 11.5 and 13.5 *dpc*). This is accompanied by changes in chromatin organization and loss of numerous histone modifications brought about by genome-wide histone replacement. This second wave of reprogramming includes the erasure of imprints allowing the establishment of sex-specific imprints during gametogenesis. Evidence exists that DNA demethylation in PGCs employs both a passive (DNA replication-dependent) and active mechanisms. Active demethylation may involve the DNA repair pathway [activation-induced cytidine deaminase (AID) and thymine DNA glycosylase (TDG) enzymes], ten-eleven translocation (TET) enzymes, and poly(ADP-ribose) polymerase (PARP) activities (Ciccarone et al 2012; reviewed in De Felici 2011).

In human PGCs, a few observations have been reported with regard to chromatin changes and DNA demethylation. Nevertheless, a timeline and some characteristics of these events are becoming to be available also thanks to the in vitro models of PGC-like formation for stem cell lines.

Immunohistochemical studies and DNA methylome analyses of global 5mC level showed that both female and male fetal germ cells are hypomethylated already at 5th–6th weeks of gestation when most of them are still migrating or have just arrived in the GRs (Gkountela et al 2013, 2015; Guo et al. 2015; Tang et al. 2015; Wermann et al. 2010). In particular, Tang et al. (2015) using base-resolution mexthylome analysis reported that at the 4th week migratory PGCs obtained from the hindgut exhibited low global 5mC levels compared to surrounding soma and at the 7th–9th week gonadal PGCs/oogonia, representing recent arrived germ cells into the GRs, remained devoid of 5mC (about 7 % methylated CpG). Like in mouse, consistent with global demethylation, UHRF1, was not detectable in proliferating PGCs/oogonia while de novo DNA methyltransferase DNMT3A and DNMT3B were repressed; as note TET1 and TET2 appeared enriched during this period. Such demethylation dynamics parallels the epigenetic remodeling occurring between 9.5 and 13.5 dpc in mouse PGCs. In human PGCs, however, demethylation of imprint controlled regions (ICRs) is completed [except for the insulin growth factor 2 receptor (*IGF2R*) and the progression-elevated gene-10 (*PEG10*) genes ICRs] before the 7th week when most of PGCs are still migrating toward or are just arrived into the GRs, indicating earlier epigenetic dynamics. Tang et al. (2015) and Gkountela et al. (2015) found that, similarly to the mouse PGCs, potentially hazardous retrotransposons such as the LINE SINE-VNTR-*Alu* (SVA) and endogenous retrovirus (ERV) elements remained methylated in gonadal human PGCs.

Remarkably, some loci associated with metabolic and neurological disorders were also resistant to DNA demethylation, revealing potential for transgenerational epigenetic inheritance that may have phenotypic consequences.

In contrast to male mouse PGCs, which remain demethylated for only a few days, male human PGCs remain hypomethylated for weeks and only showed signs of re-methylation in the II trimesters (around the 16th week onward) (Gkountela et al. 2015; Wermann et al. 2010).

Finally, human and mouse PGCs demonstrate distinct H3K9me2 and H3K27me3 dynamics. For example, in human PGCs, H3K9me2 levels were persistently lower compared to soma, but it remained detectable in the nuclei. Although at the 4th week human PGCs showed 2.5-fold enrichment of H3K27me3, the signal diminished gradually to half of that of soma by the 9th week. Importantly, H3K27me3 foci at the inactivated X chromosome were observed in soma, but not in week 7th female PGCs, indicating reactivation of X at this time. At around the 16th–20th weeks, male germ cells demonstrate absence/low levels of repressive H3K9me2, H3K27me3, and H3K9me3 modifications but high levels of permissive H3K9ac and H2A.Z marks (Almstrup et al. 2010; Bartkova et al. 2011). Interestingly, as reported above, the PGC-like cells produced in vitro from human stem cells showed ongoing removal of parental imprinting, erasure of global DNA methylation, and histone modifications similar to their in vivo PGC counterparts (Guo et al. 2015; Irie et al. 2015b; Sugawa et al. 2015; Tang et al. 2015; Wermann et al. 2010).

These recent results support the notion that human PGCs undergo a pattern of genome reprogramming similar on the whole to the murine model but with differences in timing and chromatin changes. Such difference might be related to the diversity of timing of PGC development between these two species (a few days in mouse and several weeks in human) of cell cycle duration (doubling time about 12 h and 6 days in mouse and human PGCs, respectively; Bendsen et al. 2006; De Felici and Farini 2012) and the asynchrony of the developmental events in human germ cells. The significance and importance of such differences remain, however, to be clarified.

References

Aeckerle N, Drummer C, Debowski K et al (2014) Primordial germ cell development in the marmoset monkey as revealed by pluripotency factor expression: suggestion of a novel model of embryonic germ cell translocation. Mol Hum Reprod 21:66–80

Almstrup K, Nielsen JE, Mlynarska O et al (2010) Carcinoma *in situ* testis displays permissive chromatin modifications similar to immature foetal germ cells. Br J Cancer 103:1269–1276

Alvarez-Buylla A, Merchant-Larios H (1986) Mouse primordial germ cells use fibronectin as a substrate for migration. Exp Cell Res 165:362–368

Anderson R, Fässler R, Georges-Labouesse E et al (1999) Mouse primordial germ cells lacking beta1 integrins enter the germline but fail to migrate normally to the gonads. Development 126:1655–1664

Anderson R, Copeland TK, Schöler H et al (2000) The onset of germ cell migration in the mouse embryo. Mech Dev 91:61–68
Ara T, Nakamura Y, Egawa T et al (2003) Impaired colonization of the gonads by primordial germ cells in mice lacking a chemokine, stromal cell-derived factor-1 (SDF-1). Proc Natl Acad Sci U S A 100:5319–5323
Aramaki S, Hayashi K, Kurimoto K et al (2013) A mesodermal factor, T, specifies mouse germ cell fate by directly activating germline determinants. Dev Cell 27:516–529
Avilion AA, Nicolis SK, Pevny LH et al (2003) Multipotent cell lineages in early mouse development depend on SOX2 function. Genes Dev 17:126–140
Bartkova J, Moudry P, Hodny Z et al (2011) Heterochromatin marks HP1γ, HP1α and H3K9me3, and DNA damage response activation in human testis development and germ cell tumours. Int J Androl 34:103–113
Bendel-Stenzel MR, Gomperts M, Anderson R et al (2000) The role of cadherins during primordial germ cell migration and early gonad formation in the mouse. Mech Dev 91:143–152
Bendsen E, Byskov AG, Andersen CY et al (2006) Number of germ cells and somatic cells in human foetal ovaries during the first weeks after sex differentiation. Hum Rep 21:30–35
Bortvin A, Goodheart M, Liao M et al (2004) Dppa3/Pgc7/stella is a maternal factor and is not required for germ cell specification in mice. BMC Dev Biol 2004(4):2–6
Buehr M, McLaren A, Bartley A et al (1993) Proliferation and migration of primordial germ cells in We/We mouse embryos. Dev Dyn 198:182–189
Campolo F, Gori M, Favaro R et al (2013) Essential role of Sox2 for the establishment and maintenance of the germ cell line. Stem Cells 31:1408–1421
Chambers I, Silva J, Colby D et al (2007) Nanog safeguards pluripotency and mediates germline development. Nature 450:1230–1234
Chawengsaksophak K, Svingen T, Ng ET (2012) Loss of Wnt5a disrupts primordial germ cell migration and male sexual development in mice. Biol Reprod 86:1–12
Chen SR, Zheng QS, Zhang Y et al (2013) Disruption of genital ridge development causes aberrant primordial germ cell proliferation but does not affect their directional migration. BMC Biol. doi:10.1186/1741-7007-11-22
Chiquoine AD (1954) The identification, origin, and migration of the primordial germ cells in the mouse embryo. Anat Rec 118:135–146
Chuva de Sousa Lopes SM, van den Driesche S, Carvalho RLC et al (2005) Altered primordial germ cell migration in the absence of transforming growth factor β signaling via ALK5. Dev Biol 5:194–203
Ciccarone F, Klinger FG, Catizone A et al (2012) Poly (ADP-rybosil)ation acts in DNA demethylation of mouse primordial germ cells also with DNA damage independent roles. PlosONE 7: e46927
Covello KL, Kehler J, Yu H et al (2006) HIF-2α regulates Oct-4: effects of hypoxia on stem cell function, embryonic development, and tumor growth. Genes Dev 20:557–570
De Felici M (2011) Nuclear reprogramming in mouse primordial germ cells: epigenetic contribution. Stem Cells Int. doi:10.4061/2011/425863
De Felici M (2013) Origin, migration, and proliferation of human primordial germ cells. In: Coticchio G et al (eds) Oogenesis. Springer, London, pp 19–37. doi:10.1007/978-0-85729-826-3_2
De Felici M, Dolci S (1989) In vitro adhesion of mouse fetal germ cells to extracellular matrix components. Cell Differ Dev 26:87–96
De Felici M, Farini D (2012) The control of cell cycle in mouse primordial germ cells: old and new players. Curr Pharm Des 18:233–244
De Felici M, Pesce M, Giustiniani Q et al (1998) In vitro adhesiveness of mouse primordial germ cells to cellular and extracellular matrix component substrata. Microsc Res Tech 43:258–264
De Miguel MP, Cheng L, Holland EC et al (2002) Dissection of the c-Kit signaling pathway in mouse primordial germ cells by retroviral-mediated gene transfer. Proc Natl Acad Sci U S A 99:10458–10463

Di Carlo A, De Felici M (2000) A role for E-cadherin in mouse primordial germ cell development. Dev Biol 226:209–219

Ding J, Jiang D, Kurczy M et al (2008) Inhibition of HMG CoA reductase reveals an unexpected role for cholesterol during PGC migration in the mouse. BMC Dev Biol. doi:10.1186/1471-213X-8-120

Downs KM, Inman KE, Jin DX et al (2009) The Allontoic Core Domain (ACD): new insights into development of the murine allantois and its relation to the primitive streak. Dev Dyn 238:532–553

Farini D, La Sala G, Tedesco M et al (2007) Chemoattractant action and molecular signaling pathways of Kit ligand on mouse primordial germ cells. Dev Biol 306:572–583

Felix W (1911) Die Entwicklung der Harn- und Geschlechtsorgane. In: Keibel F, Mall FP (eds) Handbuch der Entwicklungageschichte des Menshen, vol 2. Hirzel, Leipzig, pp 732–795

Ffrench-Constant F, Hollingsworth A, Heasman J et al (1991) Response to fibronectin of mouse primordial germ cells before, during and after migration. Development 113:1365–1373

Freeman B (2003) The active migration of germ cells in the embryos of mice and men is a myth. Reproduction 125:635–643

Fujimoto T, Miyayama Y, Fuyuta M (1977) The origin, migration and fine morphology of human primordial germ cells. Anat Rec 188:315–329

Fuss A (1911) Uber extraregionare Geschlechtszellen bei einem menschlichen Embryo von 4 Wochen. Anat Am 39:407–409

Fuss A (1912) Uber die Geschlechtszellen des Menschen und der Saugetiere. Anat Entw Mech 81:1–23

García-Castro MI, Anderson R, Janet Heasman J et al (1997) Interactions between germ cells and extracellular matrix glycoproteins during migration and gonad assembly in the mouse embryo. J Cell Biol 138:471–480

Ginsburg M, Snow MH, McLaren A (1990) Primordial germ cells in the mouse embryo during gastrulation. Development 110:521–528

Gkountela S, Li Z, Vincent JJ et al (2013) The ontogeny of cKIT+ human primordial germ cells proves to be a resource for human germ line reprogramming, imprint erasure and in vitro differentiation. Nat Cell Biol 15:113–122

Gkountela S, Zhang KX, Shafiq TA et al (2015) DNA demethylation dynamics in the human prenatal germline. Cell 161:1425–1436

Gomperts M, Garcia-Castro M, Wylie C et al (1994) Interactions between primordial germ cells play a role in their migration in mouse embryos. Development 120:135–141

Goto T, Salpekar A, Monk M (2001) Expression of testis specific member of the olfactory receptor gene family in human primordial germ cells. Mol Hum Reprod 7:553–558

Gu Y, Runyan C, Shoemaker A et al (2011) Membrane-bound steel factor maintains a high local concentration for mouse primordial germ cell motility, and defines the region of their migration. PLoS One 6(10):e25984. doi:10.1371/journal.pone.0025984

Guo F, Yan L, Guo H et al (2015) The transcriptome and DNA methylome landscapes of human primordial germ cells. Cell 161:1437–1452

Hara K, Kanai-Azuma M, Uemura M et al (2009) Evidence for crucial role of hindgut expansion in directing proper migration of primordial germ cells in mouse early embryogenesis. Dev Biol 330:427–439

Hayashi K, Chuva de Sousa Lopes SM, Surani MA (2007) Germ cell specification in mice. Science 316:394–396

Hu Y-C, Okumura LM, Page DC (2013) *Gata4* is required for formation of the genital ridge in mice. PLoS Genet 9(7):e1003629. doi:10.1371/journal.pgen.1003629

Irie N, Tang WWC, Surani MA (2015a) Germ cell specification and pluripotency in mammals: a perspective from early embryogenesis. Reprod Med Biol 13:203–215

Irie N, Weinberger L, Tang WWC et al (2015b) SOX17 is a critical specifier of human primordial germ cell fate. Cell 160:253–268

Kehler J, Tolkunova E, Koschorz B et al (2004) Oct4 is required for primordial germ cell survival. EMBO Rep 5:1078–1083

Kuwana T, Fujimoto T (1983) Active locomotion of human primordial germ cells in vitro. Anat Rec 205:21–26

Laird DJ, Altshuler-Keylin S, Kissner MD et al (2011) ROR2 enhances polarity and directional migration of primordial germ cells. PLoS Genet 7:e1002428

Lange UC, Saitou M, Western PS (2003) The fragilis interferon-inducible gene family of transmembrane proteins is associated with germ cell specification in mice. BMC Dev Biol 3:1–11

Lange UC, Adams DJ, Lee C et al (2008) Normal germ line establishment in mice carrying a deletion of the Ifitm/Fragilis gene family cluster. Mol Cell Biol 28:4688–4696

Lawson KA, Hage WJ (1994) Clonal analysis of the origin of primordial germ cells in the mouse. Ciba Found Symp 182:68–84

Lawson KA, Dunn NR, Roelen BA et al (1999) Bmp4 is required for the generation of primordial germ cells in the mouse embryo. Genes Dev 13:424–436

Mahakali Zama A, Hudson FP, Bedell MA (2005) Analysis of hypomorphic KitlSl mutants suggests different requirements for KITL in proliferation and migration of mouse primordial germ cells. Biol Reprod 73:639–647

Makabe S, Nottola SA, Motta PM (1989) Life history of the human female germ cell: ultrastructural aspects. In: Van Blerkom J, Motta PM (eds) Ultrastructure of human gametogenesis and early embryogenesis. Kluwer Academic, Boston, pp 33–60

Mamsen LS, Brochner CB, Byskov AG et al (2012) The migration and loss of human primordial germ stem cells from the hind gut epithelium towards the gonadal ridge. Int J Dev Biol 56:771–778

Mattiske D, Kume T, Hogan BL (2006) The mouse forkhead gene Foxc1 is required for primordial germ cell migration and antral follicle development. Dev Biol 290:447–458

McCoshen JA, McCallion DJ (1975) A study of the primordial germ cells during their migratory phase in Steel mutant mice. Experientia 31:589–590

McKay DG, Hertig AT, Adams EC et al (1953) Histochemical observations on the germ cells of human embryos. Anat Rec 17:201–219

Mikedis MM, Downs KM (2012) STELLA-positive subregions of the primitive streak cooperate to build the posterior region of the murine conceptus. Dev Biol 363:201–218

Mikedis MM, Downs KM (2013) Widespread but tissue-specific patterns of interferon-induced transmembrane protein 3 (IFITM3, FRAGILIS, MIL-1) in the mouse gastrula. Gene Expr Patterns 13:225–239

Mintz B, Russell ES (1957) Gene-induced embryological modifications of primordial germ cells in the mouse. J Exp Zool 134:207–237

Møllgard HK, Jespersen A, Lutterodt MC et al (2010) Human primordial germ cells migrate along nerve fibers and Schwann cells from the dorsal hind gut mesentery to the gonadal ridge. Mol Hum Reprod 16:621–631

Molyneaux K, Wylie C (2004) Primordial germ cell migration. Int J Dev Biol 48:537–554

Molyneaux KA, Stallock J, Schaible K et al (2001) Time-lapse analysis of living mouse germ cell migration. Dev Biol 240:488–498

Molyneaux KA, Zinszner H, Kunwar PS et al (2003) The chemokine SDF1/CXCL12 and its receptor CXCR4 regulate mouse germ cell migration and survival. Development 18:4279–4286

Motta PM, Nottola SA, Makabe S et al (2000) Mitochondrial morphology in human fetal and adult female germ cells. Hum Reprod 15:129–147

Ohinata Y, Payer B, O'Carroll D et al (2005) Blimp1 is a critical determinant of the germ cell lineage in mice. Nature 436:207–213

Ohinata Y, Ohta H, Shigeta M et al (2009) A signaling principle for the specification of the germ cell lineage in mice. Cell 137:571–584

Ohno R, Megumi Nakayama M, Naruse C et al (2013) A replication-dependent passive mechanism modulates DNA demethylation in mouse primordial germ cells. Development 140:2892–2903
Okamura D, Kimura T, Nakano T et al (2003) Cadherin-mediated cell interaction regulates germ cell determination in mice. Development 130:6423–6430
Okamura D, Tokitake Y, Niwa H et al (2008) Requirement of Oct3/4 function for germ cell specification. Dev Biol 317:576–584
Pereda J, Zorn T, Soto-Suazo M (2006) Migration of human and mouse primordial germ cells and colonization of the developing ovary: an ultrastructural and cytochemical study. Microsc Res Tech 69:386–395
Person AD, Beiraghi S, Sieben CM et al (2010) WNT5A mutations in patients with autosomal dominant Robinow syndrome. Dev Dyn 239:327–337
Pesce M, Farrace MG, Piacentini M et al (1993) Stem cell factor and leukemia inhibitory factor promote primordial germ cell survival by suppressing programmed cell death (apoptosis). Development 118:1089–1094
Pesce M, Di Carlo A, De Felici M (1997) The c-kit receptor is involved in the adhesion of mouse primordial germ cells to somatic cells in culture. Mech Dev 68:37–44
Pesce M, Klinger FG, De Felici M (2002) Derivation in culture of primordial germ cells from cells of the mouse epiblast: phenotypic induction and growth control by Bmp4 signalling. Mech Dev 12:15–24
Runyan C, Schaible K, Molyneaux K et al (2006) Steel factor controls midline cell death of primordial germ cells and is essential for their normal proliferation and migration. Development 133:4861–4869
Saitou M, Yamaji M (2012) Primordial germ cells in mice. Cold Spring Harb Perspect Biol (4): a008375
Saitou M, Barton SC, Surani MA (2002) A molecular programme for the specification of germ cell fate in mice. Nature 418:293–300
Sasaki K, Yokobayashi S, Nakamura T et al (2015) Robust in vitro induction of human germ cell fate from pluripotent stem cells. Cell Stem Cell. doi:10.1016/j.stem.2015.06.014
Scaldaferri ML, Klinger FG, Farini D et al (2015) Hematopoietic activity in putative mouse primordial germ cell populations. Mech Dev 136:53–63
Soto-Suazo M, Abrahamsohn PA, Pereda J et al (1999) Distribution and space-time relationship of proteoglycans in the extracellular matrix of the migratory pathway of primordial germ cells in mouse embryos. Tissue Cell 31:291–300
Soto-Suazo M, Abrahamsohn PA, Pereda J et al (2002) Modulation of hyaluronan in the migratory pathway of mouse primordial germ cells. Histochem Cell Biol 117:265–723
Soto-Suazo M, San Martin S, Zorn TM (2004) Collagen and tenascin-C expression along the migration pathway of mouse primordial germ cells. Histochem Cell Biol 121(2):149–153
Stallock J, Molyneaux K, Schaible K et al (2003) The pro-apoptotic gene Bax is required for the death of ectopic primordial germ cells during their migration in the mouse embryo. Development 30:6589–6597
Sugawa F, Araúzo-Bravo MJ, Yoon J et al (2015) Human primordial germ cell commitment in vitro associates with a unique PRDM14 expression profile. EMBO J 34:1009–1024
Tam PPL, Loebel DAF (2009) Specifying mouse embryonic germ cells. Cell 137:398–400
Tanaka SS, Yamaguchi YL, Tsoi B et al (2005) IFITM/Mil/fragilis family proteins IFITM1 and IFITM3 play distinct roles in mouse primordial germ cell homing and repulsion. Dev Cell 9:745–756
Tanaka SS, Yamaguchi YL, Steiner KA et al (2010) Loss of Lhx1 activity impacts on the localization of primordial germ cells in the mouse. Dev Dyn 239:2851–2859
Tanaka SS, Nakane A, Yamaguchi YL et al (2013) Dullard/Ctdnep1 modulates WNT signalling activity for the formation of primordial germ cells in the mouse embryo. PLoS One 8:e77428. doi:10.1371/journal.pone.0057428

Tang WW, Dietmann S, Irie N et al (2015) A unique gene regulatory network resets the human germline epigenome for development. Cell 161:1453–1467

Weber S, Eckert D, Nettersheim D et al (2010) Critical function of AP-2 gamma/TCFAP2C in mouse embryonic germ cell maintenance. Biol Reprod 82:214–223

Wermann H, Stoop H, Gillis AJ et al (2010) Global DNA methylation in fetal human germ cells and germ cell tumours: association with differentiation and cisplatin resistance. J Pathol 221:433–442

West JA, Viswanathan SR, Yabuuchi A et al (2009) A role for Lin28 in primordial germ-cell development and germ-cell malignancy. Nature 460:909–913

Yamaguchi S, Kurimoto K, Yabuta Y et al (2009) Conditional knockdown of Nanog induces apoptotic cell death in mouse migrating primordial germ cells. Development 136:4011–4020

Yamaguchi YL, Tanaka SS, Kumagai M et al (2015) Sall4 is essential for mouse primordial germ cell specification by suppressing somatic cell program genes. Stem Cells 33:289–300

Yamaji M, Seki Y, Kurimoto K (2008) Critical function of *Prdm14* for the establishment of the germ cell lineage in mice. Nat Genet 40:1016–1022

Ying Y, Zhao GQ (2001) Cooperation of endoderm-derived BMP2 and extraembryonic ectoderm-derived BMP4 in primordial germ cell generation in the mouse. Dev Biol 232:484–492

Ying Y, Liu XM, Marble A et al (2000) Requirement of Bmp8b for the generation of primordial germ cells in the mouse. Mol Endocrinol 14:1053–1063

Ying Y, Qi X, Zhao GQ (2001) Induction of primordial germ cells from murine epiblasts by synergistic action of BMP4 and BMP8B signaling pathways. Proc Natl Acad Sci U S A 98:7858–7862

Chapter 3
The Gonadal Supporting Cell Lineage and Mammalian Sex Determination: The Differentiation of Sertoli and Granulosa Cells

Gwenn-Aël Carré and Andy Greenfield

Abstract The supporting cell lineage plays a crucial role in nurturing the development of germ cells in the adult gonad. Sertoli cells in the testis support the progression of spermatogonial stem cells through meiosis to the production of motile spermatozoa. Granulosa cells, meanwhile, are a critical component of the ovarian follicle that produces the mature oocyte. It is a distinctive feature of the embryonic gonad that at least some of the supporting cells are derived from a single sexually bipotential precursor lineage. It is the commitment of this somatic lineage to either the Sertoli or granulosa cell fate that defines sex determination. In this chapter we review what is known about the key molecules responsible for this lineage decision in the developing mammalian gonads, relying primarily on data from studies of mice and humans. We focus on recent advances in our understanding of the mutually antagonistic interactions of testis- and ovary-determining pathways and their complexity as revealed by genetic analyses. For the sake of simplicity, we will deal with supporting cells in testis and ovary development in separate sections, but numerous points of contact exist between these accounts of gonadogenesis in male and female embryos, primarily due to the aforementioned mutual antagonisms. The final section will offer a brief synthesis of these organ-specific overviews and a summary of the key themes that emerge in this review of supporting cell differentiation in mammalian sex determination.

Gwenn-Aël Carré • A. Greenfield (✉)
Mammalian Genetics Unit, Medical Research Council, Harwell, Oxfordshire OX11 0RD, UK
e-mail: a.greenfield@har.mrc.ac.uk

R.P. Piprek (ed.), *Molecular Mechanisms of Cell Differentiation in Gonad Development*, Results and Problems in Cell Differentiation 58,
DOI 10.1007/978-3-319-31973-5_3

3.1 Sertoli Cells: Their Origin, Differentiation, and Roles in Testis Determination

3.1.1 *Gonad Formation and the Origins of Sertoli Cells in the XY Gonad*

The gonad forms on the ventromedial surface of the mesonephros at around 10.0–10.5 days *post coitum* (*dpc*) in the mouse (Kaufman 1992). This is around 36 h before commitment to the testicular or ovarian fate. This primordium, which is sexually bipotential at this stage, is often referred to as the genital ridge. Studies of mouse knockouts indicate roles for a number of genes in this process, including *Sf1 (Nr5a1)* (Luo et al. 1994), *Wt1* (Hammes et al. 2001), *Lhx9* (Birk et al. 2000), and *Emx2* (Kusaka et al. 2010). Embryos lacking these genes initiate gonad formation by thickening of the epithelial surface of the mesonephros, the coelomic epithelium, by cell proliferation, but the developing genital ridge regresses before it is completely formed. Recent analyses of embryos, in which *Gata4* has been ablated by tamoxifen-inducible gene deletion at around 8.75 *dpc*, reveal a role for this transcription factor in the initiation of gonad formation (Hu et al. 2013). The coelomic epithelium in these mutants remains a monolayer, with no signs of thickening. In contrast to wild-type embryos, the mutant epithelium lacks expression of *Sf1* and *Lhx9*, which may account for the failure to undergo normal cell proliferation. This reported role for GATA4 in gonad formation illustrates a recurrent theme in sex determination: the use of specific transcription factors at multiple stages of the process. GATA4, WT1, and SF1, all of which are implicated in early stages of gonad development, also play roles in subsequent sexual differentiation: GATA4 is implicated in the regulation of *Sry* and *Sox9* expression (Bouma et al. 2007; Tevosian et al. 2002); WT1 is required for Sertoli cell differentiation and prevents Sertoli cells from adopting a fetal Leydig cell fate (Hammes et al. 2001; Zhang et al. 2015); and SF1 positively regulates *Sox9* expression (Sekido and Lovell-Badge 2008).

The origin of germ cells in the developing gonad is well established; they migrate from the base of the allantois, along the gut mesentery, and populate the newly formed gonad from between 10.5 and 11.5 *dpc* (Ginsburg et al. 1990). However, the origins of the somatic cell lineages of the XY gonad, including Sertoli cells, peritubular myoid cells, Leydig cells, and endothelial cells, and the relationships between them, are less well understood. Sertoli cells play a central role in testis determination: they are almost exclusively XY in chromosome constitution in XX–XY mouse chimeras, in contrast to other somatic lineages (Palmer and Burgoyne 1991). This likely reflects the fact that the Y-linked testis-determining gene, *Sry* (sex-determining region of the Y), acts cell-autonomously in pre-Sertoli cells to trigger Sertoli cell differentiation and that masculinization of other lineages occurs as a consequence of Sertoli cell-derived signals. Lineage tracing experiments in XY gonads indicate that at least some Sertoli cells are derived from the coelomic epithelium and that the ability of this epithelium to act as a source of supporting cell precursors (pre-Sertoli cells) is developmentally restricted to early stages of gonad development (prior to 11.4 *dpc*) (Karl and Capel 1998). Interestingly, similar experiments indicate that coelomic-epithelium-derived cells can populate the XX gonad during the same time frame, suggesting that supporting cells have a common origin in both XY and XX gonads. However, these studies do not rule out the possibility that supporting cells might be descended from more than one progenitor cell type. Lineage tracing of somatic cell lineages in the developing ovary suggests a more complex and dynamic situation, as will be discussed in Sect. 3.2.2.

3.1.2 *SRY*: Regulation and Mode of Action

The mammalian Y chromosome is a dominant male determinant due to the consequences of SRY activity. SRY is a high-mobility group (HMG) box transcription factor that controls target gene expression by binding to and bending DNA, thereby initiating genetic pathways that drive morphogenesis of the testis. This commitment of the bipotential gonadal primordium to the testicular or ovarian fate is known as primary sex determination. *SRY* was identified as the mammalian testis-determining gene due to its location on a small 35-kilobase region of the human Y that caused male development in XX individuals with small Y-derived translocations (Sinclair et al. 1990). SRY's function is clearly conserved in mammals, as evidenced by XY gonadal sex reversal in humans and mice harboring deleterious *Sry* mutations and male development in XX individuals with an additional copy (or copies) of *Sry* (Berta et al. 1990; Koopman et al. 1991; Lovell-Badge and Robertson 1990). A number of genes required for the normal expression of *Sry* itself have been described in the mouse (reviewed by Larney et al. 2014). These include *Gata4/Fog2* (Tevosian et al. 2002), *M33 (Cbx2)* (Katoh-Fukui et al. 1998, 2012), *Six 1/4* (Fujimoto et al. 2013), *Map3k4* (Bogani et al. 2009), and *Gadd45g* (Warr et al. 2012). These genes are expressed in both XY and XX gonads prior to sex determination, which explains why a genomic *Sry* transgene is able to function normally to direct testis development in an XX embryo (Koopman et al. 1991) and underlines the similarity of the bipotential XY and XX gonads prior to sex determination. *Sry* acts within a tight developmental window to effect testis determination—any delay can result in XY ovary development, i.e., male to female gonadal sex reversal or ovotestis development (Hiramatsu et al. 2008; Warr and Greenfield 2012; Warr et al. 2014; Wilhelm et al. 2009). A number of assays indicate that the onset of *Sry* expression occurs at around 10.5 *dpc* in the mouse in Sertoli cell precursors, in the central region of the gonad; transcript levels rise before reaching a peak at around 11.5 *dpc*, and expression is extinguished by 12.5 *dpc* (Bullejos and Koopman 2001; Hacker et al. 1995; Jeske et al. 1995). This brief period of expression is consistent with SRY's role in orchestrating the initial stages of testis determination, but not subsequent differentiation, at least in the mouse. In humans, *SRY* expression is temporally and spatially more ubiquitous in embryos and adults, but it is unclear whether this profile reflects additional, including extra-gonadal, functions (Clépet et al. 1993).

Expression profiling in the early mouse gonad indicates that somatic cells exhibit a bias toward genes associated with ovary development at later stages (Jameson et al. 2012b). This has led to the idea that SRY acts to prevent this lineage priming from leading to granulosa cell development. According to this model, a delay in SRY expression results in the supporting cell lineage, already primed for ovary development, being unable to escape this prospective fate. This model also suggests that SRY acts to inhibit, directly or indirectly, ovary-determining genes, rather than simply positively promoting pro-testis genes. We will see later that much recent evidence supports this predicted antagonism between the testis- and ovary-determining genetic pathways.

How does SRY exercise control over target gene expression? SRY has evolved rapidly due to its location on the Y chromosome. This has resulted in significant variation in the structure of SRY in different species, and even within species. In mice, a long C-terminal glutamine-rich (Q-rich) domain after the HMG box DNA-binding domain acts to stabilize the protein and to activate transcription (Zhao et al. 2014). Human SRY, along with other non-rodent species, lacks this domain, suggesting that SRY in these species may activate target gene transcription by cooperating with another protein or proteins that substitute for the "missing" trans-activation domain. Despite a number of direct SRY target genes having been reported, using a variety of molecular assays (Bhandari et al. 2012; Li et al. 2014), one target stands out as the most compelling, due to its own established role in testis determination: *SOX9* (see Fig. 3.1 for an overview of the functional relationships between genes in the testis- and ovary-determining pathways).

3.1.3 SOX9: Master Regulator and Key Antagonist of Ovarian Fate

SOX9 (Sry (sex-determining region of the Y)-box9), which encodes another HMG box protein, is both necessary and sufficient for testis determination in otherwise wild-type genetic backgrounds. Loss of SOX9 in XY gonads causes sex reversal in humans and mice, and ectopic gonadal expression of SOX9 in XX gonads can cause testis determination in both species (Barrionuevo et al. 2006; Bishop et al. 2000; Chaboissier et al. 2004; Foster et al. 1994; Huang et al. 1999; Vidal et al. 2001). *Sox9* is expressed in XY and XX gonads early in mouse gonadogenesis at low levels but becomes rapidly upregulated in the XY gonad and extinguished in the XX gonad by around 11.5 *dpc* (da Silva et al. 1996; Kent et al. 1996; Sekido et al. 2004). Its expression is restricted to pre-Sertoli cells (prior to 12.5 *dpc*) and Sertoli cells (after 12.5 *dpc*) and is therefore associated with testis cords from around 12.5 *dpc* when Sertoli cells have become epithelialized and exhibit apicobasal polarity (see Fig. 3.2 for an overview of XY and XX gonad development and examples of the

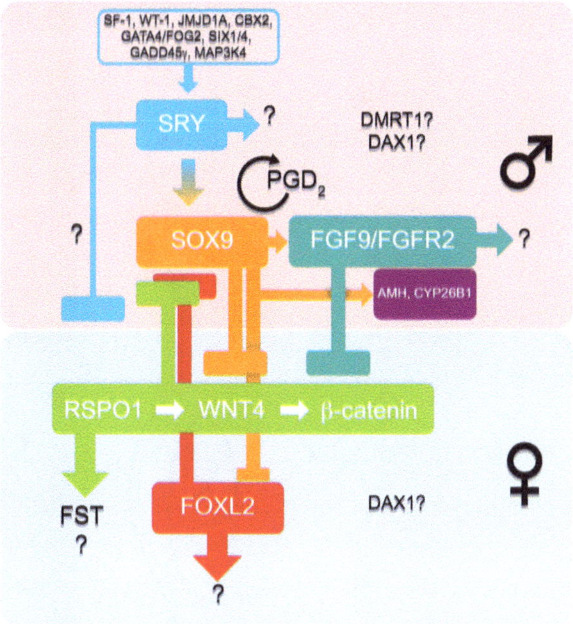

Fig. 3.1 Mutually antagonistic relationships between key testis- and ovary-determining genes in specifying supporting cell fate during mammalian sex determination. Commitment of supporting cells to the Sertoli cell fate (*upper cell*) requires the establishment of robust SOX9 expression. This is normally achieved in XY embryos by the early expression of SRY in somatic cells of the genital ridge. SOX9 acts positively to establish FGF9/FGFR2 signals, which in turn maintain SOX9 expression at high levels by inhibiting canonical WNT signals. SOX9 inhibits FOXL2 and WNT signals and positively regulates pro-testis factors such as AMH and CYP26B1. Prostaglandin D2 (PGD2) also acts to recruit cells to the Sertoli cell fate by promoting SOX9 expression. SOX9/FGF9 signals are opposed in turn by the activity of RSPO1/WNT4/β-catenin signals and FOXL2 in order to establish granulosa cell fate (*lower cell*). Positive (*arrows*) and negative (*hammered lines*) regulation of targets that remain to be established is indicated by *question marks*. Molecules required for timely SRY expression are also shown

expression patterns of markers of Sertoli and granulosa cells). Chromatin immunoprecipitation of SRY in mouse embryonic gonads revealed binding to an enhancer element around 15 kb from the transcriptional start site of *Sox9*. This binding is enhanced in the presence of SF1, and the enhancer core element, known as *TESCO* (testis-specific enhancer of *Sox9*, core), is capable of driving sexually dimorphic expression of a reporter in the embryonic gonads of transgenic mice (Sekido and Lovell-Badge 2008). SOX9 subsequently acts through its own enhancer to maintain high levels of expression in Sertoli cells. Thus, these data, and others discussed earlier, suggest that SRY activity in the supporting cells of the XY gonad acts to upregulate the expression of *Sox9* and that this transcription factor acts to "lock in" Sertoli cell differentiation and thereby promote (1) masculinization of other cell lineages and (2) morphogenetic changes associated with testis development, such as the mesonephric endothelial cell migration required for testis cord

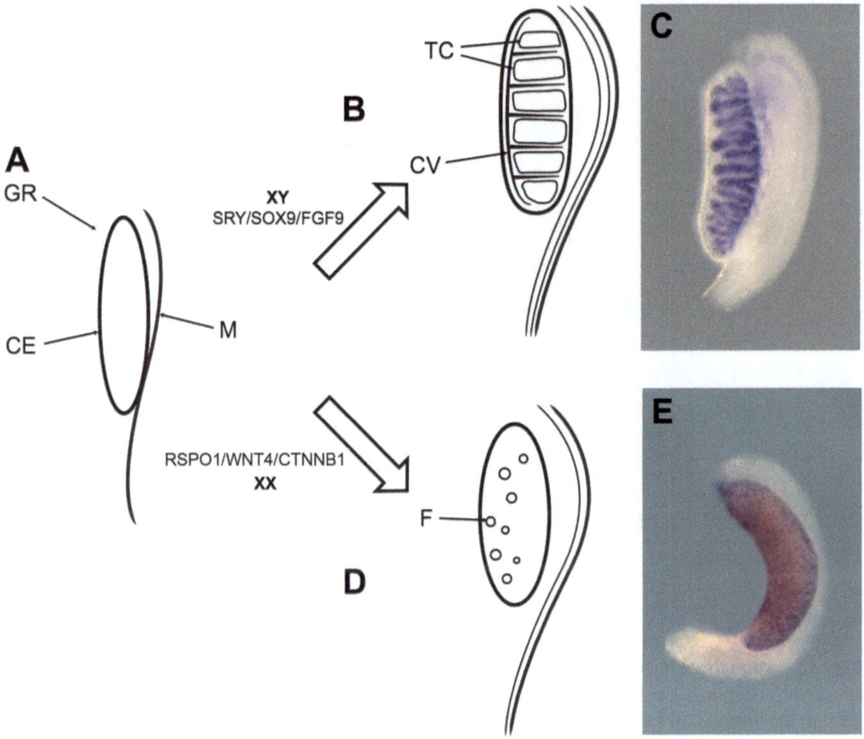

Fig. 3.2 Overview of gonadogenesis: cellular differentiation in the developing testis and ovary. (**a**) The genital ridge (GR) comprises the developing gonad and associated mesonephros (M). The coelomic epithelium (CE) is an important source of supporting cell precursors until around 11.5 *dpc*. (**b**) The presence of a Y chromosome results in SRY-dependent upregulation of SOX9 expression, leading to Sertoli cell differentiation; this causes the ingression of endothelial cells from the mesonephros to generate the coelomic vessel (CV). Sertoli cells encompass primordial germ cells and form testis cords, giving the newly formed testis its familiar architecture. (**c**) Image of whole-mount in situ hybridization (WMISH) of an XY gonad at 13.5 *dpc* showing expression of the Sertoli cell marker *Sox9* in the testis cords. (**d**) In the absence of SRY, pro-ovary pathways, including canonical WNT signals, block cell migration into the gonad and thereby prevent testis cord and coelomic vessel formation. Supporting cell precursors form granulosa cells, and these form clusters with primordial germ cells that eventually give rise to primordial follicles (F). (**e**) WMISH of an XX gonad at 13.5 *dpc* with a marker of granulosa cells, *Wnt4*. Note the absence of testis cords and prominent curvature of the gonad

formation (Combes et al. 2009; Cool et al. 2008). The phenotypic consequences of genetically ablating the *TESCO* enhancer of *Sox9* in mice are yet to be described, and there are no reported mutations in the equivalent human element associated with 46,XY disorders of sex development (DSD), a term describing disruptions to sexual development in humans (Georg et al. 2010). In this context, it is interesting to note that an additional enhancer element, termed *RevSex*, is associated with the control of human *SOX9* activity based on studies of a number of cases of 46,XY and 46,XX DSD (Benko et al. 2011; Hyon et al. 2015; Lybaek et al. 2014). This observation suggests that regulation of *SOX9* transcription may be complex, with a number of elements acting in tandem to facilitate and protect *SOX9* transcription in a supporting cell microenvironment in which pro-ovary gene products are present.

As in the case of SRY, many SOX9 targets have been identified, and these probably act in a number of pathways to promote Sertoli cell differentiation, both cell-autonomously and non-cell-autonomously. One gene whose expression depends on SOX9 is *Fgf9* (Kim et al. 2006). *Fgf9* (fibroblast growth factor, 9) encodes a secreted growth factor whose ablation in the XY mouse embryo results in complete gonadal sex reversal through its effects on *Sox9* expression (Colvin et al. 2001; Kim et al. 2006). FGF9 acts diffusibly to promote Sertoli cell differentiation along the length of the developing gonad (Hiramatsu et al. 2010), primarily by opposing pro-ovarian WNT signals (Jameson et al. 2012a). The idea that cells in the gonadal field can be recruited to the Sertoli cell fate even in the absence of SRY was clear from earlier studies of XX–XY chimeras in which a few XX Sertoli cells were present (Palmer and Burgoyne 1991). Another secreted molecule, prostaglandin D2 (PGD2), can promote the expression of *SOX9* in testis determination, independently of FGF9 (Moniot et al. 2009; Wilhelm et al. 2005). Indeed, the gene that encodes the enzyme responsible for the production of PGD2, prostaglandin D synthase (*Ptgds*), is also a reported target of SOX9 (Wilhelm et al. 2007). However, there has been no report of XY sex reversal phenotypes in mice lacking *Ptgds isoforms*. These data suggest that SOX9 acts to reinforce and amplify its testis-determining effects by the employment of diffusible molecules acting in paracrine fashion to promote its expression in other cells and thereby recruit these to the Sertoli cell fate. Such reinforcement pathways, possibly acting redundantly, may ensure that the number of Sertoli cells formed reaches a threshold required for testis development.

As previously noted, loss of *Sry* or *Sox9* in mice results in embryonic XY gonadal sex reversal; this is associated with activation of pro-ovarian genetic pathways and differentiation of ovarian cell lineages. It is worth considering that such an outcome is one of several that might have been predicted before such genetic analyses were first performed. Loss of a master testis-determining gene might have been predicted to result in a simple failure of gonadal differentiation, with an undifferentiated primordium as a consequence, or in programmed cell death and loss of the developing organ. Instead, the alternative genetic program of ovary development is employed. This "ectopic" activation of pro-ovary pathways in the XY gonad after mutation suggests that testis-determining genes such as *Sry* and *Sox9*, or their target genes, must act directly to inhibit ovary development. In Sect. 3.3 we discuss the mutual antagonism of pro-testis and pro-ovary genetic pathways in greater detail.

3.1.4 Molecules of Unclear Significance: DMRT1 and DAX1

DMRT1 (doublesex and mab-3-related transcription factor 1) is a transcription factor related to two sex-determining proteins in invertebrates: Doublesex (DSX) in *Drosophila melanogaster* and MAB-3 in *Caenorhabditis elegans*. However, its role in mammalian sex determination remains somewhat unclear. It is not required for primary sex determination in mice, since XY mice lacking *Dmrt1* are born with testes (Raymond et al. 2000). However, in the absence of this protein, testicular fate cannot be maintained, and Sertoli cells undergo postnatal reprogramming to granulosa-like cells in mutants. This role in maintenance of the Sertoli cell fate is corroborated by adult conditional gene deletion experiments (Matson et al. 2011). Studies of compound mutant mice, lacking both *Dmrt1* and pro-ovary genes such as *Foxl2*, suggest that DMRT1 acts to prevent retinoic acid (RA)-dependent reprogramming of Sertoli cells in the postnatal testis, allowing Sertoli cells to utilize RA signaling to support germ cell development (Minkina et al. 2014). However, notwithstanding its role in Sertoli cell fate maintenance, recent experiments in which *Dmrt1* is overexpressed in the mouse XX embryonic gonad, resulting in Sertoli cell development, indicate that the potential to respond to DMRT1 at the primary sex-determining stage still exists in the mouse genome (Lindeman et al. 2015; Zhao et al. 2015). The situation in humans is just as unclear. Deletions of human chromosome 9p24 are associated with cases of 46,XY DSD, including complete gonadal dysgenesis (CGD) (Ottolenghi et al. 2000). But these deletions usually include *DMRT1* and additional closely linked *DMRT* genes, which may also contribute to the phenotype. A recent report of a single, rare de novo missense mutation in *DMRT1* associated with 46,XY CGD does not aid clarity, since molecular studies suggest that the sex reversal in this patient may be partly attributed to disruption to normal DMRT binding stoichiometry at target genes, disrupting binding not just of DMRT1, but potentially other factors, in a dominant-negative fashion (Murphy et al. 2015).

DAX1 (*NR0B1*, nuclear receptor subfamily 0, group B, member 1), which encodes an orphan nuclear hormone receptor, has been reported to be an anti-testis gene and a testis-determining gene. Duplications of a region of human Xp21, which contains *DAX1*, are associated with 46,XY CGD—a phenomenon also known as dosage-sensitive sex (*DSS*) reversal (Bardoni et al. 1994; Zanaria et al. 1994). Moreover, high levels of expression of *Dax1* using transgenesis in XY mice with a highly sensitized genetic background have also been reported to cause embryonic gonadal sex reversal (Swain et al. 1998). Studies in transgenic mice and cell lines suggest that overexpression of *Dax1* can inhibit activation of *Sox9* expression through *TESCO* (Ludbrook et al. 2012). However, loss-of-function studies in mice suggest that DAX1 can also be required for testis determination on certain sensitized genetic backgrounds, possibly through its positive effects on the expression of *Sox9* (Meeks et al. 2003). Humans with Xp21 deletions develop adrenal hypoplasia congenita (AHC): testes develop, but individuals exhibit gonadal defects including disorganized testis cords and hypogonadotropic hypogonadism (Muscatelli et al. 1994). How is it possible to reconcile these contrasting roles revealed by loss- and gain-of-function studies? Firstly, it is worth noting that gain-of-function studies indicate what the phenotypic consequences of overexpression are; that is, the consequences of a gene/protein being present in a cell at levels that do not reflect the circumstances under which it was selected during evolution to its current functional status. In brief, they show what a gene *can* do, given sufficient levels, not necessarily what it *does* do. Nevertheless, the phenotypic similarities between the consequences of DAX1 loss of function and DAX1 gain of function have led to the idea that DAX1 has a positive regulatory role to play in testis determination only within a certain activity range: outside of that range, it can equally disrupt this process (Ludbrook and Harley 2004). It is anticipated that future mechanistic studies of how DAX1 exerts its effects, particularly on chromatin of its target genes, will clarify its role in these distinct models.

3.2 Granulosa Cells and Ovary Development

3.2.1 Investigating Ovary Development

We have seen that complete gonadal sex reversal in XY knockout mice has led to the discovery of several genes in the testis determination pathway and aided their characterization. In contrast, in female (XX) mice, loss of function of pro-ovary factors does not lead to complete female-to-male sex reversal (see Sects. 3.2.3–3.2.5). This might indicate that ovary determination is more modular, with loss of a specific module not sufficient to cause sex reversal (Warr and Greenfield 2012). It may also indicate that some pathways indispensable for ovarian determination are still undiscovered. The additional complexity of the ovarian supporting cell lineage (see next section) may be a contributory factor (Mork et al. 2012; Rastetter et al. 2014; Zheng et al. 2014). One of the difficulties in studying ovarian differentiation is the absence of overt morphological changes associated with the earliest steps of granulosa cell differentiation. This contrasts with testis determination, in which whole-scale morphological remodeling of the gonad occurs due to SRY activity: endothelial cells migrate and generate the coelomic vessel, while testis cords develop and form the testicular interstitium as a consequence (Cool et al. 2012). These morphological changes have facilitated the identification of distinct testicular cell lineages and the discovery of markers specifically associated with those lineages. However, recent studies of ovary development using lineage tracing and employing novel markers are revealing added complexity in the ovarian soma.

3.2.2 The Origins of Granulosa Cells

Sertoli cells are thought to be derived from bipotential precursors, which originate in the coelomic epithelium and are present in the gonad prior to sex determination (Karl and Capel 1998). This hypothesis is supported by the observation that *Sry-EGFP* transgenic XX gonads express EGFP with a spatiotemporal profile resembling endogenous *Sry* expression, showing that some XX gonadal somatic cells are able to activate the *Sry* promoter, explaining their potential to differentiate into Sertoli cells (Albrecht and Eicher 2001). Moreover, transdifferentiation of adult Sertoli cells to granulosa cells, and vice versa, in certain contexts also suggests a commonality of bipotential precursor cell. However, recent studies using lineage tracing suggests that granulosa cells in the adult XX ovary, identified by expression of *Foxl2*, have at least two distinct origins (Mork et al. 2012; Rastetter et al. 2014; Zheng et al. 2014).

Using inducible *Foxl2*-dependent Cre activity (from fetal stages 12.5 to 14.5 *dpc*) and the R26R reporter to trace the fate of early supporting cell precursors, Mork et al. (2012) show that *Foxl2*-positive cells in the fetal mouse XX gonad contribute to a population of follicles activated immediately after birth. The traced granulosa cells were only detected in the ovarian medullary region. This suggests that granulosa cells associated with primordial follicles of the cortex, which are activated postpubertally and throughout adult life, are not derived from the same *Foxl2*-expressing cells. Instead, cortical granulosa cells were derived from a second perinatal wave of ingressing cells from the ovarian surface epithelium. These cortical granulosa cell precursors, in and immediately beneath the ovarian surface epithelium, express the adult stem cell marker, *Lgr5* (leucine-rich repeat-containing G-protein-coupled receptor 5; Rastetter et al. 2014). By postnatal day 7, when follicle assembly is essentially complete, ingression of granulosa cell precursors ceases. The existence of two classes of primordial follicles, based on distinct developmental dynamics, has been independently verified using additional inducible mouse models (Zheng et al. 2014).

3.2.3 Granulosa Cell Fate and FOXL2

FOXL2 (forkhead box L2), which encodes a forkhead transcription factor, is a sex-determining gene in at least one mammal. In goats, naturally occurring (polled-intersex-syndrome (*PIS*)), or induced, loss of function of *FOXL2* leads to complete female-to-male sex reversal (Boulanger et al. 2014; Pailhoux et al. 2001). Although it is expressed from around 11.5 *dpc* in mice in gonadal somatic cells near the boundary of the gonad and mesonephros (Wilhelm et al. 2009), its deletion does not impair primary sex determination; instead, loss of FOXL2 leads to an upregulation of the pro-testis pathway later during embryonic development (Garcia-Ortiz et al. 2009), in addition to disruption of follicle activation (Schmidt et al. 2004;

Uda et al. 2004). However, postnatal deletion of *Foxl2* induces a transdifferentiation of granulosa cells to Sertoli-like cells (Ottolenghi et al. 2007; Uhlenhaut et al. 2009), indicating that FOXL2 acts as an anti-testis gene to maintain ovarian fate. Thus, the testis-determining and ovary-determining pathways are mutually antagonistic in both the embryonic and adult gonad.

The mechanism by which FOXL2 antagonizes Sertoli cell fate in the adult ovary was investigated further (Uhlenhaut et al. 2009). Firstly, *Foxl2* deletion was induced in the ovaries of mice harboring a *TESCO*-CFP fluorescent reporter, revealing activation of the reporter shortly after loss of FOXL2. This suggested that FOXL2 might inhibit *Sox9* transcription via interaction with its core gonadal enhancer. The *TESCO* sequence contains predicted FOX-binding sites, and chromatin immunoprecipitation of FOXL2 revealed enrichment of the transcription factor at *TESCO* and around the *Sox9* transcription start site in adult ovaries. Moreover, FOXL2 was shown to attenuate activation of *TESCO* by SRY/SF-1/SOX9 in transient transfection reporter assays in vitro. These inhibitory effects of FOXL2 were enhanced by the presence of interacting estrogen receptors, ESR1/2 (Uhlenhaut et al. 2009). Other FOXL2 targets, both negative and positive, have been proposed, including *Dmrt1* and *Cyp19a1* (Uhlenhaut et al. 2009).

3.2.4 Granulosa Cells and the RSPO1/WNT/β-Catenin Signaling Pathway

Studies of humans exhibiting 46,XX testicular DSD have revealed a requirement for canonical WNT signals in normal ovary development (Parma et al. 2006). The loss of R-spondin1 (*Rspo1*), *Wnt4*, or β-catenin in mice also results in varying degrees of masculinization of the mouse XX gonad. Loss of *Rspo1* reveals its role in the activation of WNT4 and its effector β-catenin in ovary differentiation (Chassot et al. 2008; Tomizuka et al. 2008). Moreover, transgenic expression of a stabilized form of β-catenin in somatic cells of the XY gonad disrupts testis determination, causing sex reversal (Maatouk et al. 2008). This suggests that canonical WNT signals act to inhibit testis development, perhaps through disruptive effects on *SOX9* transcription (Bernard et al. 2012). Consistent with their roles in ovary development, *Rspo1* and *Wnt4* transcripts are detected in the bipotential gonad of XY and XX embryos; their expression persists in the XX gonad after 11.5 *dpc* but is lost from the XY gonad at the same stage (Auguste et al. 2011; Bouma et al. 2010; Jameson et al. 2012b; Nef et al. 2005; Parma et al. 2006; Vainio et al. 1999). Deletion of *Rspo1*, *Wnt4*, or *Ctnnb1* results in similar gonadal phenotypes, characterized by partial female-to-male sex reversal, including ectopic migration of mesonephric endothelial cells leading to development of a testis-like coelomic vessel, ectopic migration of steroidogenic cell precursors from the adrenal glands inducing androgen production, and, in the case of loss of *Rspo1* and *Wnt4*, transdifferentiation of some granulosa cells into Sertoli-like cells expressing *Sox9* and formation of testis cords around the time of birth (Chassot et al. 2008; Heikkila et al. 2002; Jeays-Ward et al. 2003, 2004; Liu et al. 2009; Maatouk et al. 2008; Manuylov et al. 2008; Nicol and Yao 2015; Tomizuka et al. 2008; Vainio et al. 1999). Follistatin (*Fst*) and bone morphogenetic protein 2 (*Bmp2*), both of which exhibit enhanced expression in the XX gonad after 11.5 *dpc*, have been identified as targets of the WNT signaling pathway during ovarian development. Study of *Fst*-deficient XX gonads suggests that FST is required for the inhibition of coelomic vessel formation and survival of meiotic germ cells in female gonads (Yao et al. 2004). *Bmp2* loss-of-function mutants die around mid-gestation, preventing the examination of sexual development in these embryos, but studies in hamster ovaries in vitro suggest that BMP2 acts to facilitate germ cell entry into meiosis and prevent their apoptosis (Chakraborty and Roy 2015), consistent with observations that germ cells in *Rspo1*- and *Wnt4*-deficient mutant gonads fail to enter meiosis and instead undergo apoptosis (Auguste et al. 2011; Maatouk et al. 2013; Naillat et al. 2010; Vainio et al. 1999; Yao et al. 2004). The production of androgens in *Wnt4*-deficient mutants leads to a partial masculinization of the internal genitalia, including persistence of the Wolffian duct (Heikkila et al. 2005).

In *Rspo1*-deficient XX mutants, there is a reduction in *Foxl2* levels, but its expression is not abolished (Auguste et al. 2011). *Foxl2* expression persists in the region of the mutant gonad in close contact with the mesonephros—the medullary region—suggesting that there are two spatially distinct populations of *Foxl2*-positive somatic cells: one in which *Foxl2* expression is RSPO1 dependent and one in which it is RSPO1 independent. In contrast, *Rspo1* expression is not dependent on FOXL2 (Auguste et al. 2011). Moreover, analysis of mouse knockouts shows that *Wnt4* and *Foxl2* are also independently expressed (Ottolenghi et al. 2007). Along with data concerning the origin of granulosa cells discussed in Sect. 3.2.2, these results support the idea that ovarian differentiation in the mouse is not controlled by a simple linear genetic pathway (see Fig. 3.1). As discussed above, *Lgr5* has been identified as a marker of somatic cells in the ovarian cortex that are precursors to cortical granulosa cells. *Lgr5* encodes a G-protein-coupled receptor that is able to bind RSPO1 and activate the RSPO1/β-catenin pathway (Carmon et al. 2011; Chen et al. 2013; de Lau et al. 2011). It is expressed specifically in the XX gonad from 12.5 *dpc*, when its expression declines in XY gonads (Rastetter et al. 2014). However, in contrast to *Rspo1*- or *Wnt4*-deficient XX gonads, loss of *Lgr5* does not lead to a partial female-to-male sex reversal phenotype, suggesting possible compensation by some other LGR family members (Rastetter et al. 2014). Interestingly, *Foxl2* and *Lgr5* expression profiles are mutually exclusive at 13.5, 14.5, and 18.5 *dpc*, similar to the situation in chicken and goat, where FOXL2 and RSPO1 act in two distinct populations of cells (Kocer et al. 2008; Smith et al. 2008).

3.2.5 Loss of Both FOXL2 and RSPO1/WNT/β-Catenin Signals

Analyses of double knockouts for *Foxl2* and *Wnt4* (Ottolenghi et al. 2007) and *Rspo1* and *Foxl2* (Auguste et al. 2011) reveal a somewhat more severe XX sex reversal phenotype than the single gene knockouts, including an increase in the numbers of Leydig- and Sertoli-like cells. This suggests that the FOXL2 and canonical WNT pathways act together in order to orchestrate ovary development. However, those double knockouts do not fully sex reverse; Sertoli cell differentiation is still delayed and does not occur throughout the XX gonad. These data suggest that there may still be unidentified ovary-determining genes/pathways in the mouse that are able to compensate for the loss of FOXL2 and the RSPO1/WNT/β-catenin pathway. The distinct granulosa cell lineages discussed above may also contribute to compensatory pathways by exhibiting differential sensitivity to gene ablation. Moreover, we cannot exclude a role for germ cells in inhibiting sex reversal in pro-ovary gene knockouts at the early stage of sex determination. Earlier reports claimed that absence of germ cells in XX gonads could lead to partial masculinization of the gonad (McLaren 1991). More recently, induced ablation of female germ cells in transgenic mice has shed doubt on the idea that oocytes are required for the maintenance of somatic cell identity during ovarian development (Uhlenhaut et al. 2009). Maatouk et al (2013) showed that loss of germ cells in the XX *Wnt4*-deficient gonad occurs in an anterior to posterior wave from around 15.5 *dpc* and that in the region missing germ cells, granulosa cells exit their quiescent stage and are then able to express pro-testis genes such *Sox9* around birth. However, induction of early germ cell loss in *Wnt4*-deficient gonads prevents this disruption to granulosa cell quiescence. These observations suggest that somatic cell–germ cell interactions in the prenatal ovary at early meiotic stages are important for pregranulosa cell quiescence and germ cell survival signals, at least in the context of disruption to WNT signaling (Maatouk et al. 2013). Loss of pregranulosa cell quiescence late in fetal life, due to disruption of this signaling, is a prerequisite for subsequent activation of SOX9. The authors also predict that early disruption to the mitotic arrest of pregranulosa cells, in combination with loss of key pro-ovarian pathways, might precipitate primary female-to-male gonadal sex reversal (Maatouk et al. 2013).

In humans, duplication of *WNT4* is associated with 46,XY gonadal dysgenesis (Jordan et al. 2001). In XY mice, overexpression of *Wnt4* is not sufficient to induce ovary differentiation, although some disruption to male development is observed, including gonadal vasculature defects and lower levels of circulating testosterone (Jeays-Ward et al. 2003; Jordan et al. 2003). In addition, overexpression of the goat *RSPO1* gene in transgenic mice does not disrupt male development (Buscara et al. 2009). These results suggest that the RSPO1/WNT/β-catenin pathway is not the master regulatory pathway directing ovary differentiation in mice. It is unclear whether such a pathway exists.

3.2.6 Probing the Antagonism Between the Testis- and Ovary-Determining Pathways in More Depth

One question that has intrigued researchers is what would be the consequence of simultaneously deleting a key testis- and ovary-determining gene? Would neither testicular nor ovarian differentiation occur in the absence of two mutually antagonistic sex-determining genes? This was first tested by deletion of pro-ovarian *Rspo1* and pro-testis *Sox9* in the developing gonad, using a combination of constitutive and conditional gene targeting (Lavery et al. 2012). XY and XX embryos lacking both genes in the gonadal supporting cell lineage developed varying degrees of testicular tissue, albeit not with the same temporal profile of control XY gonads. Thus, remarkably, both SRY and SOX9 are dispensable for Sertoli cell differentiation and testicular morphogenesis in the absence of RSPO1. The authors suggest that additional factors, particularly SOX8, SOX10, and SRY (in XY embryos), are responsible for the testicular differentiation in double knockouts, highlighting redundant "backup" pathways that can be employed in the event of genetic (or other) perturbations. Such redundancy is a common theme in our understanding of sex determination. Similar observations have been made in embryonic gonads lacking *Sox9* and *Ctnnb1* (Nicol and Yao 2015) and *Fgf9* and *Wnt4* (Jameson et al. 2012a). One intriguing question remains after these observations, namely, why does the removal of two key testis- and ovary-determining genes result in a tendency toward testis development, given the initial ovarian lineage priming that exists in the bipotential gonad? Of course, it will require a more extensive dataset of double knockouts to know whether this male bias in outcome is a consistently observed phenomenon; any answer will likely depend on the gene pairs analyzed, the balance of their respective roles as either positive or inhibitory molecules, and the strength of these effects.

3.3 Conclusion and Future Perspectives

This review of the opposing genetic pathways regulating differentiation of supporting cells in male and female gonads has revealed a complex story of mutual antagonism and compensatory mechanisms that act to reinforce sex-specific cell fates and canalize the development of the bipotential gonad primordium. It seems likely that additional sex-determining genes remain to be identified: the majority of cases of 46,XY CGD in humans lack a molecular explanation. A small number of cases of 46,XX testicular DSD not attributable to *SRY* translocations also lack a genetic cause. It remains to be determined whether any combination of gene deletions will result in genuine XX testis development during early embryogenesis in mice, and if not, why not? This will require better understanding of the differences between mouse, human, and goat ovary developments. The relationship between the pro-ovarian FOXL2 and canonical WNT signaling pathways remains to be clarified, as do the consequences of granulosa cell heterogeneity for how these pathways interact—and the contribution of germ cell–somatic cell interactions.

Numerous points of mutually antagonistic contact between the testis- and ovary-determination pathways exist, mostly defined by genetics. The molecular basis of these antagonisms will shed light on fundamental mechanisms in cell fate determination. It is perhaps surprising that some key sex-determining genes appear to have primarily negative roles in inhibiting the opposite sexual fate. This is perhaps due to an incomplete picture of the whole process, but may also reflect what it is to be classified as a sex-determining gene: irrespective of any positive roles a gene has, in the absence of a role in inhibiting the opposite sex, sex reversal, even partial, would presumably not be a phenotypic consequence of its loss.

It will be interesting to compare comprehensively the pathways operating during primary sex determination and the maintenance of adult supporting cell fate and account for the unexpected plasticity of gonadal supporting cells. It is unclear whether this plasticity is a unique feature of supporting cells, but it will be interesting to determine its transcriptomic and epigenomic basis. Comparison of data generated in humans, mice, and other mammals, in addition to other vertebrates, will no doubt inform our developing understanding of sex determination and its molecular and cellular basis.

References

Albrecht KH, Eicher EM (2001) Evidence that Sry is expressed in pre-Sertoli cells and Sertoli and granulosa cells have a common precursor. Dev Biol 240:92–107

Auguste A, Chassot AA, Gregoire EP et al (2011) Loss of R-spondin1 and Foxl2 amplifies female-to-male sex reversal in XX mice. Sex Dev 5:304–317

Bardoni B, Zanaria E, Guioli S et al (1994) A dosage sensitive locus at chromosome Xp21 is involved in male to female sex reversal. Nat Genet 7:497–501

Barrionuevo F, Bagheri-Fam S, Klattig J, Kist R, Taketo MM, Englert C, Scherer G (2006) Homozygous inactivation of Sox9 causes complete XY sex reversal in mice. Biol Reprod 74:195–201

Benko S, Gordon CT, Amiel J, Lyonnet S (2011) Disruption of a long distance regulatory region upstream of SOX9 in isolated disorders of sex development. J Med Genet 48:825–830. doi:10.1136/jmedgenet-2011-100255

Bernard P, Ryan J, Sim H, Czech DP, Sinclair AH, Koopman P, Harley VR (2012) Wnt signaling in ovarian development inhibits Sf1 activation of Sox9 via the Tesco enhancer. Endocrinology 153:901–912

Berta P, Hawkins JR, Sinclair AH, Taylor A, Griffiths BL, Goodfellow PN, Fellous M (1990) Genetic evidence equating SRY and the testis-determining factor. Nature 348:448–450

Bhandari RK, Haque MM, Skinner MK (2012) Global genome analysis of the downstream binding targets of testis determining factor SRY and SOX9. PLoS One 7:e43380

Birk OS, Casiano DE, Wassif CA et al (2000) The LIM homeobox gene Lhx9 is essential for mouse gonad formation. Nature 403:909–913

Bishop CE, Whitworth DJ, Qin Y et al (2000) A transgenic insertion upstream of sox9 is associated with dominant XX sex reversal in the mouse. Nat Genet 26:490–494

Bogani D, Siggers P, Brixey R et al (2009) Loss of mitogen-activated protein kinase kinase kinase 4 (MAP3K4) reveals a requirement for MAPK signalling in mouse sex determination. PLoS Biol 7:e1000196

Boulanger L, Pannetier M, Gall L et al (2014) FOXL2 is a female sex-determining gene in the goat. Curr Biol 24:404–408

Bouma GJ, Washburn LL, Albrecht KH, Eicher EM (2007) Correct dosage of Fog2 and Gata4 transcription factors is critical for fetal testis development in mice. Proc Natl Acad Sci U S A 104:14994–14999

Bouma GJ, Hudson QJ, Washburn LL, Eicher EM (2010) New candidate genes identified for controlling mouse gonadal sex determination and the early stages of granulosa and Sertoli cell differentiation. Biol Reprod 82:380–389

Bullejos M, Koopman P (2001) Spatially dynamic expression of Sry in mouse genital ridges. Dev Dyn 221:201–205

Buscara L, Montazer-Torbati F, Chadi S et al (2009) Goat RSPO1 over-expression rescues sex-reversal in Rspo1-knockout XX mice but does not perturb testis differentiation in XY or sex-reversed XX mice. Transgenic Res 18:649–654

Carmon KS, Gong X, Lin Q, Thomas A, Liu Q (2011) R-spondins function as ligands of the orphan receptors LGR4 and LGR5 to regulate Wnt/beta-catenin signaling. Proc Natl Acad Sci U S A 108:11452–11457

Chaboissier MC, Kobayashi A, Vidal VI et al (2004) Functional analysis of Sox8 and Sox9 during sex determination in the mouse. Development 131:1891–1901

Chakraborty P, Roy SK (2015) Bone morphogenetic protein 2 promotes primordial follicle formation in the ovary. Sci Rep 5:12664

Chassot AA, Ranc F, Gregoire EP et al (2008) Activation of beta-catenin signaling by Rspo1 controls differentiation of the mammalian ovary. Hum Mol Genet 17:1264–1277

Chen PH, Chen X, Lin Z, Fang D, He X (2013) The structural basis of R-spondin recognition by LGR5 and RNF43. Genes Dev 27:1345–1350

Clépet C, Schafer AJ, Sinclair AH, Palmer MS, Lovell-Badge R, Goodfellow PN (1993) The human SRY transcript. Hum Mol Genet 2:2007–2012

Colvin JS, Green RP, Schmahl J, Capel B, Ornitz DM (2001) Male-to-female sex reversal in mice lacking fibroblast growth factor 9. Cell 104:875–889

Combes AN, Wilhelm D, Davidson T, Dejana E, Harley V, Sinclair A, Koopman P (2009) Endothelial cell migration directs testis cord formation. Dev Biol 326:112–120

Cool J, Carmona FD, Szucsik JC, Capel B (2008) Peritubular myoid cells are not the migrating population required for testis cord formation in the XY gonad. Sex Dev 2:128–133

Cool J, DeFalco T, Capel B (2012) Testis formation in the fetal mouse: dynamic and complex de novo tubulogenesis Wiley interdisciplinary reviews. Dev Biol 1:847–859

da Silva SM, Hacker A, Harley V, Goodfellow P, Swain A, Lovell-Badge R (1996) *Sox9* expression during gonadal development implies a conserved role for the gene in testis differentiation in mammals and birds. Nat Genet 14:62–67

de Lau W, Barker N, Low TY et al (2011) Lgr5 homologues associate with Wnt receptors and mediate R-spondin signalling. Nature 476:293–297

Foster JW, Dominguez-Steglich MA, Guioli S et al (1994) Campomelic dysplasia and autosomal sex reversal caused by mutations in an SRY-related gene. Nature 372:525–530

Fujimoto Y, Tanaka SS, Yamaguchi YL et al (2013) Homeoproteins six1 and six4 regulate male sex determination and mouse gonadal development. Dev Cell 26:416–430

Garcia-Ortiz JE, Pelosi E, Omari S et al (2009) Foxl2 functions in sex determination and histogenesis throughout mouse ovary development. BMC Dev Biol 9:36

Georg I, Bagheri-Fam S, Knower KC, Wieacker P, Scherer G, Harley VR (2010) Mutations of the SRY-responsive enhancer of SOX9 are uncommon in XY gonadal dysgenesis. Sex Dev 4:321–325

Ginsburg M, Snow MH, McLaren A (1990) Primordial germ cells in the mouse embryo during gastrulation. Development 110:521–528

Hacker A, Capel B, Goodfellow P, Lovell-Badge R (1995) Expression of *Sry*, the mouse testis-determining gene. Development 121:1603–1614
Hammes A, Guo JK, Lutsch G et al (2001) Two splice variants of the Wilms' tumor 1 gene have distinct functions during sex determination and nephron formation. Cell 106:319–329
Heikkila M, Peltoketo H, Leppaluoto J, Ilves M, Vuolteenaho O, Vainio S (2002) Wnt-4 deficiency alters mouse adrenal cortex function, reducing aldosterone production. Endocrinology 143:4358–4365
Heikkila M, Prunskaite R, Naillat F et al (2005) The partial female to male sex reversal in Wnt-4-deficient females involves induced expression of testosterone biosynthetic genes and testosterone production, and depends on androgen action. Endocrinology 146:4016–4023
Hiramatsu R, Matoba S, Kanai-Azuma M et al (2008) A critical time window of Sry action in gonadal sex determination in mice. Development 136:129–138
Hiramatsu R, Harikae K, Tsunekawa N, Kurohmaru M, Matsuo I, Kanai Y (2010) FGF signaling directs a center-to-pole expansion of tubulogenesis in mouse testis differentiation. Development 137:303–312
Hu YC, Okumura LM, Page DC (2013) Gata4 is required for formation of the genital ridge in mice. PLoS Genet 9:e1003629
Huang B, Wang S, Ning Y, Lamb AN, Bartley J (1999) Autosomal XX sex reversal caused by duplication of SOX9. Am J Med Genet 87:349–353
Hyon C, Chantot-Bastaraud S, Harbuz R et al (2015) Refining the regulatory region upstream of SOX9 associated with 46,XX testicular disorders of sex development (DSD). Am J Med Genet A 167:1851–1858
Jameson SA, Lin YT, Capel B (2012a) Testis development requires the repression of Wnt4 by Fgf signaling. Dev Biol 370:24–32
Jameson SA, Natarajan A, Cool J et al (2012b) Temporal transcriptional profiling of somatic and germ cells reveals biased lineage priming of sexual fate in the fetal mouse gonad. PLoS Genet 8:e1002575
Jeays-Ward K, Hoyle C, Brennan J, Dandonneau M, Alldus G, Capel B, Swain A (2003) Endothelial and steroidogenic cell migration are regulated by WNT4 in the developing mammalian gonad. Development 130:3663–3670
Jeays-Ward K, Dandonneau M, Swain A (2004) Wnt4 is required for proper male as well as female sexual development. Dev Biol 276:431–440
Jeske YWA, Bowles J, Greenfield A, Koopman P (1995) Expression of a linear Sry transcript in the mouse genital ridge. Nat Genet 10:480–482
Jordan BK, Mohammed M, Ching ST et al (2001) Up-regulation of WNT-4 signaling and dosage-sensitive sex reversal in humans. Am J Hum Genet 68:1102–1109
Jordan BK, Shen JH, Olaso R, Ingraham HA, Vilain E (2003) Wnt4 overexpression disrupts normal testicular vasculature and inhibits testosterone synthesis by repressing steroidogenic factor 1/beta-catenin synergy. Proc Natl Acad Sci U S A 100:10866–10871
Karl J, Capel B (1998) Sertoli cells of the mouse testis originate from the coelomic epithelium. Dev Biol 203:323–333
Katoh-Fukui Y, Tsuchiya R, Shiroishi T, Nakahara Y, Hashimoto N, Noguchi K, Higashinakagawa T (1998) Male-to-female sex reversal in M33 mutant mice. Nature 393:688–692
Katoh-Fukui Y, Miyabayashi K, Komatsu T et al (2012) Cbx2, a polycomb group gene, is required for Sry gene expression in mice. Endocrinology 153:913–924
Kaufman MH (1992) The atlas of mouse development. Elsevier Academic, London
Kent J, Wheatley SC, Andrews JE, Sinclair AH, Koopman P (1996) A male-specific role for SOX9 in vertebrate sex determination. Development 122:2813–2822
Kim Y, Kobayashi A, Sekido R et al (2006) Fgf9 and Wnt4 act as antagonistic signals to regulate mammalian sex determination. PLoS Biol 4:e187
Kocer A, Pinheiro I, Pannetier M et al (2008) R-spondin1 and FOXL2 act into two distinct cellular types during goat ovarian differentiation. BMC Dev Biol 8:36

Koopman P, Gubbay J, Vivian N, Goodfellow P, Lovell-Badge R (1991) Male development of chromosomally female mice transgenic for Sry. Nature 351:117–121

Kusaka M, Katoh-Fukui Y, Ogawa H et al (2010) Abnormal epithelial cell polarity and ectopic epidermal growth factor receptor (EGFR) expression induced in Emx2 KO embryonic gonads. Endocrinology 151:5893–5904

Larney C, Bailey TL, Koopman P (2014) Switching on sex: transcriptional regulation of the testis-determining gene Sry. Development 141:2195–2205

Lavery R, Chassot AA, Pauper E et al (2012) Testicular differentiation occurs in absence of R-spondin1 and Sox9 in mouse sex reversals. PLoS Genet 8:e1003170

Li Y, Zheng M, Lau YF (2014) The sex-determining factors SRY and SOX9 regulate similar target genes and promote testis cord formation during testicular differentiation. Cell Rep 8:723–733

Lindeman RE, Gearhart MD, Minkina A, Krentz AD, Bardwell VJ, Zarkower D (2015) Sexual cell-fate reprogramming in the ovary by DMRT1. Curr Biol 25:764–771

Liu CF, Bingham N, Parker K, Yao HH (2009) Sex-specific roles of beta-catenin in mouse gonadal development. Hum Mol Genet 18:405–417

Lovell-Badge R, Robertson E (1990) XY female mice resulting from a heritable mutation in the murine primary testis determining gene, Tdy. Development 109:635–646

Ludbrook LM, Harley VR (2004) Sex determination: a 'window' of DAX1 activity. Trends Endocrinol Metab 15:116–121

Ludbrook LM, Bernard P, Bagheri-Fam S et al (2012) Excess DAX1 leads to XY ovotesticular disorder of sex development (DSD) in mice by inhibiting steroidogenic factor-1 (SF1) activation of the testis enhancer of SRY-box-9 (Sox9). Endocrinology 153:1948–1958

Luo X, Ikeda Y, Parker KL (1994) A cell-specific nuclear receptor is essential for adrenal and gonadal development and sexual differentiation. Cell 77:481–490

Lybaek H, de Bruijn D, den Engelsman-van Dijk AH, Vanichkina D, Nepal C, Brendehaug A, Houge G (2014) RevSex duplication-induced and sex-related differences in the SOX9 regulatory region chromatin landscape in human fibroblasts. Epigenetics 9:416–427

Maatouk DM, Dinapoli L, Alvers A, Parker KL, Taketo MM, Capel B (2008) Stabilization of {beta}-catenin in XY gonads causes male-to-female sex-reversal. Hum Mol Genet 17:2949–2955

Maatouk DM, Mork L, Chassot AA, Chaboissier MC, Capel B (2013) Disruption of mitotic arrest precedes precocious differentiation and transdifferentiation of pregranulosa cells in the perinatal Wnt4 mutant ovary. Dev Biol 383:295–306

Manuylov NL, Smagulova FO, Leach L, Tevosian SG (2008) Ovarian development in mice requires the GATA4-FOG2 transcription complex. Development 135:3731–3743

Matson CK, Murphy MW, Sarver AL, Griswold MD, Bardwell VJ, Zarkower D (2011) DMRT1 prevents female reprogramming in the postnatal mammalian testis. Nature 476:101–104

McLaren A (1991) Development of the mammalian gonad: the fate of the supporting cell lineage. Bioessays 13:151–156

Meeks JJ, Weiss J, Jameson JL (2003) Dax1 is required for testis determination. Nat Genet 34:32–33

Minkina A, Matson CK, Lindeman RE, Ghyselinck NB, Bardwell VJ, Zarkower D (2014) DMRT1 protects male gonadal cells from retinoid-dependent sexual transdifferentiation. Dev Cell 29:511–520

Moniot B, Declosmenil F, Barrionuevo F et al (2009) The PGD2 pathway, independently of FGF9, amplifies SOX9 activity in Sertoli cells during male sexual differentiation. Development 136:1813–1821

Mork L, Maatouk DM, McMahon JA, Guo JJ, Zhang P, McMahon AP, Capel B (2012) Temporal differences in granulosa cell specification in the ovary reflect distinct follicle fates in mice. Biol Reprod 86:37

Murphy MW, Lee JK, Rojo S et al (2015) An ancient protein-DNA interaction underlying metazoan sex determination. Nat Struct Mol Biol 22:442–451

Muscatelli F, Strom TM, Walker AP et al (1994) Mutations in the *DAX-1* gene give rise to both X-linked adrenal hypoplasia congenita and hypogonadotropic hypogonadism. Nature 372:672–676

Naillat F, Prunskaite-Hyyrylainen R, Pietila I, Sormunen R, Jokela T, Shan J, Vainio SJ (2010) Wnt4/5a signalling coordinates cell adhesion and entry into meiosis during presumptive ovarian follicle development. Hum Mol Genet 19:1539–1550

Nef S, Schaad O, Stallings NR et al (2005) Gene expression during sex determination reveals a robust female genetic program at the onset of ovarian development. Dev Biol 287:361–377

Nicol B, Yao HH (2015) Gonadal identity in the absence of pro-testis factor SOX9 and pro-ovary factor beta-catenin in mice. Biol Reprod 93:35

Ottolenghi C, Veitia R, Quintana-Murci L et al (2000) The region on 9p associated with 46,XY sex reversal contains several transcripts expressed in the urogenital system and a novel doublesex-related domain. Genomics 64:170–178

Ottolenghi C, Pelosi E, Tran J et al (2007) Loss of Wnt4 and Foxl2 leads to female-to-male sex reversal extending to germ cells. Hum Mol Genet 16:2795–2804

Pailhoux E, Vigier B, Chaffaux S et al (2001) A 11.7-kb deletion triggers intersexuality and polledness in goats. Nat Genet 29:453–458

Palmer SJ, Burgoyne PS (1991) *In situ* analysis of fetal, prepuberal and adult XX–XY chimaeric mouse testes: Sertoli cells are predominantly, but not exclusively, XY. Development 112:265–268

Parma P, Radi O, Vidal V et al (2006) R-spondin1 is essential in sex determination, skin differentiation and malignancy. Nat Genet 38:1304–1309

Rastetter RH, Bernard P, Palmer JS et al (2014) Marker genes identify three somatic cell types in the fetal mouse ovary. Dev Biol 394:242–252

Raymond CS, Murphy MW, O'Sullivan MG, Bardwell VJ, Zarkower D (2000) Dmrt1, a gene related to worm and fly sexual regulators, is required for mammalian testis differentiation. Genes Dev 14:2587–2595

Schmidt D, Ovitt CE, Anlag K, Fehsenfeld S, Gredsted L, Treier AC, Treier M (2004) The murine winged-helix transcription factor Foxl2 is required for granulosa cell differentiation and ovary maintenance. Development 131:933–942

Sekido R, Lovell-Badge R (2008) Sex determination involves synergistic action of SRY and SF1 on a specific Sox9 enhancer. Nature 453:930–934

Sekido R, Bar I, Narvaez V, Penny G, Lovell-Badge R (2004) SOX9 is up-regulated by the transient expression of SRY specifically in Sertoli cell precursors. Dev Biol 274:271–279

Sinclair AH, Berta P, Palmer MS et al (1990) A gene from the human sex-determining region encodes a protein with homology to a conserved DNA-binding motif. Nature 346:240–244

Smith CA, Shoemaker CM, Roeszler KN, Queen J, Crews D, Sinclair AH (2008) Cloning and expression of R-Spondin1 in different vertebrates suggests a conserved role in ovarian development. BMC Dev Biol 8:72

Swain A, Narvaez V, Burgoyne P, Camerino G, Lovell-Badge R (1998) Dax1 antagonizes Sry action in mammalian sex determination. Nature 391:761–767

Tevosian SG, Albrecht KH, Crispino JD, Fujiwara Y, Eicher EM, Orkin SH (2002) Gonadal differentiation, sex determination and normal Sry expression in mice require direct interaction between transcription partners GATA4 and FOG2. Development 129:4627–4634

Tomizuka K, Horikoshi K, Kitada R et al (2008) R-spondin1 plays an essential role in ovarian development through positively regulating Wnt-4 signaling. Hum Mol Genet 17:1278–1291

Uda M, Ottolenghi C, Crisponi L et al (2004) Foxl2 disruption causes mouse ovarian failure by pervasive blockage of follicle development. Hum Mol Genet 13:1171–1181

Uhlenhaut NH, Jakob S, Anlag K et al (2009) Somatic sex reprogramming of adult ovaries to testes by FOXL2 ablation. Cell 139:1130–1142

Vainio S, Heikkila M, Kispert A, Chin N, McMahon AP (1999) Female development in mammals is regulated by Wnt-4 signalling. Nature 397:405–409

Vidal VP, Chaboissier MC, de Rooij DG, Schedl A (2001) Sox9 induces testis development in XX transgenic mice. Nat Genet 28:216–217

Warr N, Greenfield A (2012) The molecular and cellular basis of gonadal sex reversal in mice and humans WIREs. Dev Biol 1:559–577

Warr N, Carre GA, Siggers P et al (2012) Gadd45gamma and Map3k4 interactions regulate mouse testis determination via p38 MAPK-mediated control of Sry expression. Dev Cell 23:1020–1031

Warr N, Siggers P, Carre GA et al (2014) Transgenic expression of Map3k4 rescues T-associated sex reversal (Tas) in mice. Hum Mol Genet 23:3035–3044

Wilhelm D, Martinson F, Bradford S et al (2005) Sertoli cell differentiation is induced both cell-autonomously and through prostaglandin signaling during mammalian sex determination. Dev Biol 287:111–124

Wilhelm D, Hiramatsu R, Mizusaki H, Widjaja L, Combes AN, Kanai Y, Koopman P (2007) SOX9 regulates prostaglandin D synthase gene transcription in vivo to ensure testis development. J Biol Chem 282:10553–10560

Wilhelm D, Washburn LL, Truong V, Fellous M, Eicher EM, Koopman P (2009) Antagonism of the testis- and ovary-determining pathways during ovotestis development in mice. Mech Dev 126:324–336

Yao HH, Matzuk MM, Jorgez CJ, Menke DB, Page DC, Swain A, Capel B (2004) Follistatin operates downstream of Wnt4 in mammalian ovary organogenesis. Dev Dyn 230:210–215

Zanaria E, Muscatelli F, Bardoni B et al (1994) An unusual member of the nuclear hormone receptor superfamily responsible for X-linked adrenal hypoplasia congenita. Nature 372:635–641

Zhang L, Chen M, Wen Q et al (2015) Reprogramming of Sertoli cells to fetal-like Leydig cells by Wt1 ablation. Proc Natl Acad Sci U S A 112:4003–4008

Zhao L, Ng ET, Davidson TL et al (2014) Structure-function analysis of mouse Sry reveals dual essential roles of the C-terminal polyglutamine tract in sex determination. Proc Natl Acad Sci U S A 111:11768–11773

Zhao L, Svingen T, Ng ET, Koopman P (2015) Female-to-male sex reversal in mice caused by transgenic overexpression of Dmrt1. Development 142:1083–1088

Zheng W, Zhang H, Gorre N, Risal S, Shen Y, Liu K (2014) Two classes of ovarian primordial follicles exhibit distinct developmental dynamics and physiological functions. Hum Mol Genet 23:920–928

Chapter 4
Mesonephric Cell Migration into the Gonads and Vascularization Are Processes Crucial for Testis Development

Sarah M. Romereim and Andrea S. Cupp

Abstract Testis morphogenesis requires the integration and reorganization of multiple cell types from several sources, one of the more notable being the mesonephric-derived cell population. One of the earliest sex-specific morphogenetic events in the gonad is a wave of endothelial cell migration from the mesonephros that is crucial for (1) partitioning the gonad into domains for testis cords, (2) providing the vasculature of the testis, and (3) signaling to cells both within the gonad and beyond it to coordinately regulate testis development. In addition to endothelial cell migration, there is evidence that precursors of peritubular myoid cells migrate from the mesonephros, an event which is also important for testis cord architecture. Investigation of the mesonephric cell migration event has utilized histology, lineage tracing with mouse genetic markers, and many studies of the signaling molecules/pathways involved. Some of the more well-studied signaling molecules involved include vascular endothelial growth factor (VEGF), platelet-derived growth factor (PDGF), and neurotrophins. In this chapter, the morphogenetic events, relevant signaling pathways, mechanisms underlying the migration, and the role of the migratory cells within the testis will be discussed. Overall, the migration of mesonephric cells into the early testis is indispensable for its development and future functionality.

4.1 Introduction

Mammalian testis development requires the integration of cells from different sources into a complex structure. Cells from the coelomic epithelium and the mesonephros contribute to the gonad via migratory events at multiple points during development. Some of these migrations occur in both male and female gonads, while some are sex specific. In the male, for example, a wave of cell migration from

S.M. Romereim • A.S. Cupp (✉)
University of Nebraska-Lincoln, Lincoln, NE, USA
e-mail: sromereim2@unl.edu; acupp2@unl.edu

R.P. Piprek (ed.), *Molecular Mechanisms of Cell Differentiation in Gonad Development*, Results and Problems in Cell Differentiation 58,
DOI 10.1007/978-3-319-31973-5_4

the mesonephros into the gonad initiates key testis architecture changes. This testis-specific mesonephric migration is dependent on prior sex differentiation and occurs in response to chemotactic signaling molecules. The cells that migrate into the early testis are primarily endothelial, and these cells are required for seminiferous cord morphogenesis and testis vascularization. In this chapter, we present the advancements in understanding that have been made regarding the contribution of mesonephric migration to testis development. The studies discussed chiefly focus on mouse and rat models, as much of the work has necessitated knockout and conditional transgenic mouse lines.

4.2 The Timing of Mesonephric Cell Migration and Initial Vascularization of the Developing Gonad

The developmental context of testis-specific mesonephric cell migration is a key part of understanding the importance of this event. The gonad arises as a thickening and restructuring of the coelomic epithelium to form the genital ridge at the 5–6 tail somite developmental stage or around 10.25–10.4 days *post coitum* (*dpc*) in the mouse (Hu et al. 2013; Kusaka et al. 2010) (Fig. 4.1). The exact timing of this event in *dpc* can vary due to subtle differences in embryonic growth rate in various mouse lines, which can be caused by variable accelerated preimplantation development related to the Y chromosome (Burgoyne et al. 1995). Genital ridge formation has been reported from 9 *dpc* up to 10.5 *dpc* (Chen et al. 2012; Hacker et al. 1995; Harikae et al. 2013; Karl and Capel 1995; Nef and Parada 2000; Tanaka and Nishinakamura 2014). This event involves proliferation and ingression of cells from the coelomic epithelium to a mesenchymal compartment producing the genital ridge (Harikae et al. 2013; Karl and Capel 1998; Kusaka et al. 2010; Schmahl et al. 2000; Schmahl and Capel 2003). Coelomic epithelial migration prior to 11.5 *dpc* occurs in both XX and XY gonads, and the migratory cells will eventually become Sertoli cells and interstitial cells in the testis and follicular granulosa cells in the ovary (Karl and Capel 1998; McLaren 2000; Mork et al. 2012). The resultant gonad primordia are initially long, thin structures beside the mesonephros. Primordial germ cells must also migrate into and colonize the gonad starting at 10.0 *dpc* and continuing up to 11.5 *dpc* (Ewen and Koopman 2010; Gomperts et al. 1994; Molyneaux et al. 2001) (Fig. 4.1). Importantly, up until 11.5 *dpc*, the early gonad is morphogenetically identical in males and females as the genital ridge is bipotential and thus can give rise to either a testis or an ovary.

Male sexual differentiation is dependent on the expression of the *Sry* gene and is supported by continued expression of a set of male differentiation maintenance genes (Koopman et al. 1991; Lovell-Badge and Robertson 1990; Sinclair et al. 1990; Ungewitte and Yao 2013). A network of transcription factors including Wilms' tumor 1 (WT1) initiates *Sry* mRNA expression at ~10.5 *dpc* beginning in the center of the gonad and spreading outward in a wave (Bullejos and Koopman 2001; Bradford et al. 2009). The main events in testis development are the

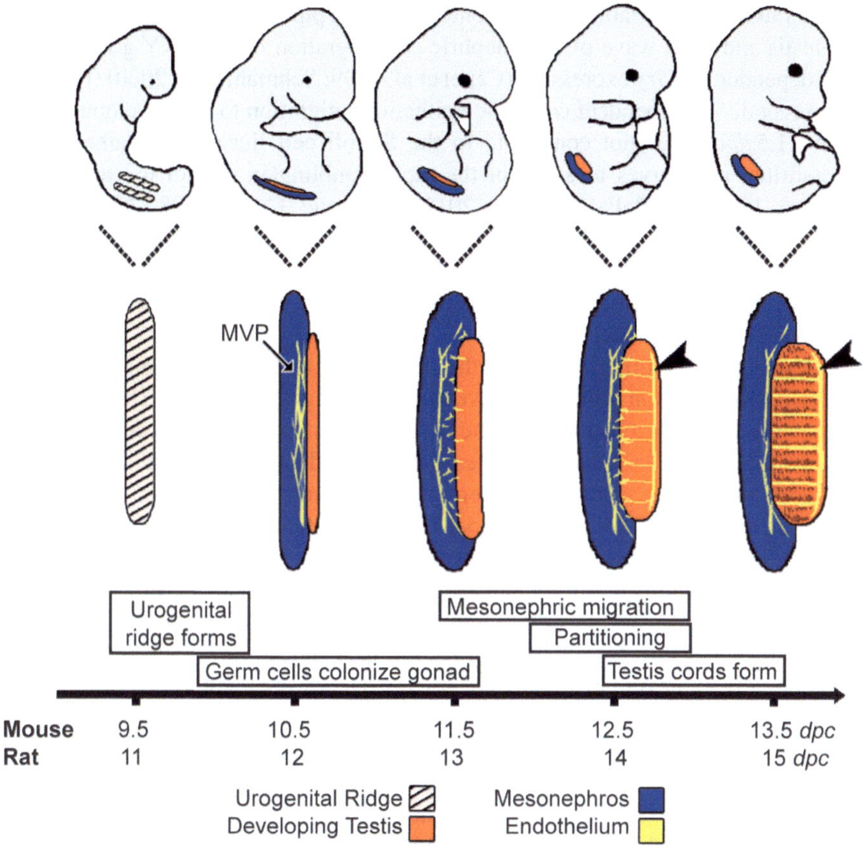

Fig. 4.1 This timeline of testis development shows a progression from the undifferentiated urogenital ridge at 9.5 days *postcoitum* (*dpc*) in the mouse (11 *dpc* in the rat) to the vascularized, compartmentalized testis complete with testis cords at 13.5 *dpc* in the mouse. Mesonephric cell migration into the male gonad begins in the mouse at 11.5 *dpc* with the breakdown of the mesonephric vascular plexus (MVP, *arrow*), and this MVP dissolution and subsequent events do not occur in female embryos. The migrating endothelial cells partition the gonad into avascular domains and form the coelomic vessel (CV, *arrowheads*) by 12.5 *dpc* in the mouse. Testis cords then begin to form between 12.5 and 13.5 *dpc*

specification of the Sertoli cell lineage, testis cord morphogenesis, and then pre-Sertoli cell expansion concurrent with testis cord elongation (Ungewitte and Yao 2013). All of these phases of development are vital to the adult testis structure and function. This chapter focuses on the second event in that series, the formation of testis cords, and the critical role that mesonephric cell migration plays in that process.

While male-specific cellular identities are established earlier, the first two morphogenetic manifestations of testis differentiation are (1) an *Sry*-dependent thickening of the gonad and the coelomic epithelium due to increased cellular proliferation in preparation for a second coelomic epithelial contribution of cells to the testis and (2) a wave of mesonephric cell migration into the XY gonad that is also dependent on *Sry* expression (Capel et al. 1999; Schmahl et al. 2000) (Fig. 4.1). The second, *Sry*-dependent coelomic epithelium migration to the developing testis after 11.5 *dpc*, does not contribute to the Sertoli cell lineage but rather to the interstitium and serves to assist in the rapid doubling in size of the testis from 11.5 to 13.5 *dpc* (DeFalco et al. 2011; Karl and Capel 1998; Nel-Themaat et al. 2009; Schmahl et al. 2000). The mesonephric cell migration, on the other hand, has less to do with increasing the overall size of the testis and is instead required for tissue patterning and testis cord morphogenesis (Bott et al. 2006; Combes et al. 2009; Cool et al. 2011).

Mesonephric cell migration into the developing testis shares some basic characteristics in common with other developmental migration events. This migration does not require direct contact with the XY gonad and can actually occur through an XX gonad placed in between a mesonephros and an XY gonad in culture at 11.5 *dpc*, suggesting that the trigger to migrate must come from a diffusible signaling molecule (Tilmann and Capel 1999). This functions similarly to morphogen gradients that promote and direct migration during embryonic development.

The events that occur during mesonephric cell migration also mirror cell behaviors seen in other migratory events and have been closely observed with time-lapse confocal microscopy from 11.5 to 13.5 *dpc* in a mouse line with GFP markers conditionally expressed in endothelial cells (Coveney et al. 2008). The majority of the migrating mesonephric cells are endothelial in nature, and thus descriptions of the behavior of the migrating population focus on that cell type (Combes et al. 2009) (Fig. 4.2). First, the endothelial cells that will migrate are released from an existing vessel network within the mesonephros in a sex-specific dissolution of the mesonephric vascular plexus. In XX embryos, the mesonephric vascular network remains intact. In XY embryos, vascular cells that are released from the plexus then undergo an endothelial-to-mesenchymal transition (EMT) by losing their extended squamous shape, detaching from the surrounding endothelial cells, and then by extending long filopodia and beginning to migrate to the coelomic domain of the developing gonad. The migrating population does not move as a unit, but rather individual cells move at different paces along the same general paths. These cells then form the coelomic vessel, the male-specific main testicular artery that persists into adulthood in rodents, and vessels branch off into the interstitium between the testis cords (Figs. 4.1 and 4.2). Ultimately, this process partitions the gonad into approximately ten avascular domains which form the testis cords and subsequently the seminiferous tubules (Coveney et al. 2008) (Fig. 4.2).

This wave of patterned endothelial migration does not occur in the XX gonad (Fig. 4.2). During the 11.5–13.5 *dpc* timeframe, the female mesonephric vascular plexus remains intact (Coveney et al. 2008). In fact, very few morphological changes are seen in the developing ovary until shortly before birth, 18.5 *dpc* in the mouse (Brennan and Capel 2004). However, there is research on ovarian development in domestic livestock such as cattle and sheep that supports a model involving early migration from the mesonephros. In this model, mesonephric

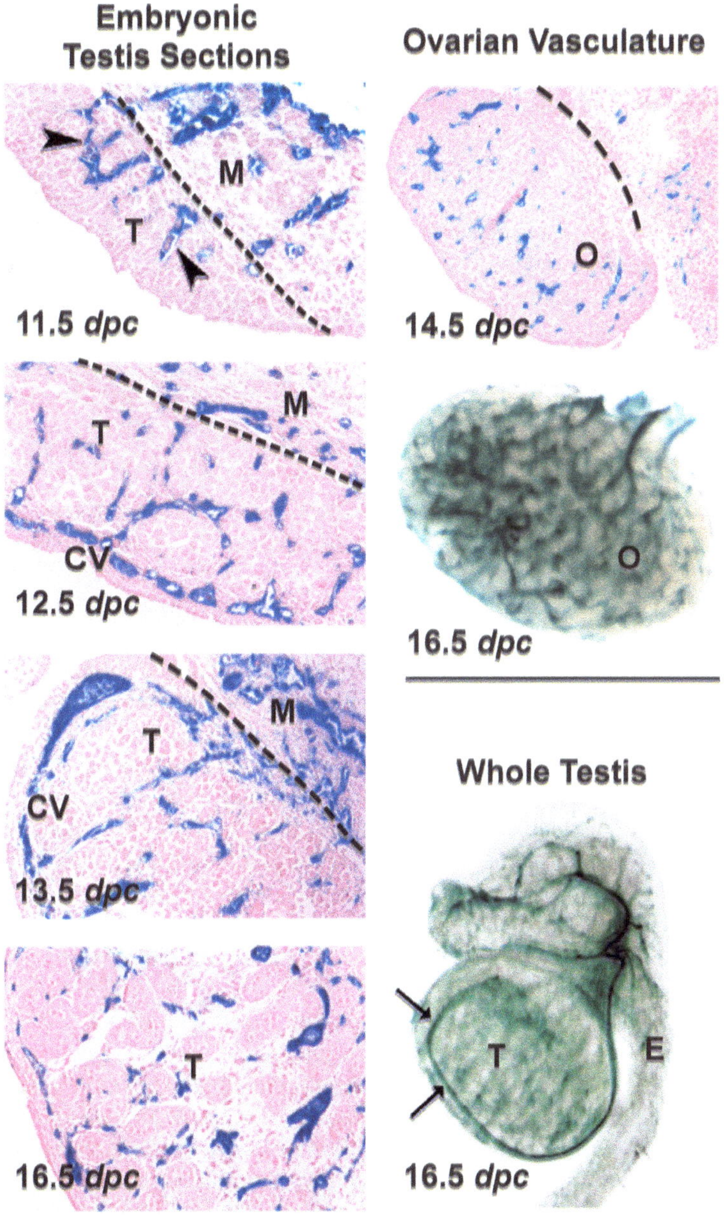

Fig. 4.2 Endothelial cells are stained blue in mesonephros (M) and gonads (*T* testis, *O* ovary) from transgenic mice expressing KDR-LacZ (a VEGF receptor-driven beta-galactosidase marker). At 11.5 *dpc* in the male, mesonephric endothelial cells are in the process of migrating into the developing testis (*arrowheads*). At 12.5 and 13.5 *dpc*, the coelomic vessel (CV) is visible in the developing testis, avascular domains have segmented, and testis cords are beginning to form. By 16.5 *dpc*, the male gonad contains seminiferous tubules as seen in the histological section and a defined vasculature with a prominent coelomic vessel (*arrows*) as seen in the whole mount stained testis with the epididymis (E) attached. Ovaries at 14.5 *dpc* as shown in histological section and at 16.5 *dpc* with whole mount staining both lack all male-specific vascular structures and instead

surface epithelial cells at 10–11.5 *dpc* in mice (days of gestation 30–50 for cattle and 22–38 for sheep) differentiate into the Gonad Ridge Epithelial-Like (GREL) cells (Smith et al. 2014). Based on data primarily from sheep and bovine models, mesonephric stromal cells, endothelial cells, and the GREL cells all migrate into the developing ovary after sex specification, but this process has not been definitively shown in rodent models (Hummitzsch et al. 2013; Smith et al. 2014). The key sex-specific differences are the relatively disorganized mesonephric migration of the developing XX gonad and that the processes are regulated by different signaling pathways (Ungewitte and Yao 2013).

4.3 How the Knowledge of Mesonephric Cell Migration Emerged

Based on histological examination, researchers suspected for years that cells from the mesonephros migrated into and contributed to the structure of both the undifferentiated gonad and sex-differentiated testes (Upadhyay et al. 1979, 1981; Wartenberg et al. 1991). Once the technological tools became available in the 1990s, studies using lineage-tracing markers directly demonstrated mesonephric migration. The first lineage study used mouse testis explants cultured with mesonephros containing a transgenic marker that could be detected by in situ hybridization to show that (1) the mesonephros contributes non-germ and non-Sertoli cells to the testes by migration, (2) this cell migration is required for normal testis architecture, and (3) both XX and XY mesonephroi can contribute cells to a developing testis (Buehr et al. 1993). While the identity of the migrating cells was inconclusive at this point, the demonstration that mesonephric cell migration is required for testis morphogenesis was a significant advancement.

A few years later, researchers used gonad explants cultured with mesonephroi of *ROSA26-LacZ* mice to show that only XY gonads promote mesonephric cell migration from 11.5 to 16.5 *dpc* (Martineau et al. 1997). The *ROSA26-LacZ* transgenic mouse line ubiquitously expresses the enzyme β-galactosidase. When *ROSA26-LacZ* mesonephric cells migrated into an XY gonad explant lacking the transgene, the substrate for β-galactosidase was provided to selectively stain those cells blue while the surrounding gonad remained unstained. This proved that those blue cells in the gonad originated in the *ROSA26-LacZ* mesonephros explant. This study also provided more insight into the cell types that undergo migration based on the final morphology and destinations of the migratory cells. The authors suggested that endothelial cells, cells that migrated to positions near areas of Sertoli cell

---

Fig. 4.2 (continued) have blood vessels spread uniformly throughout the ovary. Microscopy images reprinted with publisher (Springer) permission from Bott et al. (2010)

condensation, and other cells that migrated to locations close to endothelial cells composed the migratory population (Martineau et al. 1997).

The exact cell types that migrate from the mesonephros have been a topic of debate over the years. Early on Martineau et al. (1997) and Merchant-Larios et al. (1993) proposed that peritubular myoid cells and other interstitial cells migrate into the gonad along with endothelial cells. Peritubular myoid cells form a thin, flattened layer around the testis cords and future seminiferous tubules, defining the structure of the cords and potentially contributing to the adult functionality of the tubules by muscular contraction to aid movement of sperm through the lumen (Wilhelm et al. 2007). Interestingly, these peritubular myoid cells are the only testis cell type with no known ovarian equivalent. They also have no known cell type-specific genetic markers and can only be identified by genes common to muscle cells such as alpha-smooth muscle actin and desmin (Jeanes et al. 2005).

Later studies argue that the migrating population is primarily or even exclusively endothelial cells based on the expression of endothelial markers before/during the migratory period. In one such article, Combes et al. (2009) performed coculture from 11.5 to 12.5 *dpc* of mesonephros constitutively expressing green fluorescent protein (GFP) with XY gonads not expressing GFP and showed by immunofluorescence that >99 % of the GFP-labeled cells co-localized with the endothelial marker cadherin-5 (also designated VE-cadherin, CD144 antigen, and Cadherin 5, type 2). These data support the claim that the majority of the initial wave of migratory cells is endothelial in nature, but does not rule out subsequent migration or differentiation of other cell types. In a separate study, Cool et al. (2008) utilized lineage tracing with an EYFP alpha-smooth muscle actin (*α-Sma*) marker of interstitial cells expressed in transgenic mesonephroi cultured with wild-type XY gonads. This experiment showed that no cells already expressing alpha-smooth muscle actin were part of the migrating population, which the authors argue demonstrates that no peritubular myoid cells in the testis originate in the mesonephros (Combes et al. 2009; Cool et al. 2008). However, this finding does not rule out the possibility of precursors that do not yet express the interstitial actin marker migrating into the gonad and subsequently differentiating. Overall, the literature supports the idea that the wave of mesonephric cell migration into the gonad concurrent with male sex specification is primarily endothelial in nature but does not rule out subsequent migration of other testis cell types or their precursors. In addition to the evidence for a large migratory endothelial component, there is also support for the idea that the precursors of peritubular myoid cells migrate from the mesonephros in response to neurotrophins, though the timing of the migration is difficult to determine. As will be discussed in more detail in the section specific to neurotrophin signaling, mesonephric-derived peritubular myoid cells express the low-affinity neurotrophin receptor NGFR (formerly p75NTR), and these cells do not appear in the fetal testis unless cocultured with a mesonephros explant (Russo et al. 1999; Campagnolo et al. 2001). Neurotrophin-3 (NT3) signaling inhibition via antisense nucleotides against NT3, a pharmacological inhibitor of the receptor NGFR, and receptor-IgG fusion dominant negative expression also disrupt seminiferous tubule formation and testis architecture (Cupp et al. 2000, 2003; Levine et al. 2000). Additionally, a small proportion of GFP-expressing mesonephric cells that migrated into the gonad and subsequently were isolated by fluorescence-assisted cell sorting (FACS) were shown to express the muscle-specific intermediate filament desmin, which is a general marker for myoid cells, including the peritubular myoid cells (Nishino et al. 2001). Furthermore, there is evidence that some cells that will contribute to the Leydig cell (an interstitial cell adjacent to the mature seminiferous tubules) population migrate into the gonad from the mesonephric border alongside vascular cells (DeFalco et al. 2011). Thus, the mesonephros contributes migratory cells of both endothelial and non-endothelial cell types, though the bulk of the migratory population express endothelial cell markers during migration.

Migration from the mesonephros into the gonad after the period of sex specification also occurs in the developing ovary. After 12.5 *dpc* the transcription factor GLI1 is expressed in the mesonephros but not the ovary. GLI1+ mesonephric cells then migrate into the XX gonad starting at 18.5 *dpc*, just before birth, to contribute a small but crucial subpopulation to the theca cells of the ovarian follicles (Liu et al. 2015). Mesonephric migration into the gonad thus has the possibility to occur at multiple points during development and contribute to its architecture and function.

4.4 Signals That Promote Mesonephric Cell Migration

The signaling pathways that coordinately regulate testis development are numerous. This section will discuss those pathways that are involved in mesonephric cell migration into the XY gonad and the phenotypic results if those signals are inhibited (Table 4.1). Some signaling pathways are directly required for mesonephric cell migration, while others act in a permissive capacity or modulate the effects of other signals.

4.4.1 Sex Determination and Pleiotropic Regulators of Migration

SRY and SOX9 Male sex determination is a prerequisite for migration of mesonephric cells (Fig. 4.3). The initiator of the male-specific genetic program is *Sry* expression, which occurs from 10.5 to 12.5 *dpc* in the mouse and is necessary for cell migration into the testes from the mesonephros (Capel et al. 1999; Koopman et al. 1991). Shortly after the initiation of *Sry* expression, the SRY-box containing gene 9 (*Sox9*) is expressed in the pre-Sertoli cells via synergistic action of SRY and the transcription factor steroidogenic factor 1 (SF1, also known as NR5A1) on a *Sox9* enhancer (Luo et al. 1994; Sekido and Lovell-Badge 2008). Unlike *Sry*, *Sox9*

Table 4.1 Selected genes that play key roles in mesonephric cell migration into the developing testis

Gene symbol	Gene name	Function/role	Site and time of expression during early gonad development	Mutation/inhibition effects	References
Cdh5	Cadherin-5 (VE-cadherin, CD144)	Adhesion molecule involved in adherens junctions in vasculature	Expressed in all endothelial cells	Smaller vessels that rely on adherens junction binding cannot form, testis vascularization is inhibited, and testis cord formation prevented	Combes et al. (2009)
Cited2	cbp/p300-interacting transactivator 2	Cofactor for SF1 that has sex-specific roles in regulating gonad development	Briefly expressed in the gonad primordia at 10 *dpc* and then expressed again at 13.5 *dpc*	*Cited2*$^{-/-}$ XY gonads show delayed development and disrupted architecture including lack of testis cords and disrupted coelomic vessel formation. *Cited2*$^{-/-}$ XX gonads experience ectopic mesonephric cell migration into the gonad	Combes et al. (2010)
Dkk1	Dickkopf homolog 1	An inhibitor of the canonical WNT/RSPO signaling pathway	From 12.5 to 18.5 *dpc* protein expressed in the testicular cords but not in coelomic domain of the testis	Exogenous DKK1 inhibits testicular vascularization	Caruso et al. (2015)
Fgf9	Fibroblast growth factor 9	Upregulates and maintains *Sox9* expression after sex determination in the XY gonad	Expression is upregulated in XY gonads in the *Sox9*-expressing cells from 11.5 *dpc* onward, while it is present in the XX gonad at	*Fgf9*$^{-/-}$ mice can initiate but not maintain *Sox9* expression, resulting in impaired Sertoli cell differentiation and testis cord formation	Kim et al. (2006)

(continued)

Table 4.1 (continued)

Gene symbol	Gene name	Function/role	Site and time of expression during early gonad development	Mutation/inhibition effects	References
			11.5 *dpc* but subsequently downregulated		
Fgfr2	Fibroblast growth factor receptor 2	FGF9 signaling in the gonad acts through FGFR2	Expressed in scattered cells in the XY gonad at 11.0 *dpc* with the number of FGFR2+ cells increasing rapidly as sex determination occurs. Expressed in the XX gonad during the same time period but does not localize to the nucleus	*Fgfr2*$^{-/-}$ mice have similar phenotypes to *Fgf9*$^{-/-}$ mice and experience sex reversal	Bagheri-Fam et al. (2008); Kim et al. (2007); Schmahl et al. (2004)
Kdr	Kinase insert domain receptor	VEGFA receptor that is active during and responsible for endothelial cell migration and testis vascularization	Expressed in all endothelial cells	Pharmacological inhibition of KDR and of both KDR and FLT1 (another VEGFA receptor not expressed in the XY gonad/mesonephros until after testis cord formation) prevents endothelial migration and testis cord formation	Bott et al. (2006, 2010); Cool et al. (2011)
Ngfr	Nerve growth factor receptor	Low-affinity receptor for neurotrophin 3	Expressed in the mesonephros starting at 10.5 *dpc* and then marking peritubular myoid cells that migrate into the testis	Pharmacological inhibition inhibits seminiferous tubule formation	Campagnolo et al. (2001); Levine et al. (2000)

Nt3	Neurotrophin 3	Chemotactic factor that promotes non-endothelial mesonephric cell migration into the XY gonad and thus contributes to the population of peritubular myoid cells	In the Sertoli cells starting at 11.5 *dpc* in the mouse and 13 *dpc* in the rat	Genetic knockout of the neurotrophin ligand results in decreased testis interstitial volume	Levine et al. (2000); Russo et al. (1999)
Ntrk1	Neurotrophic tyrosine kinase, receptor, type 1	Preferentially binds nerve growth factor, but binds NT3 with lower affinity	Expressed in the interstitium of the XY gonad at 16 *dpc*	Genetic knockout resulted in a reduced number of testis cords	Cupp et al. (2000, 2002)
Ntrk3	Neurotrophic tyrosine kinase, receptor, type 1	High-affinity receptor for NT3	Expressed in the mesonephros starting at 11.5 *dpc* in the mouse and 13 *dpc* in the rat and subsequently in the peritubular myoid cells of the testis	Genetic knockout resulted in a reduced number of testis cords and a reduced interstitial area	Cupp et al. (2002)
Pdgf	Platelet-derived growth factor	Participates in endothelial-somatic cell cross talk in the XY gonad in response to VEGFA signaling, required for testis cord formation	Non-overlapping, sex-specific patterns of expression for three ligands: PDGF-A in both XX and XY gonads and in the mesonephric tubules at 11.5 *dpc*, in Sertoli cells in testis and downregulated in ovary by 12.5 *dpc*. PDGF-B in XX and XY gonads at early stages at low levels and exclusively endothelial by 13.5 *dpc*. PDGF-C in the coelomic epithelium and at the gonad/mesonephros boundary at 11.5 *dpc*, restricted to the	Pharmacological inhibition of PDGF signaling prevents testis cord formation	Brennan et al. (2003); Cool et al. (2011); Smith et al. (2005); Uzumcu et al. (2002)

(continued)

Table 4.1 (continued)

Gene symbol	Gene name	Function/role	Site and time of expression during early gonad development	Mutation/inhibition effects	References
			mesonephros/gonad boundary by 12.5 *dpc*		
Pdgfra	Alpha platelet-derived growth factor receptor	PDGF signaling promotes testis interstitial cell proliferation and communication with endothelial cells to shape testis architecture	In both sexes, expressed in the coelomic epithelium and gonad-mesonephros border at 11.5 *dpc*. By 12.5 *dpc*, strong expression in the interstitial cells of the XY gonad, almost no expression in the XX gonad	Genetic knockouts of the alpha PDGF receptor subunit show disrupted vascularization and compartmentalization	Brennan et al. (2003); Cool et al. (2011)
Rspo1	R-spondin 1	Positively regulates *Wnt4* signaling with sex-specific effects	Expressed in the fetal ovary starting at 12.5 *dpc*; also expressed in the coelomic region of the testis beginning at 12.5 *dpc*	In XX gonads, loss of RSPO1 function can lead to female-to-male sex reversal. In XY gonads, exogenous RSPO1 enhances testicular vascularization	Caruso et al. (2015); Wilhelm (2007)
Sf1; *Nr5a1*	Steroidogenic factor 1; nuclear receptor subfamily 5, group A, member 1	Transcriptional activator involved in sex determination and *Sox9* expression along with SRY	First expressed by genital ridge precursors in the coelomic epithelium at 9.5 *dpc*, ultimately marks the Sertoli cell lineage	Genetic mutation of *Sf1*/*Nr5a1* inhibits male gonad development resulting in male-to-female sex reversal	Luo et al. (1994); Sekido and Lovell-Badge (2008)
Sox9	SRY-box containing gene	Many roles in embryonic development, but in the gonad SOX9 regulates male sex determination	Expression in SF1+ cells starts shortly after 10.5 *dpc*, increases up to 11.5 *dpc* in both sexes, and then is upregulated in the testis (maintained into adulthood)	Genetic mutation in and around *Sox9* inhibits male gonad development resulting in male-to-female sex reversal; transgenic *Sox9* expression in XX	Sekido and Lovell-Badge (2008); Vidal et al. (2001); Wagner et al. (1994)

			and downregulated in the ovary by 12.5 *dpc*	gonads induces female-to-male sex reversal	
Sry	Sex determining region of Chr Y	Initiates male developmental programming by promoting *Sox9* expression	Transiently expressed in Sertoli cell precursors starting at ~10.5 *dpc* in the center of the XY gonad with expression spreading outward	Mutation of *Sry* in XY gonads results in phenotypically female gonad development; transgenic *Sry* expression in XX gonads is sufficient to induce mesonephric migration	Capel et al. (1999); Lovell-Badge and Robertson (1990); Koopman et al. (1991)
Vegfa	Vascular endothelial growth factor A	Promotes and is required for endothelial cell migration and vascularization	Expressed in Sertoli and interstitial cells starting at the time of sex differentiation 11.5 *dpc* in the mouse, 13.5 *dpc* in the rat	Pharmacological inhibition of VEGFA signaling prevents testis vascularization and testis cord formation; antibody sequestration of the anti-angiogenic isoforms of VEGFA and/or exogenous angiogenic VEGFA treatment results in increased testis vascularization	Baltes-Breitwisch et al. (2010); Bott et al. (2006); Cool et al. (2011)
Wnt4	Wingless-type MMTV integration site family, member 4	Drives the female developmental program	Expressed in the fetal ovary starting at 12.5 *dpc*	Mutation causes masculinization of the XX gonad	Vainio et al. (1999)

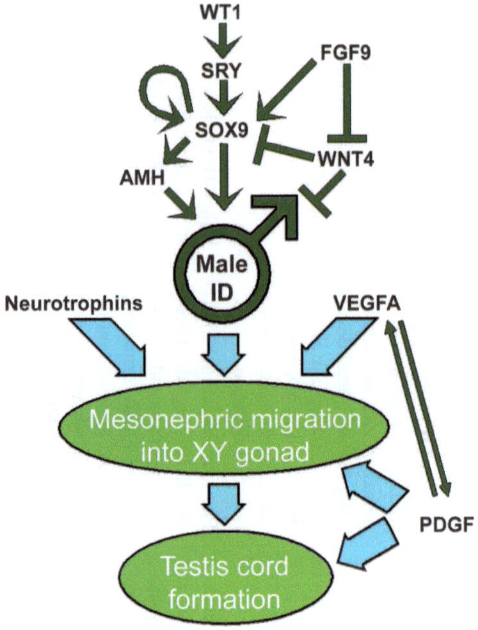

Fig. 4.3 Mesonephric cell migration and subsequent testis cord formation require the integration of multiple signaling molecules. Many of these signals promote the male gene expression profile necessary for mesonephric migration to occur; others directly participate in promoting cell migration from the mesonephros or control other aspects of testis cord morphogenesis

expression is maintained throughout embryogenesis and into adulthood (Kent et al. 1996). Mutations in *Sox9* or its regulatory regions result in male-to-female sex reversal (Wagner et al. 1994). Conversely, transgenic expression of *Sry* or *Sox9* in XX gonads causes female-to-male sex reversal (Koopman et al. 1991; Vidal et al. 2001). The transcription factors SRY and SOX9 do not directly initiate or control the mesonephric cell migration, but they put into effect a male developmental program involving several signaling molecules and pathways that result in the migratory event and seminiferous tubule morphogenesis (Fig. 4.3). These signaling factors will be discussed in more detail in this section.

FGF9 Within the developing testis, the male-specific genes expressed regulate sex specification maintenance and/or the morphogenetic changes crucial for testis architecture. Many of these genes exist in feedback loops and interacting signaling pathways. A prime example of a sex maintenance gene involved in feedback loops and pathway interaction is fibroblast growth factor 9 (FGF9), which upregulates and maintains *Sox9* expression after sex determination in the XY gonad (Fig. 4.3). *Fgf9*$^{-/-}$ mice can initiate but not maintain *Sox9* expression, resulting in impaired Sertoli cell differentiation and testis cord formation (Kim et al. 2006). This signaling occurs primarily via FGF receptor 2 (FGFR2), as *Fgfr2*$^{-/-}$ mice have similar phenotypes to *Fgf9*$^{-/-}$ mice and experience sex reversal (Bagheri-Fam et al. 2008; Kim et al. 2007; Schmahl et al. 2004). A crucial reason that FGF9 and FGFR2 are required for testis development is that FGF9 signaling actively represses the WNT4 signaling that would otherwise promote female gonad development (Jameson et al. 2012) (Fig. 4.3). Conditional knockout of either *Fgf9* or *Fgfr2* in the *Sf1*-

expressing cells of XY gonads causes male-to-female sex reversal, while testis morphology is rescued in *Fgf9/Wnt4* and *Fgfr2/Wnt4* double mutants (Jameson et al. 2012). Together these data suggest that the primary role of FGF signaling is to suppress the female developmental program.

Factors with Pleiotropic Effects in XX and XY Gonads Additionally, it is important to note that signals from the female gonad can actively inhibit mesonephric cell migration into the XX gonad. WNT4 must be downregulated in the XY gonad in order for endothelial cell migration into the developing testes and the formation of a coelomic blood vessel (Jeays-Ward et al. 2003). Interestingly, signaling pathways in gonad development are often pleiotropic and can function in both the male and female developmental programs (Fig. 4.4). One such example involves the transcription factor SF1 (which is commonly considered a marker for the Sertoli cell lineage) and its cofactor Cbp/p300-interacting transactivator 2 (CITED2). By looking at a *Cited2*$^{-/-}$ mouse and the resulting decrease in SF1 activity, researchers discovered that in XY mice SF1/CITED2 function is important for proper testis architecture, but in XX gonads SF1/CITED2 signaling is also crucial for female patterning and without it there is ectopic mesonephric cell migration (Combes et al. 2010) (Fig. 4.4).

Another such factor with pleiotropic effects is R-spondin 1 (RSPO1), which enhances canonical β-catenin/Wnt signaling to drive female gonad development (Carmon et al. 2014; Wilhelm 2007). This is required for ovarian development as masculinization and loss of female morphology is observed in XX gonads in *Rspo1* knockout mice (Chassot et al. 2008). In XY gonads of *Rspo*$^{-/-}$ mice testis morphogenesis appears normal, indicating that RSPO1 is not required *in vivo* for testis development (Chassot et al. 2008; Lavery et al. 2012). However, a recently published study shows that there may be a complex balance between RSPO1 and Dickkopf homolog 1 (DKK1), an inhibitor of the canonical WNT/RSPO signaling pathway regulating testicular angiogenesis with R-spondin 1 driving vascularization (Caruso et al. 2015) (Fig. 4.4). In cultured explants of XY gonads/mesonephroi, DKK1 treatment inhibits endothelial cell migration into the gonad while RSPO1 treatment rescues this inhibition (Caruso et al. 2015). Thus, molecules that promote mesonephric cell migration in the XY gonad such as SF1/CITED2 or RSPO1 have the opposite effect in the XX gonad and prevent ectopic vascularization.

4.4.2 VEGFA

The VEGF Family The vascular endothelial growth factor (VEGF) family contains five mammalian proteins: VEGFA, VEGFB, VEGFC, VEGFD, and PGF (placental growth factor, formerly known as PlGF). The factor for which the family is named is VEGFA, originally known simply as VEGF, and it was originally discovered as a gene vital for embryonic blood vessel development (Carmeliet et al. 1996; Ferrara and Henzel 1989; Ferrara et al. 1996). VEGFB and PGF are both

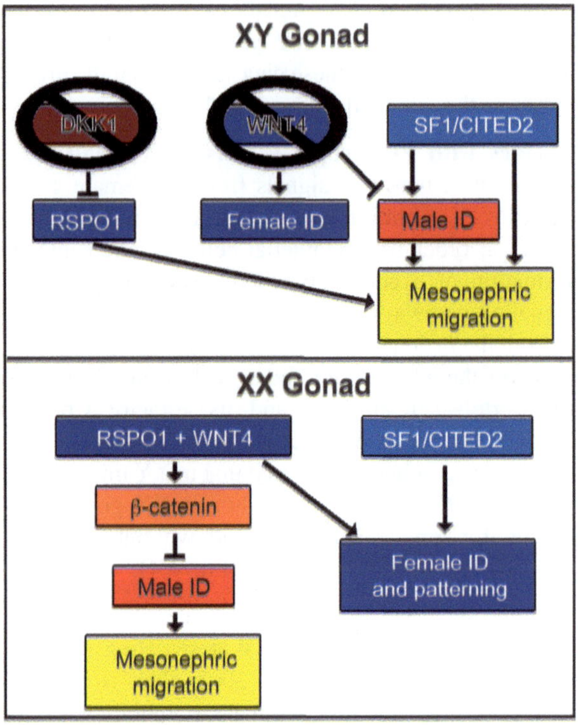

Fig. 4.4 Pleiotropic signaling factors such as RSPO1 and SF1/CITED2 have opposite effects in the XY gonad compared to the XX gonad. In the XY gonad in the absence of WNT4 and DKK1, RSPO1 and SF1/CITED2 both promote testis vascularization. In the XX gonad, RSPO1 enhances WNT4 canonical signaling via β-catenin to inhibit male sex specification and SF1/CITED2 promotes female gonad patterning

involved in blood vessel development, but do not appear to be expressed in the developing gonad at the time of mesonephric cell migration into the testis (Olofsson et al. 1998; Park et al. 1994). VEGFC and VEGFD primarily function in lymphatic angiogenesis (Achen et al. 1998; Jeltsch et al. 1997; Joukov et al. 1996; Karkkainen et al. 2002).

VEGFA A great deal of investigation has been centered on VEGFA, as it is critical for vascularization in many contexts (Ferrara et al. 2003). The function of VEGFA is also multifaceted because it undergoes alternative splicing of its eight exons that produces many isoforms with varying functions (Dehghanian et al. 2014; Olsson et al. 2006). The known human isoforms include VEGFA-111, -121, -145, -148, -165, -183, -189, and -206 (named for the length of the protein in amino acids), while the mouse isoforms are one amino-acid residue shorter than the corresponding human isoform and named accordingly (Dehghanian et al. 2014). However, there is an additional alternative splicing variation in which the previously known version of exon 8 of the VEGFA transcript, now called exon 8a, is replaced with exon 8b which gives the protein anti-angiogenic properties (Bates et al. 2002). This creates a set of anti-angiogenic isoforms corresponding to the pro-angiogenic isoforms containing exon 8: VEGFA-121b, -145b, -165b, -183b, and -189b (Dehghanian et al. 2014). Most of the research regarding mesonephric endothelial migration and the role that VEGFA plays in this process has looked at broad inhibition of VEGF receptors or other methods that are not able to distinguish the function of individual isoforms. Thus, as the field moves forward, it will be interesting to discover how much the complexity and nuance in signaling is available from so many signaling molecules factors into testis vascularization.

What is currently known is that manipulations of VEGFA can have strong effects on testis vascularization. Immunohistochemistry and explant culture in the presence of signaling inhibitors SU1498 (a selective inhibitor of one VEGF receptor), VEGFR-TKI (a general VEGFA receptor antagonist), Je-11 (competitively binds VEGFA to prevent receptor binding), and LY 294002 (a phosphoinositide 3-kinase (PI3K) pathway inhibitor) demonstrated that angiogenic isoforms of VEGFA including Vegfa120, Vegfa164, and Vegfa188 are expressed in the developing testis in the rat and that VEGFA signaling via KDR is required for vascularization and testis cord formation (Bott et al. 2006) (Fig. 4.3). Inhibition of VEGFA signaling with pharmacological treatments blocks the action of both angiogenic and antiangiogenic isoforms, showing that the vascularization and structural defects are due to impaired VEGFA signaling as a whole without distinguishing the roles of individual variants (Bott et al. 2006; Cool et al. 2011). Specific inhibition of anti-angiogenic VEGFA using antibody treatment, on the other hand, results in excessive vascularization and thus suggests that those isoforms inhibit over-vascularization (Baltes-Breitwisch et al. 2010). VEGFA signaling can thus either promote vascularization or inhibit it based on the splice variants produced.

VEGFA Signaling Transduction In addition to the variety of VEGFA isoforms possible, the ways in which VEGFA signaling can occur are quite varied and include both canonical ligand–receptor interactions and ligand-independent signaling, involving various receptor dimers and interactions with co-receptors (Domigan et al. 2015). Many of these unconventional signaling pathways are recently discovered, and the extent to which they might function in the developing gonad is not yet known. Within the early gonad, most of the research to date has focused on traditional ligand–receptor signaling. Two major receptors play important roles in transduction VEGFA signaling in the early gonad: the fms-like tyrosine kinase receptor (FLT1, sometimes cited as VEGFR-1) and the kinase insert domain receptor (KDR, alternatively called FLK1 or VEGFR-2) (de Vries et al. 1992; Shibuya et al. 1990; Terman et al. 1992). KDR and FLT1 have different modes of signal transduction and binding affinities for the VEGFA ligand. Binding analyses showed that VEGFA has a ~50-fold greater binding affinity for FLT1 than for KDR (Waltenberger et al. 1994). However, KDR is the primary receptor through which VEGFA signal transduction occurs (Kroll and Waltenberger 1997; Waltenberger et al. 1994). VEGFA binding-induced tyrosine kinase autophosphorylation of KDR results in its downstream signal transduction in endothelial cells (Waltenberger et al. 1994). When VEGFA binds to FLT1 in endothelial cells autophosphorylation and signal transduction do not occur, though in non-endothelial cells FLT1 can transduce VEGFA signaling (Caires et al. 2012).

The expression patterns of FLT1 and KDR impart critical sex-specific differences to VEGFA signaling. In the XY gonad and mesonephros, FLT1 is not expressed until after testis cord formation (18 *dpc* in the rat) (Bott et al. 2006). In XX embryos, FLT1 is expressed much earlier and may prevent ectopic endothelial cell migration by acting as a "decoy" receptor that sequesters VEGFA away from KDR to attenuate the signal transduction (Fong et al. 1999; Gille et al. 2000; Hiratsuka et al. 1998; McFee et al. 2009). Interestingly, this FLT1 decoy function/lack of kinase activity is defined by a single amino acid (Meyer et al. 2006).

There are many proteins that can modulate VEGFA signaling in addition to its receptors. VEGFA itself can bind to heparan sulfate proteoglycans within the extracellular matrix, which can alter its diffusion through tissues (Park et al. 1993; Vieira et al. 2010). In addition, the heparan sulfate proteoglycan binding domain of some VEGFA isoforms allows them to bind to co-receptors of the neuropilin family which includes neuropilin 1 (NRP1) and neuropilin 2 (NRP2) (Chen et al. 1997; Fidder 1996; Gitay-Goren et al. 1996; Kawasaki et al. 1999). The way that NRP1 functions to modulate VEGFA signaling is to bind VEGFA and form complexes with either KDR or FLT1 (without directly binding to either receptor) (Soker et al. 1998, 2002). The angiogenic isoforms of VEGFA such as Vegfa-165 can bind NRP1 and KDR, and proteins that interact with the C-terminus of NRP1 can then enhance KDR signal transduction (Wang et al. 2006). Additionally, NRP1 binding may promote KDR trafficking and recycling and thereby further enhance VEGFA signaling (Kofler and Simons 2015). The anti-angiogenic isoforms, on the other hand, cannot bind to NRP1 which may help explain why those isoforms do not have strong KDR signaling properties (Kawamura et al. 2008).

VEGFA in Mesonephric Chemotaxis Within the developing testis, the most studied VEGFA signaling pathway is the traditional angiogenic ligand–KDR receptor interaction. Based on a transgenic marker line (*KDR-LacZ*), mesonephric endothelial cells that undergo migration express KDR prior to and after migration (Bott et al. 2010) (Fig. 4.2). KDR as a lineage marker, however, may also include hematopoetic stem cells due to KDR expression beginning in a shared precursor, the hemangioblasts (Ziegler et al. 1999). Importantly, pharmacological inhibition of KDR alone with SU1498 or KDR and FLT1 both with VEGFR-TKI (a general VEGFA receptor antagonist) and VEGF-Trap (which contains domains of KDR and FLT1) prevent endothelial migration and testis cord formation (Bott et al. 2006, 2010; Cool et al. 2011). VEGFA does not act in isolation, however, as both platelet-derived growth factor (PDGF) and NRP1 act in concert with VEGFA. As will be discussed further in the next section, PDGF is crucial for coelomic vessel formation and likely participates in VEGFA-mediated cross talk between the migrating endothelial cells and the mesenchymal cells of the gonad (Brennan et al. 2003; Cool et al. 2011; Uzumcu et al. 2002). NRP1, on the other hand, can function as both a modulator of canonical VEGFA signaling and can regulate endothelial cell function independently of KDR (Kofler and Simons 2015). NRP1 can regulate adhesion and motility of endothelial cells by interacting with integrins and the non-receptor tyrosine kinase ABL1 (Fukasawa et al. 2007; Murga et al. 2005; Raimondi et al. 2014; Valdembri et al. 2009). Thus, VEGFA signaling plays a major role in testis vascularization and testis cord morphogenesis, but it is supported by other signaling pathways and potentially made redundant based on the observation that a *pDmrt1*-Cre-driven *Vegfa* knockout (excision of exon 3 specifically in the Sertoli and germ cells) does not prevent vascularization or testis cord formation (Lu et al. 2013).

4.4.3 PDGF

Platelet-derived growth factor (PDGF) is a family of signaling molecules that are made of dimers of four subunits (A, B, C, and D), resulting in PDGF-AA, -BB, -CC, -DD, and -AB (Heldin 2013). The receptors for these factors are also dimers with alpha and beta subunits corresponding to separate PDGF subunits (Heldin 2013). Within the developing testis, PDGF-AA, -BB, and -CC have been shown to be expressed and to play a role in testis cord morphogenesis (Hoch and Soriano 2003; Puglianiello et al. 2004). PDGF-BB has various functions in vasculature throughout embryonic development. In the developing testis, selective PDGF signaling inhibition with AG1296 (tyrphostin) prevents testis cord formation while treatment with PDGF factors in culture partially rescues testis cord morphogenesis, but the resulting cords are of a smaller diameter (Ricci et al. 2004; Smith et al. 2005) (Fig. 4.3). Additionally, genetic knockouts of the alpha PDGF receptor subunit show disrupted vascularization and compartmentalization prior to testis cord formation in XY gonads at 13.5 *dpc* (Brennan et al. 2003). The interactions between PDGF signaling and VEGF signaling, both of which are known regulators of testis cord morphogenesis, are a great example of cooperative signaling driving two interdependent morphological processes. Using inhibition of VEGF signaling with VEGF-Trap (which inhibits both KDR and FLT1) and mouse transgenic lineage markers for various cell types, Cool et al. (2011) showed that somatic cells of the early testis express VEGFA to promote endothelial migration and respond, by proliferation, to that endothelial migration (Fig. 4.3). Interestingly, there is some cell culture-based evidence that VEGFA can directly signal via PDGF receptors in cells lacking VEGF receptors (Ball et al. 2007). Overall, the interactions and complexities of these interconnected signaling pathways likely provide precision of control and potential redundancy in the regulation of mesonephric endothelial migration, vascularization of the testis, and testis cord morphogenesis.

4.4.4 Neurotrophins

Another chemotactic factor that has been shown to be involved in mesonephric cell migration into the early testis, but not endothelial cells, are the neurotrophins

(Fig. 4.3). Beginning at 13 *dpc* in the developing rat testis and 11.5 *dpc* in the mouse, immunohistochemistry can be used to localize neurotrophin 3 in the Sertoli cells and the neurotrophin receptors in the mesonephros and subsequently in the peritubular myoid cells (Campagnolo et al. 2001; Cupp et al. 2000; Levine et al. 2000; Russo et al. 1999). In fact, Campagnolo et al. (2001) showed that the low-affinity neurotrophin receptor NGFR can be used as a lineage marker for mesonephric-derived peritubular myoid cells and that these cells do not appear in the fetal testis unless cocultured with a mesonephros explant. Neurotrophin signaling inhibition with antisense nucleotides against NT3, a pharmacological inhibitor of the receptor NGFR, and receptor-IgG fusion dominant negative expression in rat testis explants also inhibit seminiferous tubule formation, and mouse genetic knockout of the neurotrophin ligand or its receptor results in decreased testis interstitial volume (Cupp et al. 2000, 2003; Levine et al. 2000). In addition, neurotrophin receptor knockout mice (*Ntrk1*, *Ntrk3*) show impaired seminiferous tubule formation and reduced interstitial area in the testis at 13–14 *dpc* (Cupp et al. 2002). Thus, the mesonephros contributes both endothelial cells and at least part of the peritubular myoid cell population to testis architecture.

4.4.5 Other Signaling Factors Contributing to Mesonephric Migration

In addition to the more well-studied contributors to mesonephric cell migration into the testis, there are other signaling pathways that can influence this migration. Anti-Müllerian hormone (AMH) deficient mice do not have testis abnormalities, but exogenous AMH in an organ culture can induce both mesonephric cell migration into XX gonads and testis cord formation (Ross et al. 2003; Vigier et al. 1987). *Amh* mRNA is expressed at 12.5 *dpc* in the mouse testis and thus could play a contributing role in inducing mesonephric migration into the XY gonad, though it is likely not the main initiator of the migration in vivo due to the timing of the expression (Hacker et al. 1995; Münsterberg and Lovell-Badge 1991) (Fig. 4.3).

Another pathway with an incompletely defined role in testis cord formation is Hedgehog signaling involving two ligands: desert hedgehog (DHH) and sonic Hedgehog (SHH). DHH is required for the formation of the basal lamina that surrounds the testis cords and is also involved in testicular somatic cell fate specification (Clark et al. 2000; Pierucci-Alves et al. 2001; Yao et al. 2002). In fact, DHH expression in the XY gonad shortly after *Sry* expression is one of the earliest signs of sex-specific differentiation of Sertoli cells (Bitgood et al. 1996). SHH, on the other hand, is expressed in the Wolffian duct epithelium of the mesonephros (Bitgood and McMahon 1995). The Hedgehog receptor Patched, which binds and transduces signals for both DHH and SHH, is expressed in the mesonephric mesenchyme as well as the testis interstitium (Franco and Yao 2012; Yao et al. 2002). Importantly, the migration of mesonephric endothelial cells into the early testis and its vascularization do not depend on Hedgehog signaling based on data from DHH genetic null mice and pharmacological inhibition of the Hedgehog co-receptor Smoothened (Yao et al. 2002; Yao and Capel 2002). Instead, a primary role of DHH appears to be to promote differentiation of fetal Leydig cells (Barsoum et al. 2009; Yao et al. 2002). However, DHH also plays a role in peritubular myoid cell function or differentiation, as $Dhh^{-/-}$ mice have peritubular myoid defects (Clark et al. 2000; Pierucci-Alves et al. 2001). In addition to the likely DHH signaling to peritubular myoid cells from Sertoli cells, it is possible that SHH from the mesonephric epithelium may play a role in peritubular myoid cell function either through direct interaction with myoid cells in the early testis or by its function within the mesonephros. NOTCH signaling (like DHH) is also a key regulator of the Leydig cell lineage, though NOTCH acts to promote maintenance of the progenitor line by inhibiting Leydig cell differentiation (Tang et al. 2008). Interestingly, *Notch1* receptor and the ligands *Jagged1* and *Jagged2* have been reported to be expressed in the testis vasculature during the period of mesonephric endothelial migration and testis cord formation (Brennan et al. 2002). There is also evidence that this results in active NOTCH signaling due to coincident expression of a downstream target, *Hes-5* (Brennan et al. 2002). Other cells such as Sertoli cells and the interstitium can also express *Notch2* and *Notch3* with corresponding expression of the downstream target genes *Hes-1* and *Hes-5* (Tang et al. 2008). Inhibition of NOTCH signaling with mouse genetic knockout of *Hes-1* and pharmacological treatment with DAPT (a gamma-secretase inhibitor) results in increases in differentiated Leydig cell numbers. Constitutive overexpression of *Notch* driven by the *Sf1* promoter, on the other hand, results in a loss of Leydig cells. And, ultimately, both $Hes\text{-}1^{-/-}$ and ectopic *Notch* result in smaller and irregularly shaped testis cords (Tang et al. 2008). However, the possible connection between the mesonephric-derived endothelial cell migration and regulation of Leydig cell differentiation in the testis via NOTCH signaling has not been thoroughly investigated.

4.5 Mechanisms of Mesonephric Cell Migration

Actively migrating cells must undergo several coordinated changes to detach from the original tissue, receive and interpret polarizing cues to orient their movement, translocate, and integrate into the target tissue (Ridley et al. 2003). The first step, disengaging from the original tissue, often requires the change from an epithelial state to a mesenchymal cell morphology. In the case of the mesonephric endothelial cells that migrate into the XY gonad, this is actually an endothelial-to-mesenchymal transition (EMT), though it still follows the same pattern of events based on immunofluorescence and cell shape changes (Combes et al. 2009; Thiery et al. 2009). The vascular plexus of the mesonephros and the endothelial cells that migrate into the early testis express classic endothelial cell adhesion molecules such as VE-cadherin (Combes et al. 2009). Though there is currently little direct

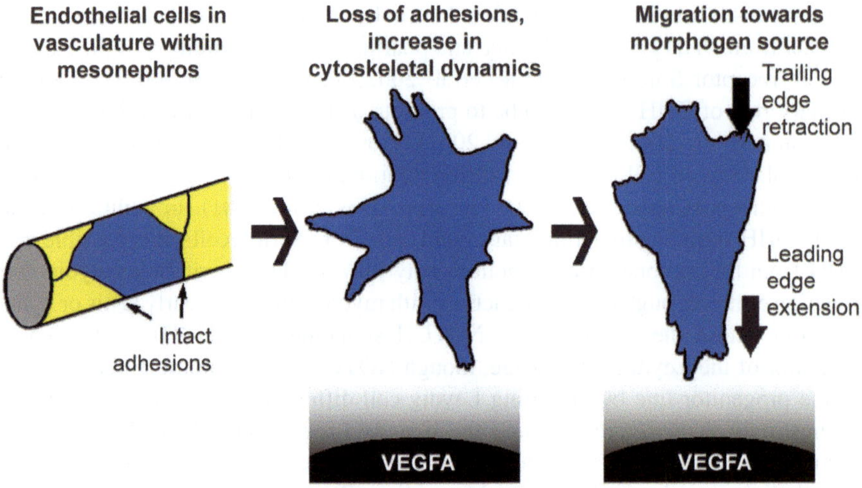

Fig. 4.5 During the process of migrating into the XY gonad, endothelial cells within the vascular plexus of the mesonephros must lose their existing adhesions and structure and acquire a mesenchymal phenotype with greater cytoskeletal dynamics and multiple filopodia-like extensions. In response to a gradient of a signaling molecule such as VEGFA, the closer filopodia extend toward the source of the morphogen and the further edges retract. This results in directional migration

evidence, it is conceivable that the endothelial-to-mesenchymal transition that allows migration would cause a change in cell–cell binding by internalizing adhesion molecules from the membrane and sequestering or degrading them temporarily. Simultaneously, the endothelial cells lose their extended squamous shape and become elongated with filopodia-like extensions (Combes et al. 2009; Coveney et al. 2008). The loss of existing cell adhesion and changes in cytoskeletal structure required for these cell shape changes permit the cell to undergo directed movements with polarized adhesion formation and cytoskeletal filament elongation (Fig. 4.5).

4.5.1 Cues to Orient Mesonephric Migration

An obvious candidate for an endothelial chemotactic signal is VEGF. Neovascularization in many contexts is controlled by VEGFA gradients, and it serves both to orient migration and to promote the necessary cellular changes to cause cell movement. For example, VEGFA signaling via human KDR expressed in porcine endothelial cells resulted in changes in cell morphology, actin reorganization, membrane ruffling, and chemotaxis while FLT1 signaling did not have the same responses (Waltenberger et al. 1994). Other data from human umbilical endothelial cells show that signaling via KDR may regulate focal adhesion complexes while FLT1 affects cytoskeletal dynamics in this context (Kanno et al. 2000). Additionally, VEGFA can promote chemotactic migration of non-endothelial human cells including osteoblasts, lung cancer cells, and mesenchymal progenitor cells (Chen et al. 2009; Fiedler et al. 2005; Mayr-Wohlfart et al. 2002). This role for VEGFA chemotaxis has also been shown for endothelial cells in other developmental systems such as the kidney (Li et al. 2005). All of this, combined with the previously demonstrated role of VEGFA in mesonephric endothelial migration into the testis, makes it highly likely to be a major directional cue to those migrating cells.

The way in which a gradient of signaling ligands is interpreted varies based on the context. Unlike cells with established adhesions in a tissue, migrating cells must respond to directional cues autonomously. Migrating endothelial tip cells in the retina and other contexts do this by extruding filopodia in multiple directions (Gerhardt et al. 2003). Further cytoskeletal polymerization is encouraged in the filopodia that receive the most signal transduction, thus elongating the cell toward the source of the signaling ligand (Fig. 4.5). VEGFA signaling gradients in the developing brain cause endothelial tip cells to migrate toward them because they express KDR and NRP1, both of which are required for migration (Fantin et al. 2013; Gerhardt et al. 2003, 2004). The mosaic studies that show that endothelial migration has a cell autonomous dependence on KDR and NRP1 signaling have not been performed in the developing testis/mesonephros, but it is likely that migration in that system occurs with a similar mechanism. It is also possible that other signaling pathways that promote mesonephric cell migration into the XY gonad do so through an analogous system of random filopodial extension, greater signaling response nearer the chemotactic source, and reinforcement of cytoskeletal dynamics in that direction.

The way in which a filopodium seeking and finding a greater source of a chemotactic factor like VEGFA responds and causes migration varies based on the cell type, but there are some common themes. Such a localized area of signaling activation results in phosphorylation/activation of downstream signal transducers and in proteins that regulate cytoskeletal dynamics. For example, endothelial cell culture assays show that a VEGF/KDR-initiated MAP-Kinase signal transduction cascade phosphorylates LIM-Kinase which inhibits cofilin, preventing cofilin from depolymerizing actin filaments and thus forming actin stress fibers that further stabilize the local filopodium (Kobayashi et al. 2006). This is just one of many potential avenues by which VEGFA or other signaling molecules regulate cytoskeletal dynamics in a complex network of interacting signal transductions and cytoskeleton-interacting proteins.

4.5.2 Narrowing the Migratory Path

One possibility that has been proposed to control the breadth of the migration path is that the individual migrating endothelial cells follow prelaid extracellular matrix scaffolding paths to partition the gonad into avascular domains prior to testis cord formation (Coveney et al. 2008). These extracellular matrix paths would be, in a sense, the path of least resistance with more matrix molecules that the leading edge of the migrating cell could bind to more easily than the surrounding matrix in order to narrow the directional movement toward a more general chemotactic signal. Another possibility is that the very first migratory endothelial cells would only have a broad directional cue to follow but would modify the surrounding tissue as they passed to guide subsequent migratory cells along the same path (Combes et al. 2009; Murphy and Gavrilovic 1999).

4.5.3 Reintegration into Target Tissue

Endothelial cells migrate individually rather than as a unit and must integrate into the target tissue (the XY gonad) and aggregate to form the coelomic vessel. Smaller vessels then branch off of the coelomic vessel between the avascular domains that will form the testis cords. In order to establish the new vasculature of the testis, VE-cadherin binding is crucial as treatment with an antibody blocking that binding prevents the formation of a nascent vascular plexus in the testis that would give rise to the coelomic vessel (Combes et al. 2009). The localization of endothelial cells between partitioned sections occurs by 12.25 *dpc*, and by 12.5 *dpc*, the endothelial cells have become interconnected again organized into vessels, so the reestablishment of cell–cell adhesion is coordinated and relatively fast (Combes et al. 2009). Both the act of partitioning the domains of the future testis cords and the presence of the endothelial cells themselves are important for the rest of the testis architecture formation.

4.6 The Role of Mesonephric-Derived Cells in the Testis

Once cells have migrated into the early XY gonad from the mesonephros, they take their place in the developing testis. Endothelial cells that were dissociated reform vasculature. The main purpose of the vasculature, of course, is to allow blood flow. Before 12 *dpc*, blood flows into the developing testis from the mesonephros vascular plexus through the gonad/mesonephros boundary (Brennan et al. 2002). After mesonephric endothelial cell migration and remodeling of the testis, blood flow is redirected into the testis via the arterial coelomic vessel (Brennan et al. 2002). Funneling blood flow in this manner is necessary for an external organ that will both need a sufficient blood supply and need to transport secreted hormones into circulation.

The importance of the testis vasculature to the development and functionality of the gonad is clear when vasculature formation is inhibited. When mesonephros–gonad organ cultures are treated with an antibody against the extracellular domain of VE-cadherin (thereby preventing endothelial cells from forming adherens junctions), mesonephric endothelial cell migration into the testis is reduced by 63 %

(Combes et al. 2009). The coelomic vessel, which like other large vessels utilizes tight junctions in addition to adherens junctions, still forms despite inhibited VE-cadherin binding. However, the branches from the coelomic vessel that separate the avascular domains do not form, and this lack of segmentation prevents testis cord development entirely (Combes et al. 2009). And, as discussed in Sect. 4.4.2, inhibition of the VEGFA signaling that is important for endothelial migration into the gonad and its vascularization also results in inhibition of testis cord formation. Thus, the testis vasculature plays a key developmental role in establishing testis architecture in addition to its role in the circulatory system. Importantly, once endothelial cells reach the gonad, they are still active in controlling testis morphogenesis. This occurs because endothelial cells secrete signaling molecules that allow cross talk with other cells such as the mesenchymal cells of the interstitium. Using fluorescent cell type markers and inhibition of signaling pathways, Cool et al. (2011) showed that VEGFA secreted from the testis interstitium recruits endothelial cells to migrate into the testis, but then PDGF from those endothelial cells also acts on the mesenchymal cells to promote their proliferation. This feed-forward signaling drives testis cord morphogenesis, and disruption of any part of the loop inhibits cord formation. This signaling loop does not represent a self-contained system by which morphogenesis occurs, as other cells from outside the gonad continue to migrate in and contribute to testis morphogenesis. Notably, fetal macrophages have migrated to the vascular plexus developing into the coelomic vessel as early as 10.5 *dpc* and appear to participate in vascularization of the testis and cord formation by promoting tissue clearance and remodeling (DeFalco et al. 2014). The recruitment of those macrophages is dependent on endothelial cell presence (DeFalco et al. 2014). Thus, the mesonephric-derived endothelial cell population plays many crucial roles in (1) partitioning the testis into domains for cords, (2) providing the vasculature of the testis, and (3) signaling to cells already present and recruiting cells from outside the gonad to coordinately regulate testis morphogenesis.

4.7 Conclusion

Testis architecture and vascularization is a complex morphogenetic event that cannot occur without a key migratory influx of cells from the mesonephros that begins at 11.5 *dpc* in the mouse. The majority of these cells are endothelial and come from the breakdown of the mesonephric vascular plexus. These cells then migrate into the XY gonad along paths that partition the early testis into avascular domains. This migration and compartmentalization are required for testis cord formation, which is necessary for the subsequent development of the seminiferous tubules. Male-specific gene expression in the gonad is a prerequisite for mesonephric migration, and many secreted signaling molecules are part of the signaling network that controls the migration including VEGFA, PDGF, neurotrophins, FGF9, AMH, DHH, and SHH. Based on the current body of evidence, the main directional cue that guides endothelial cell migration is a VEGFA morphogen gradient originating from the XY gonad. The cellular behaviors that occur during this specific migration event seem to parallel other well-studied cellular migrations during development in which the adhesions and structure of the endothelial cell are lost, a mesenchymal phenotype is adopted, and migration occurs in the direction of the target tissue via polarized cytoskeletal dynamics. Once the endothelial cell arrives, it interacts with the existing cells in the gonad to promote interstitial cell proliferation and restructure the early testis. The vasculature established during this time also serves as the foundation for the future vasculature of the testis. Thus, mesonephric cell migration into the early gonad is central to testis development and function.

References

Achen M, Jeltsch M, Kukk E, Mäkinen T, Vitali A, Wilks A, Alitalo K, Stacker S (1998) Vascular endothelial growth factor D (VEGF-D) is a ligand for the tyrosine kinases VEGF receptor 2 (Flk1) and VEGF receptor 3 (Flt4). Proc Natl Acad Sci U S A 95:548–553. doi:10.1073/pnas.95.2.548

Bagheri-Fam S, Sim H, Bernard P, Jayakody I, Taketo MM, Scherer G, Harley VR (2008) Loss of Fgfr2 leads to partial XY sex reversal. Dev Biol 314:71–83. doi:10.1016/j.ydbio.2007.11.010

Ball SG, Shuttleworth CA, Kielty CM (2007) Vascular endothelial growth factor can signal through platelet-derived growth factor receptors. J Cell Biol 177:489–500. doi:10.1083/jcb.200608093

Baltes-Breitwisch M, Artac RA, Bott RC, McFee RM, Kerl JG, Clopton DT, Cupp AS (2010) Neutralization of vascular endothelial growth factor antiangiogenic isoforms or administration of proangiogenic isoforms stimulates vascular development in the rat testis. Reproduction 140:319–329. doi:10.1530/REP-09-0456.Neutralization

Barsoum IB, Bingham NC, Parker KL, Jorgensen JS, Yao H (2009) Activation of the Hedgehog pathway in the mouse fetal ovary leads to ectopic appearance of fetal Leydig cells and female pseudohermaphroditism. Dev Biol 329:96–103. doi:10.1016/j.ydbio.2009.02.025

Bates DO, Cui T-G, Doughty JM, Winkler M, Sugiono M, Shields JD, Peat D, Gillatt D, Harper SJ (2002) VEGF165b, an inhibitory splice variant of vascular endothelial growth factor, is down-regulated in renal cell carcinoma. Cancer Res 62:4123–4131

Bitgood M, McMahon A (1995) Hedgehog and Bmp genes are coexpressed at many diverse sites of cell-cell interaction in the mouse embryo. Dev Biol 172:126–138. doi:10.1006/dbio.1995.0010

Bitgood M, Shen L, McMahon A (1996) Sertoli cell signaling by Desert hedgehog regulates the male germline. Curr Biol 6:298–304. doi:10.1016/S0960-9822(02)00480-3

Bott RC, McFee RM, Clopton DT, Toombs C, Cupp AS (2006) Vascular endothelial growth factor and kinase domain region receptor are involved in both seminiferous cord formation and vascular development during testis morphogenesis in the rat. Biol Reprod 75:56–67. doi:10.1095/biolreprod.105.047225

Bott R, Clopton D, Fuller A, McFee R, Lu N, Cupp A (2010) KDR-LacZ-expressing cells are involved in ovarian and testis-specific vascular development, suggesting a role for VEGFA in the regulation of this vasculature. Cell Tissue Res 342:117–130. doi:10.1007/s00441-010-1038-9

Bradford ST, Wilhelm D, Bandiera R, Vidal V, Schedl A, Koopman P (2009) A cell-autonomous role for WT1 in regulating Sry in vivo. Hum Mol Genet 18:3429–3438. doi:10.1093/hmg/ddp283

Brennan J, Capel B (2004) One tissue, two fates: molecular genetic events that underlie testis versus ovary development. Nat Rev Genet 5:509–521. doi:10.1038/nrg1381

Brennan J, Karl J, Capel B (2002) Divergent vascular mechanisms downstream of Sry establish the arterial system in the XY gonad. Dev Biol 244:418–428. doi:10.1006/dbio.2002.0578

Brennan J, Tilmann C, Capel B (2003) Pdgfr-α mediates testis cord organization and fetal Leydig cell development in the XY gonad. Genes Dev 17:800–810. doi:10.1101/gad.1052503

Buehr M, Gu S, McLaren A (1993) Mesonephric contribution to testis differentiation in the fetal mouse. Development 117:273–281

Bullejos M, Koopman P (2001) Spatially dynamic expression of Sry in mouse genital ridges. Dev Dyn 221:201–205. doi:10.1002/dvdy.1134

Burgoyne PS, Thornhill AR, Boudrean SK, Darling SM, Bishop CE, Evans EP, Capel B, Mittwoch U (1995) The genetic basis of XX-XY differences present before gonadal sex differentiation in the mouse [and discussion]. Philos Trans R Soc B Biol Sci 350:253–261. doi:10.1098/rstb.1995.0159

Caires KC, De Avila JM, Cupp AS, McLean DJ (2012) VEGFA family isoforms regulate spermatogonial stem cell homeostasis in vivo. Endocrinology 153:887–890. doi:10.1210/en.2011-1323

Campagnolo L, Russo MA, Puglianiello A, Favale A, Siracusa G (2001) Mesenchymal cell precursors of peritubular smooth muscle cells of the mouse testis can be identified by the presence of the p75 neurotrophin receptor. Biol Reprod 64:464–472

Capel B, Albrecht KH, Washburn LL, Eicher EM (1999) Migration of mesonephric cells into the mammalian gonad depends on Sry. Mech Dev 84:127–131. doi:10.1016/S0925-4773(99)00047-7

Carmeliet P, Ferreira V, Breier G, Pollefeyt S, Kieckens L, Gertsenstein M, Fahrig M, Vandenhoeck A, Harpal K, Eberhardt C, Declercq C, Pawling J, Moons L, Collen D, Risau W, Nagy A (1996) Abnormal blood vessel development and lethality in embryos lacking a single VEGF allele. Nature 380:435–439

Carmon KS, Gong X, Yi J, Thomas A, Liu Q (2014) RSPO-LGR4 functions via IQGAP1 to potentiate Wnt signaling. Proc Natl Acad Sci U S A 111:E1221–E1229. doi:10.1073/pnas.1323106111

Caruso M, Ferranti F, Corano Scheri K, Dobrowolny G, Ciccarone F, Grammatico P, Catizone A, Ricci G (2015) R-spondin 1/dickkopf-1/beta-catenin machinery is involved in testicular embryonic angiogenesis. PLoS One 10:e0124213. doi:10.1371/journal.pone.0124213

Chassot AA, Ranc F, Gregoire EP, Roepers-Gajadien HL, Taketo MM, Camerino G, de Rooij DG, Schedl A, Chaboissier MC (2008) Activation of β-catenin signaling by Rspo1 controls differentiation of the mammalian ovary. Hum Mol Genet 17:1264–1277. doi:10.1093/hmg/ddn016

Chen H, Chédotal A, He Z, Goodman CS, Tessier-Lavigne M (1997) Neuropilin-2, a novel member of the neuropilin family, is a high affinity receptor for the semaphorins Sema E and Sema IV but not Sema III. Neuron 19:547–559. doi:10.1016/S0896-6273(00)80371-2

Chen CH, Lai JM, Chou TY, Chen CY, Su LJ, Lee YC, Cheng TS, Hong YR, Chou CK, Whang-Peng J, Wu YC, Huang CYF (2009) VEGFA upregulates FLJ10540 and modulates migration and invasion of lung cancer via PI3K/AKT pathway. PLoS One. doi:10.1371/journal.pone.0005052

Chen H, Palmer JS, Thiagarajan RD, Dinger ME, Lesieur E, Chiu H, Schulz A, Spiller C, Grimmond SM, Little MH, Koopman P, Wilhelm D (2012) Identification of novel markers of mouse fetal ovary development. PLoS One. doi:10.1371/journal.pone.0041683

Clark AM, Garland KK, Russell LD (2000) Desert hedgehog (Dhh) gene is required in the mouse testis for formation of adult-type Leydig cells and normal development of peritubular cells and seminiferous tubules. Biol Reprod 63:1825–1838. doi:10.1095/biolreprod63.6.1825

Combes AN, Wilhelm D, Davidson T, Dejana E, Harley V, Sinclair A, Koopman P (2009) Endothelial cell migration directs testis cord formation. Dev Biol 326:112–120. doi:10.1016/j.ydbio.2008.10.040

Combes AN, Spiller CM, Harley VR, Sinclair AH, Dunwoodie SL, Wilhelm D, Koopman P (2010) Gonadal defects in Cited2-mutant mice indicate a role for SF1 in both testis and ovary differentiation. Int J Dev Biol 54:683–689. doi:10.1387/ijdb.092920ac

Cool J, Carmona FD, Szucsik JC, Capel B (2008) Peritubular myoid cells are not the migrating population required for testis cord formation in the XY gonad. Sex Dev 2:128–133. doi:10.1159/000143430

Cool J, DeFalco TJ, Capel B (2011) Vascular-mesenchymal cross-talk through Vegf and Pdgf drives organ patterning. Proc Natl Acad Sci U S A 108:167–172. doi:10.1073/pnas.1010299108

Coveney D, Cool J, Oliver T, Capel B (2008) Four-dimensional analysis of vascularization during primary development of an organ, the gonad. Proc Natl Acad Sci U S A 105:7212–7217. doi:10.1073/pnas.0707674105

Cupp AS, Kim GH, Skinner MK (2000) Expression and action of neurotropin-3 and nerve growth factor in embryonic and early postnatal rat testis development. Biol Reprod 63:1617–1628. doi:10.1095/biolreprod63.6.1617

Cupp AS, Tessarollo L, Skinner MK (2002) Testis developmental phenotypes in neurotropin receptor trkA and trkC null mutations: role in formation of seminiferous cords and germ cell survival. Biol Reprod 66:1838–1845. doi:10.1095/biolreprod66.6.1838

Cupp AS, Uzumcu M, Skinner MK (2003) Chemotactic role of neurotropin 3 in the embryonic testis that facilitates male sex determination. Biol Reprod 68:2033–2037. doi:10.1095/biolreprod.102.012617

DeFalco T, Takahashi S, Capel B (2011) Two distinct origins for Leydig cell progenitors in the fetal testis. Dev Biol 352:14–26. doi:10.1016/j.ydbio.2011.01.011

DeFalco T, Bhattacharya I, Williams AV, Sams DM, Capel B (2014) Yolk-sac-derived macrophages regulate fetal testis vascularization and morphogenesis. Proc Natl Acad Sci U S A. doi:10.1073/pnas.1400057111

Dehghanian F, Hojati Z, Kay M (2014) New insights into VEGF-A alternative splicing: key regulatory switching in the pathological process. Avicenna J Med Biotechnol 6:192–199

de Vries C, Escobedo J, Ueno H, Houck K, Ferrara N, Williams L (1992) The fms-like tyrosine kinase, a receptor for vascular endothelial growth factor. Science 255:989–991. doi:10.1126/science.1312256

Domigan CK, Ziyad S, Iruela-Arispe ML (2015) Canonical and noncanonical vascular endothelial growth factor pathways: new developments in biology and signal transduction. Arterioscler Thromb Vasc Biol 35:30–39. doi:10.1161/ATVBAHA.114.303215

Ewen KA, Koopman P (2010) Mouse germ cell development: from specification to sex determination. Mol Cell Endocrinol 323:76–93. doi:10.1016/j.mce.2009.12.013

Fantin A, Vieira JM, Plein A, Denti L, Fruttiger M, Pollard JW, Ruhrberg C (2013) NRP1 acts cell autonomously in endothelium to promote tip cell function during sprouting angiogenesis. Blood 121:2352–2362. doi:10.1182/blood-2012-05-424713

Ferrara N, Henzel WJ (1989) Pituitary follicular cells secrete a novel heparin-binding growth factor specific for vascular endothelial cells. Biochem Biophys Res Commun 161:851–858. doi:10.1016/j.bbrc.2012.08.021

Ferrara N, Carver-Moore K, Chen H, Dowd M, Lu L, O'Shea KS, Powell-Braxton L, Hillan KJ, Moore MW (1996) Heterozygous embryonic lethality induced by targeted inactivation of the VEGF gene. Nature 380:439–442

Ferrara N, Gerber H-P, LeCouter J (2003) The biology of VEGF and its receptors. Nat Med 9:669–676. doi:10.1038/nm0603-669

Fidder H (1996) Characterization of novel vascular endothelial growth factor (VEGF) receptors on tumor cells that bind VEGF[IMAGE] via its exon 7-encoded domain. J Biol Chem 271:5761–5767. doi:10.1074/jbc.271.10.5761

Fiedler J, Leucht F, Waltenberger J, Dehio C, Brenner RE (2005) VEGF-A and PlGF-1 stimulate chemotactic migration of human mesenchymal progenitor cells. Biochem Biophys Res Commun 334:561–568. doi:10.1016/j.bbrc.2005.06.116

Fong GH, Zhang L, Bryce DM, Peng J (1999) Increased hemangioblast commitment, not vascular disorganization, is the primary defect in flt-1 knock-out mice. Development 126:3015–3025

Franco HL, Yao HHC (2012) Sex and hedgehog: roles of genes in the hedgehog signaling pathway in mammalian sexual differentiation. Chromosome Res 20:247–258. doi:10.1007/s10577-011-9254-z

Fukasawa M, Matsushita A, Korc M (2007) Neuropilin-1 interacts with integrin beta1 and modulates pancreatic cancer cell growth, survival and invasion. Cancer Biol Ther 6:1173–1180. doi:10.4161/cbt.6.8.4363

Gerhardt H, Golding M, Fruttiger M, Ruhrberg C, Lundkvist A, Abramsson A, Jeltsch M, Mitchell C, Alitalo K, Shima D, Betsholtz C (2003) VEGF guides angiogenic sprouting utilizing endothelial tip cell filopodia. J Cell Biol 161:1163–1177. doi:10.1083/jcb.200302047

Gerhardt H, Ruhrberg C, Abramsson A, Fujisawa H, Shima D, Betsholtz C (2004) Neuropilin-1 is required for endothelial tip cell guidance in the developing central nervous system. Dev Dyn 231:503–509. doi:10.1002/dvdy.20148

Gille H, Kowalski J, Yu L, Chen H, Pisabarro MT, Davis-Smyth T, Ferrara N (2000) A repressor sequence in the juxtamembrane domain of Flt-1 (VEGFR-1) constitutively inhibits vascular endothelial growth factor-dependent phosphatidylinositol 3′-kinase activation and endothelial cell migration. EMBO J 19:4064–4073. doi:10.1093/emboj/19.15.4064

Gitay-Goren H, Cohen T, Tessler S, Soker S, Gengrinovitch S, Rockwell P, Klagsbrun M, Levi BZ, Neufeld G (1996) Selective binding of VEGF121 to one of the three vascular endothelial growth factor receptors of vascular endothelial cells. J Biol Chem 271:5519–5523

Gomperts M, Garcia-Castro M, Wylie C, Heasman J (1994) Interactions between primordial germ cells play a role in their migration in mouse embryos. Development 120:135–141

Hacker A, Capel B, Goodfellow P, Lovell-Badge R (1995) Expression of Sry, the mouse sex determining gene. Development 121:1603–1614

Harikae K, Miura K, Kanai Y (2013) Early gonadogenesis in mammals: significance of long and narrow gonadal structure. Dev Dyn 242:330–338. doi:10.1002/dvdy.23872

Heldin C-H (2013) Targeting the PDGF signaling pathway in tumor treatment. Cell Commun Signal 11:97. doi:10.1186/1478-811X-11-97

Hiratsuka S, Minowa O, Kuno J, Noda T, Shibuya M (1998) Flt-1 lacking the tyrosine kinase domain is sufficient for normal development and angiogenesis in mice. Proc Natl Acad Sci U S A 95:9349–9354. doi:10.1073/pnas.95.16.9349

Hoch RV, Soriano P (2003) Roles of PDGF in animal development. Development 130:4769–4784. doi:10.1242/dev.00721

Hu YC, Okumura LM, Page DC (2013) Gata4 is required for formation of the genital ridge in mice. PLoS Genet 9:1–12. doi:10.1371/journal.pgen.1003629

Hummitzsch K, Irving-Rodgers HF, Hatzirodos N, Bonner W, Sabatier L, Reinhardt DP, Sado Y, Ninomiya Y, Wilhelm D, Rodgers RJ (2013) A new model of development of the mammalian ovary and follicles. PLoS One. doi:10.1371/journal.pone.0055578

Jameson SA, Lin YT, Capel B (2012) Testis development requires the repression of Wnt4 by Fgf signaling. Dev Biol 370:24–32. doi:10.1016/j.ydbio.2012.06.009

Jeanes A, Wilhelm D, Wilson MJ, Bowles J, McClive PJ, Sinclair AH, Koopman P (2005) Evaluation of candidate markers for the peritubular myoid cell lineage in the developing mouse testis. Reproduction 130:509–516. doi:10.1530/rep.1.00718

Jeays-Ward K, Hoyle C, Brennan J, Dandonneau M, Alldus G, Capel B, Swain A (2003) Endothelial and steroidogenic cell migration are regulated by WNT4 in the developing mammalian gonad. Development 130:3663–3670. doi:10.1242/dev.00591

Jeltsch M, Kaipainen A, Joukov V, Meng X, Lakso M, Rauvala H, Swartz M, Fukumura D, Jain RK, Alitalo K (1997) Hyperplasia of lymphatic vessels in VEGF-C transgenic mice. Science 276:1423–1425. doi:10.1126/science.276.5317.1423

Joukov V, Pajusola K, Kaipainen A, Chilov D, Lahtinen I, Kukk E, Saksela O, Kalkkinen N, Alitalo K (1996) A novel vascular endothelial growth factor, VEGF-C, is a ligand for the Flt4 (VEGFR-3) and KDR (VEGFR-2) receptor tyrosine kinases. EMBO J 15:290–298

Kanno S, Oda N, Abe M, Terai Y, Ito M, Shitara K, Tabayashi K, Shibuya M, Sato Y (2000) Roles of two VEGF receptors, Flt-1 and KDR, in the signal transduction of VEGF effects in human vascular endothelial cells. Oncogene 19:2138–2146. doi:10.1038/sj.onc.1203533

Karkkainen MJ, Mäkinen T, Alitalo K (2002) Lymphatic endothelium: a new frontier of metastasis research. Nat Cell Biol 4:E2–E5. doi:10.1038/ncb0102-e2

Karl J, Capel B (1995) Three-dimensional structure of the developing mouse genital ridge. Philos Trans R Soc Lond B Biol Sci 350:235–242. doi:10.1098/rstb.1995.0157

Karl J, Capel B (1998) Sertoli cells of the mouse testis originate from the coelomic epithelium. Dev Biol 203:323–333. doi:10.1006/dbio.1998.9068

Kawamura H, Li X, Harper SJ, Bates DO, Claesson-Welsh L (2008) Vascular endothelial growth factor (VEGF)-A165b is a weak in vitro agonist for VEGF receptor-2 due to lack of coreceptor binding and deficient regulation of kinase activity. Cancer Res 68:4683–4692. doi:10.1158/0008-5472.CAN-07-6577

Kawasaki T, Kitsukawa T, Bekku Y, Matsuda Y, Sanbo M, Yagi T, Fujisawa H (1999) A requirement for neuropilin-1 in embryonic vessel formation. Development 126:4895–4902

Kent J, Wheatley SC, Andrews JE, Sinclair AH, Koopman P (1996) A male-specific role for SOX9 in vertebrate sex determination. Development 122:2813–2822

Kim Y, Kobayashi A, Sekido R, DiNapoli L, Brennan J, Chaboissier MC, Poulat F, Behringer RR, Lovell-Badge R, Capel B (2006) Fgf9 and Wnt4 act as antagonistic signals to regulate mammalian sex determination. PLoS Biol 4:1000–1009. doi:10.1007/11558958_121

Kim Y, Bingham N, Sekido R, Parker KL, Lovell-Badge R, Capel B (2007) Fibroblast growth factor receptor 2 regulates proliferation and Sertoli differentiation during male sex determination. Proc Natl Acad Sci U S A 104:16558–16563. doi:10.1073/pnas.0702581104

Kobayashi M, Nishita M, Mishima T, Ohashi K, Mizuno K (2006) MAPKAPK-2-mediated LIM-kinase activation is critical for VEGF-induced actin remodeling and cell migration. EMBO J 25:713–726. doi:10.1038/sj.emboj.7600973

Kofler NM, Simons M (2015) Angiogenesis versus arteriogenesis: neuropilin 1 modulation of VEGF signaling. F1000Prime Rep 7:26. doi:10.12703/P7-26

Koopman P, Gubbay J, Vivian N, Goodfellow P, Lovell-Badge R (1991) Male development of chromosomally female mice transgenic for Sry. Nature 351:117–121. doi:10.1038/351117a0

Kroll J, Waltenberger J (1997) The vascular endothelial growth factor receptor KDR activates multiple signal transduction pathways in porcine aortic endothelial cells. J Biol Chem 272:32521–32527

Kusaka M, Katoh-Fukui Y, Ogawa H, Miyabayashi K, Baba T, Shima Y, Sugiyama N, Sugimoto Y, Okuno Y, Kodama R, Iizuka-Kogo A, Senda T, Sasaoka T, Kitamura K, Aizawa S, Morohashi KI (2010) Abnormal epithelial cell polarity and ectopic Epidermal Growth Factor Receptor (EGFR) expression induced in Emx2 KO Embryonic Gonads. Endocrinology 151:5893–5904. doi:10.1210/en.2010-0915

Lavery R, Chassot A-A, Pauper E, Gregoire EP, Klopfenstein M, de Rooij DG, Mark M, Schedl A, Ghyselinck NB, Chaboissier M-C (2012) Testicular differentiation occurs in absence of R-spondin1 and Sox9 in mouse sex reversals. PLoS Genet 8:e1003170. doi:10.1371/journal.pgen.1003170

Levine E, Cupp AS, Skinner MK (2000) Role of neurotropins in rat embryonic testis morphogenesis (cord formation). Biol Reprod 62:132–142. doi:10.1095/biolreprod62.1.132

Li ZD, Bork JP, Krueger B, Patsenker E, Schulze-Krebs A, Hahn EG, Schuppan D (2005) VEGF induces proliferation, migration, and TGF-beta1 expression in mouse glomerular endothelial cells via mitogen-activated protein kinase and phosphatidylinositol 3-kinase. Biochem Biophys Res Commun 334:1049–1060. doi:10.1016/j.bbrc.2005.07.005

Liu C, Peng J, Matzuk MM, Yao HH-C (2015) Lineage specification of ovarian theca cells requires multicellular interactions via oocyte and granulosa cells. Nat Commun 6:6934. doi:10.1038/ncomms7934
Lovell-Badge R, Robertson E (1990) XY female mice resulting from a heritable mutation in the primary testis-determining gene, Tdy. Development 109:635–646
Lu N, Sargent KM, Clopton DT, Pohlmeier WE, Brauer VM, McFee RM, Weber JS, Ferrara N, Silversides DW, Cupp AS (2013) Loss of vascular endothelial growth factor A (VEGFA) isoforms in the testes of male mice causes subfertility, reduces sperm numbers, and alters expression of genes that regulate undifferentiated spermatogonia. Endocrinology 154:4790–4802. doi:10.1210/en.2013-1363
Luo X, Ikeda Y, Parker KL (1994) A cell-specific nuclear receptor is essential for adrenal and gonadal development and sexual differentiation. Cell 77:481–490. doi:10.1016/0092-8674(94)90211-9
Martineau J, Nordqvist K, Tilmann C, Lovell-Badge R, Capel B (1997) Male-specific cell migration into the developing gonad. Curr Biol 7:958–968. doi:10.1016/S0960-9822(06)00415-5
Mayr-Wohlfart U, Waltenberger J, Hausser H, Kessler S, Günther KP, Dehio C, Puhl W, Brenner RE (2002) Vascular endothelial growth factor stimulates chemotactic migration of primary human osteoblasts. Bone 30:472–477. doi:10.1016/S8756-3282(01)00690-1
McFee RM, Artac RA, McFee RM, Clopton DT, Smith RAL, Rozell TG, Cupp AS (2009) Inhibition of vascular endothelial growth factor receptor signal transduction blocks follicle progression but does not necessarily disrupt vascular development in perinatal rat ovaries. Biol Reprod 81:966–977. doi:10.1095/biolreprod.109.078071
McLaren A (2000) Germ and somatic cell lineages in the developing gonad. Mol Cell Endocrinol 163:3–9. doi:10.1016/S0303-7207(99)00234-8
Merchant-Larios H, Moreno-Mendoza N, Buehr M (1993) The role of the mesonephros in cell differentiation and morphogenesis of the mouse fetal testis. Int J Dev Biol 37:407–415
Meyer RD, Mohammadi M, Rahimi N (2006) A single amino acid substitution in the activation loop defines the decoy characteristic of VEGFR-1/FLT-1. J Biol Chem 281:867–875. doi:10.1074/jbc.M506454200
Molyneaux KA, Stallock J, Schaible K, Wylie C (2001) Time-lapse analysis of living mouse germ cell migration. Dev Biol 240:488–498. doi:10.1006/dbio.2001.0436
Mork L, Maatouk DM, McMahon JA, Guo JJ, Zhang P, McMahon AP, Capel B (2012) Temporal differences in granulosa cell specification in the ovary reflect distinct follicle fates in mice. Biol Reprod 86:37. doi:10.1095/biolreprod.111.095208
Münsterberg A, Lovell-Badge R (1991) Expression of the mouse anti-Müllerian hormone gene suggests a role in both male and female sexual differentiation. Development 113:613–624
Murga M, Fernandez-capetillo O, Tosato G (2005) Neuropilin-1 regulates attachment in human endothelial cells independently of vascular endothelial growth factor receptor-2. Cell 105:1992–1999. doi:10.1182/blood-2004-07-2598.Reprints
Murphy G, Gavrilovic J (1999) Proteolysis and cell migration: creating a path? Curr Opin Cell Biol 11:614–621. doi:10.1016/S0955-0674(99)00022-8
Nef S, Parada L (2000) Hormones in male sexual development. Genes Dev 14:3075–3086. doi:10.1101/gad.843800.genital
Nel-Themaat L, Vadakkan TJ, Wang Y, Dickinson ME, Akiyama H, Behringer RR (2009) Morphometric analysis of testis cord formation in sox9-egfp Mice. Dev Dyn 238:1100–1110. doi:10.1002/dvdy.21954
Nishino K, Yamanouchi K, Naito K, Tojo H (2001) Characterization of mesonephric cells that migrate into the XY gonad during testis differentiation. Exp Cell Res 267:225–232. doi:10.1006/excr.2001.5238
Olofsson B, Korpelainen E, Pepper MS, Mandriota SJ, Aase K, Kumar V, Gunji Y, Jeltsch MM, Shibuya M, Alitalo K, Eriksson U (1998) Vascular endothelial growth factor B (VEGF-B)

binds to VEGF receptor-1 and regulates plasminogen activator activity in endothelial cells. Proc Natl Acad Sci U S A 95:11709–11714. doi:10.1073/pnas.95.20.11709
Olsson A-K, Dimberg A, Kreuger J, Claesson-Welsh L (2006) VEGF receptor signalling—in control of vascular function. Nat Rev Mol Cell Biol 7:359–371. doi:10.1038/nrm1911
Park JE, Keller GA, Ferrara N (1993) The vascular endothelial growth factor (VEGF) isoforms: differential deposition into the subepithelial extracellular matrix and bioactivity of extracellular matrix-bound VEGF. Mol Biol Cell 4:1317–1326. doi:10.1091/mbc.4.12.1317
Park JE, Chen HH, Winer J, Houck KA, Ferrara N (1994) Placenta growth factor: potentiation of vascular endothelial growth factor bioactivity, in vitro and in vivo, and high affinity binding to Flt-1 but not to Flk-1/KDR. J Biol Chem 269:25646–25654. doi:10.1371/journal.pone.0018076
Pierucci-Alves F, Clark AM, Russell LD (2001) A developmental study of the Desert hedgehog-null mouse testis. Biol Reprod 65:1392–1402. doi:10.1095/biolreprod65.5.1392
Puglianiello A, Campagnolo L, Farini D, Cipollone D, Russo MA, Siracusa G (2004) Expression and role of PDGF-BB and PDGFR-beta during testis morphogenesis in the mouse embryo. J Cell Sci 117:1151–1160. doi:10.1242/jcs.00981
Raimondi C, Fantin A, Lampropoulou A, Denti L, Chikh A, Ruhrberg C (2014) Imatinib inhibits VEGF-independent angiogenesis by targeting neuropilin 1-dependent ABL1 activation in endothelial cells. J Exp Med 211:1167–1183. doi:10.1084/jem.20132330
Ricci G, Catizone A, Galdieri M (2004) Embryonic mouse testis development: role of platelet derived growth factor (PDGF-BB). J Cell Physiol 200:458–467. doi:10.1002/jcp.20035
Ridley AJ, Schwartz MA, Burridge K, Firtel RA, Ginsberg MH, Borisy G, Parsons JT, Horwitz AR (2003) Cell migration: integrating signals from front to back. Science 302:1704–1709. doi:10.1126/science.1092053
Ross AJ, Tilman C, Yao H, MacLaughlin D, Capel B (2003) AMH induces mesonephric cell migration in XX gonads. Mol Cell Endocrinol 211:1–7. doi:10.1016/j.mce.2003.09.021
Russo MA, Giustizieri ML, Favale A, Fantini MC, Campagnolo L, Konda D, Germano F, Farini D, Manna C, Siracusa G (1999) Spatiotemporal patterns of expression of neurotrophins and neurotrophin receptors in mice suggest functional roles in testicular and epididymal morphogenesis. Biol Reprod 61:1123–1132. doi:10.1095/biolreprod61.4.1123
Schmahl J, Capel B (2003) Cell proliferation is necessary for the determination of male fate in the gonad. Dev Biol 258:264–276. doi:10.1016/S0012-1606(03)00122-2
Schmahl J, Eicher EM, Washburn LL, Capel B (2000) Sry induces cell proliferation in the mouse gonad. Development 127:65–73
Schmahl J, Kim Y, Colvin JS, Ornitz DM, Capel B (2004) Fgf9 induces proliferation and nuclear localization of FGFR2 in Sertoli precursors during male sex determination. Development 131:3627–3636. doi:10.1242/dev.01239
Sekido R, Lovell-Badge R (2008) Sex determination involves synergistic action of SRY and SF1 on a specific Sox9 enhancer. Nature 453:930–934. doi:10.1038/nature07622
Shibuya M, Yamaguchi S, Yamane A, Ikeda T, Tojo A, Matsushime H, Sato M (1990) Nucleotide sequence and expression of a novel human receptor-type tyrosine kinase gene (flt) closely related to the fms family. Oncogene 5:519–524
Sinclair AH, Berta P, Palmer MS, Hawkins JR, Griffiths BL, Smith MJ, Foster JW, Frischauf AM, Lovell-Badge R, Goodfellow PN (1990) A gene from the human sex-determining region encodes a protein with homology to a conserved DNA-binding motif. Nature 346:240–244. doi:10.1038/346240a0
Smith CA, McClive PJ, Hudson Q, Sinclair AH (2005) Male-specific cell migration into the developing gonad is a conserved process involving PDGF signalling. Dev Biol 284:337–350. doi:10.1016/j.ydbio.2005.05.030
Smith P, Wilhelm D, Rodgers RJ (2014) Development of mammalian ovary. J Endocrinol. doi:10.1530/JOE-14-0062

Soker S, Takashima S, Miao HQ, Neufeld G, Klagsbrun M (1998) Neuropilin-1 is expressed by endothelial and tumor cells as an isoform-specific receptor for vascular endothelial growth factor. Cell 92:735–745. doi:10.1016/S0092-8674(00)81402-6

Soker S, Miao HQ, Nomi M, Takashima S, Klagsbrun M (2002) VEGF165 mediates formation of complexes containing VEGFR-2 and neuropilin-1 that enhance VEGF165-receptor binding. J Cell Biochem 85:357–368. doi:10.1002/jcb.10140

Tanaka SS, Nishinakamura R (2014) Regulation of male sex determination: genital ridge formation and Sry activation in mice. Cell Mol Life Sci 71:4781–4802. doi:10.1007/s00018-014-1703-3

Tang H, Brennan J, Karl J, Hamada Y, Raetzman L, Capel B (2008) Notch signaling maintains Leydig progenitor cells in the mouse testis. Development 135:3745–3753. doi:10.1242/dev.024786

Terman BI, Dougher-Vermazen M, Carrion ME, Dimitrov D, Armellino DC, Gospodarowicz D, Böhlen P (1992) Identification of the KDR tyrosine kinase as a receptor for vascular endothelial cell growth factor. Biochem Biophys Res Commun 187:1579–1586. doi:10.1016/0006-291X(92)90483-2

Thiery JP, Acloque H, Huang RYJ, Nieto MA (2009) Epithelial-Mesenchymal Transitions in Development and Disease. Cell 139:871–890. doi:10.1016/j.cell.2009.11.007

Tilmann C, Capel B (1999) Mesonephric cell migration induces testis cord formation and Sertoli cell differentiation in the mammalian gonad. Development 126:2883–2890

Ungewitte EK, Yao H (2013) How to make a gonad: cellular mechanisms governing formation of the testes and ovaries. Sex Dev 7:7–20. doi:10.1159/000338612

Upadhyay S, Luciani JM, Zamboni L (1979) The role of the mesonephros in the development of indifferent gonads and ovaries of the mouse. Ann Biol Anim Biochim Biophys 19:1179–1196. doi:10.1051/rnd:19790802

Upadhyay S, Luciani JM, Zamboni L (1981) The role of the mesonephros in the development of the mouse testis and its excurrent pathways. In: Byskov AG, Peters H (eds) Development and function of reproductive organs. Excerpta Medica, Amsterdam, pp 18–27

Uzumcu M, Dirks KA, Skinner MK (2002) Inhibition of platelet-derived growth factor actions in the embryonic testis influences normal cord development and morphology. Biol Reprod 66:745–753. doi:10.1095/biolreprod66.3.745

Vainio S, Heikkilä M, Kispert A, Chin N, McMahon AP (1999) Female development in mammals is regulated by Wnt-4 signalling. Nature 397:405–409. doi:10.1038/17068

Valdembri D, Caswell PT, Anderson KI, Schwarz JP, König I, Astanina E, Caccavari F, Norman JC, Humphries MJ, Bussolino F, Serini G (2009) Neuropilin-1/GIPC1 signaling regulates α5β1 integrin traffic and function in endothelial cells. PLoS Biol. doi:10.1371/journal.pbio.1000025

Vidal VPI, Chaboissier M-C, de Rooij DG, Schedl A (2001) Sox9 induces testis development in XX transgenic mice. Nat Genet 28:216–217. doi:10.1038/90046

Vieira JM, Ruhrberg C, Schwarz Q (2010) VEGF receptor signaling in vertebrate development. Organogenesis 6:97–106. doi:10.4161/org.6.2.11686

Vigier B, Watrin F, Magre S, Tran D, Josso N (1987) Purified bovine AMH induces a characteristic freemartin effect in fetal rat prospective ovaries exposed to it in vitro. Development 100:43–55

Wagner T, Wirth J, Meyer J, Zabel B, Held M, Zimmer J, Pasantes J, Bricarelli FD, Keutel J, Hustert E, Wolf U, Tommerup N, Schempp W, Scherer G (1994) Autosomal sex reversal and campomelic dysplasia are caused by mutations in and around the SRY-related gene SOX9. Cell 79:1111–1120. doi:10.1016/0092-8674(94)90041-8

Waltenberger J, Claesson-Welsh L, Siegbahn A, Shibuya M, Heldin CH (1994) Different signal transduction properties of KDR and Flt1, two receptors for vascular endothelial growth factor. J Biol Chem 269:26988–26995

Wang L, Mukhopadhyay D, Xu X (2006) C terminus of RGS-GAIP-interacting protein conveys neuropilin-1-mediated signaling during angiogenesis. FASEB J 20:1513–1515. doi:10.1096/fj.05-5504fje

Wartenberg H, Kinsky I, Viebahn C, Schmolke C (1991) Fine structural characteristics of testicular cord formation in the developing rabbit gonad. J Electron Microsc Tech 19:133–157. doi:10.1002/jemt.1060190203
Wilhelm D (2007) R-spondin1—discovery of the long-missing, mammalian female-determining gene? Bioessays 29:314–318. doi:10.1002/bies.20553
Wilhelm D, Palmer S, Koopman P (2007) Sex determination and gonadal development in mammals. Physiol Rev 87:1–28. doi:10.1152/physrev.00009.2006
Yao HH-C, Capel B (2002) Disruption of testis cords by cyclopamine or forskolin reveals independent cellular pathways in testis organogenesis. Dev Biol 246:356–365. doi:10.1006/dbio.2002.0663
Yao HHC, Whoriskey W, Capel B (2002) Desert Hedgehog/Patched 1 signaling specifies fetal Leydig cell fate in testis organogenesis. Genes Dev 16:1433–1440. doi:10.1101/gad.981202
Ziegler BL, Valtieri M, Porada GA, De Maria R, Müller R, Masella B, Gabbianelli M, Casella I, Pelosi E, Bock T, Zanjani ED, Peschle C (1999) KDR receptor: a key marker defining hematopoietic stem cells. Science 285:1553–1558. doi:10.1126/science.285.5433.1553

Chapter 5
Origin and Differentiation of Androgen-Producing Cells in the Gonads

Sarah J. Potter, Deepti Lava Kumar, and Tony DeFalco

Abstract Sexual reproduction is dependent on the activity of androgenic steroid hormones to promote gonadal development and gametogenesis. Leydig cells of the testis and theca cells of the ovary are critical cell types in the gonadal interstitium that carry out steroidogenesis and provide key androgens for reproductive organ function. In this chapter, we will discuss important aspects of interstitial androgenic cell development in the gonad, including: the potential cellular origins of interstitial steroidogenic cells and their progenitors; the molecular mechanisms involved in Leydig cell specification and differentiation (including Sertoli-cell-derived signaling pathways and Leydig-cell-related transcription factors and nuclear receptors); the interactions of Leydig cells with other cell types in the adult testis, such as Sertoli cells, germ cells, peritubular myoid cells, macrophages, and vascular endothelial cells; the process of steroidogenesis and its systemic regulation; and a brief discussion of the development of theca cells in the ovary relative to Leydig cells in the testis. Finally, we will describe the dynamics of steroidogenic cells in seasonal breeders and highlight unique aspects of steroidogenesis in diverse vertebrate species. Understanding the cellular origins of interstitial steroidogenic cells and the pathways directing their specification and differentiation has implications for the study of multiple aspects of development and will help us gain insights into the etiology of reproductive system birth defects and infertility.

5.1 Origins of Androgen-Producing Cells (Cell Migration and Progenitors)

Androgen-producing cells differentiate in the interstitium of the mammalian gonad, outside of the germ-cell-containing compartments of the testis (testis cords/seminiferous tubules) and ovary (follicles). While steroidogenic cells of the male gonad,

Potter and Kumar contributed equally to this work and are co-first authors.

S.J. Potter • D.L. Kumar • T. DeFalco (✉)
Division of Reproductive Sciences, Cincinnati Children's Hospital Medical Center, Cincinnati, OH 45229, USA
e-mail: sarah.potter@cchmc.org; deepti.lava.kumar@cchmc.org; tony.defalco@cchmc.org

R.P. Piprek (ed.), *Molecular Mechanisms of Cell Differentiation in Gonad Development*, Results and Problems in Cell Differentiation 58,
DOI 10.1007/978-3-319-31973-5_5

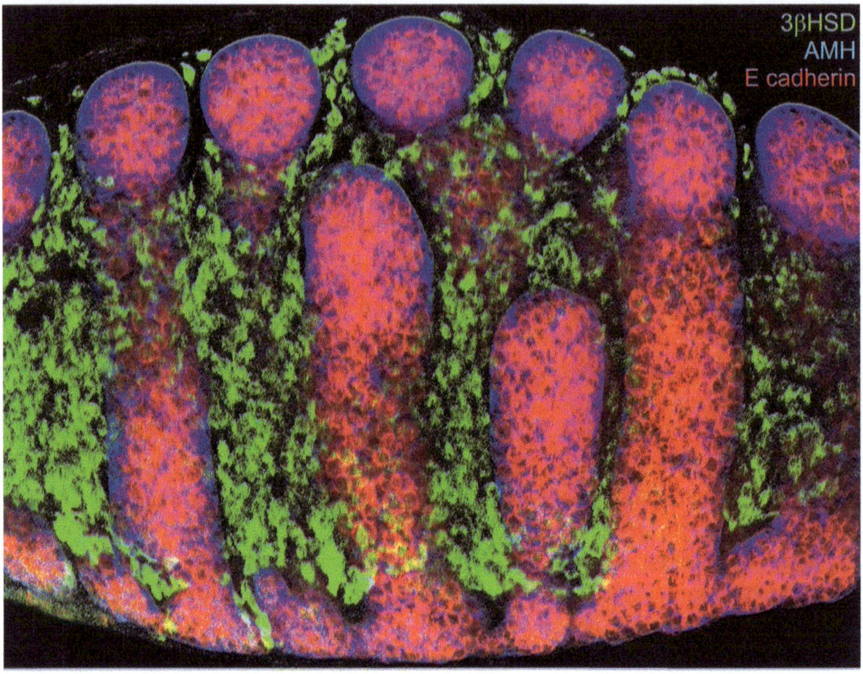

Fig. 5.1 Leydig cells differentiate in the interstitium of the fetal testis. Confocal microscopy image of a 13.5 *dpc* fetal mouse XY gonad, showing the development of two main testicular compartments: the testis cords, which contain Sertoli cells (marked in *blue* with anti-AMH antibody) and germ cells (marked in *red* with anti-Cadherin1/E-cadherin antibody), and the interstitium, which contains differentiated steroidogenic Leydig cells (marked in *green* with anti-HSD3B1/3β-HSD antibody)

termed Leydig cells (after the German anatomist Franz Leydig who first described them), arise during fetal development (Fig. 5.1), theca cells, one of the steroidogenic cell populations of the ovary which produces androgens, do not differentiate until perinatal stages when follicular development begins (Barsoum and Yao 2010; Magoffin 2005; Mannan and O'Shaughnessy 1991).

Fetal Leydig cells (FLCs), the steroidogenic population in the developing testis, arise 24 h after Sertoli cell specification at approximately 12.5 *dpc* (days *post coitum*) in the mouse fetal gonad (Barsoum and Yao 2010). The number of FLCs increases dramatically from 12.5 to 18.5 *dpc* (Barsoum and Yao 2010), after which FLC number decreases over the first week of postnatal life. FLCs are gradually replaced by adult Leydig cells (ALCs), which arise anew postnatally (i.e., not directly derived from differentiated FLCs; this is discussed in greater detail later in this chapter). As differentiated FLCs and ALCs are mitotically inactive (Byskov 1986; Migrenne et al. 2001; Orth 1982), the expansion of both cell types depends on the differentiation of progenitor cells in the interstitium. The origin of FLCs has not yet been identified definitively and has been a subject of debate for many years. Nevertheless, multiple potential sources have been speculated for the origin of FLCs (Fig. 5.2). These include neural crest cells, coelomic epithelium,

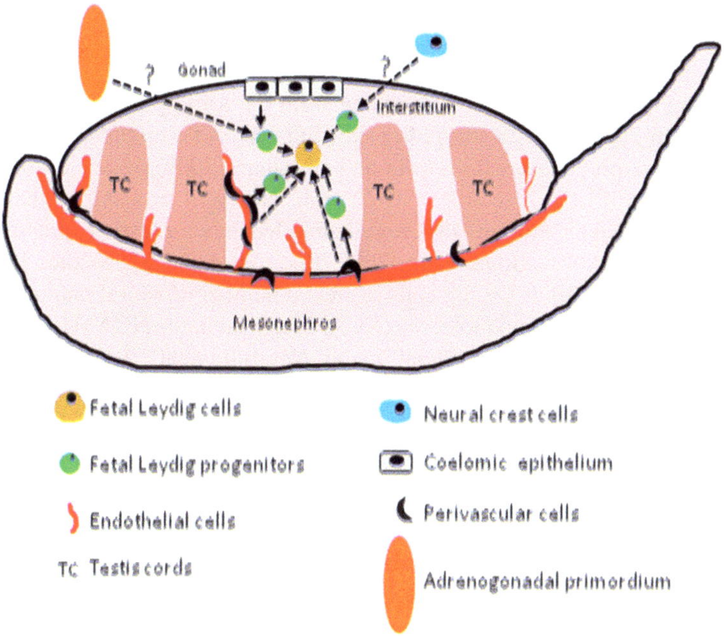

Fig. 5.2 Potential origins of fetal Leydig cells. Schematic of a fetal mouse XY gonad, showing potential cellular sources that give rise to fetal Leydig cells and their progenitors. *Arrows* indicate cellular sources that have been suggested in the literature (neural crest cells, coelomic epithelium, mesonephros, adrenogonadal primordium, and perivascular cells), while *question marks* indicate cellular sources that are still untested (adrenogonadal primordium) or are unsupported or inconsistent with data in the field (neural crest cells). It is still unclear for some cell populations whether they initially give rise to progenitors or if they directly differentiate into mature Leydig cells. The events depicted likely do not occur simultaneously, but can occur over a span of 48 h in the mouse (11.5–13.5 *dpc*)

mesonephros, adrenogonadal primordium, and perivascular cells. While it is assumed that female androgen-producing cells arise from similar progenitors as their male counterparts, not as much is known about the fetal origin of theca cells; we have discussed theca cell origin relative to Leydig cells in greater detail in Sect. 5.6. Other cell types in the gonad, such as Sertoli cells, granulosa cells, and luteal cells, also produce steroid hormones and other factors involved in steroid hormone regulation; however, we will focus on the interstitial androgen-producing cells, as the other cell types will be discussed in greater detail in other chapters in this volume.

Neural crest cells Neural crest cells (NCCs) arise from a group of precursor cells located between the neural plate and the epidermis in the developing embryo and give rise to different cell types of the body. Studies have shown that NCCs can contribute to neuroendocrine populations of the developing embryo, including the steroidogenic cells of the adrenal medulla and neuroendocrine cells in the pancreas and foregut (Huber 2006). Despite some contradictory findings with respect to the pancreas and the foregut, the hypothesis that NCCs are putative precursors for neuroendocrine cells has persisted (Andrew and Kramer 1979; Andrew et al. 1998; Kirchgessner et al. 1992; Pictet et al. 1976). Given that the adrenal gland and the gonad share a common origin, i.e., arising from the same primordium (Griswold and Behringer 2009), along with studies showing that NCCs might be precursors for neuroendocrine populations (Griswold and Behringer 2009), it was reasonable to hypothesize that FLCs arise from NCCs; additionally, it was shown that Sertoli cells and NCCs have common DNA regulatory elements for testis-expressed genes such as *Gata4* and *Sry* (Boyer et al. 2006; Pilon et al. 2008). Studies have also shown that fetal and adult Leydig cells express a number of neural markers such as Nestin, Neural cell adhesion molecule (NCAM, official name NCAM1), S-100, and Neurofilament protein 200 (official name Neurofilament, heavy polypeptide/NEFH) (Davidoff et al. 2002; Lobo et al. 2004; Middendorff et al. 1993; Mayerhofer et al. 1996). These neuroendocrine characteristics of the FLCs led to the idea that one of the potential sources of FLC origin could be NCCs. To determine if NCCs give rise to FLCs, Brennan and colleagues lineage traced NCCs using a Cre-responsive *lacZ* reporter strain driven by two distinct NCC-specific Cre lines (under the control of either the *P0* or *Wnt1* promoter), but did not find significant evidence for the contribution of NCCs toward Leydig cell lineages (Brennan et al. 2003). These findings suggested that FLCs do not originate from the neural crest cell lineage and instead acquire their neuroendocrine properties later on during development of the testis.

Coelomic epithelium Another potential source for the origin of FLCs is the coelomic epithelium of the gonad. Karl and Capel used a lipophilic dye (DiI) to label the cells of the coelomic epithelium and tracked them for 12–24 h in *ex vivo* organ explant cultures (Karl and Capel 1998). The migration of labeled cells into the testicular compartment and their subsequent differentiation were dependent upon the developmental stage at which the gonad was exposed to the dye. When the dye was injected at 11.2–11.4 *dpc* (during 15–17 tail-somite stages), the labeled cells were able to migrate into the gonad and were found in the testis cords (suggestive of a Sertoli cell fate) as well as in the interstitium (suggestive of FLC fate). In later stages of development, i.e., at 11.5 *dpc* or later (18–20 tail-somite stages), labeled cells were excluded from the testis cords and were restricted to the interstitium. Whether these labeled cells directly gave rise to FLCs was not specifically tested. However, DeFalco et al. (2011) performed similar dye-labeling experiments using a MitoTracker dye and showed that a subset of later-stage (11.5 *dpc* or later) labeled cells expressed the Leydig-cell-specific enzyme 3β-HSD (official name HSD3B1), demonstrating definitively that the coelomic epithelium is one of the cellular origins of FLCs. Further studies are required to establish the specific roles of the progeny of the coelomic epithelium at different stages with regard to FLC development.

Mesonephros The mesonephros, the organ that borders the gonad, is another tissue that is thought to contain precursors of FLCs. The mesonephros gives rise to the extra-gonadal ducts of the reproductive tract, such as the epididymis and oviduct in males and females, respectively. In mice, shortly after sex determination at around

12.5 *dpc*, one of the dramatic changes that occur in the male gonad is migration of cells from the mesonephros. This migration is induced by the male gonad (the sex of the mesonephros does not matter) and is dependent on *Sry* function (Capel et al. 1999; Martineau et al. 1997). Studies using gonad/mesonephros recombination organ cultures (male wild-type gonads cultured on top of a mesonephros carrying a ubiquitously expressed GFP or *lacZ* reporter under the *β-actin* or *ROSA26* promoter, respectively) demonstrated that cells from the adjacent mesonephros contribute to the gonadal population. GFP- and *lacZ*-positive cells from the mesonephros were shown to migrate into the gonad, but most of these cells were characterized as endothelial cells (Combes et al. 2009; Cool et al. 2008; Martineau et al. 1997). Additionally, using *kinase insert domain protein receptor* (*Kdr*)-*lacZ* mice (mice with *lacZ* expression driven by the promoter of *Kdr*, the gene encoding VEGFR2), it was shown that *Kdr-lacZ*-expressing endothelial cells from the mesonephros migrated in the XY gonad at 11.5 *dpc* (Bott et al. 2010). This migratory event occurs specifically in the male gonad and is essential for normal testis cord formation (Tilmann and Capel 1999). Further research is required to determine definitively if any non-endothelial mesonephric cells that migrate into the gonad originate from the mesenchyme or epithelia of the mesonephros, and if any cells that may migrate during earlier gonad development (i.e., prior to the stages when gonad–mesonephros recombination cultures can be set up) contribute to the FLC population.

Steroidogenic factor-1 (SF-1)-positive progenitors in the adrenogonadal primordium Studies have shown that 11.5 *dpc* (a stage prior to the migration of cells from the mesonephros) male gonads cultured in the absence of the mesonephros failed to develop testis cords, but did contain a subpopulation of cells that have FLC characteristics (Merchant-Larios et al. 1993). This indicates that the precursors that give rise to the FLC population are present in the gonad prior to the migratory event that takes place from the adjacent mesonephros. Both Sertoli cells and Leydig cells express SF-1 (Steroidogenic factor 1; official name is nuclear receptor subfamily 5, group A, member 1 [NR5A1]), and hence, it has been suggested that they arise from progenitors that express SF-1 (Luo et al. 1994, 1995). SF-1 expression begins early on in adrenogonadal development and SF-1-positive cells are known to contribute to steroidogenic cells in the adrenal cortex. Whether these SF-1-positive cells are the FLC precursors present in the adrenogonadal primordium or whether the precursors arise as a separate population and subsequently initiate SF-1 expression independently is not known and warrants further testing.

Perivascular cells and pericytes DeFalco et al. (2011) performed a detailed analysis of the cells that form the interstitial compartment of the developing mouse gonad. They found that the interstitium is comprised of a heterogeneous population of cells, some of which are closely associated with the vasculature during organ formation. Using immunofluorescence, the authors identified expression of the Maf family of basic leucine zipper transcription factors, specifically MAFB and C-MAF (official name MAF), within the interstitium. These factors are the mammalian orthologs of the *Drosophila* gene *traffic jam* known to regulate cell adhesion interactions during morphogenesis of the fly gonad (Li et al. 2003) and were examined for their potential roles in mammalian gonad development; MAFB and C-MAF were found to be early markers of the interstitial lineage. MAFB was dynamically expressed in the gonad with expression along the gonad–mesonephros border in 11.5 *dpc* embryos and in some interstitial cells in both male and female gonads. Additionally, some scattered cells at 11.5 *dpc* also showed expression of MAFB in the gonadal surface domain just beneath the coelomic epithelium only in the male gonad; the expression of MAFB was mutually exclusive with Sertoli cell markers, suggestive of an early interstitial fate. By 12.5 *dpc*, expression expanded to most of the somatic cells in the male interstitium, i.e., outside the testis cords. By 13.5 *dpc* MAFB and C-MAF were expressed in most interstitial cells. Later on in fetal life (by 14.5 *dpc*), *Mafb*-GFP knock-in reporter expression was observed specifically in the Leydig cell lineage. C-MAF expression was also observed along the gonad–mesonephros border and also within interstitial cells, although the onset of its expression was slightly delayed relative to MAFB; later expression of C-MAF was restricted to non-Leydig interstitial cells (DeFalco et al. 2011). Subsequent work revealed that some of the C-MAF-positive cells are gonadal macrophages (DeFalco et al. 2014). The MAFB/C-MAF double-positive interstitial cells likely represent Leydig progenitors during initial testis formation, and these two factors subsequently segregate into unique lineages, similar to MAFA and MAFB during alpha and beta cell specification in the pancreas (Nishimura et al. 2009).

Additionally, to determine if MAFB/C-MAF-positive perivascular cells along the gonad–mesonephric border migrate into the gonad and contribute to the interstitial population, a modified GFP recombination assay (see above) was performed by DeFalco and colleagues (DeFalco et al. 2011). Usually vascular sprouts and associated perivascular cells are disrupted during preparation of gonadal and mesonephric tissues in recombination assays, and GFP-positive migratory cells are strictly endothelial cells (Combes et al. 2009). However, in this case, vascular sprouts and associated cells were kept intact in the mesonephric component of the recombination culture. The researchers found that the GFP-positive cells also expressed VCAM1, a marker for vasculature-associated interstitial cells, and these cells migrated into the gonad. They also showed that these GFP-positive interstitial cells were adjacent to GFP-positive endothelial cells, suggesting that these two cell types migrated into the gonad together. Staining with 3β-HSD, a marker for differentiated FLCs, following a 64-h culture (time during which the cells would differentiate into Leydig cells), revealed 3β-HSD expression in GFP-positive cells (DeFalco et al. 2011). Thus, these findings indicated that migrating perivascular cells at the gonad–mesonephros border represent a progenitor population that gives rise to fetal Leydig cells.

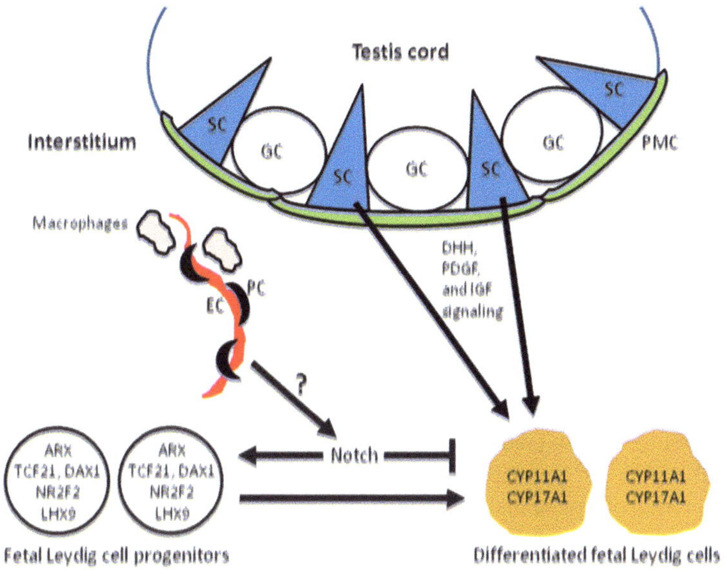

Fig. 5.3 Mechanisms regulating fetal Leydig cell differentiation. Signaling pathways and factors involved in maintaining a balance between fetal Leydig cell progenitors and differentiated fetal Leydig cells in the interstitium of the XY gonad. The *question mark* indicates the unknown relationship between vasculature/perivascular cells and Notch signaling. The transcription factors mentioned in the progenitor cell population are expressed in these cells but do not necessarily promote progenitor identity, as many are responsible for promoting differentiation. Abbreviations are as follows: *EC* endothelial cells, *GC* germ cells, *PC* pericytes, *PMC* peritubular myoid cells, *SC* Sertoli cells

5.2 Molecular Mechanisms of FLC Differentiation

5.2.1 Signaling Pathways Regulating FLC Development

Regardless of their origin, FLCs lack expression of *Sry* or *Sox9* and hence their male-specific differentiation likely depends on molecular cues from the neighboring Sertoli cells (Barsoum and Yao 2010). For example, Sertoli-cell-derived signaling molecules such as Desert hedgehog (DHH) and Platelet-derived growth factor alpha (PDGFA) have been shown to positively regulate FLC differentiation (Brennan et al. 2003; Yao et al. 2002). Balance between differentiation-promoting and differentiation-inhibiting factors regulates maintenance and differentiation of the Leydig cell population in the fetal gonad (Fig. 5.3).

Hedgehog signaling In the XY gonad, *desert hedgehog* (*Dhh*) mRNA is expressed in Sertoli cells beginning at 11.5 *dpc* and the gene encoding its corresponding receptor, *patched homolog 1* (*Ptch1*), is expressed in the interstitium of only XY gonads at around 12.5 *dpc* (Yao et al. 2002). XY gonads lacking *Dhh* show defects in the differentiation of FLCs and have disruptions in testis cord formation

(Yao et al. 2002). In prepubertal and adult stages, *Dhh*-null mice show spermatogenetic defects and lack adult Leydig cells (Clark et al. 2000). Humans with mutation in the *DHH* gene show pseudohermaphroditism and gonadal dysgenesis, a phenotype similar to *Dhh*-null mice (Canto et al. 2004; Umehara et al. 2000). *Dhh* is thought to act by regulating the expression of *Sf-1* and *cytochrome P450, family 11, subfamily a, polypeptide 1* (*Cyp11a1*, also called *side chain cleavage* [*Scc*], a steroidogenic enzyme required for androgen synthesis) genes (Barsoum and Yao 2010). These data suggest that DHH secreted from Sertoli cells signals to the interstitium, which expresses the PTCH1 receptor, to determine FLC cell fate (Yao et al. 2002).

Platelet-derived growth factor (PDGF) signaling The PDGFA ligand is secreted from Sertoli cells and is thought to signal to its receptor PDGFRA, which is expressed throughout the mesenchyme of the XY interstitium (fewer cells express PDGFRA in the XX interstitium) (Brennan et al. 2003). While both PDGFRA and PDGFRB receptors are expressed in the interstitium, only *Pdgfra* function is uniquely required for testis differentiation: analyses of *Pdgfrb*-mutant embryos revealed no gross defects in the early fetal testis (Brennan et al. 2003). The *Pdgfra*-null XY gonad showed a disruption in the formation of testis-specific vasculature and patterning of the testis cords; mesonephric cell migration and proliferation of interstitial FLC precursors were also severely reduced (Brennan et al. 2003). This indicates that PDGFA-PDGFRA signaling plays an important role in testis organogenesis and Leydig cell differentiation (Brennan et al. 2003). These studies strongly suggest that a network of signaling from the Sertoli cells to the mesenchymal cells of the interstitium is responsible for regulating the process of FLC differentiation.

Insulin-like growth factor (IGF) signaling In addition to Hedgehog and PDGF signaling pathways, Insulin-like growth factor (IGF) signaling from the Sertoli cells is also known to positively regulate FLC differentiation (Baker et al. 1996). *Insulin-like growth factor 1* (*Igf1*)-mutant mice are infertile dwarfs with dramatically reduced testosterone levels, causing a failure of masculinization and subsequent infertility (Baker et al. 1996). Further analyses of mutant Leydig cells showed that there is a developmental delay, pointing toward an essential role of *Igf1* in Leydig cell differentiation (Baker et al. 1996), although it is possible that Leydig cell differentiation is altered due to delays or defects in testis cord development in IGF pathway mutants (Nef et al. 2003; Pitetti et al. 2013).

Notch signaling While Sertoli-cell-derived DHH and PDGFA positively regulate FLC differentiation, Notch signaling was shown to negatively regulate FLC differentiation (Tang et al. 2008). Inhibition of Notch signaling in the fetal testis using DAPT (*N*-[*N*-(3,5-difluorophenacetyl)-L-alanyl]-*S*-phenylglycine *t*-butyl ester), a gamma-secretase inhibitor, or disruption of the Notch target gene *hairy and enhancer of split 1* (*Hes1*) resulted in an increase in the number of differentiated FLCs (Tang et al. 2008). Conversely, constitutive activation of Notch in the somatic gonad using a *Sf-1*-Cre-driven *Rosa*-Notch1-intracellular-domain transgene led to a decrease in FLC number (Tang et al. 2008). Thus, Notch signaling in the fetal gonad promotes the maintenance of the Leydig progenitor cell population by restricting Leydig cell differentiation within the interstitial compartment. The exact cell types engaging in Notch signaling have not been clearly defined; however, a recent study showed that the activity of *Jag1* (a gene encoding one of the Notch ligands) is required in the interstitium and perivascular cells for a proper balance between progenitors and differentiated cells, and JAG1 signaling may act through the receptor NOTCH2 (DeFalco et al. 2013).

5.2.2 Transcription Factors and Nuclear Receptors Involved in FLC Differentiation

Apart from paracrine factors, studies have shown that several transcription factors and nuclear receptors play a role in regulating the differentiation of FLCs. These factors likely act downstream of key signaling pathways (described above) cell autonomously to regulate key aspects of steroidogenic development within interstitial cells.

Transcription factor 21 (TCF21) Transcription factor 21 (TCF21) (also called Pod1/Capsulin or Epicardin), a basic helix-loop-helix transcription factor, negatively regulates FLC differentiation (Cui et al. 2004). While *Tcf21*-mutant mice die at birth due to respiratory failure caused by the absence of alveoli, the authors described feminization of the external genitalia in XY gonads from *Tcf21*-mutant mice. The effect of *Tcf21* mutation on gonad development was revealed to be an ectopic expression of SF-1; this upregulation of SF-1 resulted in more progenitor cells committing to a steroidogenic cell fate, resulting in abnormal steroidogenesis (Cui et al. 2004).

Aristaless-related homeobox (ARX) ARX is another transcription factor that plays a role in FLC differentiation (Kitamura et al. 2002; Miyabayashi et al. 2013). *Arx*-mutant testes show defective Leydig cell differentiation with a decrease in FLC number throughout fetal life. Since differentiated Leydig cells do not proliferate, this decrease in Leydig cell number is attributed to defects in the Leydig cell progenitor population. Miyabayashi et al. (2013) showed that the progenitor population that gives rise to differentiated FLCs is ARX positive. The authors' general conclusion was that the ARX transcription factor positively regulates fetal Leydig cell differentiation (Miyabayashi et al. 2013).

LIM homeobox protein 9 (LHX9) LHX9 belongs to the LIM homeobox gene family of transcription factors (Birk et al. 2000). It is known to play an important role in gonadal primordium development, as mice lacking *Lhx9* show gonadal agenesis (lack of gonads), lack male accessory sex organs, and are infertile (Birk et al. 2000). In early-stage embryos (11.5 *dpc*), *Lhx9* is highly expressed in the somatic progenitors of the coelomic epithelium and the subjacent mesenchyme. Tang et al. (2008) reported an increase in the number of LHX9-positive spindle-shaped progenitor cells and a concomitant decrease in the number of differentiated Leydig cells following constitutive activation of Notch signaling in the male fetal gonad. As spindle-shaped cells have been identified as putative adult Leydig stem cells (Davidoff et al. 2004), it is likely that these spindle-shaped LHX9-positive cells represent a progenitor pool that gives rise to differentiated FLCs. Whether these LHX9-positive progenitor cells directly give rise to differentiated FLCs is not known. Nevertheless, the authors concluded that constitutive activation of Notch signaling maintains the LHX9 progenitor population and likely prevents their differentiation into Leydig cells (Tang et al. 2008).

Nuclear receptor subfamily 0, group B, member 1 (NR0B1) Nuclear receptor subfamily 0, group B, member 1 (NR0B1), also called DAX1, is a nuclear receptor that plays an important role in testis cord organization during gonadal differentiation (Meeks et al. 2003). Gonads lacking *Dax1* show a host of testis defects, such as abnormal organization of Sertoli cells, lack of basal laminae in the testis, and disrupted peritubular myoid cell development. Both the number and organization of immature Leydig cells, as shown by the expression of CYP11A1 (a steroidogenic enzyme required for androgen synthesis), were affected. Immunohistochemical staining with SF-1 revealed that Leydig cells were restricted to the coelomic surface in *Dax1*-mutant mice (Meeks et al. 2003).

Nuclear receptor subfamily 2, group F, member 2 (NR2F2) Nuclear receptor subfamily 2, group F, member 2 (NR2F2), also called COUP-TFII, is a member of the nuclear receptor family of proteins and was also shown to play an essential role in differentiation of both fetal and adult Leydig cells (Qin et al. 2008). *Nr2f2* is highly expressed in the mesenchyme and is absolutely essential for development, as embryos lacking a functional copy of *Nr2f2* die at 10.5 *dpc* due to angiogenic and cardiovascular defects (Pereira et al. 1999; Tsai and Tsai 1997). In rats, between 15.5 and 21.5 *dpc* there is a threefold increase in FLC number, cytoplasmic volume per Leydig cell, and intratesticular testosterone levels, along with a decrease in the Leydig cell population expressing NR2F2. Exposing fetal rats to dibutyl phthalate (DBP), a drug that induces masculinization disorders, prevented this decrease in NR2F2-expressing Leydig cells and also prevented a normal age-dependent change in Leydig cell cytoplasmic volume, nuclear volume, and ITT levels, but did not have any effect on FLC number at any age. Furthermore, exposure to DBP downregulated the expression of SF-1-regulated steroidogenic genes such as *StAR*, *Cyp11a1*, and *Cyp17a1* (with binding sites for both SF-1 and NR2F2 in their promoter regions), but not *3β-HSD* (with binding sites for only SF-1 in its promoter region). The authors concluded that repression of NR2F2 may be an important mechanism by which fetal Leydig cells regulate the production of testosterone to ultimately drive masculinization (van den Driesche et al. 2012). To determine the role of *Nr2f2* in postnatal stages, conditional ablation of the *Nr2f2* gene was achieved using a ubiquitously expressed tamoxifen-inducible Cre line (Qin et al. 2008). Loss of *Nr2f2* during prepubertal stages following injection of tamoxifen at postnatal day 14 resulted in defective testosterone biosynthesis that led to arrest of spermatogenesis at the round spermatid stage, ultimately leading to infertility (Qin et al. 2008).

5.3 Adult Leydig Cell Sources and Differentiation

5.3.1 Sources of Adult Leydig Cells

Unlike most cell types, Leydig cells in the adult testis do not directly arise from Leydig cells in the fetal testis; instead they differentiate anew in postnatal life as fetal cells disappear. It has been suggested that both FLCs and ALCs originate from stem cell precursors (Benton et al. 1995; Ge et al. 2006; Hardy et al. 1989, 1990), but whether FLCs and ALCs arise from the same stem cell progenitor has not been fully resolved. Additionally, the source of the stem Leydig cell (SLC) that gives rise to the ALC is unclear; possibilities include perivascular smooth muscle and pericytes (Davidoff et al. 2004) or peritubular mesenchyme cells (Haider et al. 1995).

Davidoff et al. (2004, 2009) reported that vascular smooth muscle cells and pericytes present in the testis represent the progenitor population that gives rise to adult Leydig cells (ALCs). The putative Leydig cell progenitors (i.e., vascular smooth muscle cells and pericytes) expressed the intermediate filament protein Nestin, also known to be expressed in neural stem cells and progenitors (Davidoff et al. 2004, 2009). Furthermore, to determine if these Nestin-positive progenitors give rise to the ALC population, the authors injected ethane dimethane sulfonate (EDS) into adult rats. EDS is a cytotoxic compound that specifically eliminates the existing Leydig cell population temporarily; however, Leydig cells start regenerating within 2 weeks and are completely recovered by 1 month. Fourteen days following EDS treatment, Nestin-positive cells reappeared as clusters near the vicinity of the blood vessels (Nestin was used as a marker for Leydig progenitors). These Nestin-positive cells were identified as vascular smooth muscle cells and pericytes. Using Nestin-GFP transgenic mice it was recently demonstrated that Nestin-GFP-positive cells have the ability to undergo self-renewal in vitro and are able to differentiate into Leydig cells (Jiang et al. 2014). Additionally, when these Nestin-positive cells were transplanted into Leydig-cell-depleted mice (following injection of EDS), these cells were able to colonize the interstitium, differentiate into Leydig cells, partially rescue testosterone deficiency, and thereby improve spermatogenesis, as testosterone is known to be critical for completion of meiosis and spermiogenesis in rodents (McLachlan et al. 2002). In all, it was shown that perivascular cells and pericytes can give rise to adult Leydig cells.

SLCs are capable of differentiation into the Leydig lineage, as SLCs give rise to ALCs when transferred to postnatal mice, resulting in ALC reconstitution (Ge et al. 2006). SLCs that give rise to ALCs express growth factor receptors, such as those for Kit ligand/Stem cell factor (SCF), Leukemia inhibitory factor

(LIF), and Platelet-derived growth factor alpha (PDGFA), which allow SLCs to functionally respond to these mitogens by proliferating (Ge et al. 2006). Decreased androgen receptor (AR) expression and DBP-induced epigenetic changes both influence the SLC population, thus affecting ALC differentiation (Kilcoyne et al. 2014; O'Shaughnessy et al. 2002). Factors, such as SF-1, a Leydig-cell-related transcription factor, and DHH, a Sertoli-cell-released factor, both have roles in FLC differentiation and influence ALCs, suggesting that the presence of FLCs or their related intrinsic or extrinsic factors may be involved in ALC development (Park et al. 2007). Yang et al. (2015) showed that the development of ALCs from mouse embryonic stem (mES) cells requires SF-1, supporting the role of intrinsic FLC-related transcription factors in the development of ALCs. Alternatively, Kilcoyne et al. (2014) via lineage tracing in rats determined that NR2F2-positive cells, the putative progenitor cell population for ALCs, exist during fetal stages. Therefore, ALC stem cells might be the same progenitor population for both FLCs and ALCs. Furthermore, FLC and ALC populations overlap during the first few weeks of postnatal life in rodents, when FLCs decline and are slowly replaced with ALCs, leaving a small number of FLCs remaining in adulthood (Benton et al. 1995).

5.3.2 Differentiation of Adult Leydig Cells from Stem Cell Progenitors

In humans, there is a triphasic progression of Leydig cells: fetal (first trimester), postnatal (2–3 months), and adult stages (puberty), all of which have distinct characteristics (Forest et al. 1973; Prince 1990; Reyes et al. 1974). In humans and pigs, FLCs (discussed in earlier sections) reach their maximum number by the end of fetal development, regressing in both number and size during neonatal periods followed by the propagation of a new subset of Leydig cells. The postnatal population peaks and also regresses by the first wave of spermatogenesis. This is followed by the appearance of a more permanent population that persists throughout adulthood (Prince 1990). In rats, most studies suggest a biphasic pattern, with FLCs peaking and then declining around the time when ALCs arise (Christensen and Gillim 1969). This difference may be due to the timescale of human gestation as compared to rodents.

Differentiation Although several studies have addressed morphological and gene expression differences between FLCs and the progenitor stages of adult Leydig cells (ALC) and others have analyzed SLC and ALC gene expression (Stanley et al. 2011), the differences between the terminally differentiated populations of FLCs and the ALCs have not been fully elucidated. In the rat model, the stem Leydig cells (SLCs) arise around postnatal day 7 and differentiate through various discrete stages resulting in mature ALCs by postnatal days 28–90 (Chen et al. 2010; Ge et al. 2006; Stanley et al. 2011) (Fig. 5.4 depicts the timescale and

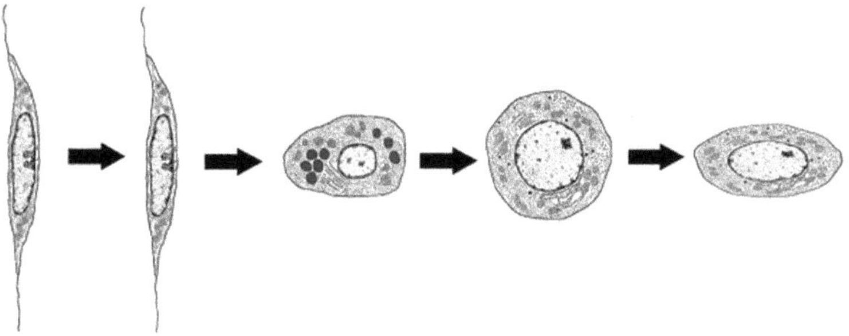

CELLULAR ATTRIBUTES	STEM LEYDIG CELL	PROGENITOR LEYDIG CELL	IMMATURE LEYDIG CELL	ADULT LEYDIG CELL	AGED LEYDIG CELL
Mitotic activity	Low	high	Once	Probably none	Probably none
Morphology	Spindle	Spindle	Round with large lipid droplets	Round with small lipid droplets	Round with small lipid droplets
Smooth ER	Low	Low	High	Highest	High
Testosterone Metabolism	None	High	Highest	Low	Low
Testosterone Synthesis	None	Low	Intermediate	High	Intermediate
Age when cell is typical	7 days	21 days	35 days	90 days	20 months
LH Receptor Number	None	Low	Intermediate	High	Intermediate
Androgen Receptor Number	?	High	High	Low	?

Fig. 5.4 Adult Leydig cell development. Diagram describing the characteristics (mitotic activity, morphology, smooth ER, testosterone metabolism, testosterone synthesis, age when cell is typical, LH receptor number, and androgen receptor number) of adult Leydig cell progenitors through differentiated cell populations in postnatal through aged rats. This figure is from Chen et al. (2009), used with permission from Elsevier and *Molecular and Cellular Endocrinology*. Abbreviation is as follows: *ER* endoplasmic reticulum. Refer to Sect. 5.3.2 in text

characteristics of each stage). The SLCs differentiate into progenitor Leydig cells (PLCs), which are characterized by their expression of luteinizing hormone receptor (LHR), 3α-HSD, and 3β-HSD. PLCs appear more spindle shaped and have negligible smooth ER, but still retain the ability to secrete androsterone as compared to SLCs. The PLCs enlarge, become round, and decrease their proliferative capacity, transitioning into immature Leydig cells (ILCs). ILCs are considered steroidogenic, containing abundant amounts of smooth ER and cytoplasmic lipid droplets. ILCs contain high levels of 5α-androstan-3α,17β-diol (3α-diol) (the inactive form of 5α-dihydrotestosterone [DHT]) and undergo a single mitotic division to produce ALCs (Chen et al. 2009). ALCs are large post-mitotic polyhedral cells, which have decreased lipid droplets and decreased levels of 5α-androstan-3α. Although the ALCs themselves do not proliferate (Teerds et al. 1989), studies have shown that, if eliminated upon administration of cytotoxic drugs like EDS, the ALC (3β-HSD-positive) population is replenished from SLCs (3β-HSD-/LHR-/PDGFRA+ cells) in as little as 10 days in rats (Ge et al. 2006).

Aging In aging adults, declines in testosterone production occur (Harman et al. 2001). There is some debate if this decrease is due to a decline in ALC numbers (Neaves et al. 1985) or due to reduced steroidogenesis in individual cells (Chen et al. 1994). Zirkin and Chen (2000) demonstrated that reactive oxygen species from the Leydig cells themselves may be a source for ALC age-related testosterone decrease. Furthermore, degeneration of ALCs in aging adults by atrophy (i.e., changes in cytoplasmic volume), rather than dedifferentiation (Neaves et al. 1984), is supported by observations of classic aging pigments (lipofuscin granules), crystalline inclusions, decreased mitochondria, and diminished smooth endoplasmic reticulum (Mori et al. 1978, 1982; Paniagua et al. 1986).

5.4 Cellular Interaction with Leydig Cells

Leydig cells have complex interactions both locally and systemically which influence their survival, differentiation, and functionality (Fig. 5.5). Here we discuss their specific interactions with cells of the interstitium (including macrophages and vascular endothelial cells), as well as with cells of the seminiferous tubules (including germ cells, Sertoli cells, and peritubular myoid cells [PMCs]). However, the potential influence of FLCs on the generation of ALCs cannot be discounted, as ALC stem cells temporally overlap with FLC populations and, thus, the microenvironment during fetal and postnatal stages may influence ALC functionality later on (Kilcoyne et al. 2014); this interaction needs to be elucidated in further detail.

5.4.1 Cells of the Interstitium

Macrophages Testicular macrophages are intimately associated with Leydig cells. The interdigitation of slender Leydig cytoplasmic processes within macrophages forms deep channels (Christensen and Gillim 1969; Hutson 1992; Miller et al. 1983) and is thought to mediate proper ultrastructure in ALCs that is required for steroidogenesis (Cohen et al. 1997). This macrophage–Leydig interdigitation has been documented as early as 20–30 days postnatally in rats, just prior to pubertal testosterone secretion (Christensen and Gillim 1969; Hutson 1992; Miller et al. 1983). Their close proximity and cellular interdigitation has been thought to modulate the production of testosterone through lipophilic factors and/or cytokines (Hales 2002; Hutson 1998; Lukyanenko et al. 1998; Watson et al. 1994). Some cytokines (including interleukin-1, tumor necrosis factor-α, interleukin-6, transforming growth factor-β (TGFβ), interferon-γ, lipopolysaccharide, reactive oxygen species, and nitric oxide) are known to decrease steroidogenesis (reviewed

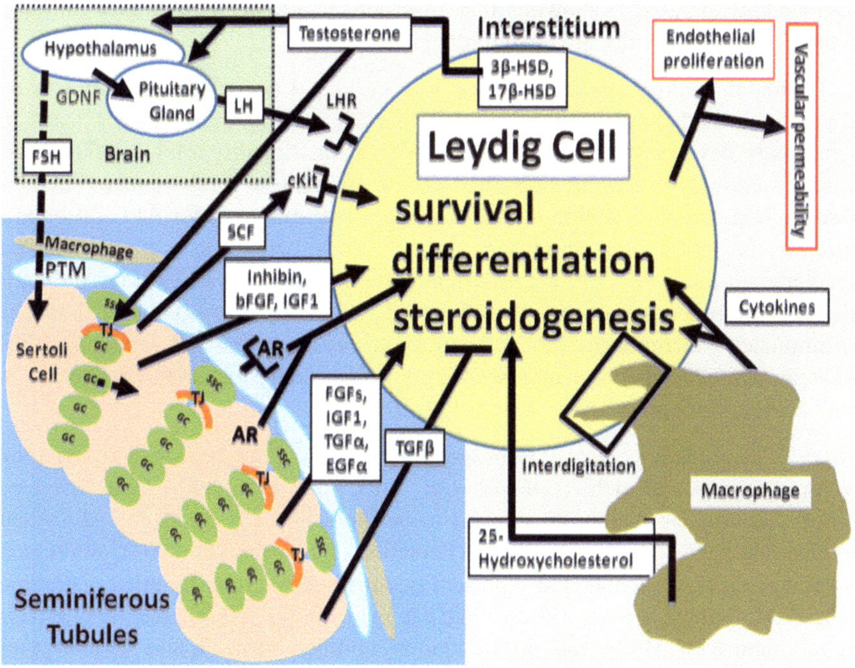

Fig. 5.5 Regulation of adult Leydig cell function. Cartoon depicting local and systemic factors that influence adult Leydig cell survival, differentiation, or steroidogenesis. *Dashed arrows* indicate long-distance or indirect interactions. Abbreviations are as follows: *AR* androgen receptor, *C-KIT* kit ligand/stem cell factor receptor, *EGFα* epidermal growth factor-α, *FGF-9* fibroblast growth factor-9, *bFGF* basic fibroblast growth factor, *GC* differentiated germ cell, *GDNF* glial cell line-derived neurotrophic factor, *IGF1* insulin-like growth factor, *LH* luteinizing hormone, *LHR* luteinizing hormone receptor, *PMC* peritubular myoid cell, *SCF* stem cell factor, *SSC* spermatogonial stem cell, *TGFα* transforming growth factor-α, *TGFβ* transforming growth factor-β, *TJ* tight junction

by Hales 2002), whereas unidentified lipophilic factors seem to increase testosterone production (Hutson 1998). Modulation of testosterone production has systemic effects and has been associated with changes in male mating behaviors, libido, and inter-male aggression (Hales 2002). It has been long known that macrophages directly secrete factors that positively influence steroidogenesis in Leydig cells (Yee and Hutson 1985); one of these candidate macrophage-derived factors has been identified as 25-hydroxycholesterol (Chen et al. 2002), an intermediate in the testosterone biosynthetic pathway (Nes et al. 2000). It is assumed that macrophage-secreted 25-hydroxycholesterol can be metabolized by Leydig cells and influences the capacity of Leydig cells to produce testosterone (Lukyanenko et al. 2001). Additionally, transient macrophage depletion (via a *Cx3cr1*-Cre-induced diphtheria toxin receptor ablation method) in the adult mouse testis does not disrupt Leydig cell numbers, but does decrease testosterone intratesticularly by half within 10 days

(DeFalco et al. 2015), showing that macrophage–Leydig cell interactions are dynamic and can rapidly affect steroidogenesis.

Testicular macrophages also play a role in adult Leydig cell development by providing growth and differentiation factors, as lack of macrophages leads to abnormally developed Leydig cells with altered functionality (Hales 2002); additionally, recovery of Leydig cells after EDS treatment is severely hindered in their absence (Gaytan et al. 1994b, c). Macrophages are important for ALC differentiation, as depletion of macrophages during critical windows in prepubertal rat development hindered Leydig cell differentiation (Gaytan et al. 1994a). Additionally, Leydig functionality may be influenced by macrophage activation status, as unstimulated macrophages inhibit testosterone in adult, but not immature, rats, whereas activated macrophages seemingly enhanced testosterone production in both adult and immature rats (Afane et al. 1998).

Vascular endothelial cells FLC progenitors are maintained via Notch signaling within a perivascular niche (DeFalco et al. 2013; Tang et al. 2008). Cells localized in the perivascular region are thought to be the stem cell source for ALCs that reconstitute the testis after Leydig cell depletion (Davidoff et al. 2004), although it is unknown if perivascular cells represent a common progenitor for other interstitial cells. Additionally, there is a knowledge gap in the field regarding the direct effects of vasculature on ALCs (e.g., differentiation, number, survival, and steroidogenesis). In the adult, Leydig cells are intimately localized to arterial and venous microvasculature, such as capillaries (Ergun et al. 1994). Conversely, Leydig cell activation via human chorionic gonadotropin (hCG) stimulation contributes to testicular vascular permeability (Collin and Bergh 1996; Setchell and Rommerts 1985) and endothelial cell proliferation (Collin and Bergh 1996). Given that ALCs localize near interstitial vasculature and that influential substances (e.g., luteinizing hormone [LH]) are delivered to ALCs from the systemic vasculature, it is likely that vasculature is critical for Leydig cell function. DeFalco et al. (2013) demonstrated that active Notch signaling, which is often associated with vasculature, was not detected in the interstitium of the adult testis; therefore, unlike FLCs, it is unlikely that vascular Notch signaling regulates ALC differentiation or function. However, alternative direct regulatory interactions have not been elucidated.

5.4.2 Cells of the Seminiferous Tubules

Germ cells Bergh (1982, 1983) previously hypothesized that germ cells influence Leydig steroidogenesis, due to differences in the size of Leydig cells associated with various stages of spermatogenesis. This correlation may be an indirect relationship, as other groups have demonstrated that germ cells do not play a role in Leydig cell steroidogenesis locally or systemically, utilizing both in vivo and cell culture germ-cell-depletion models (Chen et al. 2004; O'Shaughnessy et al. 2008; Sprando and Zirkin 1997; Wright et al. 1993). Germ cell influence on Leydig cells is probably mediated through their modulation of Sertoli cells by controlling, among other factors, 17β-estradiol, transferrin, and inhibin (Carreau 1996), all of which influence testosterone production (Boujrad et al. 1992). Conversely, Leydig cells support germ cell development (Carreau et al. 2008; Hess 2003). This effect could be mediated through testosterone, as removal of testosterone blocks spermatogenesis at meiotic stages, and post-meiotic cells detach from Sertoli cells and undergo programmed cell death at the time of testosterone blockade.

Sertoli cells Adult Sertoli cells release factors important for Leydig cell survival (Rebourcet et al. 2014a, b), such as follicle-stimulating hormone (FSH), which is important for survival of both FLCs and ALCs (Benahmed et al. 1985); furthermore, loss of FSH receptors causes a decrease in Leydig cell numbers (Baker et al. 2003). Modulating AR expression in Sertoli cells has been shown to influence Leydig cell number and differentiation (Hazra et al. 2013). Additional paracrine activity of Sertoli cells (e.g., through inhibin subunits (Hsueh et al. 1987), basic fibroblast growth factor (bFGF) (Liu et al. 2014), and IGF1 (Hu et al. 2010) affects Leydig cell differentiation and influences steroidogenesis positively through fibroblast growth factor-9 (FGF-9) (Lin et al. 2010), basic fibroblast growth factor (bFGF) (Laslett et al. 1997; Liu et al. 2014; Sordoillet et al. 1992), insulin-like growth factor-1 (IGF1) (Gelber et al. 1992), epidermal growth factor-α (EGF-α), and transforming growth factor-α (TGF-α) (Millena et al. 2004), and negatively through transforming growth factor-β (TGF-β) (Lin et al. 1987). Testosterone influences Sertoli cell expression of Claudin-3, a tight junction protein comprising part of the blood–testis barrier (BTB), as well as affecting germ cell differentiation, thus having a major influence on fertility (Meng et al. 2005).

Peritubular myoid cells PMCs, a muscle cell type surrounding the tubules, influence adult Leydig cell development, ultrastructure, and function through the activity of AR. Mice lacking AR in PMCs had Leydig cells which were split into two groups: an unaffected, apparently "normal" group and an "abnormal" group, which displayed an arrested phenotype with lower levels of Insulin-like 3 (INSL3), LHR, and steroidogenic enzymes (Welsh et al. 2012), demonstrating the influence of PMCs on Leydig cells.

5.5 Interstitial Steroidogenesis

5.5.1 Normal Interstitial Steroidogenesis in the Male

Male interstitial steroidogenic cells Steroidogenic cells are located in the brain, placenta, adrenal glands, and the gonad. Regulation of steroidogenesis within the gonad is essential for sexual differentiation (Barsoum and Yao 2006), organ maintenance, and fertility (Meng et al. 2005). Testosterone, produced and released by ALCs and fetal/immature Sertoli cells (after production of androgenic precursors by FLCs), functionally influences multiple aspects of male development: spermatogenesis (Walker 2011); the levels of gonadotropin-releasing hormone (GnRH) and luteinizing hormone (LH); male external genitalia differentiation (Tomiyama et al. 2003); and secondary sexual characteristics at puberty. Although steroidogenesis is commonly associated with local effects within the gonad, sex steroid hormones influence the maintenance of the whole body, including bone formation, metabolism, and erythropoiesis (Snyder et al. 2000).

Male interstitial steroidogenesis Steroidogenesis, the process of making steroid hormones, is performed within specialized cells, Leydig and Sertoli cells (male), and theca, granulosa, and luteal cells (female), within the gonad. Steroidogenic cells utilize enzymes to convert cholesterol into various hormones and hormone precursors. Steroidogenic acute regulatory protein (StAR), whose expression is restricted to steroidogenic organs, shuttles cholesterol across the mitochondrial membrane for processing, thereby regulating steroid production (Clark et al. 1994). This process requires LH stimulation of the Leydig cells to produce cyclic adenosine monophosphate (cAMP) to fuel movement of cholesterol across the membrane. Under the influence of gonadotropins, steroidogenesis is controlled through the production of steroidogenic enzymes, such as CYP11A1 (Hum and Miller 1993). Within the mitochondria, cholesterol is converted via CYP11A1 into cytosolic pregnenolone, which serves as a precursor to most sex hormones. Pregnenolone is further converted into progesterone via 3β-HSD (King and LaVoie 2009); the stepwise processing of the hormones from pregnenolone occurs within the smooth endoplasmic reticulum. The conversion of progesterone into androstenedione is performed by CYP17A1 (Gilep et al. 2011), followed by its final conversion into testosterone via 17β-HSD (Ge and Hardy 1998). Testosterone needs to be metabolically converted into dihydrotestosterone (DHT) via two isoforms of 5α-reductase (Ge and Hardy 1998).

Male-specific testosterone feedback loop Testosterone production is controlled systemically by the hypothalamic–pituitary–gonadal (HPG) axis. The hypothalamus produces gonadotropin-releasing hormone (GnRH), which induces the pituitary to produce the gonadotropins FSH and LH, which modulate Leydig cell function either directly or indirectly through other testicular cells, such as Sertoli cells (Krishnamurthy et al. 2001). LH signaling within Leydig cells stimulates the production of testosterone. Testosterone is then part of two feedback loops. The first is a systemic negative feedback loop, in which once a particular level of testosterone is reached, the hypothalamus will adjust the amount of GnRH produced, influencing LH, which in turn reduces testosterone output. The other feedback mechanism is local, in which Sertoli cells regulate the local pool of testosterone production in Leydig cells through releasing androgen-binding protein (ABP) which binds testosterone, thus keeping the concentration high locally; the local concentration of testosterone is 50–100-fold higher than systemic serum concentrations.

Stage-dependent male interstitial steroidogenesis Steroidogenesis in the testis occurs in a triphasic fashion in humans (Prince 2001), corresponding to the dynamics of the wavelike development and transitioning of distinct Leydig populations, thereby introducing fluctuations in testosterone production. Steroidogenesis is important for different reasons during various time periods in development. FLCs are required for the masculinization of the fetus, whereas the ALCs are essential for puberty and have important roles in spermatogenesis, libido, and overall male health. FLCs, the steroidogenic cells of the testis, appear within the mouse gonad by 12.5 *dpc*. By 13.5 *dpc*, FLCs are fully differentiated and express steroidogenic factor 1 (SF-1), which initiates steroidogenesis (Luo et al. 1994), and additionally express steroid biosynthesis enzymes, such as 3β-HSD, which are important for the production of androgens. FLC function is to convert cholesterol into androgen precursors, such as androstenedione. This is performed by acquiring exogenous cholesterol from sources outside the testis (Budefeld et al. 2009); the fact that cholesterol arrives to the testis via the vasculature is a possible reason why FLCs are localized to perivascular regions. Conversion of cholesterol to testosterone is gonadotropin-independent in rodents, as the HPG axis is not yet active during this time (O'Shaughnessy et al. 1998, 2006). FLCs do not contain a key steroidogenic enzyme, 17β-HSD, also called HSD17B3, which is required to convert androstenedione to testosterone; therefore, FLCs require the export of androstenedione to fetal Sertoli cells for the final processing of testosterone that is required for secondary sexual differentiation (O'Shaughnessy et al. 1998; Shima et al. 2013).

Postnatally, there is a transition period in which FLCs are still present, while the progenitors of ALCs are presumably present but have not started to differentiate. This period is followed by FLC degeneration, in which the production of testosterone reaches its lowest. After ALCs differentiate from ILCs, the 5α-reductase enzymes decrease, while 17β-HSD increases, switching the major output away from 5α-androstan-3α,17βdiol, a testosterone metabolite, toward testosterone within the testis. The conversion of cholesterol into testosterone in postnatal and adult stages occurs in a gonadotropin-dependent manner, as LH stimulation leads to increased testosterone production within ALCs. The importance of LH for ALC function not only lies in its stimulation of testosterone production; it also is required for the proliferation and differentiation of the ALC population (Baker and O'Shaughnessy 2001). ALCs have the capacity to produce more testosterone than FLCs, and, upon the onset of puberty, testosterone levels rise and are maintained when the organism reaches sexual maturity. This increase in testosterone is important, as testosterone deficit in hypogonadism patients is associated with multiple symptoms such as ambiguous genitalia, reduced masculinization (including decreased hair growth, decreased muscle mass, development of breasts), and infertility. In aged animals, testosterone production, both locally and systemically, starts to decrease (Harman et al. 2001), although the underlying cellular cause of this decrease is still under debate.

5.5.2 Endocrine Disruptors of Male Interstitial Steroidogenesis

Leydig cell steroidogenesis can be influenced by various environmental contaminants, including, but not limited to, pesticides, phthalates, industrial by-products, metals, and estrogenic compounds. The perturbation of Leydig cell function by these endocrine-disrupting compounds has relevance for human reproductive health, and uncovering their mechanism of action is a major area of research.

Pesticides One major class of pesticides, organochlorines, includes compounds such as *p,p′*-DDT (and its derivatives) and lindane (also known as gamma-hexachlorocyclohexane [γ-HCH]); these agents influence steroidogenesis (Enangue Njembele et al. 2014) by preventing cholesterol transport across the mitochondrial membrane via StAR (Walsh and Stocco 2000), conversion of the initial stages of testosterone synthesis from cholesterol to pregnenolone (Murono et al. 2006), or later stages by affecting 3β-, 17β-, and 11β-HSD enzymes (methoxychlor and 2,2-bis(*p*-hydroxyphenyl)-1,1,1-trichloroethane, also known as HPTE) (Guo et al. 2013; Hu et al. 2011). Some pesticides (e.g., procymidone, vinclozolin, linuron) modulate their effects on steroidogenesis via AR (Svechnikov et al. 2010). Pesticides can also influence StAR function by decreasing *StAR* expression levels transcriptionally (dimethoate and lindane) and through posttranscriptional modification (Glyphosate; commonly known as the commercial pesticide Roundup) (Walsh et al. 2000a, b; Walsh and Stocco 2000).

Phthalates Phthalate exposure can induce changes in FLC function (testosterone production), mitosis (number), and behavior (aggregation) or SLC proliferation and differentiation, all of which are hypothesized to lead to testicular dysgenesis syndrome (TDS) (Hu et al. 2009). Phthalates, commonly used in plastic manufacturing, influence Leydig cells in a stage-specific manner (Akingbemi et al. 2001; Rodriguez-Sosa et al. 2014) and in a species-specific manner (rat FLCs are sensitive, but mouse and human FLCs seem to be resistant) (Johnson et al. 2012). In rats, di(2-ethylhexyl)phthalate (DEHP) given postnatally (at 21–35 days old) caused significantly reduced 17β-HSD function and decreased testosterone output, whereas when given to adults (at 62–89 days old) there seemed to be no effect on androgen biosynthesis (Akingbemi et al. 2001).

Phthalate esters (such as DEHP and di-*n*-octyl phthalate [DOP]) perturb both the structure of Leydig cells and their testosterone output (Jones et al. 1993). Although some metabolites of phthalates (monophthalates) do not have any effect on Leydig cells (such as mono-*n*-butyl phthalate and mono-*n*-benzyl phthalate), others, such as mono-2-ethylhexyl phthalate (MEHP; the monoester of DEHP), have been demonstrated to influence mitochondrial function and stimulate basal steroidogenesis via StAR in postnatal mice (Savchuk et al. 2015). Others have shown that dibutyl phthalate (DBP) causes epigenetic modifications (via histone methylation) at the *StAR* promoter (Kilcoyne et al. 2014).

Industrial by-products Industrial by-products, such as dioxin, decrease size and number of Leydig cells (Johnson et al. 1994) and can influence cAMP levels (Lai et al. 2005) and cytochrome activity (Moore et al. 1991), thereby influencing steroidogenesis. Differential changes in steroidogenesis occur in a dose-dependent manner, as low doses of 2,3,7,8-tetrachlorodibenzo-*p*-dioxin (TCDD) (0.3 μg/kg) affect intratesticular testosterone, whereas higher concentrations (1 μg/kg) influence systemic feedback loops in rats (Adamsson et al. 2009). Another industrial by-product is benzophenones (BP). BP-1 is known to selectively inhibit 17β-HSD and therefore influences testosterone production (Nashev et al. 2010).

Metals Some metal cations (such as cadmium, cobalt, copper, mercury, nickel, and zinc) suppress hCG-induced or cAMP-induced testosterone production in Leydig cells (Laskey and Phelps 1991). This suppression may be through steroidogenic enzymes, such as 3β-HSD (arsenic, cadmium, chromium, lead, and mercury), 17β-HSD (arsenic and cadmium), CYP11A1 (lead), or via blocking StAR protein activity (cadmium, lead) (Payne and Hardy 2007).

Estrogenic compounds Testosterone can be metabolized into DHT (via 5-α-reductase) and estrogen (via aromatase). Since these two metabolites have opposing effects, disruption in the androgen:estrogen ratio can lead to infertility; therefore, a delicate balance is necessary for testicular function. Estrogen, acting through estrogen receptor alpha (ER-α), causes a reduction in *StAR*, *CYP17A1*, and *17β-HSD* steroidogenic gene expression, thereby influencing male-specific steroidogenesis (Strauss et al. 2009). Estrogen can be detrimental to Leydig cell development, as it blocks Leydig progenitor cell division when given between 5 and 30 days after EDS-induced Leydig cell depletion in rats; however, this blockade depends on the stage given, as when it is given at day 0–5 after EDS treatment it had no effect (Abney and Myers 1991). The xenoestrogen compound bisphenol A (BPA) has seemingly different effects on Leydig cells depending on the stage or species (N'Tumba-Byn et al. 2012). BPA disrupts immature Leydig steroidogenesis by affecting the conversion of testosterone into its metabolite 5α-androstan-3α,17βdiol, but does not affect basal androgen production (Savchuk et al. 2013) as it does in mature ALCs (Akingbemi et al. 2004). Differences were observed between rodent and human testis cultures: *INSL3* mRNA and testosterone production were reduced in humans, but not for rodents at low concentrations of 10(-8) M BPA (N'Tumba-Byn et al. 2012). Distinct signaling mechanisms between rodents and humans have been demonstrated, as nuclear estrogen receptor alpha, ERα, plays a role in BPA-reduced testosterone in rodents only, as fetal humans lack ERα (Gaskell et al. 2003) and seemingly signal through G-protein coupled receptor 30 and estrogen-related receptor gamma (Bouskine et al. 2009; N'Tumba-Byn et al. 2012).

5.6 Differing Properties of Male and Female Androgen-Producing Cells

The gonad initially arises as a bipotential organ during fetal development; thereafter, gonadal programming is required for sexual differentiation into the ovary and the testis. While some studies have suggested that the supporting cell types in the male and female gonad, the Sertoli and granulosa cells, respectively, arise from a common progenitor (Albrecht and Eicher 2001), not as much is known about the interstitial compartment of the gonad. The analogous interstitial androgen-producing cell type in the ovary corresponding to Leydig cells is the theca cell, which is localized around developing follicles. Similar to the origins of the adult testicular Leydig cells, theca cells are thought to arise from an adult stem cell progenitor, likely mesenchymal stromal precursor cells or steroidogenic cells with copious amounts of smooth endoplasmic reticulum. However, while Leydig cells commence steroidogenesis soon after initial testis formation, theca cells do not produce androgens until after birth upon LH stimulation when folliculogenesis begins (Mannan and O'Shaughnessy 1991; Fortune and Armstrong 1977).

While it is assumed that interstitial cells are recruited around the developing follicle due to paracrine signals from granulosa cells, the embryonic origin of theca cells is less clearly defined. Recent work has shown that, similar to Leydig cells, theca cells likely have multiple cellular origins: a subset of indigenous embryonic ovarian cells expressing *Wt1* (*Wilms tumor 1 homolog*) and migrating mesonephric mesenchymal cells expressing *Gli1* (encoding a member of the Hedgehog signaling pathway), which respond to Hedgehog signaling from granulosa cells (Liu et al. 2015). An early marker of fetal interstitial cells is the transcription factor MAFB; MAFB-expressing cells appear in the mouse gonad regardless of sexual differentiation around 12.5 *dpc* (DeFalco et al. 2011). Within the female, these cells reside near blood vessels, whereas their male counterparts not only localize to perivascular regions but also spread throughout the entire interstitial region (DeFalco et al. 2011). An additional putative marker for theca progenitors includes NR2F2, which is expressed in a unique subset of interstitial mesenchyme in the fetal ovary (Rastetter et al. 2014). Interestingly, NR2F2 is also expressed in Leydig cells and is required for Leydig cell differentiation during prepubertal stages (Qin et al. 2008), revealing a potential similarity between Leydig and theca cells.

Upon the development of the primary follicle, theca cells migrate and proliferate in the areas adjacent to the follicle (Eshkol and Lunenfeld 1972; Peters 1969). Granulosa-derived factors are thought to contribute to the androgen production and morphology of the theca interna cells (Kotsuji et al. 1990; Tajima et al. 2007). Although likely of similar cellular origin, different signaling pathways and genes likely regulate the differentiation of theca and Leydig cells; for example, *Gdf9* is required for theca cells alone, while Hedgehog signaling is required for proper function (i.e., steroid production) of both cell types (Clark et al. 2000; Dong et al. 1996; Liu et al. 2015). In the preantral follicle, Parrott and Skinner (1997) have shown that granulosa-derived SCF activates theca cell growth and androstenedione production in cows, compared to a survival factor in Leydig cells. Although this is not an exhaustive list of factors, these examples demonstrate the heterogeneity of signaling influences between male and female androgen cells.

Both theca and Leydig cell androgen production are under the control of LH (Baird et al. 1981; Campbell et al. 1998; Palermo 2007; Ryan et al. 2008). Although LH/hCG and estrogen can stimulate theca cell steroidogenesis, additional factors, such as granulosa-derived activin, inhibin, BMPs, TGF-β, Gdf-9, and follistatin, can modulate hormone production (Tajima et al. 2007; Young and McNeilly 2010). Unlike ALCs, theca cells lack 17β-HSD, which is essential for conversion of androstenedione into testosterone, but do contain 3β-HSD (Sawetawan et al. 1994). Upon reaching the age of sexual maturity, Leydig cells reside in a post-mitotic phase with consistent numbers, whereas theca cells during ovulation become theca luteal cells, which produce massive amounts of progesterone until luteolysis (reviewed in Magoffin 2005). During and after luteolysis, theca luteal cells undergo apoptosis and become reabsorbed with the rest of the cells within the corpus albicans. An alternative fate of theca cells within atretic follicles is that they become hypertrophied and eventually lose their steroidogenic capacity (Himelstein-Braw et al. 1976; reviewed in Magoffin 2005).

5.7 Origin and Differentiation of Interstitial Steroidogenic Cells in Other Vertebrates

Vertebrates encompass a diverse set of animals, including amniotes (e.g., mammals, birds, and reptiles) and anamniotes (e.g., fish and amphibians), both of which possess reproductive organs but which have distinct testicular morphology. Amniotes exhibit a tubular structure interspersed with clumps of vasculature and interstitial cells, whereas the anamniotes have a cystic structure, either in the form of cysts or lobes. Originally thought to be lacking in anamniotes, Leydig cells have been observed in areas between the cystic regions (Norris and Carr 2013). Within the amniotes, seasonal and continuous/opportunistic breeders both have Leydig cells. Within the Leydig cells, the internal steroidogenic machinery, e.g., StAR, is evolutionarily conserved across species (i.e., piscine, avian, amphibian, and mammalian species) (Bauer et al. 2000).

Even among mammalian species, there are diverse gonadal phenotypes in terms of sex-specific steroidogenic cells. European mole (*Talpa occidentalis*) females have ovotestes that exhibit major fluctuations in steroid production and steroidogenic cell number in breeding versus nonbreeding seasons (Jimenez et al. 1993), likely to promote different behaviors appropriate for mating versus protection of offspring or territory. Similarly, spotted hyena females have high levels of androgen production (Browne et al. 2006), which drives female aggression that is evolutionarily advantageous due to the scavenger lifestyle of the hyena. However, this masculinization comes at great reproductive costs, as high levels of testosterone result in a pseudoscrotum and elongated peniform clitoris which are ill-suited for the birthing process, so many females die during parturition (Browne et al. 2006; Dloniak et al. 2006; Lindeque and Skinner 1982).

Among certain species, Leydig cell function and breeding patterns vary seasonally. Rainfall, temperature, light, and nutrition all affect seasonal breeding patterns, likely through influencing androgen production. Changes in androgen biosynthesis by altered light cycle in hamsters are caused by changes in synthesis and secretion of testosterone, as well as direct regulation of circulating LH and Leydig cell responsiveness to LH signaling (Desjardins et al. 1971; Hardy et al. 1987). Opportunistic breeders, such as some rodents and humans, do not have seasonal changes in testicular weight once they reach sexual maturity (Suzuki and Racey 1978). Therefore, opportunistic breeders have relatively little change in their testis cellular populations until aging occurs. In contrast, seasonal breeders have months of recrudescence and intervening periods of regression, in which each period differs in testicular weight and spermatogenic activity. Within seasonal breeders, there are variations in how testis function is modulated: bats show changes in individual Leydig cell volume and Leydig cell granule number (Loh and Gemmell 1980), while stallions, birds, and amphibians exhibit changes in Leydig and Sertoli cell numbers instead of individual cell size (Johnson and Nguyen 1986; Johnson and Thompson 1986).

5.8 Conclusions and Future Areas of Research in the Field

While much research has been done regarding steroidogenic cells of the gonad, there are several outstanding questions in the field that are likely to be fruitful areas of future research. The first major topic deals with the origins of FLCs and ALCs. Indirect evidence points to several potential sources for the origin of FLCs and ALCs; however, which cell type(s) definitively gives rise to FLCs and ALCs still remains somewhat unresolved. Additionally, the physiological significance of the presence of two distinct populations of Leydig cells (fetal and adult) is unclear and requires further investigation. Another major unanswered question in the field is whether FLCs and ALCs arise from the same stem cell precursors or have distinct cellular origins. Finally, another topic of great interest to human health concerns the effects of endocrine-disrupting compounds on the differentiation and function of Leydig cells, as well as the subsequent impacts on male development. Answering these questions will provide a more comprehensive understanding of the molecular mechanisms of steroidogenic cell differentiation and its cellular origins in fetal and adult life. This knowledge will surely be of significance as it will help reveal the etiology of disorders of sexual development and will have implications for our understanding of fertility and other aspects of human health that are dependent on or influenced by gonadal steroid hormones.

Acknowledgments We would like to thank Haolin Chen and Barry Zirkin for permission to use a figure from their previous publication (Chen et al. 2009) in *Molecular and Cellular Endocrinology*. This work was supported by Cincinnati Children's Hospital Medical Center (CCHMC) developmental funds, a CCHMC Trustee Award, and a March of Dimes Basil O'Connor Starter Scholar Award (#5-FY14-32) to T.D.; S.J.P. was supported by a CancerFree KIDS Research Grant.

References

Abney TO, Myers RB (1991) 17 beta-estradiol inhibition of Leydig cell regeneration in the ethane dimethylsulfonate-treated mature rat. J Androl 12:295–304

Adamsson A, Simanainen U, Viluksela M, Paranko J, Toppari J (2009) The effects of 2,3,7,8-tetrachlorodibenzo-p-dioxin on foetal male rat steroidogenesis. Int J Androl 32:575–585

Afane M, Dubost JJ, Sauvezie B, Issoual D, Dosgilbert A, Grizard G, Boucher D (1998) Modulation of Leydig cell testosterone production by secretory products of macrophages. Andrologia 30:71–78

Akingbemi BT, Youker RT, Sottas CM, Ge R, Katz E, Klinefelter GR, Zirkin BR, Hardy MP (2001) Modulation of rat Leydig cell steroidogenic function by di(2-ethylhexyl)phthalate. Biol Reprod 65:1252–1259

Akingbemi BT, Sottas CM, Koulova AI, Klinefelter GR, Hardy MP (2004) Inhibition of testicular steroidogenesis by the xenoestrogen bisphenol A is associated with reduced pituitary luteinizing hormone secretion and decreased steroidogenic enzyme gene expression in rat Leydig cells. Endocrinology 145:592–603

Albrecht KH, Eicher EM (2001) Evidence that Sry is expressed in pre-Sertoli cells and Sertoli and granulosa cells have a common precursor. Dev Biol 240:92–107

Andrew A, Kramer B (1979) An experimental investigation into the possible origin of pancreatic islet cells from rhombencephalic neurectoderm. J Embryol Exp Morphol 52:23–38

Andrew A, Kramer B, Rawdon BB (1998) The origin of gut and pancreatic neuroendocrine (APUD) cells-the last word? J Pathol 186:117–118

Baird DT, Swanston IA, McNeilly AS (1981) Relationship between LH, FSH, and prolactin concentration and the secretion of androgens and estrogens by the preovulatory follicle in the ewe. Biol Reprod 24:1013–1025

Baker PJ, O'Shaughnessy PJ (2001) Role of gonadotrophins in regulating numbers of Leydig and Sertoli cells during fetal and postnatal development in mice. Reproduction 122:227–234

Baker J, Hardy MP, Zhou J, Bondy C, Lupu F, Bellve AR, Efstratiadis A (1996) Effects of an Igf1 gene null mutation on mouse reproduction. Mol Endocrinol 10:903–918

Baker PJ, Pakarinen P, Huhtaniemi IT, Abel MH, Charlton HM, Kumar TR, O'Shaughnessy PJ (2003) Failure of normal Leydig cell development in follicle-stimulating hormone (FSH) receptor-deficient mice, but not FSHbeta-deficient mice: role for constitutive FSH receptor activity. Endocrinology 144:138–145

Barsoum I, Yao HH (2006) The road to maleness: from testis to Wolffian duct. Trends Endocrinol Metab 17:223–228

Barsoum IB, Yao HH (2010) Fetal Leydig cells: progenitor cell maintenance and differentiation. J Androl 31:11–15

Bauer MP, Bridgham JT, Langenau DM, Johnson AL, Goetz FW (2000) Conservation of steroidogenic acute regulatory (StAR) protein structure and expression in vertebrates. Mol Cell Endocrinol 168:119–125

Benahmed M, Reventos J, Tabone E, Saez JM (1985) Cultured Sertoli cell-mediated FSH stimulatory effect on Leydig cell steroidogenesis. Am J Physiol 248:E176–E181

Benton L, Shan LX, Hardy MP (1995) Differentiation of adult Leydig cells. J Steroid Biochem Mol Biol 53:61–68
Bergh A (1982) Local differences in Leydig cell morphology in the adult rat testis: evidence for a local control of Leydig cells by adjacent seminiferous tubules. Int J Androl 5:325–330
Bergh A (1983) Paracrine regulation of Leydig cells by the seminiferous tubules. Int J Androl 6:57–65
Birk OS, Casiano DE, Wassif CA, Cogliati T, Zhao L, Zhao Y, Grinberg A, Huang S, Kreidberg JA, Parker KL, Porter FD, Westphal H (2000) The LIM homeobox gene Lhx9 is essential for mouse gonad formation. Nature 403:909–913
Bott RC, Clopton DT, Fuller AM, McFee RM, Lu N, McFee RM, Cupp AS (2010) KDR-LacZ-expressing cells are involved in ovarian and testis-specific vascular development, suggesting a role for VEGFA in the regulation of this vasculature. Cell Tissue Res 342(1):117–130
Boujrad N, Guillaumin JM, Bardos P, Hochereau De Reviers MT, Drosdowsky MA, Carreau S (1992) Germ cell-Sertoli cell interactions and production of testosterone by purified Leydig cells from mature rat. J Steroid Biochem Mol Biol 41:677–681
Bouskine A, Nebout M, Brucker-Davis F, Benahmed M, Fenichel P (2009) Low doses of bisphenol A promote human seminoma cell proliferation by activating PKA and PKG via a membrane G-protein-coupled estrogen receptor. Environ Health Perspect 117:1053–1058
Boyer A, Pilon N, Raiwet DL, Lussier JG, Silversides DW (2006) Human and pig SRY 5′ flanking sequences can direct reporter transgene expression to the genital ridge and to migrating neural crest cells. Dev Dyn 235(3):623–632
Brennan J, Tilmann C, Capel B (2003) Pdgfr-alpha mediates testis cord organization and fetal Leydig cell development in the XY gonad. Genes Dev 17:800–810
Browne P, Place NJ, Vidal JD, Moore IT, Cunha GR, Glickman SE, Conley AJ (2006) Endocrine differentiation of fetal ovaries and testes of the spotted hyena (Crocuta crocuta): timing of androgen-independent versus androgen-driven genital development. Reproduction 132:649–659
Budefeld T, Jezek D, Rozman D, Majdic G (2009) Initiation of steroidogenesis precedes expression of cholesterologenic enzymes in the fetal mouse testes. Anat Histol Embryol 38:461–466
Byskov AG (1986) Differentiation of mammalian embryonic gonad. Physiol Rev 66:71–117
Campbell BK, Baird DT, Webb R (1998) Effects of dose of LH on androgen production and luteinization of ovine theca cells cultured in a serum-free system. J Reprod Fertil 112:69–77
Canto P, Soderlund D, Reyes E, Mendez JP (2004) Mutations in the desert hedgehog (DHH) gene in patients with 46, XY complete pure gonadal dysgenesis. J Clin Endocrinol Metab 89:4480–4483
Capel B, Albrecht KH, Washburn LL, Eicher EM (1999) Migration of mesonephric cells into the mammalian gonad depends on Sry. Mech Dev 84:127–131
Carreau S (1996) Paracrine control of human Leydig cell and Sertoli cell functions. Folia Histochem Cytobiol 34:111–119
Carreau S, De Vienne C, Galeraud-Denis I (2008) Aromatase and estrogens in man reproduction: a review and latest advances. Adv Med Sci 53:139–144
Chen H, Hardy MP, Huhtaniemi I, Zirkin BR (1994) Age-related decreased Leydig cell testosterone production in the brown Norway rat. J Androl 15:551–557
Chen JJ, Lukyanenko Y, Hutson JC (2002) 25-hydroxycholesterol is produced by testicular macrophages during the early postnatal period and influences differentiation of Leydig cells in vitro. Biol Reprod 66:1336–1341
Chen H, Irizarry RA, Luo L, Zirkin BR (2004) Leydig cell gene expression: effects of age and caloric restriction. Exp Gerontol 39:31–43
Chen H, Ge RS, Zirkin BR (2009) Leydig cells: from stem cells to aging. Mol Cell Endocrinol 306:9–16
Chen H, Stanley E, Jin S, Zirkin BR (2010) Stem Leydig cells: from fetal to aged animals. Birth Defects Res C Embryo Today 90:272–283

Christensen AK, Gillim SW (1969) The correlation of fine structure and function in steroid-secreting cells, with emphasis on those of the gonads. In: Mckerns KW (ed) The gonads. Appleton-Century-Crofts, New York
Clark BJ, Wells J, King SR, Stocco DM (1994) The purification, cloning, and expression of a novel luteinizing hormone-induced mitochondrial protein in MA-10 mouse Leydig tumor cells. Characterization of the steroidogenic acute regulatory protein (StAR). J Biol Chem 269:28314–28322
Clark AM, Garland KK, Russell LD (2000) Desert hedgehog (Dhh) gene is required in the mouse testis for formation of adult-type Leydig cells and normal development of peritubular cells and seminiferous tubules. Biol Reprod 63:1825–1838
Cohen PE, Hardy MP, Pollard JW (1997) Colony-stimulating factor-1 plays a major role in the development of reproductive function in male mice. Mol Endocrinol 11:1636–1650
Collin O, Bergh A (1996) Leydig cells secrete factors which increase vascular permeability and endothelial cell proliferation. Int J Androl 19:221–228
Combes AN, Wilhelm D, Davidson T, Dejana E, Harley V, Sinclair A, Koopman P (2009) Endothelial cell migration directs testis cord formation. Dev Biol 326:112–120
Cool J, Carmona FD, Szucsik JC, Capel B (2008) Peritubular myoid cells are not the migrating population required for testis cord formation in the XY gonad. Sex Dev 2:128–133
Cui S, Ross A, Stallings N, Parker KL, Capel B, Quaggin SE (2004) Disrupted gonadogenesis and male-to-female sex reversal in Pod1 knockout mice. Development 131:4095–4105
Davidoff MS, Middendorff R, Kofuncu E, Muller D, Jezek D, Holstein AF (2002) Leydig cells of the human testis possess astrocyte and oligodendrocyte marker molecules. Acta Histochem 104:39–49
Davidoff MS, Middendorff R, Enikolopov G, Riethmacher D, Holstein AF, Muller D (2004) Progenitor cells of the testosterone-producing Leydig cells revealed. J Cell Biol 167:935–944
Davidoff MS, Middendorff R, Muller D, Holstein AF (2009) The neuroendocrine Leydig cells and their stem cell progenitors, the pericytes. Adv Anat Embryol Cell Biol 205:1–107
DeFalco T, Takahashi S, Capel B (2011) Two distinct origins for Leydig cell progenitors in the fetal testis. Dev Biol 352:14–26
DeFalco T, Saraswathula A, Briot A, Iruela-Arispe ML, Capel B (2013) Testosterone levels influence mouse fetal Leydig cell progenitors through notch signaling. Biol Reprod 88:91
DeFalco T, Bhattacharya I, Williams AV, Sams DM, Capel B (2014) Yolk-sac-derived macrophages regulate fetal testis vascularization and morphogenesis. Proc Natl Acad Sci U S A 111: E2384–E2393
DeFalco T, Potter SJ, Williams AV, Waller B, Kan MJ, Capel B (2015) Macrophages contribute to the spermatogonial niche in the adult testis. Cell Rep 12(7):1107–1119
Desjardins C, Ewing LL, Johnson BH (1971) Effects of light deprivation upon the spermatogenic and steroidogenic elements of hamster testes. Endocrinology 89:791–800
Dloniak SM, French JA, Holekamp KE (2006) Rank-related maternal effects of androgens on behaviour in wild spotted hyaenas. Nature 440:1190–1193
Dong J, Albertini DF, Nishimori K, Kumar TR, Lu N, Matzuk MM (1996) Growth differentiation factor-9 is required during early ovarian folliculogenesis. Nature 383:531–535
Enangue Njembele AN, Bailey JL, Tremblay JJ (2014) In vitro exposure of Leydig cells to an environmentally relevant mixture of organochlorines represses early steps of steroidogenesis. Biol Reprod 90:118
Ergun S, Stingl J, Holstein AF (1994) Microvasculature of the human testis in correlation to Leydig cells and seminiferous tubules. Andrologia 26:255–262
Eshkol A, Lunenfeld B (1972) Gonadotropic regulation of ovarian development in mice during infancy. In: Saxena BB, Boling O, Gandy HM (eds) Gonadotropins. Wiley Interscience, New York
Forest MG, Cathiard AM, Bertrand JA (1973) Evidence of testicular activity in early infancy. J Clin Endocrinol Metab 37:148–151

Fortune JE, Armstrong DT (1977) Androgen production by theca and granulosa isolated from proestrous rat follicles. Endocrinology 100:1341–1347
Gaskell TL, Robinson LL, Groome NP, Anderson RA, Saunders PT (2003) Differential expression of two estrogen receptor beta isoforms in the human fetal testis during the second trimester of pregnancy. J Clin Endocrinol Metab 88:424–432
Gaytan F, Bellido C, Aguilar E, Van Rooijen N (1994a) Requirement for testicular macrophages in Leydig cell proliferation and differentiation during prepubertal development in rats. J Reprod Fertil 102:393–399
Gaytan F, Bellido C, Morales C, Reymundo C, Aguilar E, Van Rooijen N (1994b) Effects of macrophage depletion at different times after treatment with ethylene dimethane sulfonate (EDS) on the regeneration of Leydig cells in the adult rat. J Androl 15:558–564
Gaytan F, Bellido C, Morales C, Reymundo C, Aguilar E, Van Rooijen N (1994c) Selective depletion of testicular macrophages and prevention of Leydig cell repopulation after treatment with ethylene dimethane sulfonate in rats. J Reprod Fertil 101:175–182
Ge RS, Hardy MP (1998) Variation in the end products of androgen biosynthesis and metabolism during postnatal differentiation of rat Leydig cells. Endocrinology 139:3787–3795
Ge RS, Dong Q, Sottas CM, Papadopoulos V, Zirkin BR, Hardy MP (2006) In search of rat stem Leydig cells: identification, isolation, and lineage-specific development. Proc Natl Acad Sci U S A 103:2719–2724
Gelber SJ, Hardy MP, Mendis-Handagama SM, Casella SJ (1992) Effects of insulin-like growth factor-I on androgen production by highly purified pubertal and adult rat Leydig cells. J Androl 13:125–130
Gilep AA, Sushko TA, Usanov SA (2011) At the crossroads of steroid hormone biosynthesis: the role, substrate specificity and evolutionary development of CYP17. Biochim Biophys Acta 1814:200–209
Griswold SL, Behringer RR (2009) Fetal Leydig cell origin and development. Sex Dev 3:1–15
Guo J, Deng H, Li H, Zhu Q, Zhao B, Chen B, Chu Y, Ge RS (2013) Effects of methoxychlor and its metabolite 2,2-bis(p-hydroxyphenyl)-1,1,1-trichloroethane on 11beta-hydroxysteroid dehydrogenase activities in vitro. Toxicol Lett 218:18–23
Haider SG, Laue D, Schwochau G, Hilscher B (1995) Morphological studies on the origin of adult-type Leydig cells in rat testis. Ital J Anat Embryol 100(Suppl 1):535–541
Hales DB (2002) Testicular macrophage modulation of Leydig cell steroidogenesis. J Reprod Immunol 57:3–18
Hardy MP, Mendis-Handagama SM, Zirkin BR, Ewing LL (1987) Photoperiodic variation of Leydig cell numbers in the testis of the golden hamster: a possible mechanism for their renewal during recrudescence. J Exp Zool 244:269–276
Hardy MP, Zirkin BR, Ewing LL (1989) Kinetic studies on the development of the adult population of Leydig cells in testes of the pubertal rat. Endocrinology 124:762–770
Hardy MP, Kelce WR, Klinefelter GR, Ewing LL (1990) Differentiation of Leydig cell precursors in vitro: a role for androgen. Endocrinology 127:488–490
Harman SM, Metter EJ, Tobin JD, Pearson J, Blackman MR (2001) Longitudinal effects of aging on serum total and free testosterone levels in healthy men. Baltimore Longitudinal Study of Aging. J Clin Endocrinol Metab 86:724–731
Hazra R, Jimenez M, Desai R, Handelsman DJ, Allan CM (2013) Sertoli cell androgen receptor expression regulates temporal fetal and adult Leydig cell differentiation, function, and population size. Endocrinology 154:3410–3422
Hess RA (2003) Estrogen in the adult male reproductive tract: a review. Reprod Biol Endocrinol 1:52
Himelstein-Braw R, Byskov AG, Peters H, Faber M (1976) Follicular atresia in the infant human ovary. J Reprod Fertil 46:55–59
Hsueh AJ, Dahl KD, Vaughan J, Tucker E, Rivier J, Bardin CW, Vale W (1987) Heterodimers and homodimers of inhibin subunits have different paracrine action in the modulation of luteinizing hormone-stimulated androgen biosynthesis. Proc Natl Acad Sci U S A 84:5082–5086

Hu GX, Lian QQ, Ge RS, Hardy DO, Li XK (2009) Phthalate-induced testicular dysgenesis syndrome: Leydig cell influence. Trends Endocrinol Metab 20:139–145
Hu GX, Lin H, Chen GR, Chen BB, Lian QQ, Hardy DO, Zirkin BR, Ge RS (2010) Deletion of the Igf1 gene: suppressive effects on adult Leydig cell development. J Androl 31:379–387
Hu GX, Zhao B, Chu Y, Li XH, Akingbemi BT, Zheng ZQ, Ge RS (2011) Effects of methoxychlor and 2,2-bis(p-hydroxyphenyl)-1,1,1-trichloroethane on 3beta-hydroxysteroid dehydrogenase and 17beta-hydroxysteroid dehydrogenase-3 activities in human and rat testes. Int J Androl 34:138–144
Huber K (2006) The sympathoadrenal cell lineage: specification, diversification, and new perspectives. Dev Biol 298:335–343
Hum DW, Miller WL (1993) Transcriptional regulation of human genes for steroidogenic enzymes. Clin Chem 39:333–340
Hutson JC (1992) Development of cytoplasmic digitations between Leydig cells and testicular macrophages of the rat. Cell Tissue Res 267:385–389
Hutson JC (1998) Interactions between testicular macrophages and Leydig cells. J Androl 19:394–398
Jiang MH, Cai B, Tuo Y, Wang J, Zang ZJ, Tu X, Gao Y, Su Z, Li W, Li G, Zhang M, Jiao J, Wan Z, Deng C, Lahn BT, Xiang AP (2014) Characterization of Nestin-positive stem Leydig cells as a potential source for the treatment of testicular Leydig cell dysfunction. Cell Res 24:1466–1485
Jimenez R, Burgos M, Sanchez A, Sinclair AH, Alarcon FJ, Marin JJ, Ortega E, Diaz De La Guardia R (1993) Fertile females of the mole Talpa occidentalis are phenotypic intersexes with ovotestes. Development 118:1303–1311
Johnson L, Nguyen HB (1986) Annual cycle of the Sertoli cell population in adult stallions. J Reprod Fertil 76:311–316
Johnson L, Thompson DL Jr (1986) Seasonal variation in the total volume of Leydig cells in stallions is explained by variation in cell number rather than cell size. Biol Reprod 35:971–979
Johnson L, Wilker CE, Safe SH, Scott B, Dean DD, White PH (1994) 2,3,7,8-Tetrachlorodibenzo-p-dioxin reduces the number, size, and organelle content of Leydig cells in adult rat testes. Toxicology 89:49–65
Johnson KJ, Heger NE, Boekelheide K (2012) Of mice and men (and rats): phthalate-induced fetal testis endocrine disruption is species-dependent. Toxicol Sci 129:235–248
Jones HB, Garside DA, Liu R, Roberts JC (1993) The influence of phthalate esters on Leydig cell structure and function in vitro and in vivo. Exp Mol Pathol 58:179–193
Karl J, Capel B (1998) Sertoli cells of the mouse testis originate from the coelomic epithelium. Dev Biol 203:323–333
Kilcoyne KR, Smith LB, Atanassova N, Macpherson S, Mckinnell C, Van Den Driesche S, Jobling MS, Chambers TJ, De Gendt K, Verhoeven G, O'Hara L, Platts S, Renato De Franca L, Lara NL, Anderson RA, Sharpe RM (2014) Fetal programming of adult Leydig cell function by androgenic effects on stem/progenitor cells. Proc Natl Acad Sci U S A 111:E1924–E1932
King SR, Lavoie HA (2009) Regulation of the early steps in gonadal steroidogenesis. In: Chedrese PJ (ed) Reproductive endocrinology: a molecular approach. Springer, New York
Kirchgessner AL, Adlersberg MA, Gershon MD (1992) Colonization of the developing pancreas by neural precursors from the bowel. Dev Dyn 194:142–154
Kitamura K, Yanazawa M, Sugiyama N, Miura H, Iizuka-Kogo A, Kusaka M, Omichi K, Suzuki R, Kato-Fukui Y, Kamiirisa K, Matsuo M, Kamijo S, Kasahara M, Yoshioka H, Ogata T, Fukuda T, Kondo I, Kato M, Dobyns WB, Yokoyama M, Morohashi K (2002) Mutation of ARX causes abnormal development of forebrain and testes in mice and X-linked lissencephaly with abnormal genitalia in humans. Nat Genet 32:359–369
Kotsuji F, Kamitani N, Goto K, Tominaga T (1990) Bovine theca and granulosa cell interactions modulate their growth, morphology, and function. Biol Reprod 43:726–732

Krishnamurthy H, Kats R, Danilovich N, Javeshghani D, Sairam MR (2001) Intercellular communication between Sertoli cells and Leydig cells in the absence of follicle-stimulating hormone-receptor signaling. Biol Reprod 65:1201–1207
Lai KP, Wong MH, Wong CK (2005) Inhibition of CYP450scc expression in dioxin-exposed rat Leydig cells. J Endocrinol 185:519–527
Laskey JW, Phelps PV (1991) Effect of cadmium and other metal cations on in vitro Leydig cell testosterone production. Toxicol Appl Pharmacol 108:296–306
Laslett AL, McFarlane JR, Risbridger GP (1997) Developmental response by Leydig cells to acidic and basic fibroblast growth factor. J Steroid Biochem Mol Biol 60:171–179
Li MA, Alls JD, Avancini RM, Koo K, Godt D (2003) The large Maf factor Traffic Jam controls gonad morphogenesis in Drosophila. Nat Cell Biol 5(11):994–1000
Lin T, Blaisdell J, Haskell JF (1987) Transforming growth factor-beta inhibits Leydig cell steroidogenesis in primary culture. Biochem Biophys Res Commun 146:387–394
Lin YM, Tsai CC, Chung CL, Chen PR, Sun HS, Tsai SJ, Huang BM (2010) Fibroblast growth factor 9 stimulates steroidogenesis in postnatal Leydig cells. Int J Androl 33:545–553
Lindeque M, Skinner JD (1982) Fetal androgens and sexual mimicry in spotted hyaenas (Crocuta crocuta). J Reprod Fertil 65:405–410
Liu H, Yang Y, Zhang L, Liang R, Ge RS, Zhang Y, Zhang Q, Xiang Q, Huang Y, Su Z (2014) Basic fibroblast growth factor promotes stem Leydig cell development and inhibits LH-stimulated androgen production by regulating microRNA expression. J Steroid Biochem Mol Biol 144(Pt B):483–491
Liu C, Peng J, Matzuk MM, Yao HH (2015) Lineage specification of ovarian theca cells requires multicellular interactions via oocyte and granulosa cells. Nat Commun 6:6934
Lobo MV, Arenas MI, Alonso FJ, Gomez G, Bazan E, Paino CL, Fernandez E, Fraile B, Paniagua R, Moyano A, Caso E (2004) Nestin, a neuroectodermal stem cell marker molecule, is expressed in Leydig cells of the human testis and in some specific cell types from human testicular tumours. Cell Tissue Res 316:369–376
Loh HS, Gemmell RT (1980) Changes in the fine structure of the testicular Leydig cells of the seasonally-breeding bat, Myotis adversus. Cell Tissue Res 210:339–347
Lukyanenko YO, Carpenter AM, Brigham DE, Stocco DM, Hutson JC (1998) Regulation of Leydig cells through a steroidogenic acute regulatory protein-independent pathway by a lipophilic factor from macrophages. J Endocrinol 158:267–275
Lukyanenko YO, Chen JJ, Hutson JC (2001) Production of 25-hydroxycholesterol by testicular macrophages and its effects on Leydig cells. Biol Reprod 64:790–796
Luo X, Ikeda Y, Parker KL (1994) A cell-specific nuclear receptor is essential for adrenal and gonadal development and sexual differentiation. Cell 77:481–490
Luo X, Ikeda Y, Schlosser DA, Parker KL (1995) Steroidogenic factor 1 is the essential transcript of the mouse Ftz-F1 gene. Mol Endocrinol 9:1233–1239
Magoffin DA (2005) Ovarian theca cell. Int J Biochem Cell Biol 37:1344–1349
Mannan MA, O'Shaughnessy PJ (1991) Steroidogenesis during postnatal development in the mouse ovary. J Endocrinol 130:101–106
Martineau J, Nordqvist K, Tilmann C, Lovell-Badge R, Capel B (1997) Male-specific cell migration into the developing gonad. Curr Biol 7:958–968
Mayerhofer A, Lahr G, Seidl K, Eusterschulte B, Christoph A, Gratzl M (1996) The neural cell adhesion molecule (NCAM) provides clues to the development of testicular Leydig cells. J Androl 17:223–230
McLachlan RI, O'Donnell L, Meachem SJ, Stanton PG, De Kretser DM, Pratis K, Robertson DM (2002) Identification of specific sites of hormonal regulation in spermatogenesis in rats, monkeys, and man. Recent Prog Horm Res 57:149–179
Meeks JJ, Crawford SE, Russell TA, Morohashi K, Weiss J, Jameson JL (2003) Dax1 regulates testis cord organization during gonadal differentiation. Development 130:1029–1036
Meng J, Holdcraft RW, Shima JE, Griswold MD, Braun RE (2005) Androgens regulate the permeability of the blood-tests barrier. Proc Natl Acad Sci U S A 102:16696–16700

Merchant-Larios H, Moreno-Mendoza N, Buehr M (1993) The role of the mesonephros in cell differentiation and morphogenesis of the mouse fetal testis. Int J Dev Biol 37:407–415

Middendorff R, Davidoff M, Holstein AF (1993) Neuroendocrine marker substances in human Leydig cells—changes by disturbances of testicular function. Andrologia 25:257–262

Migrenne S, Pairault C, Racine C, Livera G, Geloso A, Habert R (2001) Luteinizing hormone-dependent activity and luteinizing hormone-independent differentiation of rat fetal Leydig cells. Mol Cell Endocrinol 172:193–202

Millena AC, Reddy SC, Bowling GH, Khan SA (2004) Autocrine regulation of steroidogenic function of Leydig cells by transforming growth factor-alpha. Mol Cell Endocrinol 224:29–39

Miller SC, Bowman BM, Rowland HG (1983) Structure, cytochemistry, endocytic activity, and immunoglobulin (Fc) receptors of rat testicular interstitial-tissue macrophages. Am J Anat 168:1–13

Miyabayashi K, Katoh-Fukui Y, Ogawa H, Baba T, Shima Y, Sugiyama N, Kitamura K, Morohashi K (2013) Aristaless related homeobox gene, Arx, is implicated in mouse fetal Leydig cell differentiation possibly through expressing in the progenitor cells. PLoS One 8: e68050

Moore RW, Jefcoate CR, Peterson RE (1991) 2,3,7,8-Tetrachlorodibenzo-p-dioxin inhibits steroidogenesis in the rat testis by inhibiting the mobilization of cholesterol to cytochrome P450scc. Toxicol Appl Pharmacol 109:85–97

Mori H, Fukunishi R, Fujii M, Hataji K, Shiraishi T, Matsumoto K (1978) Stereological analysis of Reinke's crystals in human Leydig cells. Virchows Arch A Pathol Anat Histol 380:1–9

Mori H, Hiromoto N, Nakahara M, Shiraishi T (1982) Stereological analysis of Leydig cell ultrastructure in aged humans. J Clin Endocrinol Metab 55:634–641

Murono EP, Derk RC, Akgul Y (2006) In vivo exposure of young adult male rats to methoxychlor reduces serum testosterone levels and ex vivo Leydig cell testosterone formation and cholesterol side-chain cleavage activity. Reprod Toxicol 21:148–153

Nashev LG, Schuster D, Laggner C, Sodha S, Langer T, Wolber G, Odermatt A (2010) The UV-filter benzophenone-1 inhibits 17beta-hydroxysteroid dehydrogenase type 3: virtual screening as a strategy to identify potential endocrine disrupting chemicals. Biochem Pharmacol 79:1189–1199

Neaves WB, Johnson L, Porter JC, Parker CR Jr, Petty CS (1984) Leydig cell numbers, daily sperm production, and serum gonadotropin levels in aging men. J Clin Endocrinol Metab 59:756–763

Neaves WB, Johnson L, Petty CS (1985) Age-related change in numbers of other interstitial cells in testes of adult men: evidence bearing on the fate of Leydig cells lost with increasing age. Biol Reprod 33:259–269

Nef S, Verma-Kurvari S, Merenmies J, Vassalli JD, Efstratiadis A, Accili D, Parada LF (2003) Testis determination requires insulin receptor family function in mice. Nature 426 (6964):291–295

Nes WD, Lukyanenko YO, Jia ZH, Quideau S, Howald WN, Pratum TK, West RR, Hutson JC (2000) Identification of the lipophilic factor produced by macrophages that stimulates steroidogenesis. Endocrinology 141:953–958

Nishimura W, Bonner-Weir S, Sharma A (2009) Expression of MafA in pancreatic progenitors is detrimental for pancreatic development. Dev Biol 333:108–120

Norris DO, Carr JA (2013) Vertebrate endocrinology. Academic (Elsevier), China

N'tumba-Byn T, Moison D, Lacroix M, Lecureuil C, Lesage L, Prud'homme SM, Pozzi-Gaudin S, Frydman R, Benachi A, Livera G, Rouiller-Fabre V, Habert R (2012) Differential effects of bisphenol A and diethylstilbestrol on human, rat and mouse fetal leydig cell function. PLoS One 7:e51579

O'Shaughnessy PJ, Baker P, Sohnius U, Haavisto AM, Charlton HM, Huhtaniemi I (1998) Fetal development of Leydig cell activity in the mouse is independent of pituitary gonadotroph function. Endocrinology 139:1141–1146

O'Shaughnessy PJ, Johnston H, Willerton L, Baker PJ (2002) Failure of normal adult Leydig cell development in androgen-receptor-deficient mice. J Cell Sci 115:3491–3496
O'Shaughnessy PJ, Baker PJ, Johnston H (2006) The foetal Leydig cell—differentiation, function and regulation. Int J Androl 29:90–95
O'Shaughnessy PJ, Hu L, Baker PJ (2008) Effect of germ cell depletion on levels of specific mRNA transcripts in mouse Sertoli cells and Leydig cells. Reproduction 135:839–850
Orth JM (1982) Proliferation of Sertoli cells in fetal and postnatal rats: a quantitative autoradiographic study. Anat Rec 203:485–492
Palermo R (2007) Differential actions of FSH and LH during folliculogenesis. Reprod Biomed Online 15:326–337
Paniagua R, Amat P, Nistal M, Martin A (1986) Ultrastructure of Leydig cells in human ageing testes. J Anat 146:173–183
Park SY, Tong M, Jameson JL (2007) Distinct roles for steroidogenic factor 1 and desert hedgehog pathways in fetal and adult Leydig cell development. Endocrinology 148:3704–3710
Parrott JA, Skinner MK (1997) Direct actions of kit-ligand on theca cell growth and differentiation during follicle development. Endocrinology 138:3819–3827
Payne A, Hardy M (2007) The Leydig cell in health and disease. Humana, Totowa, NJ
Pereira FA, Qiu Y, Zhou G, Tsai MJ, Tsai SY (1999) The orphan nuclear receptor COUP-TFII is required for angiogenesis and heart development. Genes Dev 13:1037–1049
Peters H (1969) The development of the mouse ovary from birth to maturity. Acta Endocrinol (Copenh) 62:98–116
Pictet RL, Rall LB, Phelps P, Rutter WJ (1976) The neural crest and the origin of the insulin-producing and other gastrointestinal hormone-producing cells. Science 191:191–192
Pilon N, Raiwet D, Viger RS, Silversides DW (2008) Novel pre- and post-gastrulation expression of Gata4 within cells of the inner cell mass and migratory neural crest cells. Dev Dyn 237 (4):1133–1143
Pitetti JL, Calvel P, Romero Y, Conne B, Truong V, Papaioannou MD, Schaad O, Docquier M, Herrera PL, Wilhelm D, Nef S (2013) Insulin and IGF1 receptors are essential for XX and XY gonadal differentiation and adrenal development in mice. PLoS Genet 9(1):e1003160
Prince FP (1990) Ultrastructural evidence of mature Leydig cells and Leydig cell regression in the neonatal human testis. Anat Rec 228:405–417
Prince FP (2001) The triphasic nature of Leydig cell development in humans, and comments on nomenclature. J Endocrinol 168:213–216
Qin J, Tsai MJ, Tsai SY (2008) Essential roles of COUP-TFII in Leydig cell differentiation and male fertility. PLoS One 3:e3285
Rastetter RH, Bernard P, Palmer JS, Chassot AA, Chen H, Western PS, Ramsay RG, Chaboissier MC, Wilhelm D (2014) Marker genes identify three somatic cell types in the fetal mouse ovary. Dev Biol 394:242–252
Rebourcet D, O'Shaughnessy PJ, Monteiro A, Milne L, Cruickshanks L, Jeffrey N, Guillou F, Freeman TC, Mitchell RT, Smith LB (2014a) Sertoli cells maintain Leydig cell number and peritubular myoid cell activity in the adult mouse testis. PLoS One 9:e105687
Rebourcet D, O'Shaughnessy PJ, Pitetti JL, Monteiro A, O'Hara L, Milne L, Tsai YT, Cruickshanks L, Riethmacher D, Guillou F, Mitchell RT, Van't Hof R, Freeman TC, Nef S, Smith LB (2014b) Sertoli cells control peritubular myoid cell fate and support adult Leydig cell development in the prepubertal testis. Development 141:2139–2149
Reyes FI, Boroditsky RS, Winter JS, Faiman C (1974) Studies on human sexual development. II. Fetal and maternal serum gonadotropin and sex steroid concentrations. J Clin Endocrinol Metab 38:612–617
Rodriguez-Sosa JR, Bondareva A, Tang L, Avelar GF, Coyle KM, Modelski M, Alpaugh W, Conley A, Wynne-Edwards K, Franca LR, Meyers S, Dobrinski I (2014) Phthalate esters affect maturation and function of primate testis tissue ectopically grafted in mice. Mol Cell Endocrinol 398:89–100

Ryan KE, Glister C, Lonergan P, Martin F, Knight PG, Evans AC (2008) Functional significance of the signal transduction pathways Akt and Erk in ovarian follicles: in vitro and in vivo studies in cattle and sheep. J Ovarian Res 1:2

Savchuk I, Soder O, Svechnikov K (2013) Mouse leydig cells with different androgen production potential are resistant to estrogenic stimuli but responsive to bisphenol a which attenuates testosterone metabolism. PLoS One 8:e71722

Savchuk I, Soder O, Svechnikov K (2015) Mono-2-ethylhexyl phthalate stimulates androgen production but suppresses mitochondrial function in mouse leydig cells with different steroidogenic potential. Toxicol Sci 145:149–156

Sawetawan C, Milewich L, Word RA, Carr BR, Rainey WE (1994) Compartmentalization of type I 17 beta-hydroxysteroid oxidoreductase in the human ovary. Mol Cell Endocrinol 99:161–168

Setchell BP, Rommerts FF (1985) The importance of the Leydig cells in the vascular response to hCG in the rat testis. Int J Androl 8:436–440

Shima Y, Miyabayashi K, Haraguchi S, Arakawa T, Otake H, Baba T, Matsuzaki S, Shishido Y, Akiyama H, Tachibana T, Tsutsui K, Morohashi K (2013) Contribution of Leydig and Sertoli cells to testosterone production in mouse fetal testes. Mol Endocrinol 27:63–73

Snyder PJ, Peachey H, Berlin JA, Hannoush P, Haddad G, Dlewati A, Santanna J, Loh L, Lenrow DA, Holmes JH, Kapoor SC, Atkinson LE, Strom BL (2000) Effects of testosterone replacement in hypogonadal men. J Clin Endocrinol Metab 85:2670–2677

Sordoillet C, Savona C, Chauvin MA, De Peretti E, Feige JJ, Morera AM, Benahmed M (1992) Basic fibroblast growth factor enhances testosterone secretion in cultured porcine Leydig cells: site(s) of action. Mol Cell Endocrinol 89:163–171

Sprando RL, Zirkin BR (1997) Is Leydig cell steroidogenic function affected by the germ cell content of the seminiferous tubules? J Androl 18:424–430

Stanley EL, Johnston DS, Fan J, Papadopoulos V, Chen H, Ge RS, Zirkin BR, Jelinsky SA (2011) Stem Leydig cell differentiation: gene expression during development of the adult rat population of Leydig cells. Biol Reprod 85:1161–1166

Strauss L, Kallio J, Desai N, Pakarinen P, Miettinen T, Gylling H, Albrecht M, Makela S, Mayerhofer A, Poutanen M (2009) Increased exposure to estrogens disturbs maturation, steroidogenesis, and cholesterol homeostasis via estrogen receptor alpha in adult mouse Leydig cells. Endocrinology 150:2865–2872

Suzuki F, Racey PA (1978) The organization of testicular interstitial tissue and changes in the fine structure of the Leydig cells of European moles (Talpa europaea) throughout the year. J Reprod Fertil 52:189–194

Svechnikov K, Izzo G, Landreh L, Weisser J, Soder O (2010) Endocrine disruptors and Leydig cell function. J Biomed Biotechnol. doi:10.1155/2010/684504

Tajima K, Orisaka M, Mori T, Kotsuji F (2007) Ovarian theca cells in follicular function. Reprod Biomed Online 15:591–609

Tang H, Brennan J, Karl J, Hamada Y, Raetzman L, Capel B (2008) Notch signaling maintains Leydig progenitor cells in the mouse testis. Development 135:3745–3753

Teerds KJ, De Rooij DG, Rommerts FF, Van Der Tweel I, Wensing CJ (1989) Turnover time of Leydig cells and other interstitial cells in testes of adult rats. Arch Androl 23:105–111

Tilmann C, Capel B (1999) Mesonephric cell migration induces testis cord formation and Sertoli cell differentiation in the mammalian gonad. Development 126:2883–2890

Tomiyama H, Hutson JM, Truong A, Agoulnik AI (2003) Transabdominal testicular descent is disrupted in mice with deletion of insulinlike factor 3 receptor. J Pediatr Surg 38:1793–1798

Tsai SY, Tsai MJ (1997) Chick ovalbumin upstream promoter-transcription factors (COUP-TFs): coming of age. Endocr Rev 18:229–240

Umehara F, Tate G, Itoh K, Yamaguchi N, Douchi T, Mitsuya T, Osame M (2000) A novel mutation of desert hedgehog in a patient with 46, XY partial gonadal dysgenesis accompanied by minifascicular neuropathy. Am J Hum Genet 67:1302–1305

Van Den Driesche S, Walker M, McKinnell C, Scott HM, Eddie SL, Mitchell RT, Seckl JR, Drake AJ, Smith LB, Anderson RA, Sharpe RM (2012) Proposed role for COUP-TFII in regulating fetal Leydig cell steroidogenesis, perturbation of which leads to masculinization disorders in rodents. PLoS One 7:e37064

Walker WH (2011) Testosterone signaling and the regulation of spermatogenesis. Spermatogenesis 1:116–120

Walsh LP, Stocco DM (2000) Effects of lindane on steroidogenesis and steroidogenic acute regulatory protein expression. Biol Reprod 63:1024–1033

Walsh LP, McCormick C, Martin C, Stocco DM (2000a) Roundup inhibits steroidogenesis by disrupting steroidogenic acute regulatory (StAR) protein expression. Environ Health Perspect 108:769–776

Walsh LP, Webster DR, Stocco DM (2000b) Dimethoate inhibits steroidogenesis by disrupting transcription of the steroidogenic acute regulatory (StAR) gene. J Endocrinol 167:253–263

Watson ME, Newman RJ, Payne AM, Abdelrahim M, Francis GL (1994) The effect of macrophage conditioned media on Leydig cell function. Ann Clin Lab Sci 24:84–95

Welsh M, Moffat L, Belling K, De Franca LR, Segatelli TM, Saunders PT, Sharpe RM, Smith LB (2012) Androgen receptor signalling in peritubular myoid cells is essential for normal differentiation and function of adult Leydig cells. Int J Androl 35:25–40

Wright WW, Fiore C, Zirkin BR (1993) The effect of aging on the seminiferous epithelium of the brown Norway rat. J Androl 14:110–117

Yang Y, Su Z, Xu W, Luo J, Liang R, Xiang Q, Zhang Q, Ge RS, Huang Y (2015) Directed mouse embryonic stem cells into leydig-like cells rescue testosterone-deficient male rats in vivo. Stem Cells Dev 24:459–470

Yao HH, Whoriskey W, Capel B (2002) Desert Hedgehog/Patched 1 signaling specifies fetal Leydig cell fate in testis organogenesis. Genes Dev 16:1433–1440

Yee JB, Hutson JC (1985) Effects of testicular macrophage-conditioned medium on Leydig cells in culture. Endocrinology 116:2682–2684

Young JM, McNeilly AS (2010) Theca: the forgotten cell of the ovarian follicle. Reproduction 140:489–504

Zirkin BR, Chen H (2000) Regulation of Leydig cell steroidogenic function during aging. Biol Reprod 63:977–981

Chapter 6
Germ Cell Commitment to Oogenic Versus Spermatogenic Pathway: The Role of Retinoic Acid

Kellie S. Agrimson and Cathryn A. Hogarth

Abstract The core of the decision to commit to either oogenesis or spermatogenesis lies in the timing of meiotic entry. Primordial germ cells within the fetal ovary become committed to the female pathway prior to birth and enter meiosis during embryonic development. In the fetal testis, however, the germ cells are protected from this signal before birth and instead receive this trigger postnatally. There is a growing body of evidence to indicate that RA is the meiosis-inducing factor in both sexes, with the gender-specific timing of meiotic entry controlled via degradation of this molecule only within the fetal testis. This chapter will review our current understanding of how RA controls germ cell fate in both the embryonic ovary and postnatal testis, highlighting the key studies that have led to the hypothesis that RA can drive the commitment to meiosis in both sexes and discussing the current debate over whether RA truly is the meiosis-inducing factor in the fetal ovary.

Abbreviations

AKT	Thymoma viral proto-oncogene
ALDH	Aldehyde dehydrogenase
BrdU	5-bromo-2′-deoxyuridine
CYP26	Cytochrome p450 family 26
DHRS3	Dehydrogenase/reductase (SDR family) member 3
DMC1	DMC1 dosage suppressor of mck1 homolog
dpc	Days *post coitum*
dpp	Days *postpartum*
H2AFX	H2A histone family, member X

K.S. Agrimson • C.A. Hogarth (✉)
School of Molecular Biosciences, Washington State University, Pullman, WA 99164, USA

The Center for Reproductive Biology, Washington State University, Pullman, WA 99164, USA
e-mail: chogarth@vetmed.wsu.edu

R.P. Piprek (ed.), *Molecular Mechanisms of Cell Differentiation in Gonad Development*, Results and Problems in Cell Differentiation 58,
DOI 10.1007/978-3-319-31973-5_6

HDACs	Histone deacetylase complexes
KIT	Kit oncogene
LRAT	Lecithin-retinol acyltransferase
MKI67	Antigen identified by monoclonal antibody Ki 67
mTORC	Mouse target of rapamycin complex
NANOS2	Nanos homolog 2
PDK1	Pyruvate dehydrogenase kinase, isoenzyme 1
PGC	Primordial germ cell
PI3K	Phosphatidylinositol 3-kinase
RA	Retinoic acid
RAR	Retinoic acid receptor
RARE	Retinoic acid response element
RBP	Retinol-binding protein
RDH	Retinol dehydrogenase
REC8	REC8 meiotic recombination protein
RXR	Retinoid X receptor
SOHLH1	Spermatogenesis- and oogenesis-specific basic helix-loop-helix 1
SPO11	SPO11 meiotic protein covalently bound to DSB
STRA6	Stimulated by retinoic acid gene 6
STRA8	Stimulated by retinoic acid gene 8
SYCP3	Synaptonemal complex protein 3
TUNEL	Terminal deoxynucleotidyl transferase dUTP nick end labeling
VAD	Vitamin A deficient
ZBTB16	Zinc finger and BTB domain containing 16

6.1 Introduction

The hallmark of vertebrate reproduction is the joining of a male and female haploid cell to create a new diploid embryo and the next generation. The germline of both sexes arises from a common cell type, the primordial germ cell (PGC), and these cells must then decide whether to follow the male (spermatogenesis) or female (oogenesis) pathway of development (Fig. 6.1). In mammals, this decision is made in utero, after the PGCs have migrated into the developing gonad. PGCs that find themselves within a fetal ovary are triggered to become oogonia and enter meiosis, arresting at the diplotene stage of meiosis I just prior to birth. PGCs that populate a developing testis, however, continue to proliferate both prior to and after birth, separated by a species-specific period of quiescence, allowing meiosis to be continually, yet asynchronously, initiated from the onset of puberty throughout reproductive life.

It has been known for quite some time that the timing of meiotic entry for germ cells from either sex is not intrinsic, but requires specific signals from ovarian or testicular somatic cells. The key to understanding this phenomenon has been to determine which factors are present and which genes are expressed within the ovary, and not in the testis, during embryonic development, but that arise within

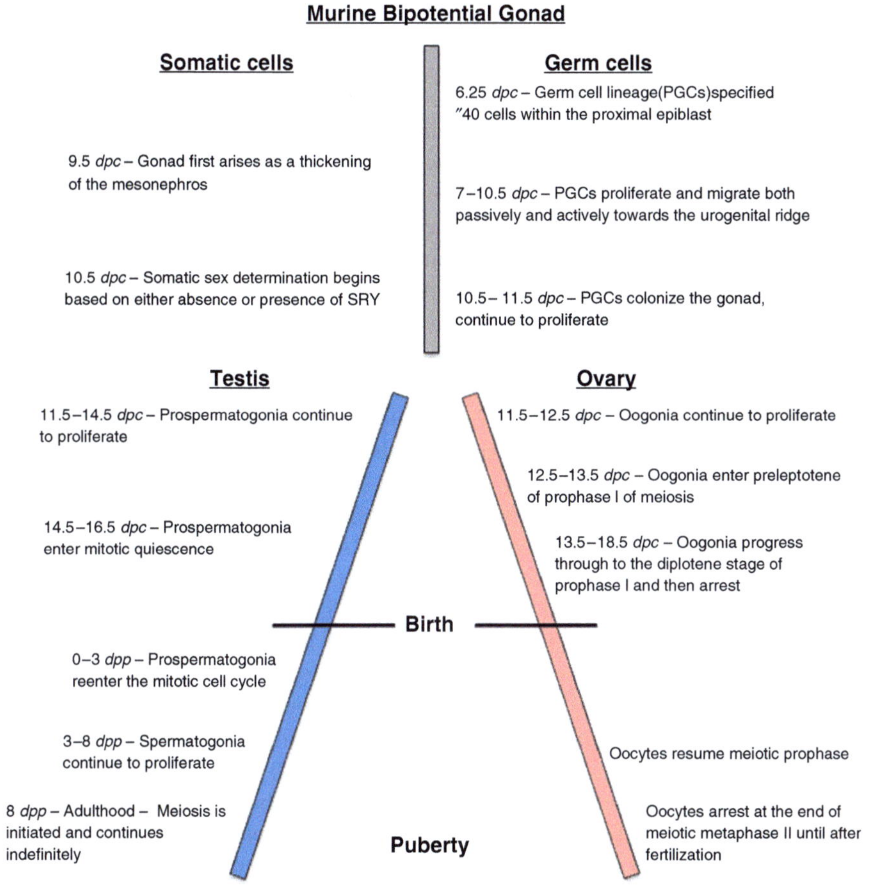

Fig. 6.1 Sex-specific timing of germ cell development. Timeline comparing murine male and female germ cell development. The *gray bar* designates events in either somatic (*left*) or germ (*right*) cells while the gonad and the primordial germ cells (PGCs) are bipotential. The *blue* and *pink bars* then highlight the events specific to either the testis or the ovary, respectively. The *black line* separates events occurring before or after birth in both sexes

the postnatal testis during puberty. Studies performed over the last decade have indicated that the active metabolite of vitamin A, all-trans retinoic acid (RA), is likely the driving force behind the sex-specific timing of meiotic entry, although debate still remains over whether this molecule is truly the meiosis-inducing factor in the embryonic ovary. This chapter will discuss the data relevant to the role of RA in the commitment to oogenesis versus spermatogenesis. It will focus mostly on the murine model, as much of the data regarding the molecular mechanisms driving meiotic entry has been generated in vivo via studies with genetically modified mice or in vitro via culture of primary mouse tissue with various chemicals, but will also include brief discussions of evidence generated from other vertebrate species.

6.2 Regulation of RA Synthesis, Signaling, and Degradation

Regulation of physiological RA concentrations occurs via complex interactions between retinoid binding and metabolizing enzymes (Napoli 2012) (Fig. 6.2). Due to its short half-life, estimated to be 30 min (Kane et al. 2005), and hydrophobicity, RA is not readily transported around the body via the circulation, but instead is synthesized within target organs. Vitamin A, in the form of retinol and bound to retinol binding protein (RBP), is delivered to target organs via the circulation. Once inside cells, retinol is metabolized to RA via a two-step oxidative process (Fig. 6.2), controlled by the retinol dehydrogenase (RDH) and aldehyde dehydrogenase (ALDH) enzymes. For each step, there are several isozymes of varying efficiency (Arnold et al. 2015a, b; Napoli 2012) capable of catalyzing the reaction and most display tissue-specific expression patterns (Alnouti and Klaassen 2008; Nishimura and Naito 2006), thereby allowing for precise regulation of RA levels in different organs. Within these enzyme families, of significant interest to reproduction are RDH10 and three ALDH enzymes, ALDH1A1, ALDH1A2, and ALDH1A3. Data derived from developmental studies indicate that ALDH1A1 and ALDH1A3 are mostly active within tissues of ectodermal origin, i.e., the brain and eyes, whereas ALDH1A2 is more broadly present throughout the mesodermal tissues, including the gonad. Transcript and protein localization studies and global and conditional mouse mutants, which will be described in the following sections, have demonstrated a role for RDH10 and all three ALDH enzymes in the synthesis of RA within the testis (Raverdeau et al. 2012; Sugimoto et al. 2012; Tong et al. 2013; Vernet et al. 2006b), yet there is significantly less known about which isozymes are expressed or essential for germ cell development in the fetal ovary.

The current dogma suggests that RA acts in a paracrine manner: synthesized by one cell and then passed to a neighboring cell to induce a response (Kumar and Duester 2011). RA drives changes in gene expression via binding to a heterodimer of a retinoic acid receptor (RAR) and a retinoid X receptor (RXR) and the interaction of this ligand-bound heterodimer with retinoic acid response elements (RAREs) within gene regulatory regions (reviewed in Bastien and Rochette-Egly 2004). In the absence of ligand, the RAR/RXR heterodimer is often bound at RAREs and recruits histone deacetylase complexes (HDACs) to aid in gene silencing. Upon association with RA, the RAR/RXR heterodimer releases the HDACs and instead attracts co-activation complexes (e.g., histone acetylases and methyltransferases), to open chromatin, and finally draws in the transcriptional machinery to drive production of RA-responsive mRNAs. For each of the two types of RA receptors, there are three different isoforms: alpha, beta, and gamma. As is the case for the RA synthesis enzymes, there is a plethora of information regarding which of the RA receptors are essential for male germ cell development and almost nothing known about either the expression or function of these receptors in the fetal ovary. Of the six receptors, RARα, RARγ, and RXRβ are all essential for spermatogenesis (Gely-Pernot et al. 2012; Lufkin et al. 1993; Vernet et al. 2006a, 2008)

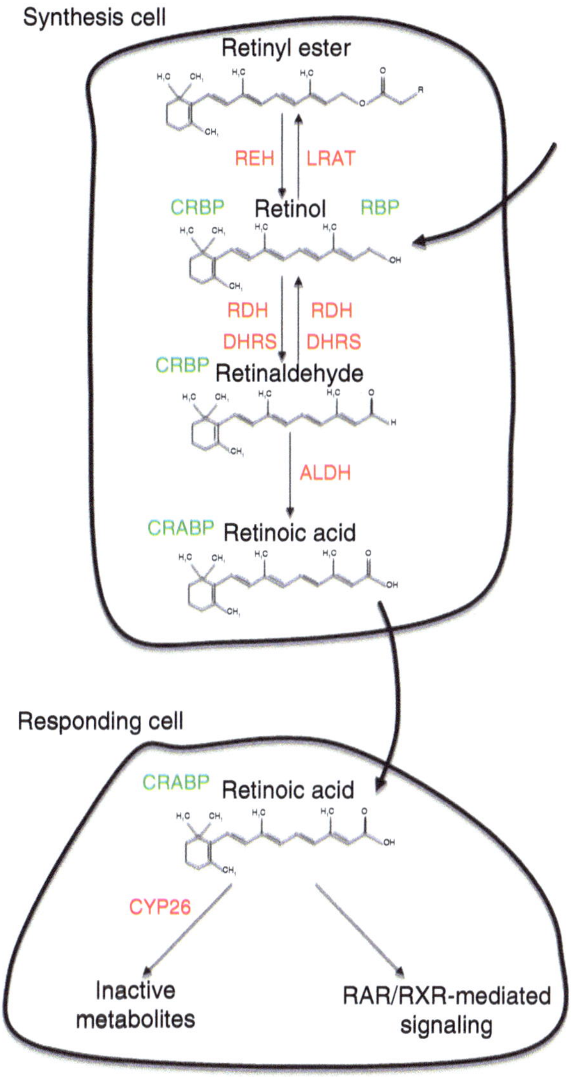

Fig. 6.2 Mammalian vitamin A metabolism, signaling, and degradation. Schematic outlining the paracrine mechanism of vitamin A metabolism and signaling or degradation. The enzymes are highlighted in *red* and the transport/binding proteins are highlighted in *green*. Dietary vitamin A is transported through the circulation to target organs as retinol bound to retinol-binding protein (RBP). Once inside cells, retinol is bound by cellular retinol-binding proteins (CRBPs) and follows one of two fates: (1) converted to retinyl esters by lecithin-retinol acyltransferase (LRAT) or (2) oxidized to RA via a two-step enzymatic process. The storage versus oxidation decision is chaperoned by CRBP, as unbound CRBP can inhibit LRAT activity (Napoli 2012), and the mobilization of cellular retinyl ester stores is driven by the activity of retinyl ester hydrolase (REH). In addition to retinol, the CRBPs also bind retinal within cells. The conversion of retinol to retinal is reversible, with these reactions catalyzed by the retinol dehydrogenases (RDHs) or short chain dehydrogenases (DHRS). The retinal to RA step is irreversible and regulated by the aldehyde dehydrogenases (ALDHs). Once synthesized, RA usually acts in a paracrine manner, being delivered to a neighboring cell to induce a response. Within cells, RA is bound by cellular retinoic acid-binding proteins (CRABPs) and then either directed toward the retinoic acid receptors

with RARα and RXRβ appearing to function mostly within Sertoli cells (Vernet et al. 2006a, 2008) and RARγ important within the germ cells (Gely-Pernot et al. 2012).

To avoid overstimulation or to protect genes from responding to RA, cells produce the RA-degrading cytochrome p450 family 26 (CYP26) enzymes. These proteins catabolize RA into inactive metabolites, thereby either blocking or stopping the expression of RA-responsive genes. There are three CYP26 family members, two of which are expressed within the testis and/or are important for normal mammalian reproduction, CYP26A1 and CYP26B1. The localization of the mRNAs encoding these enzymes to the peritubular myoid cells surrounding the seminiferous epithelium (Vernet et al. 2006b; Wu et al. 2008) suggests that the CYP26 proteins create a metabolic barrier to the influx of RA from the serum into the testis tubule, thereby allowing the germ and Sertoli cells to independently regulate their own RA concentrations. Radiolabeled RA was used to demonstrate that the testis does not readily take up RA from the serum (Kurlandsky et al. 1995; Smith et al. 1973; Zhuang et al. 1995), further data supporting the presence of a metabolic barrier; however, this hypothesis has yet to be tested genetically. In the embryo, however, localization, chemical inhibition, and genetic ablation studies have clearly demonstrated CYP26B1 expression in the Sertoli cells of the fetal testis and that this protein acts as the meiotic inhibitory factor.

6.3 Does RA Induce Meiosis in the Fetal Ovary?

The timing of meiotic initiation is a sex-specific event. Evidence has now been collected for numerous vertebrate species to indicate that meiotic onset in females occurs embryonically or during the larval stage of development whereas meiosis is not initiated in males until after birth or following metamorphosis (Childs et al. 2011; Koubova et al. 2006; Lau et al. 2013; Piprek et al. 2013; Rodriguez-Mari et al. 2013; Smith et al. 2008; Wallacides et al. 2009). These studies also all allude to RA as being the driver of the timing of this sex-specific mechanism, but the role of RA in meiotic onset was first described using the mouse model. Work published in 2006, independently by two different laboratories, claimed to have identified RA as the meiosis-inducing factor through the detection of testis-specific expression of CYP26B1 beginning in mouse at 12.5 *dpc* (days *post coitum*) (Bowles et al. 2006; Koubova et al. 2006). These studies generated a model (Fig. 6.3) whereby the synthesis of RA takes place in the mesonephros of the developing urogenital ridge in both sexes and then diffuses into both testes and ovaries in an anterior-to-posterior direction, beginning around 11.5 *dpc*. Within the ovary, this

---

Fig. 6.2 (continued) (RARs) to drive classical RA signaling or targeted for degradation by the cytochrome p450 family 26 (CYP26) enzymes

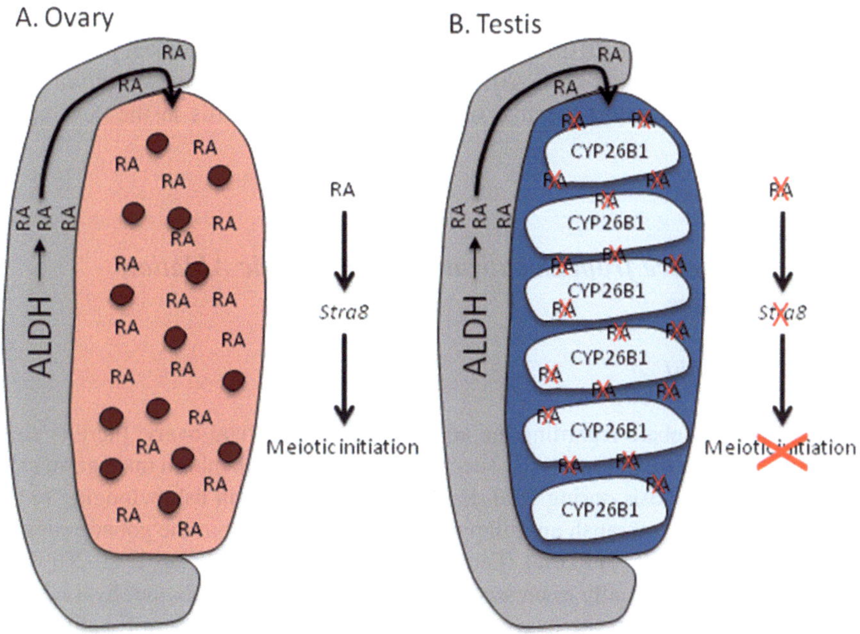

Fig. 6.3 RA and CYP26B1 function in the mouse embryonic gonad. Diagram summarizes the hypothesis generated by the studies performed in 2006 by Bowles et al. (2006) and Koubova et al. (2006). (**a**) In the female urogenital ridge, RA generated by aldehyde dehydrogenase (ALDH) activity in the mesonephros drives the expression of *Stra8* and the onset of meiosis in the oogonia (*red circles*). (**b**) In the male urogenital ridge, RA is still produced in the mesonephros; however, the action of CYP26B1 within the testis cords acts to degrade this RA (marked by *red crosses*), thereby preventing *Stra8* expression and the onset of meiosis

RA can then drive the initiation of meiosis via triggering stimulated by retinoic acid gene 8 (*Stra8*) expression, a gene known to be essential for meiotic progression (Anderson et al. 2008; Baltus et al. 2006; Mark et al. 2008) and the classic marker of an RA response. This is followed soon after by the production of meiotic-specific proteins, including REC8 meiotic recombination protein (REC8) and synaptonemal complex protein 3 (SYCP3), and by 14.5 *dpc*, the first germ cells containing meiotic chromosomal figures are seen toward the anterior end of the ovary. Within the testis, however, structural rearrangements and CYP26B1 prevent RA from triggering germ cells to enter meiosis. Between 10.5 and 12.5 *dpc*, the cells of the testis rearrange so that small groups of germ cells become encased by somatic cells, forming the testis cords. These somatic cells, the pre-Sertoli cells, express CYP26B1 and form a metabolic barrier between the incoming RA from the mesonephros and the germ cells. As a result, germ cells in a fetal testis are never exposed to RA during the window of plasticity that would allow them to enter meiosis and are instead set upon the male program. While data now exist to support this model in multiple vertebrate species, although the location of RA synthesis does differ (Feng et al. 2015; Piprek et al. 2013; Rodriguez-Mari et al. 2013; Smith et al. 2008) and the action of STRA8 does not appear to be conserved in all vertebrates (Feng et al. 2015; Rodriguez-Mari et al. 2013), this model has recently been challenged (Kumar et al. 2011) and the following sections will discuss the evidence for and against the role of RA in initiating meiosis in the developing ovary.

6.3.1 *Evidence from Localization and Genetic Ablation Studies*

6.3.1.1 *Cyp26b1*

While there is debate regarding the action of RA in the fetal ovary, there is clear evidence that CYP26B1 is a testis-specific factor required to inhibit male germ cells from entering meiosis during fetal development, although in fish belonging to the teleost class, e.g., zebrafish and tilapia, RA degradation within the gonad relies on CYP26A1 instead of CYP26B1 (Feng et al. 2015; Rodriguez-Mari et al. 2013). In mice, *Cyp26b1* is initially expressed in the gonads of both sexes, but by 12.5 *dpc*, expression is lost from the ovary and becomes specific to the Sertoli cells of the fetal testis (Bowles et al. 2006). A similar localization pattern is also present in anuran amphibians, with CYP26B1 protein present in the developing testis but not ovary (Piprek et al. 2013), whereas in zebrafish, *Cyp26a1* expression is upregulated in somatic cells of the testes during the developmental period critical for sex determination (Rodriguez-Mari et al. 2013). Global deletion of *Cyp26b1* in mice leads to late embryonic lethality or the animals die shortly after birth (Abu-Abed et al. 2001; Sakai et al. 2001; Yashiro et al. 2004), therefore allowing for the examination of fetal testis development in the absence of CYP26B1 enzymatic activity. Based on culture of control and mutant testis tissue extracts onto RARE reporter cells, fetal testes from *Cyp26b1*$^{-/-}$ mutant animals displayed threefold higher RA levels when compared to either *Cyp26b1*$^{+/+}$ or *Cyp26b1*$^{+/-}$ littermates and germ cell numbers and testis size were found to be reduced upon comparison of newborn male mutant and wild-type mice (MacLean et al. 2007). Analysis of fetal testes at multiple timepoints revealed that in the absence of CYP26B1 activity the germ cells enter meiosis, with some cells progressing through to the pachytene stage of prophase I (MacLean et al. 2007). Additionally, the numbers of apoptotic germ cells, distinguished via terminal deoxynucleotidyl transferase dUTP nick end labeling (TUNEL) and caspase-3 staining, were significantly upregulated in the *Cyp26b1*$^{-/-}$ mutant testes (MacLean et al. 2007).

Conditional elimination of CYP26B1 activity specifically from Sertoli cells demonstrated that fetal male germ cells respond to hypothesized increases in RA levels by entering meiosis even after exiting the cell cycle. In wild-type mouse fetal testes, prospermatogonia proliferate until approximately 14.5–16.5 *dpc* when they enter G0, with the timing being very strain specific (Western et al. 2008). Via conditional deletion of *Cyp26b1* using the *Amh*-Cre line, Li and Clagett-Dame

(2009) were able to demonstrate that elimination of CYP26B1 activity in a Sertoli cell-specific manner from 15.5 *dpc* onward either prevented the germ cells from undergoing quiescence or induced them to exit G0 and instead initiate meiotic prophase (Li and Clagett-Dame 2009). As CYP26B1 currently has no other known substrates (Topletz et al. 2012; Thatcher and Isoherranen 2009), these observations support RA as the trigger for meiotic entry in the fetal gonad, suggesting that CYP26B1 is responsible for forming a barrier to prevent the exposure of prospermatogonia to RA throughout fetal testis development and acts as a meiotic inhibitory factor in the testis. What remains to be determined is whether RA acts on germ cells directly or whether it induces the somatic cells to secrete a different factor that in turn triggers meiotic entry.

6.3.1.2 *Stra8*

Further genetic evidence supporting a role for RA in the initiation of meiosis in females has been derived from the *Stra8* mutant mice. *Stra8* was identified in a screen for RA-responsive genes following treatment of the embryonic carcinoma cell line P19 (Oulad-Abdelghani et al. 1996). *Stra8* transcript and protein is expressed at the onset of meiosis in both male and female germ cells in multiple vertebrate strains, including humans (Jorgensen et al. 2012), rodents (Zhou et al. 2008b), chickens (Smith et al. 2008), goats (Tang et al. 2014; Yao et al. 2014), pigs (Wang et al. 2014), and catfish (Dong et al. 2013). Multiple studies have now shown that *Stra8* is likely a direct target of RA action. *Stra8* mRNAs can be detected as quickly as 2–4 h following RA treatment (Evans et al. 2014; Oulad-Abdelghani et al. 1996) and two RARE elements can be detected within a 400 bp region upstream from the *Stra8* gene start codon (Giuili et al. 2002).

Global deletion of *Stra8* results in both female and male infertility (Baltus et al. 2006). Examination of *Stra8*-deficient fetal ovaries revealed that the germ cells do not progress past their pre-meiotic stage of development, cannot undergo the chromatin condensation process indicative of meiotic entry, nor do they execute the pre-meiotic round of DNA replication (Baltus et al. 2006). The inability of *Stra8*-deficient oogonia to undergo meiosis was also shown via inappropriate expression or localization of known meiotic gene products, REC8, SYCP3, and H2A histone family, member X (H2AFX), within these cells. These defects lead to adult female mice displaying a marked reduction in ovarian size and a lack of follicles or oocytes, although oocytes that could be fertilized but did not generate viable embryos were detected in pure background C57BL/6 *Stra8*-deficient female mice in a follow-up study (Dokshin et al. 2013).

Additional evidence for STRA8 as the key meiosis-inducing molecule has been derived from studies of dual *Cyp26b1/Stra8*-knockout male embryos (Saba et al. 2014). As described above, male germ cells enter meiosis in the absence of CYP26B1 activity; however, this phenotype is reversed when STRA8 is also eliminated from these cells (Saba et al. 2014). Taken together, these results demonstrate that the classic marker of an RA response, STRA8, is essential for the initiation of meiotic prophase in mouse germ cells of both sexes. However, meiosis still appears to be directed by RA even in vertebrates that lack a *Stra8* homolog, e.g., teleost fishes (Rodriguez-Mari et al. 2013; Feng et al. 2015), but the downstream targets of RA signaling that drive meiotic onset in the germ cells of these species have yet to be determined. Collectively, the localization and genetic ablation studies demonstrate a requirement for RA-responsive and RA-degrading gene products in the ovary and also support the hypothesis that RA is the meiotic-inducing molecule within the fetal ovary.

6.3.1.3 *Aldh1a* Genes

The main crux of the argument against RA driving meiotic initiation in the fetal ovary has been derived from studies of the *Aldh1a2*- and *Aldh1a2/Aldh1a3*-deficient mice (Kumar et al. 2011). Usually, *Aldh1a2/Aldh1a3*-deficient embryos die as they do not synthesize enough RA to complete early embryogenesis, preventing examination of RA-dependent processes later in fetal development. To circumvent this, Kumar et al. (2011) used maternal low-dose RA supplementation from 6.75 to 9.25 *dpc* to rescue the early embryonic lethality and allow the deficient animals to progress through to 14.5 *dpc*. As the supplemented RA is cleared within 12–24 h (Mic et al. 2002), the treatment scheme permitted the animals to reach an age at which the early events of gonad development could be investigated in the absence of ALDH1A2 and ALDH1A3 activity. Intriguingly, in the rescued, *Aldh1a2/Aldh1a3*-deficient embryos, robust *Stra8* expression could be detected in oogonia, indicative of meiotic initiation (Kumar et al. 2011). Examination of ovaries lacking just *Aldh1a2* activity revealed that other markers of meiosis, *Sycp3* and H2AFX, were also expressed in a pattern similar to wild-type controls in the absence of this enzyme (Kumar et al. 2011). However, the same analyses were not performed on the dual *Aldh1a2/Aldh1a3*-deficient embryos so it is unclear whether ALDH1A3 activity could have influenced the expression of these other meiotic markers. Kumar et al. (2011) then utilized a *RARE-hsplacZ* reporter transgene mouse model (Rossant et al. 1991), which enables the visualization of cells that have triggered RA signaling owing to β-galactosidase expression in response to RA transactivation, to show that RA activity was present within the mesonephros of wild-type ovaries and absent within the urogenital ridge of the *Aldh1a2* mutant animals and that wild-type ovaries are able to respond to treatment with physiological levels of RA. Based on these genetic ablation and reporter gene expression studies, Kumar et al. (2011) concluded that (1) *Stra8* expression in the fetal ovary does not require RA signaling, (2) meiotic initiation can take place within fetal ovaries devoid of RA, and (3) even though RA is synthesized in the mesonephros of female fetal gonads, not enough diffuses into the ovary to stimulate signaling.

The study by Kumar et al. (2011) highlights that further analysis of gene knockouts and more sensitive retinoid measurements and enzyme activity assays are required before RA will be definitively proven as the meiosis-inducing substance in the fetal ovary. There are enzymes other than ALDH1A2 and ALDH1A3 that can synthesize RA, namely ALDH1A1, and no enzyme activity analyses were performed on the knockout mice in the Kumar et al. (2011) study to conclusively demonstrate a lack of ALDH activity in mutant ovaries. *Aldh1a1* is known to be present within wild-type fetal testes and ovaries and it is possible that *Aldh1a1* expression is upregulated in the absence of ALDH1A2 and ALDH1A3. This possibility was not addressed by the Kumar et al. (2011) study. In fact, quantitative PCR analysis of human fetal ovaries and testes (Childs et al. 2011; Le Bouffant et al. 2010) and analysis of human ALDH isozyme efficiency (Arnold et al. 2015b) imply that ALDH1A1 is the most important of the three ALDH enzymes for RA synthesis in humans. However, if redundancy between ALDH isozymes occurs, then we would have expected to see RA activity in the mutant mice carrying the *RARE-hsplacZ* reporter transgene in the Kumar et al. (2011) study. A plausible explanation for why this was not the case is the fragile nature of this reporter gene. The *RARE-hsplacZ* line is often prone to losing activity (Sakai and Drager 2010), even when the breeding scheme involves males and females who are both positive for the transgene (Snyder, Agrimson, and Hogarth, unpublished observations). It is therefore possible that the breeding of this reporter line into the different genetic background of the ALDH mutant could have disrupted the expression of the transgene.

ALDH gene expression and protein localization in vertebrate species other than mice also imply a role for RA in driving meiotic onset, although the site of RA synthesis differs. *Aldh1a2* is expressed in fetal or developing ovaries in humans (Childs et al. 2011), chickens (Smith et al. 2008), anuran and urodele amphibians (Piprek et al. 2013; Wallacides et al. 2009), and teleost fishes (Feng et al. 2015; Rodriguez-Mari et al. 2013), rather than in the mesonephros as is the case for mice. In addition, these studies all show that the timing of *Aldh1a2* expression is dependent upon exactly when meiotic initiation occurs. These observations imply that perhaps the mouse model is an outlier with respect to where the RA required for meiotic onset is synthesized. Protein localization in both zebrafish (Rodriguez-Mari et al. 2013) and anurans (Piprek et al. 2013) demonstrated that the somatic cells of the gonads contain ALDH1A2 and are likely the site of RA production for the initiation of meiosis in most vertebrate species.

6.3.2 Evidence from Chemical Inhibitors and Deficiency Studies

As genetic ablation studies tend to only allow for the examination of one or two genes at a time, the emergence of heterogeneous phenotypes in these models is often explained by functional redundancy between protein family members. The development of chemical antagonists and agonists targeting the various retinoid metabolizing and signaling proteins families has allowed for the circumvention of this issue and generated valuable tools for the investigation of the role of RA in meiotic initiation in both the testis and ovary. In addition, the study of vitamin A- and RA-deficient rodents has provided significant insight into the processes under RA control during oogenesis and spermatogenesis.

6.3.2.1 Induction of Meiosis via Exogenous Retinoids

One of the first indicators for the action of RA on the induction of meiosis was the observation that germ cells in the fetal rat ovary could enter meiosis prematurely upon exposure to RA prior to what would normally occur in vivo (Livera et al. 2000). This was also true for fetal ovaries treated with an agonist of RARα signaling (Livera et al. 2000), implying that stimulation of RA signaling is sufficient to drive female germ cells to enter meiosis. Hallmarks of meiotic entry, including induction of *Stra8*, *Rec8*, and *Sycp3* expression, have also been observed following the incubation of whole mouse male urogenital ridges, larval stage amphibian testes, or isolated fetal male germ cells with RA (Bowles et al. 2006, 2010; Koubova et al. 2006; Ohta et al. 2010; Piprek et al. 2013), although Kumar et al. (2011) reported that RA signaling could not be induced within the fetal testis via treatment with physiologically relevant levels of exogenous RA (25–100 nM) due to CYP26B1 activity. The majority of ex vivo organ culture experiments have been performed with 0.7–1 μM RA added to the culture media, which is 25- to 50-fold higher than endogenous levels (Horton and Maden 1995; Mic et al. 2003), and hence it is possible that any cellular response seen is due only to such high levels of RA. However, incubation of germ cells in as little as 1 nM of RA is adequate to induce *Stra8* expression (Bowles et al. 2010), supportive of the conclusion that exogenous RA can induce meiotic entry.

6.3.2.2 Inhibition of RA Synthesis

Aside from the published ALDH ablation genetic model (Kumar et al. 2011), RA deficiency has been produced via either dietary vitamin A restriction or chemical inhibition of ALDH activity and both experimental designs have provided evidence to support a role for RA in triggering female meiosis. Examination of ovaries in severely vitamin A-deficient (VAD) female rat embryos revealed that the germ cells failed to express *Stra8* and initiate meiosis, indicative of a role for a metabolite of vitamin A in meiotic entry (Li and Clagett-Dame 2009). Additionally, in rat embryos with a more moderate deficiency in vitamin A, approximately 30 % of the oogonia, compared to 75 % in control embryos, properly initiated meiosis, as evidenced by cells stained positive for SYCP3, and the expression of *Stra8* was reduced by half (Li and Clagett-Dame 2009). In addition to blocking its synthesis via restriction of substrate, RA activity can also be abolished via chemical inhibition of the ALDH enzymes. The bis(dichloroacetyl) diamine WIN 18,446, an irreversible inhibitor of ALDH1A2 and reversible inhibitor of ALDH1A1 and ALDH1A3 (Arnold et al. 2015a, b), has been used both ex vivo and in vivo to study the effects of eliminating RA synthesis on oogenesis and spermatogenesis (discussed in more detail below). Treatment of female urogenital ridges with WIN 18,446 not only blocked the expression of *Stra8* in these germ cells, but parallel treatment of WIN 18,446 with vitamin A, in the form of retinol, also blocked *Stra8* expression, indicating that it was a lack of RA and not a different vitamin A metabolite that induced the WIN 18,446 effect (Hogarth et al. 2011). A second ALDH inhibitor, citral, has also been shown to block meiotic entry in mouse, amphibian, and human fetal tissues (Le Bouffant et al. 2010; Pietruszko et al. 1999; Piprek et al. 2013). Taken together, these data suggest that meiosis does not initiate in RA-deficient ovaries, in complete contradiction with the ALDH genetic ablation study (Kumar et al. 2011). Clearly, an analysis of fetal ovaries when all three ALDH enzymes are rendered nonfunctional genetically is required to determine whether the RA-synthesizing enzymes are essential for meiotic initiation in females.

6.3.2.3 Inhibition of RA Signaling and Degradation

RAR panantagonists have been utilized to generate evidence for a role of RA in meiotic initiation in the ovary. Two different RAR panantagonists, BMS-204493 and AGN193109, were shown to inhibit the expression of *Stra8*, *Sycp3*, and DMC1 dosage suppressor of mck1 homolog (*Dmc1*) following incubation of mouse E11.5 female urogential ridges with these chemicals (Bowles et al. 2006; Koubova et al. 2006), indicative of a role for RA signaling in triggering meiosis in female germ cells. In addition, the RAR panantagonist BMS453 also blocked meiotic entry in cultures of anuran larval ovaries (Piprek et al. 2013). Two different chemicals have been utilized to provide additional evidence that the degradation of RA by the CYP26 enzymes is essential to stop male germ cells within the fetal testis from entering meiosis. Treatment of mouse male urogenital ridges or anuran larval testes in ex vivo organ culture with either ketoconazole (Bowles et al. 2006; Koubova et al. 2006; Piprek et al. 2013), a nonspecific cytochrome p450 inhibitor, or the CYP26-specific inhibitor R115866 (Koubova et al. 2006), also known as talarozole, induced the germ cell expression of *Stra8*, *Sycp3*, and *Dmc1*, all hallmarks of meiotic initiation, and triggered meiosis-like morphological changes within the nuclei of these cells. Talarozole treatment has been shown to increase RA levels in various organs, although not directly within either the testis or ovary (Stoppie et al. 2000). The treatment of male urogenital ridges simultaneously with either of the CYP26 inhibitors and a RAR panantagonist indicated that RA signaling is necessary for the meiotic response seen following CYP26B1 elimination (Bowles et al. 2006; Koubova et al. 2006).

These observations clearly suggest that germ cell meiotic entry requires RA signaling and that this action is opposed by CYP26B1 activity, yet there is currently a lack of genetic evidence supporting a role for the RA signaling pathway in the initiation of meiosis in the female. Global knockout mice have been constructed for all six receptors, none of which reported evidence for female infertility, though fertility was not tested in the RXRa mutant (Ghyselinck et al. 1997; Kastner et al. 1996; Krezel et al. 1996; Lohnes et al. 1993; Lufkin et al. 1993). While based on the current genetic evidence it would seem that RXRα is the most likely candidate to be important for female meiosis, there is still the possibility of redundancy between members of the RAR protein family. As a result, the key to definitively demonstrating a role for RA signaling in female meiosis lies in the development of a conditional allele that can specifically delete RXRα within the fetal ovary and a genetic model that can block signaling via every RAR receptor, as a means of eliminating the possibility of functional redundancy.

6.3.3 Conclusions

When all the current data is taken into consideration, there is no definitive answer to the question "Does RA trigger female meiosis?" The evidence does point to the CYP26 enzymes, known at this point to only degrade RA, as the key enzymes in protecting fetal male germ cells from entering meiosis. Either vitamin A deficiency or chemicals that inhibit RA synthesis prevent the expression of meiotic markers in fetal ovaries and both exogenous RA or chemical agonists of the RA signaling pathway can induce male fetal germ cells to enter meiosis. However, the observation that meiosis is induced in fetal ovaries in the absence of two of the three ALDH enzymes and the fact that there is currently no genetic evidence to link the RA signaling pathway with triggering female meiosis would suggest that perhaps an alternative substrate of CYP26B1 is the meiosis-inducing factor. Clearly future studies to help answer this critical question should focus on compound mutations of the RA synthesis and signaling proteins and more sensitive assays to then measure RA concentrations and enzyme and receptor activity following genetic modification.

6.4 The Role of RA in Triggering Male Meiosis

In contrast to females, RA regulation of murine male meiosis begins in the postnatal testis. The cords that arise during fetal development become the seminiferous tubules by birth, with the Sertoli cells underlined by the tubule basement membrane and the male germ cells, the prospermatogonia, located at the center. Over the course of the next 3–4 days following birth, these prospermatogonia undergo significant morphological changes, reenter the cell cycle, migrate to the basement membrane, and follow one of two fates: (1) either transition directly to differentiating spermatogonia and initiate the first round of spermatogenesis or (2) become part of the undifferentiated A spermatogonial population, a small portion of which will also develop into the spermatogonial stem cell pool. Following the first round of spermatogenesis, undifferentiated A spermatogonia are continually triggered to

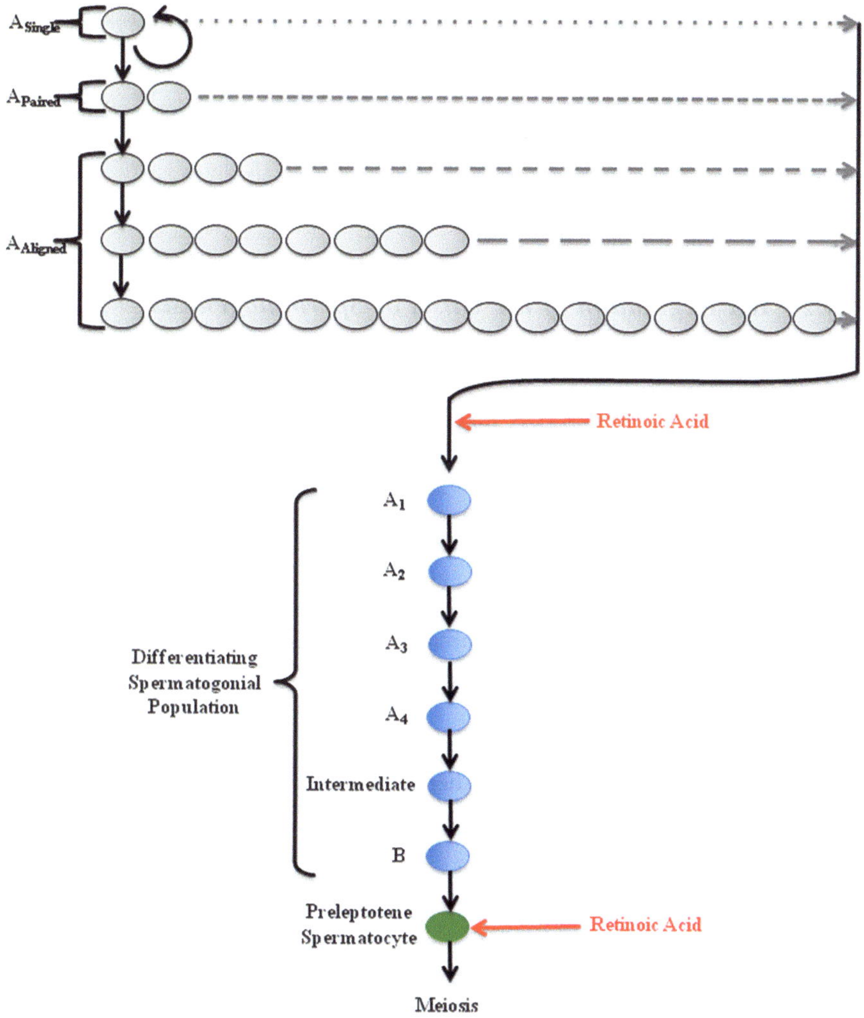

Fig. 6.4 The role of RA during spermatogonial differentiation and the onset of male meiosis in mice. The undifferentiated spermatogonial population (*gray circles*) can exist as A_{single} spermatogonia or, following mitotic divisions and incomplete cytokinesis, as chains of A_{paired} and $A_{aligned}$ spermatogonia. While germ cells from all three of these subpopulations can differentiate in response to RA, the $A_{aligned}$ spermatogonia are far more likely to differentiate compared to the A_{single} spermatogonia. This is represented by the incremental increases in size of the dashed lines. Once differentiating spermatogonia (*blue circles*) are formed following exposure to RA, six mitotic divisions occur to form preleptotene spermatocytes (*green circle*). RA signaling is believed to also be required in preleptotene spermatocytes for entry into meiosis

differentiate and become A1 spermatogonia, known as the A-to-A1 transition. This transition marks the commitment of spermatogonia to progression through meiosis and becoming spermatozoa (Fig. 6.4). There is now extensive evidence to indicate that the critical hormone controlling the A-to-A1 transition is RA, yet less is known about the molecular events that (1) allow for a spermatogonia to differentiate and (2) occur following the RA trigger. The following sections will review our current understanding of how RA regulates spermatogonial differentiation and the commitment to and initiation of meiosis and the control of these processes at the molecular level.

6.4.1 RA Regulation of the Onset of Spermatogenesis and Spermatogonial Differentiation

The onset of spermatogenesis occurs in an asynchronous manner in patches along the length of neonatal murine testis tubules. This event establishes the cycle of the seminiferous epithelium and eventually leads to the continual differentiation of spermatogonia throughout reproductive life. Evidence has been collected from numerous genetic and chemical studies to demonstrate that RA is the hormone responsible for triggering the onset of and regulating the timing of spermatogenesis. However, the molecular mechanisms and cell types responsible for regulating RA levels within the testis and the downstream consequences of RA signaling are still being elucidated.

6.4.1.1 Evidence Derived from Either an Absence of or Excess RA

Vitamin A Deficiency

The first indication that RA drives spermatogonial differentiation was deduced from studies examining testes of VAD rodents (Huang and Hembree 1979; Ismail et al. 1990; Mitranond et al. 1979; Morales and Griswold 1987; Van Beek and Meistrich 1990; van Pelt and de Rooij 1990; van Pelt et al. 1995; Wang and Kim 1993). The placement of wild-type adult rodents with normal spermatogenesis on a VAD diet resulted in the sloughing of advanced germ cells into the lumen of the testis tubules and loss of fertility in approximately 9–11 weeks. Morphological analysis of these testes revealed only Sertoli cells and undifferentiated A spermatogonia remaining within the seminiferous epithelium (Morales and Griswold 1987; van Pelt and de Rooij 1990). Interestingly, treatment of VAD rodents with exogenous retinoids and addition of vitamin A back into the diet induced spermatogonial differentiation and the re-initiation of spermatogenesis (Morales and Griswold 1987; van Pelt and de Rooij 1990), demonstrating that vitamin A plays a crucial role during spermatogonial differentiation and controls the proper progression of spermatogenesis in rodents. Upon further investigation of the seminiferous epithelium in these animals following a 2-month recovery period, only a few stages of the cycle were present in the testis suggesting spermatogenesis had become synchronized (Morales and Griswold 1987; van Pelt and de Rooij 1990). Spermatogenic synchrony is evidence of an alteration to the timing of spermatogonial differentiation, i.e., indicating that the introduction of retinoids following deficiency induced nearly all spermatogonia to differentiate simultaneously, thereby eliminating the asynchronous nature of spermatogenesis. The adult VAD and synchronized rodent models have helped shape our understanding of how RA controls spermatogonial differentiation.

While studies of spermatogenesis in the adult animal can be performed with the VAD model, analysis of wild-type neonatal testis development in a VAD environment is not possible due to the length of time required to induce vitamin A deficiency in rodents (~28 weeks in mice). To circumvent this issue, lecithin-retinol acyltransferase (*Lrat*) knockout mice were fed a VAD diet (Li et al. 2011). This mutation prevents the storage of retinyl esters (Batten et al. 2004) and hence *Lrat* knockout animals become deficient very quickly when fed a VAD diet (Li et al. 2011; Ruiz et al. 2007; Vernet et al. 2008), allowing for the analysis of the juvenile testis in a VAD environment (Li et al. 2011). Quantitative RT-PCR analysis of VAD *Lrat*-deficient mice demonstrated that *Stra8* expression levels were significantly decreased within their testes and the development of meiotic germ cells rarely occurred (Li et al. 2011). Therefore, this study provided further support that RA is required for the onset of spermatogenesis, spermatogonial differentiation, and meiotic progression.

WIN 18,446

WIN 18,446 was first investigated in 1961 as a potential reversible male contraceptive (Heller et al. 1961). Treatment of men with this compound displayed successfully suppressed sperm production but also exhibited acute side effects when WIN 18,446 was taken in conjunction with alcohol. As a result, the mechanism driving WIN 18,446 suppression of spermatogenesis was never investigated (Heller et al. 1961). Further studies with WIN 18,446 confirmed a reversible inhibition of spermatogenesis in mice (Brooks and van der Horst 2003) and the seminiferous epithelium in the testes of rabbits exposed to WIN 18,446 mimicked the phenotype reported in VAD rodent testes, i.e., only Sertoli cells and undifferentiated spermatogonia remained following treatment (Amory et al. 2011). Based on these observations and the structural similarities between the enzymes required to metabolize alcohol and vitamin A within the body, the RA synthesis enzymes were hypothesized to be the testicular target of WIN 18,446 activity (Amory et al. 2011). More recent studies have confirmed that WIN 18,446 is a suicide inhibitor of ALDH1A2 and a reversible inhibitor of ALDH1A1 (Arnold et al. 2015a). Combined these two enzymes are likely responsible for synthesizing more than 95 % of the total RA generated within the testis (Arnold et al. 2015b), indicating that WIN 18,446 should be able to nearly completely block testicular RA synthesis. Quantitative liquid chromatography/mass spectrometry assays revealed that RA levels in the testis were reduced by 90 % when exposed to WIN 18,446

(Arnold et al. 2015a) so although WIN 18,446 is not feasible for use as a male contraceptive, the compound is an excellent tool for the study of spermatogenesis in animal models.

Utilizing WIN 18,446 to generate testes lacking RA resulted in blocking spermatogonial differentiation both in vitro and in vivo. Active RA signaling could not be detected in cultured 2 days *postpartum* (*dpp*) *RARE-hsplacZ* mouse testes exposed to a combination of retinol and WIN 18,446, yet β-galactosidase activity, indicative of an RA response, was robustly detected when testes were exposed to RA alone (Hogarth et al. 2011). In addition, *Stra8* levels were significantly decreased in either primary neonatal testis cultures or isolated prospermatogonia following treatment with WIN 18,446 when compared to testes or cells exposed to retinol or RA (Hogarth et al. 2011). These observations were also consistent with the decrease in *Stra8* expression seen in testes from rabbits treated with WIN 18,446 (Amory et al. 2011). To investigate the in vivo effect of this compound, neonatal mice were treated daily for seven consecutive days and their testes harvested for morphology and cell marker analysis 24 h following the final treatment. The detection of zinc finger and BTB domain containing 16 (ZBTB16) protein, a marker of undifferentiated spermatogonia (Buaas et al. 2004), in almost every germ cell and the complete lack of STRA8-positive cells indicated that the WIN 18,446-treated testes contained an enriched population of undifferentiated spermatogonia (Hogarth et al. 2013). In addition, the same treatment scheme performed on *RARE-hsplacZ* mice demonstrated that WIN 18,446 was also able to completely inhibit the classical RA signaling mechanism (Hogarth et al. 2013). Similarly, when 0 *dpp* mice were exposed to WIN 18,446 for 4 days, no kit oncogene (KIT) protein, a known marker of male germ cell differentiation, could be detected within spermatogonia (Busada et al. 2015). Taken together, these observations show that blocking the production of RA in the neonatal testis inhibits spermatogonial differentiation.

Similar to rescued spermatogenesis in VAD mice, RA exposure following WIN 18,446 treatment initiated spermatogonial differentiation and synchronized spermatogenesis. STRA8 was present in the majority of spermatogonia 24 h following RA exposure (Hogarth et al. 2013) and morphological and immunohistochemistry analysis revealed that spermatogenesis had been synchronized within testes taken from animals treated with WIN 18,446 and RA, i.e., all testis tubule cross sections displayed the same arrangement of germ cells (Hogarth et al. 2013). As a result, the WIN 18,446/RA co-treatment protocol has provided further data to demonstrate the necessary role of RA for spermatogonial differentiation. Furthermore, these data have indicated that the timing of male germ cell differentiation can be controlled by manipulating RA availability within the testis.

Exogenous Retinoids

Multiple independent studies have investigated the effects of excess RA on the neonatal testis. The first, Snyder et al. (2011), treated 2 *dpp* *RARE-hsplacZ* mice with a single dose of RA to examine whether dramatic increases in testicular RA concentrations could disturb RA signaling within either germ or Sertoli cells at the onset of spermatogenesis. RA exposure resulted in almost every germ cell staining positive for β-galactosidase activity and STRA8, indicating a robust spermatogonial response to RA, and the early induction of SYCP3 expression in germ cells suggested that these cells may have been preparing to enter meiosis (Snyder et al. 2011). Interestingly, Sertoli cells were rarely β-galactosidase positive in testes isolated from either the RA- or vehicle control-treated animals (Snyder et al. 2010, 2011). This observation implies that (1) the Sertoli cell response to RA is below the level of detection for this assay, (2) the *RARE-hsplacZ* response element cannot be stimulated within Sertoli cells, or (3) perhaps classical RA signaling does not take place in the neonatal Sertoli cells. Additionally, the number of caspase-3-positive germ cells was significantly increased following RA treatment, suggesting that germ cell apoptosis was occurring (Snyder et al. 2011).

The second study, Busada et al. (2014), exposed 1 *dpp* mice to RA and found that this induced premature synthesis of STRA8, KIT, and spermatogenesis- and oogenesis-specific basic helix-loop-helix 1 (SOHLH1) proteins, all markers of spermatogonial differentiation, and an increase in the number of antigen identified by monoclonal antibody Ki 67 (MKI67)-positive germ cells, suggesting increased proliferation (Busada et al. 2014). However, in contrast to the first study, the authors found no evidence of excessive programmed cell death following RA exposure (Busada et al. 2014). Interestingly, both studies did find that premature exposure to RA resulted in an immediate increase in *Stra8* (24 h postinjection) followed by a decrease (48 and 72 h postinjection), most likely due to the increase in *Cyp26a1* and *Cyp26b1* mRNAs and therefore RA degradation to counteract the effect (Busada et al. 2014; Snyder et al. 2011). Following exogenous exposure to RA, both studies extended their analysis by allowing animals to recover for different periods of time to investigate the downstream effects on spermatogenesis. The onset of meiosis was delayed by a period of 2 days in testes from RA-treated neonatal mice when compared to vehicle-treated controls (Busada et al. 2014). While the reason for this meiotic delay is unknown, a plausible explanation is that while the germ cells prematurely exposed to RA express various markers of differentiating spermatogonia, the immediate upregulation of CYP26 activity to clear all available RA halts these cells from continuing to progress through their differentiation pathway, thereby delaying the onset of meiosis. Although there was a delay in the timing, spermatogenesis was synchronous within testes of animals left to recover to adulthood following treatment (Snyder et al. 2011), showing that RA exposure in excess at the onset of spermatogenesis overrides the asynchronous differentiation of spermatogonia, presumably by exposing all germ cells to RA simultaneously.

While a single dose of RA at 2 *dpp* resulted in synchronous spermatogenesis, RA treatment of adult mice did not induce the same outcome (Snyder et al. 2011). A possible explanation for this could be that while the mechanism for controlling testicular RA synthesis is likely in place in the adult testis, it may not be established in the neonatal testis. To investigate this, postnatal animals aged 4, 6, and 8 *dpp* were injected with a single dose of RA and grown to adulthood (Davis et al. 2013) to determine whether there was a point during juvenile testis development at which spermatogenesis could no longer be synchronized following RA exposure. While spermatogenesis was synchronized in the testes of animals treated with RA at 2 and 4 *dpp*, the degree of synchrony was reduced at 6 *dpp* compared with the younger timepoints and no synchrony was seen following RA exposure at 8 *dpp* (Davis et al. 2013; Snyder et al. 2011). The current hypothesis explaining this "window of synchrony" following RA treatment is that the emergence of preleptotene spermatocytes between 6 and 8 *dpp* alters whether spermatogonia can respond to exogenous RA. This conclusion is also supported by recent observations that preleptotene spermatocytes are likely able to synthesize RA (Raverdeau et al. 2012; Wu et al. 2008) and perhaps are responsible for providing the RA that drives spermatogonial differentiation following the first round of spermatogenesis. Taken together, these exogenous RA studies show that there is a period of RA sensitivity present in the postnatal testis during which exposure can manipulate the asynchronous initiation of spermatogenesis.

6.4.1.2 Evidence from Gene Expression Studies

Due to the asynchronous nature of the first round of spermatogenesis, the spermatogonial population is very heterogeneous. During the initiation of spermatogenesis, most prospermatogonia transit into undifferentiated spermatogonia that will support all subsequent rounds of spermatogenesis; however, the stem cell pool and the germ cells that first commit to undergoing spermatogenesis also arise from the prospermatogonial population. As a result, investigating and distinguishing between these spermatogonial populations has proven difficult. Recently Hermann et al. (2015) examined spermatogonial heterogeneity within the neonatal murine testis by utilizing several cell isolation techniques and single cell quantitative RT-PCR and was able to distinguish three different gene clusters. Although there was heterogeneity even within these clusters, it is likely that the three groups represent the spermatogonial stem cell pool, the progenitor spermatogonia, and the differentiating spermatogonia (Hermann et al. 2015). It is not fully understood what controls the fate decisions and induces this heterogeneity, but it is likely that exposure to RA plays a role.

A second recent study has also determined that in addition to direct effects on gene transcription, RA may regulate recruitment of mRNAs to polysomes for translation during the onset of spermatogenesis (Chappell et al. 2013). Microarray studies have shown that between 1 and 4 *dpp*, approximately 50 genes are upregulated within the testis (Shima et al. 2004). This is a relatively small number considering the extensive changes to germ cell morphology, differentiation status, and positioning within the tubules that also occur during that time. Utilizing polysome gradients, Chappell et al. (2013) were able to show that both housekeeping and germ cell-specific genes were more efficiently translated, rather than observing significant increases in transcript levels, between 1 and 4 *dpp*, the time during which RA is triggering the onset of spermatogenesis. In a continuation of this study, Busada et al. (2015) demonstrated that rather than inducing *Kit* transcription, KIT protein levels were increased via translation of stored *Kit* mRNAs following premature exposure of neonatal mice to RA. The authors were also able to show that RA-regulated KIT translation was driven via the interaction of RA receptors with the phosphatidylinositol 3-kinase (PI3K)/thymoma viral proto-oncogene (AKT)/mouse target of rapamycin complex (mTORC) signaling pathway, as chemical inhibition of PI3K, pyruvate dehydrogenase kinase, isoenzyme 1 (PDK1), or AKT resulted in a loss of KIT expression in the testis (Busada et al. 2015). Taken together, these studies all demonstrate the crucial but complex role that RA plays during the onset of spermatogenesis, i.e., RA regulates both transcription and translation, and highlight the need for further research into the various downstream effects of RA and interactions of this molecule with other signaling pathways.

WIN 18,446 has been a valuable tool for the examination of gene expression changes in the absence of RA. Microarray analyses demonstrated that WIN 18,446 treatment of neonatal testes had significant effects on gene expression (Hogarth et al. 2011), and this analysis was compared to array data generated via RA exposure of 2 *dpp* prospermatogonia (Zhou et al. 2008a). *Stra8*, dehydrogenase/reductase (SDR family) member 3 (*Dhrs3*) (a retinol reductase responsible for the conversion of retinaldehyde to retinol), and stimulated by retinoic acid gene 6 (*Stra6*) (a membrane protein that can aid in the transport of retinol into the cell) were all significantly downregulated following WIN 18,446 treatment and upregulated following RA treatment, implying perhaps that RA has direct effects on the transcription of genes encoding proteins responsible for vitamin A metabolism (Hogarth et al. 2011; Zhou et al. 2008a). In addition, the WIN 18,446/RA synchronization protocol combined with the unique RiboTag mouseline (Sanz et al. 2009) has allowed for the investigation of cell type-specific gene expression occurring during spermatogonial differentiation (Evans et al. 2014). The RiboTag mouseline was developed to enable the collection of polyribosomes, and subsequent isolation of associated mRNAs, in a cell-specific manner (Sanz et al. 2009). Testes were collected from RiboTag/*Stra8*-iCre and RiboTag/*Amh*-Cre animals to allow for the investigation of germ and Sertoli cells, respectively. Microarray analysis revealed that while both Sertoli and germ cells contain the machinery required to synthesize RA and trigger the RA signaling pathway, these cell types expressed distinct isoforms of the different family members. For example, *Rarb*, *Rarg*, and *Rxrg* were enriched within germ cells, while *Rxrb* was enriched in Sertoli cells (Evans et al. 2014), and these observations perfectly reflect conclusions drawn from gene knockout and localization studies (Gely-Pernot et al. 2012; Vernet et al. 2006a, 2008). In addition, Evans et al. (2014) also investigated which transcripts were associated with polyribosomes during spermatogonial differentiation and meiotic initiation in the synchronized postnatal testis. Interestingly, it appears that germ cells experience reorganization of chromatin and histone modifications during the first round of spermatogenesis, as transcripts encoding different histones were found to vary in response to RA exposure (Evans et al. 2014), potentially in preparation for entry into meiotic prophase. This hypothesis was also supported by the observation that meiotic genes were associated with polysomes well before the appearance of preleptotene spermatocytes (Evans et al. 2014). These data provide evidence that once RA triggers spermatogonial differentiation, germ cells are committed to entering meiosis in the postnatal testis.

6.4.1.3 Evidence from Transgenic and Genetic Ablation Mouse Models

RARE-hsplacZ

Evidence that RA is the trigger for the initial differentiation of spermatogonia within the neonatal testis was derived from the use of the *RARE-hsplacZ* transgenic mouse model (Rossant et al. 1991). Multiple studies have shown that RA signaling first occurs in at approximately 2–3 *dpp* in the neonatal testis and induces the synthesis of STRA8 and KIT protein in the newly formed differentiating spermatogonia (Busada et al. 2014, 2015; Snyder et al. 2010; Zhou et al. 2008b) (Fig. 6.5). Lower magnification views of these stained testis cross sections revealed nonuniform expression of STRA8, implying that the onset of spermatogonial differentiation occurs in an asynchronous manner. β-galactosidase staining of testes isolated from 3 *dpp RARE-hsplacZ* transgenic mice revealed that RA signaling takes place within patches along the tubules (Snyder et al. 2010). In addition, STRA8 co-localized with β-galactosidase activity in differentiating spermatogonia, but was absent from the patches of tubule that lacked RA signaling (Snyder et al. 2010). Interestingly, a single dose of RA eliminated the patches that were β-galactosidase negative and staining was present along the length of whole tubules (Snyder et al. 2010). These observations suggest that RA signaling initiates the onset of spermatogenesis via asynchronously triggering the first wave of spermatogenesis.

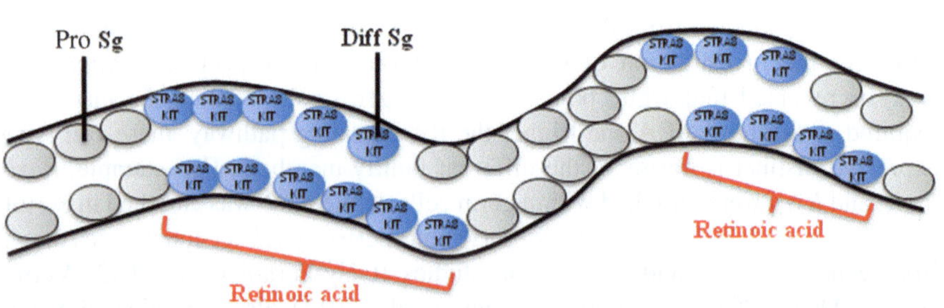

Fig. 6.5 The asynchronous initiation of spermatogenesis. Schematic depicts the patchy nature of the onset of spermatogenesis along the length of the seminiferous tubules (Snyder et al. 2010). The *gray cells* represent the prospermatogonia and *blue cells* represent the newly differentiating A1 spermatogonia. Spermatogonial differentiation occurs only in the patches of seminiferous tubule that are exposed to RA signaling (*red*), thereby triggering STRA8 and KIT expression (markers of differentiating spermatogonia) and establishing the asynchronous initiation of spermatogenesis

Rdh10

The RA-synthesizing enzyme RDH10 is required for juvenile spermatogenesis (Tong et al. 2013). Elimination of *Rdh10* specifically within Sertoli cells resulted in a partial block of the initiation of spermatogenesis in the juvenile testis. However, this block became more complete following the additional conditional mutation of this gene within germ cells (Tong et al. 2013), indicating that RDH10 activity is most important within Sertoli cells but that some compensation within germ cells does occur. Interestingly, analysis of testes taken from 28 *dpp* Sertoli- and germ cell-specific *Rdh10* knockout mice revealed that spermatogenesis had eventually initiated, even in the absence of RDH10 activity within the seminiferous epithelium, resulting in fertile male mice. This recovery of function following RDH10 ablation is puzzling but could be the result of compensation from another retinaldehyde dehydrogenase over time or that the testis eventually receives RA from a different source.

Aldh1a Enzymes

Evidence to demonstrate that the Sertoli cells provide the RA required for the initial differentiation of spermatogonia in the neonatal testis has been derived from the analysis of mutant mice. *Aldh1a1*, *Aldh1a2*, and *Aldh1a3* were conditionally deleted specifically within Sertoli cells by utilizing the *Amh*-Cre transgenic mouse line (Lecureuil et al. 2002; Raverdeau et al. 2012), herein referred to as $Aldh1a^{Ser-/-}$. The resulting animals displayed juvenile testes containing only Sertoli cells and undifferentiated spermatogonia (Raverdeau et al. 2012), demonstrating that Sertoli cells synthesize the RA required to drive the onset of spermatogenesis. To investigate whether this block in spermatogonial differentiation could be rescued, *Aldh1a* $^{Ser-/-}$ mice were treated with a single dose of exogenous RA (Raverdeau et al. 2012). Twenty-four hours following treatment, KIT-positive differentiating spermatogonia had developed (Raverdeau et al. 2012), emphasizing the necessity of RA for spermatogonial differentiation during the onset of spermatogenesis and that these spermatogonia could respond normally when eventually exposed to RA.

6.4.2 RA Regulation of Male Meiotic Initiation

There are several lines of evidence suggesting that RA signaling is necessary for meiotic progression within the juvenile testis. The main source of data supporting this hypothesis has been the *Stra8* knockout mouse model, with two independent studies investigating how the elimination of this protein affects male meiotic progression (Anderson et al. 2008; Mark et al. 2008). Analysis following *Stra8* ablation would suggest that RA appears to control meiotic initiation through the induction of this protein. Further evidence supporting a role for RA in meiotic initiation has been derived utilizing the *Aldh1a*$^{Ser-/-}$ mice and localization analyses to determine which cells within the testis can synthesize and degrade RA.

6.4.2.1 *Stra8* Knockout Model

A role for RA in driving male meiotic initiation has mostly been inferred from the study of STRA8-deficient male mice. STRA8 protein is present in differentiating A spermatogonia and preleptotene spermatocytes in the postnatal testis (Hogarth et al. 2015; Zhou et al. 2008b). Nuclear and cytoplasmic localization of STRA8 have both been reported, suggesting that the protein can shuttle between different cellular compartments (Hogarth et al. 2015; Tedesco et al. 2009), and while a putative DNA binding domain has been identified within its amino acid sequence (Tedesco et al. 2009; and Michael Griswold, personal communication), the function of STRA8 remains unknown. Three independent studies investigated the effects of a complete loss of *Stra8* in the testis (Anderson et al. 2008; Endo et al. 2015; Mark et al. 2008). *Stra8*-null male mice are infertile, due to a block during meiotic prophase (Anderson et al. 2008; Mark et al. 2008) and reduced numbers of differentiating spermatogonia compared to controls (Endo et al. 2015), but two of the studies came to different conclusions regarding what the most advanced germ cell type present was within the seminiferous epithelium. Anderson et al. (2008) only observed preleptotene spermatocytes within the testes of *Stra8*-null male mice (Anderson et al. 2008), while Mark et al. (2008) reported that the majority of seminiferous tubules were filled with preleptotene and leptotene spermatocytes while very few zygotene and pachytene spermatocytes were present. Mark et al. (2008) concluded that preleptotene spermatocytes formed properly in the testes of *Stra8*-null mice based on histology and by detection of *Rec8* transcript expression (Mark et al. 2008). Anderson et al. (2008) also investigated REC8 expression and localization within preleptotene spermatocyte and concluded that while REC8 was present, the localization was abnormal due to a lack of association with chromosomes. In addition, a lack of H2AFX expression implied that DNA double-strand breaks were not properly occurring within STRA8-deficient spermatocytes. DNA replication appeared to proceed normally in the preleptotene spermatocytes, based on 5-bromo-2′-deoxyuridine (BrdU) incorporation, suggesting that loss of *Stra8* did not have an effect on the process of replication in these cells (Mark et al. 2008; Anderson et al. 2008). As a result, these two studies combined have revealed that although STRA8 does not control progression into replication, it is necessary for proper meiotic prophase.

Further evidence for the role of STRA8 in driving meiotic initiation was collected during the examination of the nanos homolog 2 (*Nanos2*)-null animal. RA is known to downregulate the synthesis of NANOS2 in the testis (Barrios et al. 2010) and NANOS2 is expressed within embryonic and neonatal male germ cells at a time when RA signaling does not occur in the testis (Barrios et al. 2010; Bowles et al. 2006; Snyder et al. 2010; Suzuki et al. 2007; Tsuda et al. 2003, 2006), implying a link between RA and NANOS2 function. *Stra8* expression was induced within prospermatogonia in the embryonic testis of the *Nanos2* global knockout mice (Suzuki et al. 2007; Suzuki and Saga 2008), implying that these cells have inappropriately entered meiosis. While the *Stra8* mutation resulted in a block during meiotic prophase, meiosis was induced in the *Nanos2* global knockout mice (Anderson et al. 2008; Suzuki et al. 2007; Tedesco et al. 2009). This study implied that NANOS2 may play a role in the repression of STRA8 and thus assist in the prevention of premature meiotic initiation in the testis.

Of interest is that while mutating *Stra8* does block meiotic progression at a similar point in male and female germ cells (Anderson et al. 2008; Mark et al. 2008), it is possible that STRA8 function is not completely conserved between the sexes. In the embryonic ovary, the first appearance of STRA8 in germ cells corresponds with the onset of meiotic prophase. However, in the testis STRA8 can be detected in two different cell types: (1) newly differentiating spermatogonia and (2) the preleptotene spermatocytes. As a result, it is the second appearance of STRA8 during spermatogenesis that coincides with the physical entry of that cell into meiosis and not the first, as the newly differentiated spermatogonia undergo multiple rounds of mitotic division before entering meiotic prophase. As mentioned above, the function of STRA8 is unknown so it remains to be determined why there is a sex-specific response when *Stra8* is first expressed. The most likely explanation is that while RA activity may be conserved in males and females, the downstream targets in spermatogonia differ from those in oogonia. For example, RA triggers the translation of KIT protein in spermatogonia (Busada et al. 2015), an event that does not take place in female germ cells. In addition, there is also the possibility that other cues from the somatic cell environment interplay with RA in either sex and influence the response.

6.4.2.2 *Aldh1a* Knockout Model

Further evidence to support a role for RA during meiotic initiation has been derived from the *Aldh1a*$^{Ser-/-}$ mice. Spermatogenesis was rescued in *Aldh1a*$^{Ser-/-}$ mice testes following a single RA injection, recovering spermatogonial differentiation and resulting in preleptotene and leptotene spermatocyte formation at 8 and 9 days postinjection, respectively (Raverdeau et al. 2012). STRA8-positive preleptotene spermatocytes were present in all tubules at 8 days post-RA treatment and SPO11 meiotic protein covalently bound to DSB (*Spo11*) transcript levels increased significantly 9 days post-RA (Raverdeau et al. 2012). These results suggest that a cell type other than the Sertoli cells synthesizes the RA required to induce *Stra8* expression within preleptotene spermatocytes and drive the subsequent rounds of spermatogonial differentiation. To further investigate whether RA is necessary for meiotic initiation, BMS493, a pan-RAR inverse agonist, was utilized to prevent RA signaling in the testis (Raverdeau et al. 2012). STRA8-positive preleptotene spermatocytes did not form following exposure to BMS493; however, when animals were exposed to a combination of RA and BMS493 meiotic initiation occurred normally (Raverdeau et al. 2012). Therefore, it was hypothesized that the

RAR/RXR heterodimer responsible for inducing *Stra8* expression in germ cells is activated by RA and blocking RAR signaling in germ cells prevented meiotic initiation (Raverdeau et al. 2012). This study provided evidence to suggest RA synthesis is not required within Sertoli cells to either form or induce *Stra8* expression in preleptotene spermatocytes and that germ cells provide the alternative source of RA needed during meiotic initiation.

6.5 Concluding Remarks

Our understanding of how germ cells make the decision to commit to oogenesis or spermatogenesis really began to take shape over the last decade. The core of this decision is in the timing of meiotic entry, occurring embryonically in females but after birth in males. The focus of research in this area has been on deciphering the mechanisms in place that allow for the sex-specific timing of meiosis and the identification of either inhibitory or inductive factors in the testis and ovary, respectively. There is a large body of evidence to support the hypothesis that RA is the key meiotic-inducing molecule in the embryonic ovary, with the activity of CYP26B1 blocking the RA signaling cascade and preventing the onset of meiosis in the fetal testis. Data generated by both genetic and chemical ablation studies demonstrate that in the absence of CYP26B1 or its activity, male germ cells in fetal testes display markers of meiotic entry. What is less clear is whether RA truly is the meiotic-inducing factor in the ovary. While chemical inhibition of the RA signaling pathway blocks female fetal germ cells from entering meiosis, genetic ablation of two of the three RA synthesis enzymes does not stop the expression of meiotic markers in the embryonic ovary. The challenge now will be to devise novel and more sensitive methods for analyzing potential meiotic-inducing factors in fetal gonads, including retinoids, and performing more complete genetic ablation studies to definitively prove that RA is required for the onset of meiosis in females.

There is significantly less doubt with regard to the action of RA in committing male germ cells to undergo spermatogenesis. The A-to-A1 spermatogonial transition, now known to be controlled by RA, marks an irreversible step toward the differentiation of male germ cells to become sperm. Elimination of RA within the testis, induced genetically, chemically, or via the diet, leads to the accumulation of undifferentiated spermatogonia within the seminiferous epithelium, indicating that spermatogonial differentiation does not take place in the absence of RA. Less is known about the direct role of RA at the onset of meiosis in testis. However, the observation that meiosis is blocked following elimination of the classical marker of an RA response, STRA8, would suggest that RA is also important for meiotic progression in males. Research efforts in this area should now turn to how RA levels are regulated within the seminiferous epithelium and how RA triggers the A-to-A1 transition at the molecular level.

References

Abu-Abed S, Dolle P, Metzger D, Beckett B, Chambon P, Petkovich M (2001) The retinoic acid-metabolizing enzyme, CYP26A1, is essential for normal hindbrain patterning, vertebral identity, and development of posterior structures. Genes Dev 15(2):226–240

Alnouti Y, Klaassen CD (2008) Tissue distribution, ontogeny, and regulation of aldehyde dehydrogenase (Aldh) enzymes mRNA by prototypical microsomal enzyme inducers in mice. Toxicol Sci 101(1):51–64. doi:10.1093/toxsci/kfm280, kfm280 [pii]

Amory JK, Muller CH, Shimshoni JA, Isoherranen N, Paik J, Moreb JS, Amory DW Sr, Evanoff R, Goldstein AS, Griswold MD (2011) Suppression of spermatogenesis by bisdichloroacetyldiamines is mediated by inhibition of testicular retinoic acid biosynthesis. J Androl 32(1):111–119. doi:10.2164/jandrol.110.010751, jandrol.110.010751 [pii]

Anderson EL, Baltus AE, Roepers-Gajadien HL, Hassold TJ, de Rooij DG, van Pelt AM, Page DC (2008) Stra8 and its inducer, retinoic acid, regulate meiotic initiation in both spermatogenesis and oogenesis in mice. Proc Natl Acad Sci U S A 105(39):14976–14980

Arnold SL, Kent T, Hogarth CA, Griswold MD, Amory JK, Isoherranen N (2015a) Pharmacological inhibition of ALDH1A in mice decreases all-trans retinoic acid concentrations in a tissue specific manner. Biochem Pharmacol 95(3):177–192. doi:10.1016/j.bcp.2015.03.001

Arnold SL, Kent T, Hogarth CA, Schlatt S, Prasad B, Haenisch M, Walsh T, Muller CH, Griswold MD, Amory JK, Isoherranen N (2015b) Importance of ALDH1A enzymes in determining human testicular retinoic acid concentrations. J Lipid Res 56(2):342–357. doi:10.1194/jlr.M054718

Baltus AE, Menke DB, Hu YC, Goodheart ML, Carpenter AE, de Rooij DG, Page DC (2006) In germ cells of mouse embryonic ovaries, the decision to enter meiosis precedes premeiotic DNA replication. Nat Genet 38(12):1430–1434

Barrios F, Filipponi D, Pellegrini M, Paronetto MP, Di Siena S, Geremia R, Rossi P, De Felici M, Jannini EA, Dolci S (2010) Opposing effects of retinoic acid and FGF9 on Nanos2 expression and meiotic entry of mouse germ cells. J Cell Sci 123(Pt 6):871–880. doi:10.1242/jcs.057968, jcs.057968 [pii]

Bastien J, Rochette-Egly C (2004) Nuclear retinoid receptors and the transcription of retinoid-target genes. Gene 328:1–16. doi:10.1016/j.gene.2003.12.005

Batten ML, Imanishi Y, Maeda T, Tu DC, Moise AR, Bronson D, Possin D, Van Gelder RN, Baehr W, Palczewski K (2004) Lecithin-retinol acyltransferase is essential for accumulation of all-trans-retinyl esters in the eye and in the liver. J Biol Chem 279(11):10422–10432. doi:10.1074/jbc.M312410200

Bowles J, Knight D, Smith C, Wilhelm D, Richman J, Mamiya S, Yashiro K, Chawengsaksophak K, Wilson MJ, Rossant J, Hamada H, Koopman P (2006) Retinoid signaling determines germ cell fate in mice. Science 312(5773):596–600. doi:10.1126/science.1125691, 1125691 [pii]

Bowles J, Feng CW, Spiller C, Davidson TL, Jackson A, Koopman P (2010) FGF9 suppresses meiosis and promotes male germ cell fate in mice. Dev Cell 19(3):440–449. doi:10.1016/j.devcel.2010.08.010

Brooks NL, van der Horst G (2003) Short-term effects of N′N-bis(dichloroacetyl)-1,8-octamethylenediamine (WIN 18446) on the testes, selected sperm parameters and fertility of male CBA mice. Lab Anim 37(4):363–373. doi:10.1258/002367703103051921

Buaas FW, Kirsh AL, Sharma M, McLean DJ, Morris JL, Griswold MD, de Rooij DG, Braun RE (2004) Plzf is required in adult male germ cells for stem cell self-renewal. Nat Genet 36(6):647–652. doi:10.1038/ng1366, ng1366 [pii]

Busada JT, Kaye EP, Renegar RH, Geyer CB (2014) Retinoic acid induces multiple hallmarks of the prospermatogonia-to-spermatogonia transition in the neonatal mouse. Biol Reprod 90(3):64. doi:10.1095/biolreprod.113.114645

Busada JT, Chappell VA, Niedenberger BA, Kaye EP, Keiper BD, Hogarth CA, Geyer CB (2015) Retinoic acid regulates Kit translation during spermatogonial differentiation in the mouse. Dev Biol 397(1):140–149. doi:10.1016/j.ydbio.2014.10.020

Chappell VA, Busada JT, Keiper BD, Geyer CB (2013) Translational activation of developmental messenger RNAs during neonatal mouse testis development. Biol Reprod 89(3):61. doi:10.1095/biolreprod.113.109819

Childs AJ, Cowan G, Kinnell HL, Anderson RA, Saunders PT (2011) Retinoic acid signalling and the control of meiotic entry in the human fetal gonad. PLoS One 6(6):e20249. doi:10.1371/journal.pone.0020249

Davis JC, Snyder EM, Hogarth CA, Small C, Griswold MD (2013) Induction of spermatogenic synchrony by retinoic acid in neonatal mice. Spermatogenesis 3(1):e23180. doi:10.4161/spmg.23180, 2012SPGEN0033R1 [pii]

Dokshin GA, Baltus AE, Eppig JJ, Page DC (2013) Oocyte differentiation is genetically dissociable from meiosis in mice. Nat Genet 45(8):877–883. doi:10.1038/ng.2672

Dong R, Yang S, Jiao J, Wang T, Shi H, Zhou L, Zhang Y, Wang D (2013) Characterization of Stra8 in Southern catfish (Silurus meridionalis): evidence for its role in meiotic initiation. BMC Mol Biol 14:11. doi:10.1186/1471-2199-14-11

Endo T, Romer KA, Anderson EL, Baltus AE, de Rooij DG, Page DC (2015) Periodic retinoic acid-STRA8 signaling intersects with periodic germ-cell competencies to regulate spermatogenesis. Proc Natl Acad Sci U S A 112(18):E2347–E2356. doi:10.1073/pnas.1505683112

Evans E, Hogarth C, Mitchell D, Griswold M (2014) Riding the spermatogenic wave: profiling gene expression within neonatal germ and Sertoli cells during a synchronized initial wave of spermatogenesis in mice. Biol Reprod 90(5):108. doi:10.1095/biolreprod.114.118034

Feng R, Fang L, Cheng Y, He X, Jiang W, Dong R, Shi H, Jiang D, Sun L, Wang D (2015) Retinoic acid homeostasis through aldh1a2 and cyp26a1 mediates meiotic entry in Nile tilapia (Oreochromis niloticus). Sci Rep 5:10131. doi:10.1038/srep10131

Gely-Pernot A, Raverdeau M, Celebi C, Dennefeld C, Feret B, Klopfenstein M, Yoshida S, Ghyselinck NB, Mark M (2012) Spermatogonia differentiation requires retinoic acid receptor gamma. Endocrinology 153(1):438–449. doi:10.1210/en.2011-1102, en.2011-1102 [pii]

Ghyselinck NB, Dupe V, Dierich A, Messaddeq N, Garnier JM, Rochette-Egly C, Chambon P, Mark M (1997) Role of the retinoic acid receptor beta (RARbeta) during mouse development. Int J Dev Biol 41(3):425–447

Giuili G, Tomljenovic A, Labrecque N, Oulad-Abdelghani M, Rassoulzadegan M, Cuzin F (2002) Murine spermatogonial stem cells: targeted transgene expression and purification in an active state. EMBO Rep 3(8):753–759. doi:10.1093/embo-reports/kvf149

Heller CG, Moore DJ, Paulsen CA (1961) Suppression of spermatogenesis and chronic toxicity in men by a new series of bis(dichloroacetyl) diamines. Toxicol Appl Pharmacol 3:1–11

Hermann BP, Mutoji KN, Velte EK, Ko D, Oatley JM, Geyer CB, McCarrey JR (2015) Transcriptional and translational heterogeneity among neonatal mouse spermatogonia. Biol Reprod 92(2):54. doi:10.1095/biolreprod.114.125757

Hogarth CA, Evanoff R, Snyder E, Kent T, Mitchell D, Small C, Amory JK, Griswold MD (2011) Suppression of Stra8 expression in the mouse gonad by WIN 18,446. Biol Reprod 84(5):957–965. doi:10.1095/biolreprod.110.088575, biolreprod.110.088575 [pii]

Hogarth CA, Evanoff R, Mitchell D, Kent T, Small C, Amory JK, Griswold MD (2013) Turning a spermatogenic wave into a tsunami: synchronizing murine spermatogenesis using WIN 18,446. Biol Reprod 88(2):40. doi:10.1095/biolreprod.112.105346, biolreprod.112.105346 [pii]

Hogarth CA, Arnold S, Kent T, Mitchell D, Isoherranen N, Griswold MD (2015) Processive pulses of retinoic acid propel asynchronous and continuous murine sperm production. Biol Reprod 92(2):37. doi:10.1095/biolreprod.114.126326

Horton C, Maden M (1995) Endogenous distribution of retinoids during normal development and teratogenesis in the mouse embryo. Dev Dyn 202(3):312–323. doi:10.1002/aja.1002020310

Huang HF, Hembree WC (1979) Spermatogenic response to vitamin A in vitamin A deficient rats. Biol Reprod 21(4):891–904
Ismail N, Morales C, Clermont Y (1990) Role of spermatogonia in the stage-synchronization of the seminiferous epithelium in vitamin-A-deficient rats. Am J Anat 188(1):57–63. doi:10.1002/aja.1001880107
Jorgensen A, Nielsen JE, Blomberg Jensen M, Graem N, Rajpert-De Meyts E (2012) Analysis of meiosis regulators in human gonads: a sexually dimorphic spatio-temporal expression pattern suggests involvement of DMRT1 in meiotic entry. Mol Hum Reprod 18(11):523–534. doi:10.1093/molehr/gas030
Kane MA, Chen N, Sparks S, Napoli JL (2005) Quantification of endogenous retinoic acid in limited biological samples by LC/MS/MS. Biochem J 388(Pt 1):363–369. doi:10.1042/BJ20041867
Kastner P, Mark M, Leid M, Gansmuller A, Chin W, Grondona JM, Decimo D, Krezel W, Dierich A, Chambon P (1996) Abnormal spermatogenesis in RXR beta mutant mice. Genes Dev 10(1):80–92
Koubova J, Menke DB, Zhou Q, Capel B, Griswold MD, Page DC (2006) Retinoic acid regulates sex-specific timing of meiotic initiation in mice. Proc Natl Acad Sci U S A 103(8):2474–2479
Krezel W, Dupe V, Mark M, Dierich A, Kastner P, Chambon P (1996) RXR gamma null mice are apparently normal and compound RXR alpha +/-/RXR beta -/-/RXR gamma -/- mutant mice are viable. Proc Natl Acad Sci U S A 93(17):9010–9014
Kumar S, Duester G (2011) SnapShot: retinoic acid signaling. Cell 147(6):1422–1422.e1421. doi:10.1016/j.cell.2011.11.034
Kumar S, Chatzi C, Brade T, Cunningham TJ, Zhao X, Duester G (2011) Sex-specific timing of meiotic initiation is regulated by Cyp26b1 independent of retinoic acid signalling. Nat Commun 2:151. doi:10.1038/ncomms1136, ncomms1136 [pii]
Kurlandsky SB, Gamble MV, Ramakrishnan R, Blaner WS (1995) Plasma delivery of retinoic acid to tissues in the rat. J Biol Chem 270(30):17850–17857
Lau EL, Lee MF, Chang CF (2013) Conserved sex-specific timing of meiotic initiation during sex differentiation in the protandrous black porgy Acanthopagrus schlegelii. Biol Reprod 88 (6):150. doi:10.1095/biolreprod.112.105536
Le Bouffant R, Guerquin MJ, Duquenne C, Frydman N, Coffigny H, Rouiller-Fabre V, Frydman R, Habert R, Livera G (2010) Meiosis initiation in the human ovary requires intrinsic retinoic acid synthesis. Hum Reprod 25(10):2579–2590. doi:10.1093/humrep/deq195
Lecureuil C, Fontaine I, Crepieux P, Guillou F (2002) Sertoli and granulosa cell-specific Cre recombinase activity in transgenic mice. Genesis 33(3):114–118. doi:10.1002/gene.10100
Li H, Clagett-Dame M (2009) Vitamin A deficiency blocks the initiation of meiosis of germ cells in the developing rat ovary in vivo. Biol Reprod 81(5):996–1001. doi:10.1095/biolreprod.109.078808, biolreprod.109.078808 [pii]
Li H, Palczewski K, Baehr W, Clagett-Dame M (2011) Vitamin A deficiency results in meiotic failure and accumulation of undifferentiated spermatogonia in prepubertal mouse testis. Biol Reprod 84(2):336–341. doi:10.1095/biolreprod.110.086157, biolreprod.110.086157 [pii]
Livera G, Rouiller-Fabre V, Valla J, Habert R (2000) Effects of retinoids on the meiosis in the fetal rat ovary in culture. Mol Cell Endocrinol 165(1-2):225–231
Lohnes D, Kastner P, Dierich A, Mark M, LeMeur M, Chambon P (1993) Function of retinoic acid receptor gamma in the mouse. Cell 73(4):643–658, doi:0092-8674(93)90246-M [pii]
Lufkin T, Lohnes D, Mark M, Dierich A, Gorry P, Gaub MP, LeMeur M, Chambon P (1993) High postnatal lethality and testis degeneration in retinoic acid receptor alpha mutant mice. Proc Natl Acad Sci U S A 90(15):7225–7229
MacLean G, Li H, Metzger D, Chambon P, Petkovich M (2007) Apoptotic extinction of germ cells in testes of Cyp26b1 knockout mice. Endocrinology 148(10):4560–4567. doi:10.1210/en.2007-0492, en.2007-0492 [pii]
Mark M, Jacobs H, Oulad-Abdelghani M, Dennefeld C, Feret B, Vernet N, Codreanu CA, Chambon P, Ghyselinck NB (2008) STRA8-deficient spermatocytes initiate, but fail to

complete, meiosis and undergo premature chromosome condensation. J Cell Sci 121 (Pt 19):3233–3242. doi:10.1242/jcs.035071, 121/19/3233 [pii]

Mic FA, Haselbeck RJ, Cuenca AE, Duester G (2002) Novel retinoic acid generating activities in the neural tube and heart identified by conditional rescue of Raldh2 null mutant mice. Development 129(9):2271–2282

Mic FA, Molotkov A, Benbrook DM, Duester G (2003) Retinoid activation of retinoic acid receptor but not retinoid X receptor is sufficient to rescue lethal defect in retinoic acid synthesis. Proc Natl Acad Sci U S A 100(12):7135–7140. doi:10.1073/pnas.1231422100

Mitranond V, Sobhon P, Tosukhowong P, Chindaduangrat W (1979) Cytological changes in the testes of vitamin-A-deficient rats. I. Quantitation of germinal cells in the seminiferous tubules. Acta Anat (Basel) 103(2):159–168

Morales C, Griswold MD (1987) Retinol-induced stage synchronization in seminiferous tubules of the rat. Endocrinology 121(1):432–434

Napoli JL (2012) Physiological insights into all-trans-retinoic acid biosynthesis. Biochim Biophys Acta 1821(1):152–167. doi:10.1016/j.bbalip.2011.05.004

Nishimura M, Naito S (2006) Tissue-specific mRNA expression profiles of human phase I metabolizing enzymes except for cytochrome P450 and phase II metabolizing enzymes. Drug Metab Pharmacokinet 21(5):357–374. doi:10.2133/dmpk.21.357 [pii]

Ohta K, Lin Y, Hogg N, Yamamoto M, Yamazaki Y (2010) Direct effects of retinoic acid on entry of fetal male germ cells into meiosis in mice. Biol Reprod 83(6):1056–1063. doi:10.1095/biolreprod.110.085787

Oulad-Abdelghani M, Bouillet P, Decimo D, Gansmuller A, Heyberger S, Dolle P, Bronner S, Lutz Y, Chambon P (1996) Characterization of a premeiotic germ cell-specific cytoplasmic protein encoded by Stra8, a novel retinoic acid-responsive gene. J Cell Biol 135(2):469–477

Pietruszko R, Abriola DP, Izaguirre G, Kikonyogo A, Dryjanski M, Ambroziak W (1999) Aldehyde inhibitors of aldehyde dehydrogenases. Adv Exp Med Biol 463:79–87

Piprek RP, Pecio A, Laskowska-Kaszub K, Kloc M, Kubiak JZ, Szymura JM (2013) Retinoic acid homeostasis regulates meiotic entry in developing anuran gonads and in Bidder's organ through Raldh2 and Cyp26b1 proteins. Mech Dev 130(11-12):613–627. doi:10.1016/j.mod.2013.09.001

Raverdeau M, Gely-Pernot A, Feret B, Dennefeld C, Benoit G, Davidson I, Chambon P, Mark M, Ghyselinck NB (2012) Retinoic acid induces Sertoli cell paracrine signals for spermatogonia differentiation but cell autonomously drives spermatocyte meiosis. Proc Natl Acad Sci U S A 109(41):16582–16587. doi:10.1073/pnas.1214936109, 1214936109 [pii]

Rodriguez-Mari A, Canestro C, BreMiller RA, Catchen JM, Yan YL, Postlethwait JH (2013) Retinoic acid metabolic genes, meiosis, and gonadal sex differentiation in zebrafish. PLoS One 8(9):e73951. doi:10.1371/journal.pone.0073951

Rossant J, Zirngibl R, Cado D, Shago M, Giguere V (1991) Expression of a retinoic acid response element-hsplacZ transgene defines specific domains of transcriptional activity during mouse embryogenesis. Genes Dev 5(8):1333–1344

Ruiz A, Ghyselinck NB, Mata N, Nusinowitz S, Lloyd M, Dennefeld C, Chambon P, Bok D (2007) Somatic ablation of the Lrat gene in the mouse retinal pigment epithelium drastically reduces its retinoid storage. Invest Ophthalmol Vis Sci 48(12):5377–5387. doi:10.1167/iovs.07-0673

Saba R, Wu Q, Saga Y (2014) CYP26B1 promotes male germ cell differentiation by suppressing STRA8-dependent meiotic and STRA8-independent mitotic pathways. Dev Biol 389 (2):173–181. doi:10.1016/j.ydbio.2014.02.013

Sakai Y, Drager UC (2010) Detection of retinoic acid catabolism with reporter systems and by in situ hybridization for CYP26 enzymes. Methods Mol Biol 652:277–294. doi:10.1007/978-1-60327-325-1_16

Sakai Y, Meno C, Fujii H, Nishino J, Shiratori H, Saijoh Y, Rossant J, Hamada H (2001) The retinoic acid-inactivating enzyme CYP26 is essential for establishing an uneven distribution of retinoic acid along the anterio-posterior axis within the mouse embryo. Genes Dev 15 (2):213–225

Sanz E, Yang L, Su T, Morris DR, McKnight GS, Amieux PS (2009) Cell-type-specific isolation of ribosome-associated mRNA from complex tissues. Proc Natl Acad Sci U S A 106 (33):13939–13944. doi:10.1073/pnas.0907143106, 0907143106 [pii]

Shima JE, McLean DJ, McCarrey JR, Griswold MD (2004) The murine testicular transcriptome: characterizing gene expression in the testis during the progression of spermatogenesis. Biol Reprod 71(1):319–330

Smith JE, Milch PO, Muto Y, Goodman DS (1973) The plasma transport and metabolism of retinoic acid in the rat. Biochem J 132(4):821–827

Smith CA, Roeszler KN, Bowles J, Koopman P, Sinclair AH (2008) Onset of meiosis in the chicken embryo; evidence of a role for retinoic acid. BMC Dev Biol 8:85. doi:10.1186/1471-213X-8-85

Snyder EM, Small C, Griswold MD (2010) Retinoic acid availability drives the asynchronous initiation of spermatogonial differentiation in the mouse. Biol Reprod 83(5):783–790. doi:10.1095/biolreprod.110.085811, biolreprod.110.085811 [pii]

Snyder EM, Davis JC, Zhou Q, Evanoff R, Griswold MD (2011) Exposure to retinoic acid in the neonatal but not adult mouse results in synchronous spermatogenesis. Biol Reprod 84 (5):886–893. doi:10.1095/biolreprod.110.089755, biolreprod.110.089755 [pii]

Stoppie P, Borgers M, Borghgraef P, Dillen L, Goossens J, Sanz G, Szel H, Van Hove C, Van Nyen G, Nobels G, Vanden Bossche H, Venet M, Willemsens G, Van Wauwe J (2000) R115866 inhibits all-trans-retinoic acid metabolism and exerts retinoidal effects in rodents. J Pharmacol Exp Ther 293(1):304–312

Sugimoto R, Nabeshima Y, Yoshida S (2012) Retinoic acid metabolism links the periodical differentiation of germ cells with the cycle of Sertoli cells in mouse seminiferous epithelium. Mech Dev 128(11–12):610–624. doi:10.1016/j.mod.2011.12.003, S0925-4773(11)00116-X [pii]

Suzuki A, Saga Y (2008) Nanos2 suppresses meiosis and promotes male germ cell differentiation. Genes Dev 22(4):430–435. doi:10.1101/gad.1612708

Suzuki A, Tsuda M, Saga Y (2007) Functional redundancy among Nanos proteins and a distinct role of Nanos2 during male germ cell development. Development 134(1):77–83. doi:10.1242/dev.02697

Tang F, Yao X, Zhu H, Mu H, Niu Z, Yu M, Yang C, Peng S, Li G, Hua J (2014) Expression pattern of Ngn3 in dairy goat testis and its function in promoting meiosis by upregulating Stra8. Cell Prolif 47(1):38–47. doi:10.1111/cpr.12077

Tedesco M, La Sala G, Barbagallo F, De Felici M, Farini D (2009) STRA8 shuttles between nucleus and cytoplasm and displays transcriptional activity. J Biol Chem 284 (51):35781–35793. doi:10.1074/jbc.M109.056481, M109.056481 [pii]

Thatcher JE, Isoherranen N (2009) The role of CYP26 enzymes in retinoic acid clearance. Expert Opin Drug Metab Toxicol 5(8):875–886. doi:10.1517/17425250903032681

Tong MH, Yang QE, Davis JC, Griswold MD (2013) Retinol dehydrogenase 10 is indispensible for spermatogenesis in juvenile males. Proc Natl Acad Sci U S A 110(2):543–548. doi:10.1073/pnas.1214883110, 1214883110 [pii]

Topletz AR, Thatcher JE, Zelter A, Lutz JD, Tay S, Nelson WL, Isoherranen N (2012) Comparison of the function and expression of CYP26A1 and CYP26B1, the two retinoic acid hydroxylases. Biochem Pharmacol 83(1):149–163. doi:10.1016/j.bcp.2011.10.007

Tsuda M, Sasaoka Y, Kiso M, Abe K, Haraguchi S, Kobayashi S, Saga Y (2003) Conserved role of nanos proteins in germ cell development. Science 301(5637):1239–1241. doi:10.1126/science.1085222, 301/5637/1239 [pii]

Tsuda M, Kiso M, Saga Y (2006) Implication of nanos2-3′UTR in the expression and function of nanos2. Mech Dev 123(6):440–449. doi:10.1016/j.mod.2006.04.002

Van Beek ME, Meistrich ML (1990) A method for quantifying synchrony in testes of rats treated with vitamin A deprivation and readministration. Biol Reprod 42(3):424–431

van Pelt AM, de Rooij DG (1990) Synchronization of the seminiferous epithelium after vitamin A replacement in vitamin A-deficient mice. Biol Reprod 43(3):363–367 van Pelt AM, van Dissel-Emiliani FM, Gaemers IC, van der Burg MJ, Tanke HJ, de Rooij DG (1995) Characteristics of A spermatogonia and preleptotene spermatocytes in the vitamin A-deficient rat testis. Biol Reprod 53(3):570–578

Vernet N, Dennefeld C, Guillou F, Chambon P, Ghyselinck NB, Mark M (2006a) Prepubertal testis development relies on retinoic acid but not rexinoid receptors in Sertoli cells. EMBO J 25 (24):5816–5825. doi:10.1038/sj.emboj.7601447, 7601447 [pii]

Vernet N, Dennefeld C, Rochette-Egly C, Oulad-Abdelghani M, Chambon P, Ghyselinck NB, Mark M (2006b) Retinoic acid metabolism and signaling pathways in the adult and developing mouse testis. Endocrinology 147(1):96–110. doi:10.1210/en.2005-0953, en.2005-0953 [pii]

Vernet N, Dennefeld C, Klopfenstein M, Ruiz A, Bok D, Ghyselinck NB, Mark M (2008) Retinoid X receptor beta (RXRB) expression in Sertoli cells controls cholesterol homeostasis and spermiation. Reproduction 136(5):619–626. doi:10.1530/REP-08-0235, REP-08-0235 [pii]

Wallacides A, Chesnel A, Chardard D, Flament S, Dumond H (2009) Evidence for a conserved role of retinoic acid in urodele amphibian meiosis onset. Dev Dyn 238(6):1389–1398. doi:10.1002/dvdy.21948

Wang Z, Kim KH (1993) Vitamin A-deficient testis germ cells are arrested at the end of S phase of the cell cycle: a molecular study of the origin of synchronous spermatogenesis in regenerated seminiferous tubules. Biol Reprod 48(5):1157–1165

Wang X, Chen T, Song C, Gao B, Zhang Y (2014) Cloning and expression characteristics of the pig Stra8 gene. Int J Mol Sci 15(7):12480–12494. doi:10.3390/ijms150712480

Western PS, Miles DC, van den Bergen JA, Burton M, Sinclair AH (2008) Dynamic regulation of mitotic arrest in fetal male germ cells. Stem Cells 26(2):339–347. doi:10.1634/stemcells.2007-0622, 2007-0622 [pii]

Wu JW, Wang RY, Guo QS, Xu C (2008) Expression of the retinoic acid-metabolizing enzymes RALDH2 and CYP26b1 during mouse postnatal testis development. Asian J Androl 10 (4):569–576. doi:10.1111/j.1745-7262.2008.00408.x

Yao X, Tang F, Yu M, Zhu H, Chu Z, Li M, Liu W, Hua J, Peng S (2014) Expression profile of Nanos2 gene in dairy goat and its inhibitory effect on Stra8 during meiosis. Cell Prolif 47 (5):396–405. doi:10.1111/cpr.12128

Yashiro K, Zhao X, Uehara M, Yamashita K, Nishijima M, Nishino J, Saijoh Y, Sakai Y, Hamada H (2004) Regulation of retinoic acid distribution is required for proximodistal patterning and outgrowth of the developing mouse limb. Dev Cell 6(3):411–422

Zhou Q, Li Y, Nie R, Friel P, Mitchell D, Evanoff RM, Pouchnik D, Banasik B, McCarrey JR, Small C, Griswold MD (2008a) Expression of stimulated by retinoic acid gene 8 (Stra8) and maturation of murine gonocytes and spermatogonia induced by retinoic acid in vitro. Biol Reprod 78(3):537–545

Zhou Q, Nie R, Li Y, Friel P, Mitchell D, Hess RA, Small C, Griswold MD (2008b) Expression of stimulated by retinoic acid gene 8 (Stra8) in spermatogenic cells induced by retinoic acid: an in vivo study in vitamin A-sufficient postnatal murine testes. Biol Reprod 79(1):35–42

Zhuang YH, Sainio EL, Sainio P, Vedeckis WV, Ylikomi T, Tuohimaa P (1995) Distribution of all-trans-retinoic acid in normal and vitamin A deficient mice: correlation to retinoic acid receptors in different tissues of normal mice. Gen Comp Endocrinol 100(2):170–178. doi:10.1006/gcen.1995.1146

Chapter 7
Ovarian Folliculogenesis

Nitzan Rimon-Dahari, Lia Yerushalmi-Heinemann, Liat Alyagor, and Nava Dekel

Abstract The ovary, the female gonad, serves as the source for the germ cells as well as the major supplier of steroid sex hormones. During embryonic development, the primordial germ cells (PGCs) are specified, migrate to the site of the future gonad, and proliferate, forming structures of germ cells nests, which will eventually break down to generate the primordial follicles (PMFs). Each PMF contains an oocyte arrested at the first prophase of meiosis, surrounded by a flattened layer of somatic pre-granulosa cells. Most of the PMFs are kept dormant and only a selected population is activated to join the growing pool of follicles in a process regulated by both intra- and extra-oocyte factors. The PMFs will further develop into secondary pre-antral follicles, a stage which depends on bidirectional communication between the oocyte and the surrounding somatic cells. Many of the signaling molecules involved in this dialog belong to the transforming growth factor β (TGF-β) superfamily. As the follicle continues to develop, a cavity called antrum is formed. The resulting antral follicles relay on the pituitary gonadotropins, follicle-stimulating hormone (FSH), and luteinizing hormone (LH) for their development. Most of the follicles undergo atretic degeneration and only a subset of the antral follicles, known as the dominant follicles, will reach the preovulatory stage at each reproductive cycle, respond to LH, and subsequently ovulate, releasing a fertilizable oocyte. The remaining somatic cells in the raptured follicle will undergo terminal differentiation and form the corpus luteum, which secretes progesterone necessary to maintain pregnancy.

7.1 Introduction

The ovary, the female gonad, serves as the source for the germ cells as well as the major supplier of steroid sex hormones. During early embryonic development, at 7.25–7.5 *dpc* (days *post coitum*) primordial germ cells (PGCs), which are the origin

N. Rimon-Dahari • L. Yerushalmi-Heinemann • L. Alyagor • N. Dekel (✉)
Department of Biological Regulation, Weizmann Institute of Science, Rehovot, Israel
e-mail: nava.dekel@weizmann.ac.il

R.P. Piprek (ed.), *Molecular Mechanisms of Cell Differentiation in Gonad Development*, Results and Problems in Cell Differentiation 58,
DOI 10.1007/978-3-319-31973-5_7

of both oocytes and spermatozoa, are first identified in the extraembryonic mesoderm as a small cluster of alkaline phosphatase (AP)-positive cells. These cells migrate into the posterior portion of the primitive streak (Ginsburg et al. 1990; reviewed by McLaren 2003) and move by active locomotion directly into the endoderm of the developing hindgut (Anderson et al. 2000). Beginning at 9.0–9.5 *dpc*, PGCs undergo an active, directed migration from the dorsal axis of the hindgut toward the developing genital ridges (Molyneaux et al. 2001), where they reach at around 10.5–11.5 *dpc*. Throughout their migration, PGCs proliferate rapidly (Ginsburg et al. 1990). In female mice, the PGCs continue to proliferate until 13.5 *dpc*, forming germ cells clusters known as cysts or nests. At approximately 13.5 *dpc*, they embark on meiosis, thus becoming oocytes (Jagarlamudi and Rajkovic 2012). The oocytes precede their meiotic division up to the diplotene of the first prophase. Between 17.5 *dpc* to postnatal day 5 *dpp* (*days post partum*), the germ cell cysts break down to form primordial follicles (PMFs), a process accompanied by a massive loss of germ cells (Pepling and Spradling 2001; Menke et al. 2003).

Each PMF contains a prophase-arrested oocyte surrounded by a flattened epithelium (pre-granulosa cells) that will differentiate to form granulosa cells (GC). The PMFs constitute the ovarian quiescent follicle reserve. From this reserve, follicles will be recruited to the growing pool. The transition from a primordial to a primary follicle is characterized by a morphological change in the GC that turn from flattened to cuboidal. During further development to secondary follicle, the oocyte continues to grow, the GC proliferate, and an additional layer of theca cells forms outside the basement membrane that surrounds the follicle. These stages of follicle growth are gonadotropin independent, but require a complex bidirectional communication between the oocyte and the somatic cells (Eppig 2001; reviewed by Edson et al. 2009).

In the subsequent stages of folliculogenesis, small fluid-filled cavities are formed within the follicle, ultimately joining to a single cavity, known as the antrum, thus forming the antral follicle. The antrum defines two separate populations of GC: the cumulus GC, which are adjacent to the oocyte, and the mural GC that line the follicle wall and serve as the major source for steroid hormones (reviewed by Edson et al. 2009). Folliculogenesis at this stage depends on the presence of the pituitary gonadotropins, follicle-stimulating hormone (FSH), and luteinizing hormone (LH). FSH is required for GC survival, proliferation, estradiol production, and LH receptor expression (reviewed by Edson et al. 2009; Richards and Pangas 2010a). Most of the antral follicles will undergo atretic degeneration, whereas only a subset of them will continue to develop to the preovulatory stage. The mural GC of these selected follicles express high concentration of LH receptors, subjecting them to respond to the preovulatory LH surge, which activates a sequence of events culminating in ovulation of the dominant follicles. These events include oocyte meiotic resumption, cumulus expansion, follicle rupture, and the release of a cumulus–oocyte complex that contains a fertilizable oocyte. Once the oocyte is released, the remaining granulosa and theca cells undergo terminal differentiation to create the corpus luteum (CL) (reviewed by Edson et al. 2009). The CL is essential for establishing and maintaining pregnancy, mainly through secretion of progesterone. In case pregnancy has not occurred, the CL degenerates (Stocco et al. 2007).

7.2 Primordial Germ Cells' Specification and Migration

Prior to gastrulation at 6.25 *dpc*, the primordial germ cells (PGCs) split off from the proximal region of the pluripotent epiblast and arise, at around 7.25 *dpc*, as a cluster in the extraembryonic mesoderm (Ginsburg et al. 1990). This process of PGC formation and specification depends on bone morphogenetic protein (BMP) family signaling from the extraembryonic ectoderm (ExE) and visceral endoderm (VE), which surround the epiblast cells. Specifically, BMP4 and BMP8B originate from the ExE and BMP2 is secreted from the VE. Loss of any of these genes results in a reduction or even the absence of PGC (reviewed by Edson et al. 2009). This BMP signal is elicited through Smad1/5/8, which translocate into the nucleus with the common mediator, Smad4 (Hayashi et al. 2007; Ohinata et al. 2009; Saitou et al. 2012). At ~5.5 *dpc*, BMP4 signaling activates the transcriptional regulators PRDM1 (PRDI-BF1-RIZ domain containing 1; BLIMP1) and PRDM14 in the most proximal epiblast cells. Both BLIMP1 and PRDM14 are essential for PGC specification. BLIMP1 is required for the repression of the somatic program (for example, by turning off Hox gene expression), and both BLIMP1 and PRDM14 are involved in the re-expression of factors of pluripotency (such as *Sox2*, *Oct4*, and *Nanog*) and in preparation for epigenetic reprogramming (Ohinata et al. 2009; Saitou et al. 2012). Interestingly, the abovementioned BMP4 signaling from the ExE is antagonized by signals from the anterior visceral endoderm (AVE). BMP8b from the ExE restricts AVE development, thereby allowing the BMP4 signaling. In addition, *Wnt3* expression in the epiblast was shown to be essential for the responsiveness of the epiblast cells to BMP4, thus allowing them to determine the germ cell fate (Ohinata et al. 2009).

One of the earliest genes induced by BMPs is *Ifitm3* (interferon-induced transmembrane protein 3; Fragilis), which serves as an early marker for PGCs but does not have an essential role in their differentiation. Additional early markers of PGCs which are not essential for their formation are ALPL (alkaline phosphatase) and STELLA (reviewed by Edson et al. 2009). Starting at ~7.5 *dpc*, the specified PGCs migrate from the base of the yolk sac through the hindgut and dorsal mesentery to colonize the genital ridge (10.5 *dpc*; Jagarlamudi and Rajkovic 2012; Saitou et al. 2012). This process probably relays on chemoattractant receptor ligand interactions. One such example is the Kit ligand (reviewed by Edson et al. 2009).

Between 7.5 and 8.5 *dpc*, PGCs express markers of pluripotency (including POU5F1, SOX2, NANOG) and undergo chromatin changes, resembling the patterns of pluripotent stem cells (specifically, increasing the levels of H3K27me3 and erasing the H3K9me2 methylation). During their migration, they arrest in G2 stage of the cell cycle and transiently become transcriptionally quiescent (Seki et al. 2007). Additional epigenetic events in PGCs are the genome-wide DNA demethylation, which is completed in both sexes by E13.5, erasure of imprinting, and reactivation of the inactive X chromosome in females. Starting ~9.5 *dpc*, throughout their migration, PGCs undergo active proliferation (for a detailed review, see Saitou et al. 2012). Interestingly, it was recently demonstrated that this process of PGC expansion and migration is done in an incoherent mode, which involves physical mixing of the progenies of dividing PGCs as they migrate before their allocation to different gonads (Reizel et al. 2012).

Upon their arrival at the genital ridge where the future gonad will form, PGCs are subjected to a global change in gene expression, in which the pluripotency program is turned off and sexual specific genes are turned on (reviewed by Hu et al. 2015). Interestingly, the sexual fate of the PGCs is determined by the sexual identity of the fetal gonad to which the PGCs migrate and not by the PGCs' own identity (Gill et al. 2011). In addition, it was recently shown that the ability of PGCs to undertake sexual specialization requires an active developmental transition to allow them to respond to the surrounding gonad, a process called germ cell licensing (Gill et al. 2011). This process was recently shown to require the formation and signaling of the genital ridges, which induce deleted in azoospermia-like (DAZL) expression in the PGCs (Hu et al. 2015).

7.3 Male and Female Germ Cell Specification

After the entrance of the PGCs to the genital ridges, the bipotential gonad will continue its development in a sexually dependent manner, to form either ovaries or testes. The somatic supporting cells in the gonad, which arise within the urogenital ridge from a common mesodermal progenitor (Jameson et al. 2012), play a critical role in this process. In the XY gonad, expression of *Sry* gene triggers upregulation of *Sox9* and *Fgf9*, which activate the male pathway and repress signals that promote the female pathway (i.e., *Wnt4*/*Rspo1* and β-catenin). In the XX gonad, in the absence of *Sry*, WNT4 and RSPO1 promote the female pathway by maintaining β-catenin signaling. Supporting cell precursors commit to granulosa cell fate and control ovarian development by promoting theca cell differentiation and regulating meiotic entry in germ cells (Lin and Capel 2015).

The time of germ cells' entry into meiosis is sex specific, as XY germ cells arrest in mitosis and do not divide again until postnatally as spermatogonia, whereas the XX germ cells continue to divide and then enter meiosis at approximately 13.5 *dpc* (Koubova et al. 2006; Lei and Spradling 2013). Therefore, a complex network of signals is required to coordinate this process; the mesonephroi adjacent to the developing ovary and testes contain several aldehyde dehydrogenases that convert retinaldehyde to all-trans-retinoic acid (RA). The enzyme CYP26B1, which degrades RA, is downregulated in the somatic cells of the developing ovary but is upregulated in the somatic cells of the developing testis. Therefore, RA is accessible only to XX germ cells and not to XY germ cells. The binding of RA to its receptor RAR in XX germ cells induces *Stra8* expression, which, in turn, induces SYCP3, a component of the synaptonemal complex. SYCP3 is stably translated in the presence of DAZL and loaded on chromosomes, facilitating the pairing of homologous chromosomes, which is required in meiosis (reviewed by Edson et al. 2009).

7.4 Primordial Follicle Formation

After the female PGCs enter the genital ridges and differentiate to oogonia, they proliferate with incomplete cytokinesis, creating clusters (syncytia) of cells connected by intercellular bridges, through which organelles and mitochondria can move. The germ cells in the cysts are synchronized; thus, the number of cells in each cyst is in the power of two (Epifano and Dean 2002; Lei and Spradling 2013; Pepling and Spradling 1998). However, it has been recently demonstrated that these cysts are partially segmented prior to meiosis into smaller cysts and aggregate with cysts from different PGCs to form a nest. From 13.5 *dpc*, the germ cells within the clusters enter meiosis in an anterior to posterior wave and become oocytes (Lei and Spradling 2013; Menke et al. 2003). Oocytes remain arrested in the diplotene stage of meiosis I, and only after the female reaches sexual maturation, selected oocytes will resume their meiotic division and ovulate (see hereinafter).

Between 17.5 *dpc* to postnatal day 5 *dpp*, cyst clusters begin to break down to form PMFs, a process which is accompanied with massive cell lost (Pepling and Spradling 2001). A single layer of squamous somatic pre-granulosa cells encapsulates individual oocytes in the process of follicle formation.

Several factors and signaling pathways were shown to affect follicle assembly:

- Factor in the germline α (FIGLA), a germ cell-specific bHLH transcription factor. Primordial follicles fail to form in *Figlα*-null mice, suggesting it is required for initial follicle formation (Soyal et al. 2000).
- The transcription factor Forkhead box L2 (FOXL2). In mice null for *Foxl2*, the somatic pre-granulosa cells fail to develop around the growing oocyte, which initiate its growth prematurely and thus undergo atresia (Uda et al. 2004).
- Members of the TGF-β superfamily. Activin A was reported to increase the number of PMFs (reviewed by Knight et al. 2012); AMH inhibits PMF assembly (Nilsson et al. 2011); Gremlin, a BMP antagonist, was also shown to affect PMF assembly (Myers et al. 2011); however, there is no sufficient indication that BMPs play a role in cyst breakdown (Pangas 2012).
- Neurotrophins (NTs) and their tyrosine kinase receptor (NTRK). The amount of PMFs is decreased in the absence of nerve growth factor (NGF) and its receptor, NTRK1, and in the absence of NTRK2, which is the receptor of neurotrophin 4 (NTF4) and brain-derived neurotrophic factor (BDNF) (Dissen et al. 2001; Kerr et al. 2009).

- KITL/KIT signaling. Inhibition of KIT in ovary organ cultures caused a reduction in cyst breakdown and an increase in oocyte numbers, accompanied with reduced cell death. Kit ligand increased levels of phosphorylated MAP kinase. Thus, KIT signaling has a role in cyst breakdown, which may be mediated by MAP kinase (Jones and Pepling 2013).
- Notch signaling. In the neonatal murine ovary, the Notch ligands, Jagged1 and Jagged2, are expressed in germ cells, whereas the Notch receptors, Notch1 and Notch2, are expressed in pre-granulosa cells. Disruption of Jagged1 or Notch2 leads to altered follicular composition and aberrant follicle formation such as multi-oocytic follicles, which indicates a failure in cyst breakdown (Vanorny et al. 2014).
- Steroid hormones. An inhibitory involvement of estrogen and progesterone has also been suggested in follicle formation (Chen et al. 2007; Kezele and Skinner 2003).

Although there is mounting evidence to the contrary (Tilly and Telfer 2009), the current consensus is that female germ cells are not formed later in life; therefore, the PMFs constitute a fixed pool known as the ovarian reserve.

7.5 Primordial Follicle Activation

The PMFs are expected to endure one of three possible fates: (1) remain quiescent, (2) be activated and join the growing pool of follicles and then either go through atresia or finish their developmental process up till ovulation, and (3) die directly from their dormant state (Zheng et al. 2012). Most of the PMFs are kept dormant and only a selected subpopulation is activated and joins the growing pool of follicles throughout life until menopause. The transition from primordial to primary follicle is marked histologically by a change in the GC morphology, from squamous to cuboidal and the initiation of oocyte growth (Pedersen and Hannah 1968). By utilizing transgenic mice models, several studies from the last years have demonstrated that inhibitory signals from the oocyte itself and signals from the somatic cells maintain the PMF pool in its dormant state.

7.5.1 Suppression of PMF Activation by Intra-oocyte Inhibitory Signals

The most central signaling pathway identified to this date that controls PMFs activation is the **PI3K pathway**: oocyte deletion of *Pten*, a PI3K-negative regulator, caused activation of the entire PMF pool and thus premature ovarian failure (POF) (Reddy et al. 2008). A similar phenotype was also observed in oocyte deletion of FOXO3a, a substrate of AKT and a central downstream target of the of the PI3K pathway (Castrillon et al. 2003). In addition, hyper-phosphorylation and nuclear export of FOXO3a were observed in *Pten*-null oocytes (John et al. 2009). Therefore, it was suggested that FOXO3a acts downstream of PI3K/PTEN–AKT to negatively regulate follicular activation (Adhikari et al. 2009a; John et al. 2009; Reddy et al. 2010).

Overactivation of the PMF pool was also demonstrated in oocyte-specific knockout of either TSC1 or TSC2 (Adhikari et al. 2009a, b). The TSC1–TSC2 complex suppresses the activation of mTORC1. Indeed, it was shown that in both mutants, elevated activation of mTORC1 increased the activation of p70 S6 kinase 1 (S6K1)–ribosomal protein S6 (rpS6) signaling that promoted protein translation and ribosome biogenesis in oocytes (Reddy et al. 2010). Interestingly, although both PTEN and TSC1/2 regulate the activation of the PMFs by negatively regulating S6K1 and rpS6 activation in oocytes, it appears that they do so by phosphorylating different residues of S6K1. Thus, their similar regulation on the PMF pool is carried in distinct ways (Adhikari et al. 2009b).

An important regulator of the PI3K signaling pathway is PDK1, which can phosphorylate AKT and S6K1. In mice with specific deletion of PDK1 in oocytes, PMFs were depleted directly from the dormant state, causing premature ovarian failure (POF) in early adulthood. This indicates that the PDK1–Akt–S6K1–rpS6 pathway also plays a central role in maintaining the survival of PMFs (Reddy et al. 2009, 2010).

Premature ovarian failure was also observed in mice null to $p27^{\mathrm{Kip1}}$, an inhibitor of the cell cycle, which is reported to be expressed in the nuclei of oocytes of primordial, primary, and secondary follicles (Rajareddy et al. 2007). However, loss of *p27* and *Foxo3a* leads to synergistically accelerated primordial follicle activation, indicating that they maintain PMF dormancy by different signaling pathways (Rajareddy et al. 2007).

7.5.2 Activation of PMFs by Extra-oocyte Signals

A recently published article shed light on the important role that the pre-GC of the PMF play in its activation. Inhibition of mTORC1 signaling in the GC of PMFs prevents their differentiation, arresting the dormant oocytes in their quiescent state, leading to oocyte death. Overactivation of mTORC1 signaling in GC of PMF by specific TSC1 deletion accelerates their differentiation and causes premature activation of all PMF. It was further demonstrated that the differentiation of the GC in the PMF triggers the awakening of dormant oocytes through KIT ligand (KITL), which subsequently activates the PI3K signaling pathway in the oocytes. Hence, a communication network exists between the GC and the germ cells, which is based on mTORC1-KITL signaling that initiates the KIT-PI3K signaling pathway in oocytes, inducing their activation (Zhang et al. 2014).

This model accounts for a previous report that KITL can activate PMF (Hutt et al. 2006). Additional reports demonstrated the influence of other paracrine factors on primordial follicle activation, including leukemia inhibiting factor (LIF) (Nilsson et al. 2002), basic fibroblast growth factor (BFGF) (Nilsson et al. 2001), keratinocyte growth factor (KGF) (Kezele et al. 2005), platelet-derived growth factor (PDGF) (Nilsson et al. 2006), connective tissue growth factor (CTGF) (Schindler et al. 2010), and Neurotrophins (Nilsson et al. 2009). Interestingly, at least part of these factors are known to activate the PI3K signaling cascade (Pangas 2012).

Another paracrine factor, which affects PMF activation, is anti-Müllerian hormone (AMH), a TGF-β superfamily member, which is expressed by GC of growing follicles. In AMH-null mice, the PMF pool was depleted in an earlier stage, indicating AMH suppresses or prevents PMFs from being activated (Durlinger et al. 1999). In addition, aberrant premature activation of PMFs was also recently indicated in the absence of Notch signaling. However, a link between the Notch and the PI3K signaling is yet to be examined (Vanorny et al. 2014). Finally, BMP4 (Nilsson and Skinner 2003) and BMP7 (Lee et al. 2004) have also been shown to be important for the transition from primordial to primary follicles (Pangas 2012).

Interestingly, although most PMFs lie dormant until they are activated, PMFs in the medulla of the ovary are activated immediately after their assembly, shortly after birth. It was recently demonstrated that these follicles contain GC originating from surface epithelium cells of the gonad at 11.5 *dpc*. In contrast, primordial follicles in the cortex of the ovary, which are activated during adult life, are populated by GC that arise from the surface epithelium around birth (Mork et al. 2012).

7.6 Pre-antral Folliculogenesis

Primary follicles further develop to form secondary follicles, which contain oocytes in midgrowth stages surrounded by two or more layers of GC. A basement membrane forms around the outermost GC layer and an additional layer of somatic cells called theca cells encapsulates the follicle. These cells differentiate into two cell layers, the theca interna and externa, which have ultrastructural features, such as mitochondria with tubular cristae, which allows them to function as the source of androgens for the conversion to estrogens executed by GC (Magoffin 2005). The inner theca layer (theca interna) becomes increasingly vascularized as the follicle continues to develop, while the GC layer remains avascular (Knight et al. 2012). The oocyte relays on the GC to support its development, but it also has a critical role in regulating the rate of follicular growth by controlling proliferation and differentiation of granulosa and theca cells (Edson et al. 2009; Eppig et al. 2002; Liu et al. 2006, 2015). Indeed, although growth of pre-antral follicles is gonadotropin independent, it is governed by bidirectional communication between the oocyte and the surrounding somatic cells. Many of the signaling molecules involved in this dialog belong to the transforming growth factor β (TGF-β) superfamily (Harlow et al. 2002; Ingman and Robertson 2009; Knight et al. 2012; Matzuk and Burns 2012; Pangas and Matzuk 2004; reviewed by Richards and Pangas 2010a).

7.6.1 TGF-β Superfamily and Their Role in Pre-antral Folliculogenesis

The TGF-β superfamily consists of a structurally conserved group of proteins ubiquitously expressed, which function as extracellular ligands in various physiological processes, such as cell proliferation, differentiation, apoptosis, and cell migration (Wotton and Massague 2000). This superfamily is comprised of several subfamilies, which include TGF-β, bone morphogenetic protein (BMP), growth and differentiation factor (GDF), activin/inhibin, glial cell-derived neurotrophic factor (GDNF), and several additional members such as AMH, also known as Müllerian-inhibiting substance (MIS) (Knight et al. 2012). Members of the TGF-β family bind to specific receptor complexes composed of type I and type II serine/threonine receptor kinases, initiating a signaling cascade that causes the nuclear translocation of the SMAD proteins, resulting in transcriptional activation of target genes. Based on their activated signaling pathways, the proteins can be divided into two major groups: the BMPs, which activate SMAD1-5-8, and activins/TGF-βs/GDF9, which activate SMAD2–3 (Chang et al. 2002; Myers et al. 2009). Several members of the TGF-β superfamily are produced by different cell types in the follicle: GC produce activins, theca cells BMP4 and BMP7, granulosa and theca cells both express TGF-β, and oocytes express GDF9 and BMP15 (Oktem and Urman 2010). The latter two factors, which are selectively expressed by oocytes, have a central role in the dialog between the oocyte and the surrounding somatic cells. In GDF9-null mice, folliculogenesis is blocked at the primary stage (Dong et al. 1996). Furthermore, it was recently demonstrated that GDF9 originating from the oocyte is responsible for the production of protein signals by the GC, which in turn induce theca cell differentiation (Liu et al. 2015). BMP15 is not required during pre-antral folliculogenesis in mice, though BMP15 and GDF9 may have redundant roles (Yan et al. 2001). However, the importance of GDF9 and BMP15 in pre-antral folliculogenesis may be species specific, as BMP15 missense mutation in sheep causes similar phenotypes as in mouse GDF9-null mutation (Hanrahan et al. 2004).

Activin A produced by GC was shown to promote pre-antral follicle growth (Oktem and Urman 2010). TGF-β promotes proliferation of GC and pre-antral follicle growths (Liu et al. 1999) as well as progesterone production and FSH-induced estrogen synthesis (Dodson and Schomberg 1987). It also appears that TGF-β suppresses steroidogenesis in thecal cells from most species (Juengel et al. 2005).

Another signaling pathway involved in the progress from primary to secondary follicles involves the neurotrophins NTF5 and BDNF, which signal thorough NTRK2.

In null *Ntrk2* mice, as well as in double knockout of *Ntf5* and *Bdnf*, there is a significant reduction in the number of secondary follicles, which is apparently caused by a signaling pathway other than the GDF9 (Edson et al. 2009).

In addition to paracrine signaling, direct cell-to-cell communication via gap junctions also plays a significant role in pre-antral growth, as evident from mice with deficiencies of connexin43 (CX43) and connexin37 (CX37), two of the core proteins that form the gap junctions in the ovary (Granot and Dekel 1994; Teilmann 2005). In CX43-deficient ovaries, follicles fail to develop beyond the primary stage (Ackert et al. 2001), whereas CX37-deficient mice fail to develop mature (Graafian) follicles (Simon et al. 1997).

7.7 Folliculogenesis Toward Ovulation

As the follicle continues to develop, an antrum is formed and separates the GC population into two functional cell types, the mural GC, which are responsible for steroidogenesis, and the cumulus cells, which are adjacent to the oocyte (reviewed by Edson et al. 2009). Antral follicles relay on the pituitary gonadotropins FSH and LH for their development and ovulation. FSH is required for antral follicle survival, GC proliferation, LH receptor expression, and estradiol production. FSH binds to its G-protein coupled receptor, FSHR, and initiates the classical adenylyl cyclase (AC)/cAMP/protein kinase A (PKA) pathway that results in phosphorylation and activation of the transcription factor cAMP-response element-binding protein (CREB), thus upregulating a number of target genes such as aromatase and the LH receptor. FSH also activates several PKA-independent pathways such as AKT via PI3K and SRC tyrosine kinase-dependent pathway (Ulloa-Aguirre et al. 2007). FSH and estradiol affect GC proliferation by modulating their cell cycle (Edson et al. 2009).

7.7.1 *Follicular Steroidogenesis: The Two-Cell Two-Gonadotropin System*

The theca cells are capable of de novo production of androgens from cholesterol. The key enzymes of the synthesis, namely StAR (transporter of cholesterol to the inner mitochondrial membrane), CYP11A1 (which converts the cholesterol to pregnenolone), CYP17A1 (which converts pregnenolone to dehydroepiandrosterone—DHEA), and 3β-hydroxysteroid dehydrogenase (3β-HSD, which converts DHEA into androstenedione), are under LH control (reviewed by Magoffin 2005). However, the enzymes necessary for the conversion of androstenedione to estradiol, CYP19A1 (aromatase) and 17β-hydroxysteroid dehydrogenase (17β-HSD), are expressed by the GC and are regulated by FSH (reviewed by

Magoffin 2005). Theca cells express LH receptor from the secondary follicle stage, before estradiol is necessary for follicular growth. Therefore, androgen biosynthesis is modulated in pre-antral and small antral follicles by secretion of inhibitory factors such as activins by the GC as well as autocrine inhibition of factors secreted by the theca cells themselves, including TGF-β, TGF-α, and FGF7 (Magoffin 2005; reviewed by Edson et al. 2009). In addition to signaling pathways activated by FSH and LH, locally produced factors such as insulin-like growth factor 1 (IGF-1) also take part in the GC proliferation and differentiation and thus are necessary for the survival of antral follicles and promoting ovulation (Ulloa-Aguirre et al. 2007).

Importantly, the majority of follicles in the growing pool will undergo atresia. Only a subset population of the antral follicles will reach the preovulatory stage and will respond to LH by activating downstream signal transduction pathways mediated by AC and cAMP and will eventually ovulated (Salomon et al. 1977). These are the dominant follicles.

7.8 Dominant Follicle Selection

In mice, dominant follicle (DF) selection is known to occur 36 h after the release of FSH (Salomon et al. 1977). The DF possess several characteristic features. It produces more estrogen than the subordinate follicle (SF) and by that is more effective in eliciting the negative feedback on FSH production, which results in reduced circulation of FSH. Additionally, the DF contains more GC and a higher level of FSHR as compared to the SF (Louvet and Vaitukaitis 1976; Rao et al. 1978; Van Santbrink et al. 1995). Upon stimulation by FSH, GC express anti-apoptotic factors such as x-linked inhibitor of apoptosis (XIAP) and flice-like inhibitory protein (FLIP). In the absence of FSH, GC express the death receptor and ligand FAS and FAS ligand (Jiang et al. 2003). Therefore, the DF is able to thrive in a declining FSH environment, while the SF cannot. Furthermore, the estrogen production in DF mediates the effect of FSH on the expression of LH receptors on granulosa cells. While SF only have LH receptors on theca cells, DF become less dependent on FSH and acquire further responsiveness to LH by expressing LH receptors on GC. In addition, insulin-like growth factor II mRNA levels and bioavailability are increased in the DF, stimulating steroidogenesis (Baerwald et al. 2012). A leading hypothesis regarding the mechanism of DF selection suggests that its advantage over the SF is achieved by the development of richer vascularization that allows the exposure to higher gonadotropin concentrations.

7.8.1 The Involvement of Angiogenesis in the Ovary and DF Selection

The ovary is one of the only organs in the adult that conserves the capacity to remodel vascular networks under tight regulation during each reproductive cycle. Primordial and primary follicles lack their own vasculature and use diffusion for oxygen and metabolite transfer. During their development, follicles acquire a two-capillary vascular network located in the theca interna and externa layers. At each given time point, hundreds of growing follicles recruit and remodel a vascular bed necessary to sustain their changing metabolic needs (Brown and Russell 2014). However, the follicular blood vessels will never penetrate to the GC layers due to a barrier of basement membrane that is located between them and the theca cells. After ovulation, upon the formation of the CL, the basement membrane breaks down and a massive angiogenic process, involving penetration of blood vessels into the inner parts of the follicle, takes place.

Several lines of evidence from different animal model systems suggest that DFs as compared to SFs are highly vascularized, thus being effectively exposed to higher gonadotropins level. Specifically, increased vascularization of individual follicles in rhesus monkeys resulted in preferential delivery of gonadotropins (Zeleznik et al. 1981). Moreover, in vivo studies in cattle showed that the maintenance of follicular vasculature and appropriate blood supply to follicles are essential for the establishment of follicular dominance (Acosta 2007). In mares, differential blood flow and blood flow velocity are early parameters for identification of DF (Acosta and Miyamoto 2004).

Several known pro- and anti-angiogenic factors were shown to participate during folliculogenesis and may be involved in DF selection:

- Vascular endothelial growth factor A (VEGFA) is expressed in ovaries of numerous species including human, bovine, ovine, monkeys, and rodents. It has been shown that follicular growth caused by hormonal stimulation induces inner hypoxia that causes an elevation in VEGFA expression (Neeman et al. 1997). Along this line, injection of VEGFA to rat ovaries increased the number of antral follicles (Danforth et al. 2003) and elevated the number of ovulating oocytes (Iijima et al. 2005). Additionally, the blockage of VEGF receptor 2 (VEGFR2) by neutralizing antibodies attenuated the formation of pre-antral follicles (Zimmermann et al. 2003). VEGF receptor 1 (VEGFR1), which delivers the VEGF signal intracellularly, is upregulated in theca cells of large follicles after pregnant mare serum gonadotropin (PMSG) administration in gilts (Shimizu et al. 2002). Interestingly, it has been shown that soluble Flt (sFlt), the truncated and soluble form of the same receptor, which inhibits VEGF activity, is expressed by the bovine dominant follicle (Macias et al. 2012).
- The PI3K/AKT signaling pathway, known to mediate angiogenesis, proliferation metabolism, and other cellular processes, also seems to participate in

follicular growth and DF selection. In cattle, the PI3K/AKT signaling pathway is known to be an early marker for follicular dominance (Ryan et al. 2007).

- Angiopoietin 2 (ANG2), a competitive inhibitor of the pro-angiogenic factor angiopoietin 1 (ANG1) for the TIE2 receptor, is expressed in large but not small follicles of the ewe (Chowdhury et al. 2010).
- Thrombospondin 1 (TSP1), an anti-angiogenic factor that induces endothelial apoptosis, known to promote follicular atresia in the primate and in the rat ovary (Garside et al. 2010; Thomas et al. 2008), is expressed in pre-antral and early-antral follicles of the rat (Petrik et al. 2002). The knockout of TSP1 receptor, CD36, in mice resulted in hypervascularized ovaries, increased levels of VEGF and VEGFR2, an elevation on the number of preovulatory follicles, a decrease in the number of CL, and a reduction in the number of offspring (Osz et al. 2014). Treatment with a mimetic protein of TSP1 in marmoset monkeys did not block the emergence of DF, but reduced the pre-antral and early-antral follicles survival and angiogenesis (Garside et al. 2010).

7.9 Ovulation

Dominant follicle selection is followed by ovulation. This multistep process occurs in response to the preovulatory LH surge, which is generated by the hypothalamus/pituitary/ovary feedback system as follows. As mentioned above, FSH induces antral follicle growth, associated with elevated estradiol production. The high estradiol levels produced by the follicular GC increase the pulses of the hypothalamic gonadotropin-releasing hormone (GnRH) that trigger the LH surge (reviewed by Richards and Pangas 2010b).

The ovulatory response to LH consists of the resumption of meiosis in the oocyte, the expansion/mucification of the cumulus, the rupture of the follicle, and the release of a cumulus–oocyte complex (COC) that contains a fertilizable oocyte. Once the oocyte is released, the residual follicle cells, the granulosa and thecal cells, undergo reprogramming of terminal differentiation to create the corpus luteum (CL) through a process defined as luteinization. Consequently, the gene expression program initiated by FSH is turned off and is replaced by genes that control matrix formation and luteinization (reviewed by Edson et al. 2009; Richards and Pangas 2010a, b).

Several transcriptional regulators, necessary for ovulation, are induced in response to LH. The expression of the progesterone receptor (PR) is rapidly induced in the GC of the preovulatory follicle and is responsible for the expression of several key enzymes, including Adamts1, cathepsin L (*Ctsl*) proteases, endothelin 2 (*Edn2*), and Snap25, which regulates vesicle secretion and seems to be important for the release of potent cytokines from the GC as part of the inflammatory and immune-like response of ovulation (see hereinafter). Additional transcriptional regulators that mediate the ovulatory response to LH include C/EBP, the nuclear receptor NR5A2 also known as liver receptor homolog 1 (LRH1), and NR5A1 also known as steroidogenic factor 1 (SF1) (reviewed by Edson et al. 2009).

The binding of LH to its G-protein coupled receptor in the GC activates the AC/cAMP/PKA pathway, as well as the PI3K/AKT and RAS signaling cascades, each of which is critical for ovulation (Richards and Pangas 2010b). Downstream to the AC/cAMP/PKA pathway, the CG express the EGF-like factors amphiregulin (AREG), beta-cellulin (BTC), and epiregulin (EREG) in GC. These EGF-like factors undergo proteolytic cleavage—by ADAM17/TACE in porcine (Yamashita and Shimada 2012) or by matrix metalloproteinases MMP2 and MMP9 in mice (Light and Hammes 2015). Subsequently, the EGFR is activated and initiates the ERK1/2 pathway, thus inducing expression of downstream target genes responsible for cumulus expansion, steroidogenesis, oocyte maturation, and ovulation. Disruption of the EGF ligand/receptor signaling pathway in mice caused impaired cumulus expansion and inhibited ovulation (Hsieh et al. 2007). Disruption of ERK1/2 in GC resulted in failure of the follicles to ovulate and luteinize (Fan et al. 2009). Interestingly, ERK1/2 were also shown to be essential in the female mouse pituitary to initiate the LH surge (Bliss et al. 2009).

7.9.1 Oocyte Resumption of Meiosis

As described extensively above, in rodents the germ cells, oogonia, enter the meiotic prophase asynchronously starting on 13.5 *dpc* and proceed until 18.5 *dpc*, through leptotene, zygotene, pachytene, and diplotene, a stage at which they arrest. Meiotically arrested oocytes are characterized by the large nucleus known as "germinal vesicle" (GV). Alongside with the progress of folliculogenesis described above, the oocyte grows in size but persists in the GV stage throughout infancy and for variable periods beyond puberty. In sexually mature females, during each reproductive cycle a number of fully grown oocytes, characteristic of the species, reenter meiosis as manifested by "germinal vesicle breakdown" (GVBD). These oocytes complete the first round of meiosis with extrusion of the first polar body (PBI) and immediately thereafter progress to the second meiotic metaphase. Resumption of meiosis and its progression to the metaphase of the second meiotic division is usually referred to as "oocyte maturation." At this stage, the meiotic process is arrested again, being completed upon fertilization, by the extrusion of the second polar body (reviewed by Tsafriri and Dekel 2010).

Fully grown oocytes remain arrested in the first prophase due to low activity of the maturation-promoting factor (MPF), which is a complex consisting of cyclin-dependent kinase 1 CDK1 and cyclin B that regulates the G2/M transition of the cell cycle. Low activity of CDK1 in meiotically arrested oocytes is maintained by relatively high concentrations of cAMP. Under these conditions, the active PKA stimulates the Wee1 kinase, inactivating, in turn, the Cdc25B phosphatase, required for CDK1 dephosphorylation and its resultant activation. In addition, in meiotically arrested oocytes, APC^{Cdh1} constantly degrades cyclin B1 (reviewed by Adhikari and Liu 2014).

The intra-oocyte cAMP is contributed by the somatic compartment of the ovarian follicle that supplies this inhibitor to the oocyte by cell-to-cell communication mediated by gap junctions (Dekel et al. 1981). In addition, local production of cAMP by the oocyte itself has been suggested (Mehlmann et al. 2002; Kalinowski et al. 2004). Maintenance of the high levels of cAMP in the oocyte is further executed by the prevention of its degradation by phosphodiesterase 3A (PDE3A). Inhibition of PDE3A is achieved by cGMP, which is synthesized in the cumulus cells by the guanylyl cyclase natriuretic peptide receptor 2 (NPR2) and enters the oocyte via gap junctions (reviewed by Zhang and Xia 2012). Interestingly, NPR2, which is the natriuretic peptide precursor type C (NPPC) receptor, is activated in response to NPPC generated by mural granulosa cells (Zhang et al. 2010).

Activation of CDK1 at the onset of meiosis is dependent on the reduction in intra-oocyte cAMP. The preovulatory LH surge closes the gap junctions, thus preventing entrance of both cAMP and cGMP to the oocyte. Cessation of its supply, together with its increased hydrolysis by PDE3A, lowers the intra-oocyte cAMP levels (Tsafriri and Dekel 2010). The resulting inactivated PKA can no longer inhibit Cdc25B, which now dephosphorylates CDK1, turning on its catalytic activity (Adhikari and Liu 2014).

7.9.2 Cumulus Expansion

Cumulus expansion, also known as cumulus mucification, represents the secretion and apposition of hyaluronan-rich extracellular matrix in the COC (Dekel and Kraicer 1978; Eppig 1980). Hyaluronan, which forms the structural backbone of the COC extracellular matrix, is the product of hyaluronan synthase 2 (Has2), the expression of which is upregulated by LH. This gonadotropin also elevates the expression of TNF-α-induced protein 6 (Tnfaip6), which links the HA covalently to the heavy chain of serum-derived inter-α-trypsin inhibitor (IαI) upon its interaction with multimers of pentraxin 3 (PTX3). Another protein that interacts with HA to stabilize the COC matrix is Versican, a complex matricellular proteoglycan, which is also rapidly induced in GC by the LH surge (Dunning et al. 2015). Prostaglandin synthase 2 (PTGS2; also known as COX2), the rate-limiting enzyme in the synthesis of prostaglandins, is also essential for cumulus expansion and may function upstream of TNFAIP6 (reviewed by Edson et al. 2009).

Interestingly, many of these matrix-associated genes that are upregulated in the follicle after the LH surge are also found at sites of inflammation. In fact, ovulation and inflammation share several common attributes, including increased vascular permeability and prostaglandin synthesis as well as massive generation of reactive oxygen species (ROS), which were shown to be indispensable for ovulation (Shkolnik et al. 2011). Moreover, in the process of follicle rupture, there is enhanced ovarian blood flow and participation of proteinases, bradykinins, and histamine, which are all features of acute inflammatory reaction. Furthermore, the involvement of immune cells such as dendritic cells was also demonstrated in the ovulatory process (Cohen-Fredarow et al. 2014).

7.10 Corpus Luteum Formation

Ovulation culminates in the release of an ovum from the ovarian follicle to the oviduct, where it can be fertilized by sperm. The remaining granulosa and theca cells in the ruptured follicle undergo terminal differentiation to luteinizing cells and create the CL. The CL is essential for establishing and maintaining pregnancy mainly through secretion of progesterone. In case pregnancy has not occurred, the CL degenerates.

As opposed to the FSH-stimulated GC proliferation, LH-induced luteinization is characterized by cell cycle arrest. This involves the elevated expression of endogenous Cdk inhibitors (e.g., p21cip1 and p27kip1) and loss of positive cell cycle regulators, including cyclins and Cdk2 (Stocco et al. 2007).

During luteinization, the cellular responsiveness to external hormonal signals is altered due to changes in the expression pattern of the associated receptors. The FSHR is silenced and an activation and thereafter desensitization of the LHR occur. A sustained stimulation of the prolactin (PRL) receptor (PRLR) as well as a short rapid increase in progesterone receptor (PR) expression and a shift in the expression of the estrogen receptor (ER) from the predominance of ERβ to that of ERα also take place (Stocco et al. 2007).

The formation of the CL involves extensive tissue remodeling, including changes in the ECM that allows cell migration and neovascularization. The CL is extremely vascularized and has one of the highest rates of blood flow in the organism. The dense capillary network that is formed is important for efficient supply of nutrients, hormones, and lipoprotein-bound cholesterol to the luteal cells facilitates the efficient production of progesterone and its secretion. The process of CL vascularization is greatly controlled by VEGF, which promotes endothelial cell migration and proliferation, as well as angiopoietins (Ang), which appear to be important for vessel maturation and/or stabilization (Ferrara et al. 1998; Wulff et al. 2000). The transcription factor induced by LH, C/EBPβ , has a central role in the process of luteinization, as evident from the fact that in response to gonadotropin stimulation, C/EBPβ-null mice ovulate fertilizable eggs, but luteinization does not take place, and the CL is not formed (Sterneck et al. 1997). Double knockout of CEBPα and CEBPβ specifically in GC (*Cebpα/β*$^{gc-/-}$) resulted in sterility, and microarray analyses identified numerous gene targets, including genes associated with neovascularization and endothelial cell functions (Fan et al. 2011).

As mentioned previously, the central role of the CL is the production and secretion of steroids, mainly progesterone. The major source of cholesterol used for steroid production is plasma lipoproteins (HDL and LDL) (Azhar and Menon

1982; Tureck and Strauss 1982). The HDL receptor scavenger receptor class B type I (SR-BI, *Scarb1*) is considered the central receptor responsible for cholesterol uptake in these cells. However, SR-BI-null mice synthesize normal amounts of progesterone during pseudopregnancy, suggesting the presence of additional pathways, such as de novo synthesis (Miettinen et al. 2001; Trigatti et al. 1999). Cholesterol is transported to the inner mitochondrial membrane by the sterol carrier protein 2 (SCP2) and steroidogenic acute regulatory protein (StAR). There, it is converted to pregenolone by cholesterol side chain cleavage P450 (P450scc), which is later converted in the endoplasmic reticulum to progesterone by 3β hydroxysteroid dehydrogenase (HSD). Both enzymes are highly expressed in the CL throughout pregnancy (Bachelot and Binart 2005; Stocco et al. 2007).

In mice, progesterone levels drop before parturition due to expression of the enzyme 20αHSD that catabolizes progesterone into the inactive progestin, 20α-DHP, and the enzyme P450c26, which catalyzes the conversion of cholesterol to 26-hydroxycholesterol thus reduces cholesterol availability to be turned into progesterone (Stocco et al. 2007).

In rodents and humans, the CL also produces androgens and estrogens in addition to progesterone. The weak androgen, androstenedione, is produced from progesterone by the enzyme P45017α-hydroxylase/C17–20 lyase (P450c17 or CYP17). Androstenedione is converted to estradiol by Cyp19 and 17βHSD-7 (Stocco et al. 2007).

To allow normal reproductive function, the CL must be regularly eliminated. In rodents, regression of the CL occurs in two phases. The first phase is functional and is associated with marked decrease in progesterone production. This is followed by a structural regression phase, characterized by programmed cell death of the luteal cells. The CL decreases in size and weight and eventually becomes a scar known as corpus albicans (Stocco et al. 2007).

7.11 Conclusion

The ovary is a highly dynamic organ, which undergoes massive developmental changes up to sexual maturation, which are followed by periodic functional modifications at each reproductive cycle. The ovary executes a diverse repertoire of major roles. It houses the entire reservoir of the female oocytes encapsulated in dormant PMFs constantly generating growing follicles from which selected ones will ovulate a mature oocyte ready for fertilization. It also serves as the source for female sex hormones, thus functioning as an endocrine gland. We herein present the strict and complex regulatory mechanisms that orchestrate the successful accomplishment of these functions, as schematically described in Fig. 7.1.

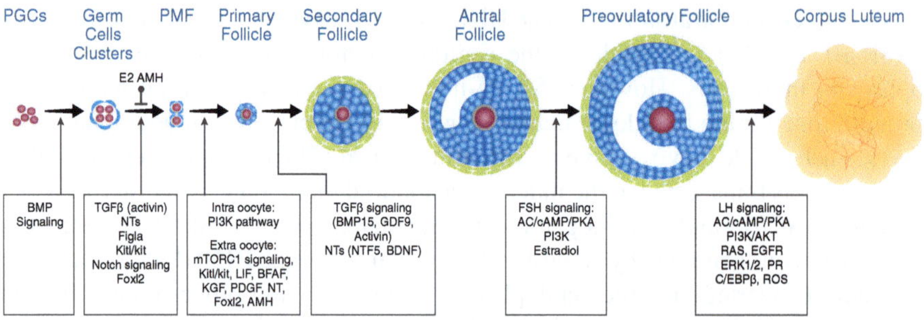

Fig. 7.1 Ovarian folliculogenesis in the mouse: a schematic presentation of the stages in ovarian folliculogenesis and the major factors and signaling pathways involved in this process. Primordial germ cells (PGCs) are specified by the BMP signaling pathway, then proliferate and migrate to the gonad, and form germ cells clusters. These clusters break down to form primordial follicles (PMFs), each containing a single oocyte (*purple*) surrounded by a flattened layer of pre-granulosa cells (*blue*). This process is negatively regulated by estradiol (E2) and AMH and is controlled by activin, neurotrophins (NTs), Figlα, Kitl/kit, FOXL2, and notch signaling (see text for details). Most of the PMFs remain quiescent, but a selected subpopulation is activated to further develop into primary follicles in a process regulated by both intra- and extra-oocyte factors (see text for details). The transition from primary to secondary follicle is accompanied by GC proliferation and encapsulation of the follicle by the theca cells (*green*). Secondary pre-antral follicles depend greatly on TGF-β signaling. Antral follicles depend on the gonadotropin hormones FSH and LH for their survival and ovulation. The LH surge triggers ovulation, while the remaining GC undergo terminal differentiation to form the CL (see text for details)

References

Ackert CL, Gittens JE, O'Brien MJ et al (2001) Intercellular communication via connexin43 gap junctions is required for ovarian folliculogenesis in the mouse. Dev Biol 233:258–270. doi:10.1006/dbio.2001.0216

Acosta TJ (2007) Studies of follicular vascularity associated with follicle selection and ovulation in cattle. J Reprod Dev 53:39–44. doi:10.1262/jrd.18153

Acosta TJ, Miyamoto A (2004) Vascular control of ovarian function: ovulation, corpus luteum formation and regression. Anim Reprod Sci 82–83:127–140. doi:10.1016/j.anireprosci.2004.04.022

Adhikari D, Liu K (2014) The regulation of maturation promoting factor during prophase I arrest and meiotic entry in mammalian oocytes. Mol Cell Endocrinol 382:480–487. doi:10.1016/j.mce.2013.07.027

Adhikari D, Flohr G, Gorre N et al (2009a) Disruption of Tsc2 in oocytes leads to overactivation of the entire pool of primordial follicles. Mol Hum Reprod 15:765–770. doi:10.1093/molehr/gap092

Adhikari D, Zheng W, Shen Y et al (2009b) Tsc/mTORC1 signaling in oocytes governs the quiescence and activation of primordial follicles. Hum Mol Genet 19:397–410. doi:10.1093/hmg/ddp483

Anderson R, Copeland TK, Scholer H et al (2000) The onset of germ cell migration in the mouse embryo. Mech Dev 91:61–68. doi:10.1016/S0925-4773(99)00271-3

Azhar S, Menon KM (1982) Receptor mediated gonadotropin action in gonadal tissues: relationship between blood cholesterol levels and gonadotropin stimulated steroidogenesis in isolated rat Leydig and luteal cells. J Steroid Biochem 16:175–184. doi:10.1016/0022-4731(82)90165-0

Bachelot A, Binart N (2005) Corpus luteum development: lessons from genetic models in mice. Curr Top Dev Biol 68:49–84. doi:10.1016/S0070-2153(05)68003-9
Baerwald AR, Adams GP, Pierson RA (2012) Ovarian antral folliculogenesis during the human menstrual cycle: a review. Hum Reprod Update 18:73–91. doi:10.1093/humupd/dmr039
Bliss SP, Miller A, Navratil AM et al (2009) ERK signaling in the pituitary is required for female but not male fertility. Mol Endocrinol 23:1092–1101. doi:10.1210/me.2009-0030
Brown HM, Russell DL (2014) Blood and lymphatic vasculature in the ovary: development, function and disease. Hum Reprod Update 20:29–39. doi:10.1093/humupd/dmt049
Castrillon DH, Miao L, Kollipara R et al (2003) Suppression of ovarian follicle activation in mice by the transcription factor Foxo3a. Science 301:215–218. doi:10.1126/science.1086336
Chang H, Brown CW, Matzuk MM (2002) Genetic analysis of the mammalian transforming growth factor-beta superfamily. Endocr Rev 23:787–823. doi:10.1210/er.2002-0003
Chen Y, Jefferson WN, Newbold RR et al (2007) Estradiol, progesterone, and genistein inhibit oocyte nest breakdown and primordial follicle assembly in the neonatal mouse ovary in vitro and in vivo. Endocrinology 148:3580–3590. doi:10.1210/en.2007-0088
Chowdhury MWH, Scaramuzzi RJ, Wheeler-Jones CPD, Khalid M (2010) The expression of angiogenic growth factors and their receptors in ovarian follicles throughout the estrous cycle in the ewe. Theriogenology 73:856–872
Cohen-Fredarow A, Tadmor A, Raz T et al (2014) Ovarian dendritic cells act as a double-edged pro-ovulatory and anti-inflammatory sword. Mol Endocrinol 28:1039–1054. doi:10.1210/me.2013-1400
Danforth DR, Arbogast LK, Ghosh S et al (2003) Vascular endothelial growth factor stimulates preantral follicle growth in the rat ovary. Biol Reprod 68:1736–1741. doi:10.1095/biolreprod.101.000679
Dekel N, Kraicer PF (1978) Induction in vitro of mucification of rat cumulus oophorus by gonadotrophins and adenosine 3′,5′-monophosphate. Endocrinology 102:1797–1802
Dekel N, Lawrence TS, Gilula NB, Beers WH (1981) Modulation of cell-to-cell communication in the cumulus-oocyte complex and the regulation of oocyte maturation by LH. Dev Biol 86:356–362. doi:10.1016/0012-1606(81)90193-7
Dissen GA, Romero C, Hirshfield AN, Ojeda SR (2001) Nerve growth factor is required for early follicular development in the mammalian ovary. Endocrinology 142:2078–2086. doi:10.1210/endo.142.5.8126
Dodson WC, Schomberg DW (1987) The effect of transforming growth factor-beta on follicle-stimulating hormone-induced differentiation of cultured rat granulosa cells. Endocrinology 120(2):512–516
Dong J, Albertini DF, Nishimori K et al (1996) Growth differentiation factor 9 is required during early ovarian folliculogenesis. Nature 383(6600):531–535
Dunning KR, Watson LN, Zhang VJ et al (2015) Activation of mouse cumulus-oocyte complex maturation in vitro through EGF-like activity of versican. Biol Reprod 92:1–10. doi:10.1095/biolreprod.114.127274
Durlinger AL, Kramer P, Karels B et al (1999) Control of primordial follicle recruitment by anti-Mullerian hormone in the mouse ovary. Endocrinology 140:5789–5796. doi:10.1210/en.140.12.5789
Edson MA, Nagaraja AK, Matzuk MM et al (2009) The mammalian ovary from genesis to revelation. Endocr Rev 30:624–712. doi:10.1210/er.2009-0012
Epifano O, Dean J (2002) Genetic control of early folliculogenesis in mice. Trends Endocrinol Metab 13:169–173. doi:10.1016/S1043-2760(02)00576-3
Eppig JJ (1980) Regulation of cumulus oophorus expansion by gonadotropins in vivo and in vitro. Biol Reprod 23:545–552. doi:10.1095/biolreprod23.3.545
Eppig JJ (2001) Review Oocyte control of ovarian follicular development and function in mammals. Reproduction 122(6):829–838

Eppig JJ, Wigglesworth K, Pendola FL (2002) The mammalian oocyte orchestrates the rate of ovarian follicular development. Proc Natl Acad Sci U S A 99:2890–2894. doi:10.1073/pnas.052658699

Fan H, Liu Z, Shimada M et al (2009) MAPK3/1 (ERK1/2) in ovarian granulosa cells are essential for female fertility. Science 324:1–4

Fan H-Y, Liu Z, Johnson PF, Richards JS (2011) CCAAT/enhancer-binding proteins (C/EBP)-α and -β are essential for ovulation, luteinization, and the expression of key target genes. Mol Endocrinol 25:253–268. doi:10.1210/me.2010-0318

Ferrara N, Chen H, Davis-Smyth T et al (1998) Vascular endothelial growth factor is essential for corpus luteum angiogenesis. Nat Med 4:336–340

Garside SA, Henkin J, Morris KD et al (2010) A thrombospondin-mimetic peptide, ABT-898, suppresses angiogenesis and promotes follicular atresia in pre- and early-antral follicles in vivo. Endocrinology 151:5905–5915. doi:10.1210/en.2010-0283

Gill ME, Hu Y-C, Lin Y, Page DC (2011) Licensing of gametogenesis, dependent on RNA binding protein DAZL, as a gateway to sexual differentiation of fetal germ cells. Proc Natl Acad Sci U S A 108:7443–7448. doi:10.1016/j.juro.2011.10.069

Ginsburg M, Snow MH, McLaren A (1990) Primordial germ cells in the mouse embryo during gastrulation. Development 110:521–528

Granot I, Dekel N (1994) Phosphorylation and expression of connexin-43 ovarian gap junction protein are regulated by luteinizing hormone. J Biol Chem 269(48):30502–30509

Hanrahan JP, Gregan SM, Mulsant P et al (2004) Mutations in the genes for oocyte-derived growth factors GDF9 and BMP15 are associated with both increased ovulation rate and sterility in Cambridge and Belclare sheep (Ovis aries). Biol Reprod 70:900–909. doi:10.1095/biolreprod.103.023093

Harlow CR, Davidson L, Burns KH et al (2002) FSH and TGF-beta superfamily members regulate granulosa cell connective tissue growth factor gene expression in vitro and in vivo. Endocrinology 143:3316–3325. doi:10.1210/en.2001-211389

Hayashi K, de Sousa Lopes SM, Surani MA (2007) Germ cell specification in mice. Science 316(5823):394–396

Hsieh M, Lee D, Panigone S et al (2007) Luteinizing hormone-dependent activation of the epidermal growth factor network is essential for ovulation. Mol Cell Biol 27:1914–1924. doi:10.1128/MCB.01919-06

Hu Y-C, Nicholls PK, Soh YQS et al (2015) Licensing of primordial germ cells for gametogenesis depends on genital ridge signaling. PLoS Genet 11:e1005019. doi:10.1371/journal.pgen.1005019

Hutt KJ, McLaughlin EA, Holland MK (2006) Kit ligand and c-Kit have diverse roles during mammalian oogenesis and folliculogenesis. Mol Hum Reprod 12:61–69. doi:10.1093/molehr/gal010

Iijima K, Jiang J-Y, Shimizu T et al (2005) Acceleration of follicular development by administration of vascular endothelial growth factor in cycling female rats. J Reprod Dev 51:161–168. doi:10.1262/jrd.51.161

Ingman WV, Robertson SA (2009) The essential roles of TGFB1 in reproduction. Cytokine Growth Factor Rev 20:233–239. doi:10.1016/j.cytogfr.2009.05.003

Jagarlamudi K, Rajkovic A (2012) Molecular and cellular endocrinology oogenesis: transcriptional regulators and mouse models. Mol Cell Endocrinol 356:31–39. doi:10.1016/j.mce.2011.07.049

Jameson SA, Natarajan A, Cool J et al (2012) Temporal transcriptional profiling of somatic and germ cells reveals biased lineage priming of sexual fate in the fetal mouse gonad. PLoS Genet 8:e1002575. doi:10.1371/journal.pgen.1002575

Jiang J-Y, Cheung CKM, Wang Y, Tsang BK (2003) Regulation of cell death and cell survival gene expression during ovarian follicular development and atresia. Front Biosci 8:222–237

John GB, Gallardo TD, Shirley LJ, Castrillon DH (2009) Foxo3 is a PI3K-dependent molecular switch controlling the initiation of oocyte growth. Dev Biol 321:197–204. doi:10.1016/j.ydbio.2008.06.017.Foxo3

Jones RL, Pepling ME (2013) KIT signaling regulates primordial follicle formation in the neonatal mouse ovary. Dev Biol 382:186–197. doi:10.1016/j.ydbio.2013.06.030

Juengel JL, Mcnatty KP, Animal W et al (2005) The role of proteins of the transforming growth factor-b superfamily in the intraovarian regulation of follicular development. Hum Reprod Update 11:144–161. doi:10.1093/humupd/dmh061

Kalinowski RR, Berlot CH, Jones TLZ et al (2004) Maintenance of meiotic prophase arrest in vertebrate oocytes by a Gs protein-mediated pathway. Dev Biol 267:1–13. doi:10.1016/j.ydbio.2003.11.011

Kerr B, Garcia-Rudaz C, Dorfman M et al (2009) NTRK1 and NTRK2 receptors facilitate follicle assembly and early follicular development in the mouse ovary. Reproduction 138:131–140. doi:10.1530/REP-08-0474

Kezele P, Skinner MK (2003) Regulation of ovarian primordial follicle assembly and development by estrogen and progesterone: endocrine model of follicle assembly. Endocrinology 144: 3329–3337. doi:10.1210/en.2002-0131

Kezele P, Nilsson EE, Skinner MK (2005) Keratinocyte growth factor acts as a mesenchymal factor that promotes ovarian primordial to primary follicle transition. Biol Reprod 73:967–973. doi:10.1095/biolreprod.105.043117

Knight PG, Satchell L, Glister C (2012) Intra-ovarian roles of activins and inhibins. Mol Cell Endocrinol 359:53–65. doi:10.1016/j.mce.2011.04.024

Koubova J, Menke DB, Zhou Q et al (2006) Retinoic acid regulates sex-specific timing of meiotic initiation in mice. Proc Natl Acad Sci U S A 103:2474–2479. doi:10.1073/pnas.0510813103

Lee W-S, Yoon S-J, Yoon T-K et al (2004) Effects of bone morphogenetic protein-7 (BMP-7) on primordial follicular growth in the mouse ovary. Mol Reprod Dev 69:159–163. doi:10.1002/mrd.20163

Lei L, Spradling AC (2013) Mouse primordial germ cells produce cysts that partially fragment prior to meiosis. Development 140:2075–2081. doi:10.1242/dev.093864

Light A, Hammes SR (2015) LH-induced steroidogenesis in the mouse ovary, but not testis, requires matrix metalloproteinase 2 and 9-mediated cleavage of upregulated EGF receptor ligands. Biol Reprod. doi:10.1095/biolreprod.115.130971

Lin Y-T, Capel B (2015) Cell fate commitment during mammalian sex determination. Curr Opin Genet Dev 32:144–152. doi:10.1016/j.gde.2015.03.003

Liu X, Andoh K, Abe Y et al (1999) A comparative study on transforming growth factor-beta and activin A for preantral follicles from adult, immature, and diethylstilbestrol-primed immature mice. Endocrinology 140:2480–2485. doi:10.1210/en.140.6.2480

Liu K, Rajareddy S, Liu L et al (2006) Control of mammalian oocyte growth and early follicular development by the oocyte PI3 kinase pathway: new roles for an old timer. Dev Biol 299:1–11. doi:10.1016/j.ydbio.2006.07.038

Liu C, Peng J, Matzuk MM, Yao HH-C (2015) Lineage specification of ovarian theca cells requires multicellular interactions via oocyte and granulosa cells. Nat Commun 6:6934. doi:10.1038/ncomms7934

Louvet JP, Vaitukaitis JL (1976) Induction of follicle stimulating hormone (FSH) receptors in rat ovaries by estrogen priming. Endocrinology 99:758–764

Macias V, Pinzón C, Fierro F et al (2012) Identification of soluble forms of vascular endothelial growth factor receptors, sVEGFR-1 and sVEGFR-2, in bovine dominant follicles. Reprod Domest Anim 47:e39–e42. doi:10.1111/j.1439-0531.2011.01919.x

Magoffin DA (2005) Ovarian theca cell. Int J Biochem Cell Biol 37:1344–1349. doi:10.1016/j.biocel.2005.01.016

Matzuk MM, Burns KH (2012) Genetics of mammalian reproduction: modeling the end of the germline. Annu Rev Physiol 74:503–528. doi:10.1146/annurev-physiol-020911-153248

McLaren A (2003) Primordial germ cells in the mouse. Dev Biol 262:1–15. doi:10.1016/S0012-1606(03)00214-8
Mehlmann LM, Jones TLZ, Jaffe LA (2002) Meiotic arrest in the mouse follicle maintained by a Gs protein in the oocyte. Science 297:1343–1345. doi:10.1126/science.1073978
Menke DB, Koubova J, Page DC (2003) Sexual differentiation of germ cells in XX mouse gonads occurs in an anterior-to-posterior wave. Dev Biol 262:303–312. doi:10.1016/S0012-1606(03)00391-9
Miettinen HE, Rayburn H, Krieger M (2001) Abnormal lipoprotein metabolism and reversible female infertility in HDL receptor (SR-BI)-deficient mice. J Clin Invest 108:1717–1722. doi:10.1172/JCI200113288
Molyneaux KA, Stallock J, Schaible K, Wylie C (2001) Time-lapse analysis of living mouse germ cell migration. Dev Biol 240:488–498. doi:10.1006/dbio.2001.0436
Mork L, Maatouk DM, McMahon JA et al (2012) Temporal differences in granulosa cell specification in the ovary reflect distinct follicle fates in mice. Biol Reprod 86:37. doi:10.1095/biolreprod.111.095208
Myers M, Middlebrook BS, Matzuk MM, Pangas SA (2009) Loss of inhibin alpha uncouples oocyte-granulosa cell dynamics and disrupts postnatal folliculogenesis. Dev Biol 334:458–467. doi:10.1016/j.ydbio.2009.08.001
Myers M, Tripurani SK, Middlebrook B et al (2011) Loss of gremlin delays primordial follicle assembly but does not affect female fertility in mice. Biol Reprod 85:1175–1182. doi:10.1095/biolreprod.111.091728
Neeman M, Abramovitch R, Schiffenbauer YS, Tempel C (1997) Regulation of angiogenesis by hypoxic stress: from solid tumours to the ovarian follicle. Int J Exp Pathol 78(2):57–70
Nilsson EE, Skinner MK (2003) Bone morphogenetic protein-4 acts as an ovarian follicle survival factor and promotes primordial follicle development. Biol Reprod 69:1265–1272. doi:10.1095/biolreprod.103.018671
Nilsson E, Parrott JA, Skinner MK (2001) Basic fibroblast growth factor induces primordial follicle development and initiates folliculogenesis. Mol Cell Endocrinol 175:123–130. doi:10.1016/S0303-7207(01)00391-4
Nilsson EE, Kezele P, Skinner MK (2002) Leukemia inhibitory factor (LIF) promotes the primordial to primary follicle transition in rat ovaries. Mol Cell Endocrinol 188:65–73. doi:10.1016/S0303-7207(01)00746-8
Nilsson EE, Detzel C, Skinner MK (2006) Platelet-derived growth factor modulates the primordial to primary follicle transition. Reproduction 131:1007–1015. doi:10.1530/rep.1.00978
Nilsson E, Dole G, Skinner MK (2009) Neurotrophin NT3 promotes ovarian primordial to primary follicle transition. Reproduction 138:697–707. doi:10.1530/REP-09-0179
Nilsson EE, Schindler R, Savenkova MI, Skinner MK (2011) Inhibitory actions of anti-Müllerian hormone (AMH) on ovarian primordial follicle assembly. PLoS One. doi:10.1371/journal.pone.0020087
Ohinata Y, Ohta H, Shigeta M et al (2009) A signaling principle for the specification of the germ cell lineage in mice. Cell 137:571–584. doi:10.1016/j.cell.2009.03.014
Oktem O, Urman B (2010) Understanding follicle growth in vivo. Hum Reprod 25:2944–2954. doi:10.1093/humrep/deq275
Osz K, Ross M, Petrik J (2014) The thrombospondin-1 receptor CD36 is an important mediator of ovarian angiogenesis and folliculogenesis. Reprod Biol Endocrinol 12:1–10. doi:10.1186/1477-7827-12-21
Pangas SA (2012) Regulation of the ovarian reserve by members of the transforming growth factor beta family. Mol Reprod Dev 79(10):666–679. doi:10.1002/mrd.22076
Pangas SA, Matzuk MM (2004) Genetic models for transforming growth factor beta superfamily signaling in ovarian follicle development. Mol Cell Endocrinol 225:83–91. doi:10.1016/j.mce.2004.02.017
Pedersen T, Hannah P (1968) Proposal for a classification of oocytes and follicles in the mouse ovary. J Reprod Fertil 17:555–557

Pepling ME, Spradling AC (1998) Female mouse germ cells form synchronously dividing cysts. Development 125:3323–3328
Pepling ME, Spradling AC (2001) Mouse ovarian germ cell cysts undergo programmed breakdown to form primordial follicles. Dev Biol 234(2):339–351. doi:10.1006/dbio.2001.0269
Petrik JJ, Gentry PA, Feige J et al (2002) Expression and localization of thrombospondin-1 and -2 and their cell-surface receptor, CD36, during rat follicular development and formation of the corpus luteum. Biol Reprod 67:1522–1531. doi:10.1095/biolreprod.102.007153
Rajareddy S, Reddy P, Du C et al (2007) p27kip1 (cyclin-dependent kinase inhibitor 1B) controls ovarian development by suppressing follicle endowment and activation and promoting follicle atresia in mice. Mol Endocrinol 21:2189–2202. doi:10.1210/me.2007-0172
Rao MC, Midgley AR, Richards JS (1978) Proliferation of ovarian cellular. Cell 14:71–78
Reddy P, Liu L, Adhikari D et al (2008) Oocyte-specific deletion of Pten causes premature activation of the primordial follicle pool. Science 319:611–613. doi:10.1126/science.1152257
Reddy P, Adhikari D, Zheng W et al (2009) PDK1 signaling in oocytes controls reproductive aging and lifespan by manipulating the survival of primordial follicles. Hum Mol Genet 18: 2813–2824. doi:10.1093/hmg/ddp217
Reddy P, Zheng W, Liu K (2010) Mechanisms maintaining the dormancy and survival of mammalian primordial follicles. Trends Endocrinol Metab 21:96–103. doi:10.1016/j.tem.2009.10.001
Reizel Y, Itzkovitz S, Adar R et al (2012) Cell lineage analysis of the mammalian female germline. PLoS Genet 8:e1002477. doi:10.1371/journal.pgen.1002477
Richards JS, Pangas SA (2010a) Fertility control. Handb Exp Pharmacol 198:3–27. doi:10.1007/978-3-642-02062-9
Richards JS, Pangas SA (2010b) The ovary: basic biology and clinical implications, Review series. J Clin Invest 120:963–972. doi:10.1172/JCI41350, Review
Ryan KE, Casey SM, Canthy MJ et al (2007) Akt and Erk signal transduction pathways are early markers of differentiation in dominant and subordinate ovarian follicles in cattle. Reproduction 133:617–626. doi:10.1530/REP-06-0130
Saitou M, Kagiwada S, Kurimoto K (2012) Epigenetic reprogramming in mouse pre-implantation development and primordial germ cells. Development 139:15–31. doi:10.1242/dev.050849
Salomon Y, Yanovsky A, Mintz Y et al (1977) Synchronous generation of ovarian hCG binding sites and LH-sensitive adenylate cyclase in immature rats following treatment with pregnant mare serum gonadotropin. J Cyclic Nucleotide Res 3:163–176
Schindler R, Nilsson E, Skinner MK (2010) Induction of ovarian primordial follicle assembly by connective tissue growth factor CTGF. PLoS One 5:e12979. doi:10.1371/journal.pone.0012979
Seki Y, Yamaji M, Yabuta Y et al (2007) Cellular dynamics associated with the genome-wide epigenetic reprogramming in migrating primordial germ cells in mice. Development 134: 2627–2638. doi:10.1242/dev.005611
Shkolnik K, Tadmor A, Ben-Dor S et al (2011) Reactive oxygen species are indispensable in ovulation. Proc Natl Acad Sci U S A 108(4):1462–1467. doi:10.1073/pnas.1017213108
Shimizu T, Jiang J, Sasada H, Sato E (2002) Changes of messenger RNA expression of angiogenic factors and related receptors during follicular development in gilts. Biol Reprod 67:1846–1852. doi:10.1095/biolreprod.102.006734
Simon AM, Goodenough DA, Li E, Paul DL (1997) Female infertility in mice lacking connexin 37. Nature 385:525–529
Soyal SM, Amleh A, Dean J (2000) FIGalpha, a germ cell-specific transcription factor required for ovarian follicle formation. Development 127:4645–4654
Sterneck E, Tessarollo L, Johnson PF (1997) An essential role for C/EBPbeta in female reproduction. Genes Dev 11:2153–2162. doi:10.1101/gad.11.17.2153
Stocco C, Telleria C, Gibori G (2007) The molecular control of corpus luteum formation, function, and regression. Endocr Rev 28:117–149. doi:10.1210/er.2006-0022

Teilmann SC (2005) Differential expression and localisation of connexin-37 and connexin-43 in follicles of different stages in the 4-week-old mouse ovary. Mol Cell Endocrinol 234:27–35. doi:10.1016/j.mce.2004.10.014

Thomas FH, Wilson H, Silvestri A, Fraser HM (2008) Thrombospondin-1 expression is increased during follicular atresia in the primate ovary. Endocrinology 149:185–192. doi:10.1210/en. 2007-0835

Tilly JL, Telfer EE (2009) Purification of germline stem cells from adult mammalian ovaries: a step closer towards control of the female biological clock? Mol Hum Reprod 15:393–398. doi:10.1093/molehr/gap036

Trigatti B, Rayburn H, Viñals M et al (1999) Influence of the high density lipoprotein receptor SR-BI on reproductive and cardiovascular pathophysiology. Proc Natl Acad Sci U S A 96: 9322–9327. doi:10.1073/pnas.96.16.9322

Tsafriri A, Dekel N (2010) Intra- and intercellular molecular mechanisms in regulation of meiosis in murid rodents. In: Tosti E, Boni R (eds) Oocyte maturation and fertilization: a long history for a short event. Bentham, Dubai, pp 38–63

Tureck RW, Strauss JF (1982) Progesterone synthesis by luteinized human granulosa cells in culture: the role of de novo sterol synthesis and lipoprotein-carried sterol. J Clin Endocrinol Metab 54:367–373. doi:10.1210/jcem-54-2-367

Uda M, Ottolenghi C, Crisponi L et al (2004) Foxl2 disruption causes mouse ovarian failure by pervasive blockage of follicle development. Hum Mol Genet 13:1171–1181. doi:10.1093/hmg/ ddh124

Ulloa-Aguirre A, Zariñán T, Pasapera AM et al (2007) Multiple facets of follicle-stimulating hormone receptor function. Endocrine 32:251–263. doi:10.1007/s12020-008-9041-6

Van Santbrink E, Hop W, Van Dessel T et al (1995) Decremental follicle-stimulating hormone and dominant follicle development during the normal menstrual cycle. Fertil Steril 1(64):37–43

Vanorny DA, Prasasya RD, Chalpe AJ et al (2014) Notch signaling regulates ovarian follicle formation and coordinates follicular growth. Mol Endocrinol 28:499–511. doi:10.1210/me. 2013-1288

Wotton D, Massague J (2000) Transcriptional control by the TGF-b/Smad signaling system. EMBO J 19:1745–1754

Wulff C, Wilson H, Largue P et al (2000) Angiogenesis in the human corpus luteum: localization and changes in angiopoietins, tie-2, and vascular endothelial growth factor messenger ribonucleic acid. J Clin Endocrinol Metab 85:4302–4309. doi:10.1210/jcem.85.11.6942

Yamashita Y, Shimada M (2012) The release of EGF domain from EGF-like factors by a specific cleavage enzyme activates the EGFR-MAPK3/1 pathway in both granulosa cells and cumulus cells during the ovulation process. J Reprod Dev 58(5):510–514

Yan C, Wang P, DeMayo J et al (2001) Synergistic roles of bone morphogenetic protein 15 and growth differentiation factor 9 in ovarian function. Mol Endocrinol 15:854–866. doi:10.1210/ mend.15.6.0662

Zeleznik AJ, Schuler HM, Reichert LE (1981) Gonadotropin-binding sites in the rhesus monkey ovary: role of the vasculature in the selective distribution of human chorionic gonadotropin to the preovulatory follicle. Endocrinology 109:356–362

Zhang M, Xia G (2012) Hormonal control of mammalian oocyte meiosis at diplotene stage. Cell Mol Life Sci 69:1279–1288. doi:10.1007/s00018-011-0867-3

Zhang M, Su Y-Q, Sugiura K et al (2010) Granulosa cell ligand NPPC and its receptor NPR2 maintain meiotic arrest in mouse oocytes. Science 330:366–370

Zhang H, Risal S, Gorre N et al (2014) Somatic cells initiate primordial follicle activation and govern the development of dormant oocytes in mice. Curr Biol 24(21):2501–2508. doi:10. 1016/j.cub.2014.09.023

Zheng W, Nagaraju G, Liu Z, Liu K (2012) Functional roles of the phosphatidylinositol 3-kinases (PI3Ks) signaling in the mammalian ovary. Mol Cell Endocrinol 356:24–30. doi:10.1016/j. mce.2011.05.027

Zimmermann RC, Hartman T, Kavic S et al (2003) Vascular endothelial growth factor receptor 2—mediated angiogenesis is essential for gonadotropin-dependent follicle development. J Clin Invest 112:659–669. doi:10.1172/JCI200318740

Chapter 8
Control of Oocyte Growth and Development by Intercellular Communication Within the Follicular Niche

Stephany El-Hayek and Hugh J. Clarke

Abstract In the mammalian ovary, each oocyte grows and develops within its own structural and developmental niche—the follicle. Together with the female germ cell in the follicle are somatic granulosa cells, specialized companion cells that surround the oocyte and provide support to it, and an outer layer of thecal cells that serve crucial roles including steroid synthesis. These follicular compartments function as a single physiological unit whose purpose is to produce a healthy egg, which upon ovulation can be fertilized and give rise to a healthy embryo, thus enabling the female germ cell to fulfill its reproductive potential. Beginning from the initial stage of follicle formation and until terminal differentiation at ovulation, oocyte and follicle growth depend absolutely on cooperation between the different cellular compartments. This cooperation synchronizes the initiation of oocyte growth with follicle activation. During growth, it enables metabolic support for the follicle-enclosed oocyte and allows the follicle to fulfill its steroidogenic potential. Near the end of the growth period, intra-follicular interactions prevent the precocious meiotic resumption of the oocyte and ensure its nuclear differenti-

S. El-Hayek
Department of Obstetrics and Gynecology, McGill University, Montreal, QC, Canada

Department of Biology, McGill University, Montreal, QC, Canada

Research Institute of the McGill University Health Centre, 1001 Decarie Blvd, Block E-M0.2218, Montreal, QC, Canada, H4A 3J1

H.J. Clarke (✉)
Department of Obstetrics and Gynecology, McGill University, Montreal, QC, Canada

Department of Biology, McGill University, Montreal, QC, Canada

Department of Experimental Medicine, McGill University, Montreal, QC, Canada

Research Institute of the McGill University Health Centre, 1001 Decarie Blvd, Block E-M0.2218, Montreal, QC, Canada, H4A 3J1
e-mail: hugh.clarke@mcgill.ca

R.P. Piprek (ed.), *Molecular Mechanisms of Cell Differentiation in Gonad Development*, Results and Problems in Cell Differentiation 58,
DOI 10.1007/978-3-319-31973-5_8

ation. Finally, cooperation enables the events of ovulation, including meiotic maturation of the oocyte and expansion of the cumulus granulosa cells. In this chapter, we discuss the cellular interactions that enable the growing follicle to produce a healthy oocyte, focusing on the communication between the germ cell and the surrounding granulosa cells.

8.1 Introduction

Each human female is born with a population of oocytes estimated to range from several hundred thousand to several million (Wallace and Kelsey 2010). Although this may seem like a large number, it pales in comparison to the millions of sperm produced daily by a young adult male (Oatley and Brinster 2012). Moreover, although some controversy remains, the weight of available evidence indicates that, unlike males, female mammals do not possess a population of germ-line stem cells that give rise to new oocytes postnatally (Eggan et al. 2006; Lei and Spradling 2013; Zhang et al. 2014a; Zhang et al. 2015). Oocytes are thus relatively rare, and each presents a precious reproductive opportunity. It is not surprising that each has been allocated its own structural and developmental niche—a follicle. Each follicle contains a single oocyte together with somatic granulosa cells, which are specialized companion cells that surround the oocyte and provide support to it as well as producing steroid hormones, and an outer layer of thecal cells that serve crucial roles in steroid synthesis and in ovulation. The follicular compartment functions as a single physiological unit whose purpose is to produce a healthy egg that upon ovulation can be fertilized and will give rise to a new organism, thus enabling the female germ cell to fulfill its reproductive potential. Here, based mainly on findings using the murine model, we will discuss the cellular interactions that enable the follicle to achieve this goal, focusing on the communication between the germ and somatic cells,

8.2 Overview of Oogenesis and Folliculogenesis

During embryonic development in the female, oogonia proliferate to generate a population of germ cells. These synchronously enter meiosis and progress to diplotene of prophase I where they become arrested (Borum 1961). By the time of birth, somatic pre-granulosa cells within the ovary have surrounded oocytes to generate primordial follicles, each consisting of one oocyte (~15 μm diameter in the mouse) enclosed by a single layer of squamous granulosa cells (Findlay et al. 2015; Grive and Freiman 2015; Jorgensen 2013; Pepling 2012). However, for reasons not yet fully understood, a significant proportion of the oocyte population is lost by apoptosis during late embryogenesis (Ene et al. 2013; Hutt 2015; Malki et al. 2014), reducing the number of primordial follicles that form or remain at birth. Primordial follicles thus constitute the ovarian reserve pool, from which follicles are continually recruited during reproductive life into the growth phase to produce fertilizable eggs. Recruitment progressively depletes the ovarian reserve and when, as a consequence of aging, disease, or environmental toxins, the supply of primordial follicles drops below a functional threshold, the female is no longer fertile.

When a primordial follicle enters the growth phase, the enclosed oocyte begins to increase in size. The granulosa cells undergo a change in shape and proliferate mitotically so that they continue to fully enclose the surface of the expanding oocyte. Such follicles, consisting of a newly growing oocyte surrounded by a single layer of granulosa cells, are termed primary. During this stage, the oocyte begins to secrete a family of glycoproteins that assemble into an extracellular coat, termed the *zona pellucida*, which physically separates the oocyte from the bodies of the adjacent granulosa cells. As the granulosa cells continue to proliferate, they generate multiple layers around the growing oocyte. In addition, thecal cells become deposited around the outside of the basement membrane surrounding the follicle. These follicles, now termed secondary, also begin to acquire their own vasculature (Suzuki et al. 1998). Eventually, a fluid-filled cavity termed the antrum appears within the follicle, defining it as antral. Once the antrum expands, it separates the granulosa cells into two populations—the mural granulosa cells that line the inside of the follicular wall and the cumulus granulosa cells that surround the oocyte (Fig. 8.1). Near the time of antral formation, the oocyte, having increased >100-fold in volume and accumulated enormous reserves of transcripts, proteins, and organelles needed to sustain early embryo development, stops increasing in size. The oocyte nonetheless continues to undergo molecular differentiation and it is only at the late antral stage that it acquires the ability—known as developmental competence—to develop as an embryo following ovulation and fertilization.

The final stage of oocyte development is meiotic maturation. Maturation is physiologically triggered by the preovulatory surge of luteinizing hormone (LH) that also triggers cumulus expansion and ovulation, thus coordinating the completion of oocyte development with its deposition into the reproductive tract to be fertilized (Eppig 1979; Espey 1980; Meinecke and Meinecke-Tillmann 1979). During maturation, the cell cycle, which has been arrested at late prophase I throughout growth, resumes and in most mammals, the oocyte completes meiosis I before becoming arrested at metaphase of meiosis II (Fig. 8.1). In addition to these nuclear events, many cytoplasmic changes also occur during maturation. These include changes in the ultrastructure of the endoplasmic reticulum (Mehlmann et al. 1995; Payne and Schatten 2003), migration of the cortical granules to just underneath the plasma membrane (Liu et al. 2003; Nicosia et al. 1977), and accumulation of the mitochondria around the meiotic spindle (Dalton and Carroll 2013; Van Blerkom et al. 2003). Many previously silent mRNAs become translationally activated, whereas others become silenced and in some cases

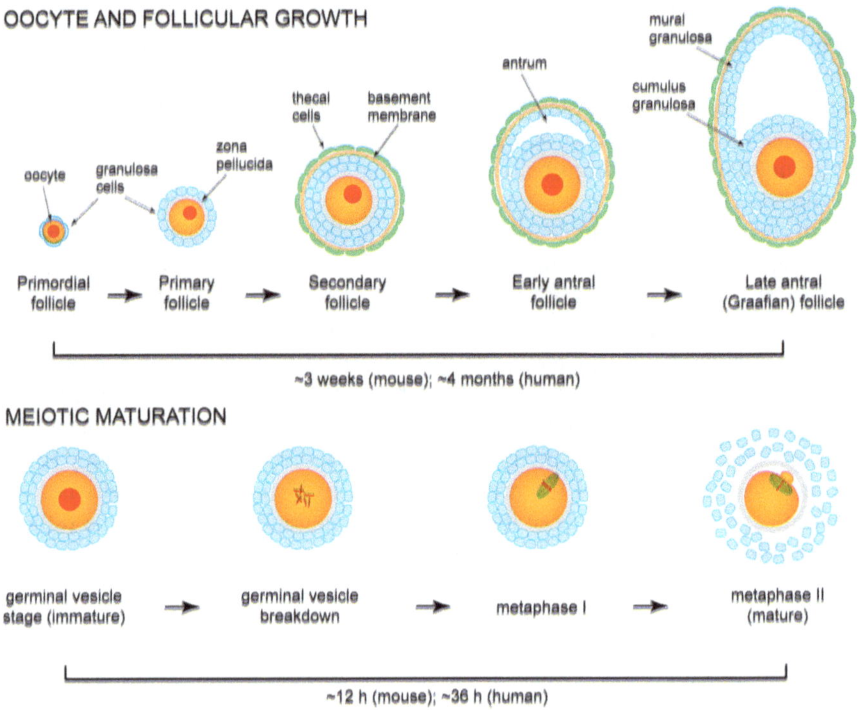

Fig. 8.1 Postnatal oocyte development. *Upper*: Oocyte and follicular growth. Oocytes in primordial follicles are surrounded by a small number of squamous granulosa cells. Incompletely understood signals cause primordial follicles to transition to the primary stage, characterized by a growing oocyte enclosed by cuboidal, mitotically active granulosa cells. When multiple granulosa cell layers have been generated and a basement membrane and thecal cell layer have been recruited, the follicles are termed secondary. The appearance of a small cavity, termed the antrum, within the granulosa cell layer defines antral follicles; oocytes at this stage are (nearly) fully grown. As the antrum enlarges, the mural and cumulus granulosa cell populations become clearly separated, and the fully grown Graafian follicle is ready to ovulate. *Lower*: Meiotic maturation. Upon the initiation of maturation, the nucleus (germinal vesicle) breaks down and the chromosomes condense and migrate to the periphery of the cell where they become assembled on the first meiotic spindle. Half of the products of meiosis I are discarded in the first polar body, and the oocyte chromosomes are assembled on the second meiotic spindle in preparation for meiosis II, which will occur after fertilization. The maturing oocyte also undergoes many cytoplasmic changes, not illustrated. During maturation, in a process termed cumulus expansion, the cumulus cells secrete proteins that generate an extracellular matrix that separates them from each other and from the oocyte and facilitates fertilization

degraded (Clarke 2012; Kang and Han 2011). During maturation, physical contact between the oocyte and granulosa cells, which has persisted since primordial follicle formation, is irreversibly lost. By the end of maturation, the mature germ cell, now termed an egg, is ovulated and can be fertilized.

8.3 Intra-follicular Cell Contact and Bidirectional Communication

The ovarian follicle is among the few physiological structures in mammals whose development can be sustained for a relatively long period of time in vitro and for which the quality of the cells that grow or differentiate in vitro can be rigorously tested—in this case, by the ability of the oocyte to give rise to an embryo. Thus, the ovarian follicle is an ideal model system to study mechanisms that regulate the formation and function of complex tissues. The relatively simple structure and cellular composition of the ovarian follicle, together with the development of culture conditions that enable oocyte and follicular growth to be reproduced in vitro, have provided researchers with an unparalleled opportunity to discover how the different cell types within the follicle cooperate to produce a healthy egg. We will focus on postnatal growth and development of the oocyte and follicle. We refer readers to recent reviews for studies of pre- and perinatal stages (Findlay et al. 2015; Grive and Freiman 2015; Jorgensen 2013; Pepling 2012).

Experimental studies have revealed a dynamic bidirectional communication network that links the germ-line and somatic cells of the follicular niche and is governed by both direct cell–cell contact and paracrine signaling. This communication is absolutely indispensable for the production of a healthy oocyte. When granulosa–oocyte complexes at an early stage of growth are removed from the ovarian follicle and placed in culture, the oocyte can complete normal development, as witnessed by its ability to give rise to offspring following fertilization (Eppig and Schroeder 1989; Hasegawa et al. 2006; Yamamoto et al. 1999). This remarkable result demonstrates that, with the possible exception of the earliest stages of growth, which have not been tested, the granulosa cells alone are sufficient to provide all of the external support needed by the oocyte to acquire developmental competence. If the granulosa cells are removed, however, the oocyte fails to grow, even when the two cell types are cocultured. Thus, physical contact with the granulosa is essential for oocyte growth.

Because the *zona pellucida* becomes assembled around the oocyte at an early stage of growth, thereby physically separating it from the bodies of the granulosa cells, an adaptation is required to allow the two cellular compartments to remain in physical contact. This is achieved by structures termed transzonal projections (TZPs). These narrow cytoplasmic fingers extend from the granulosa cells to the oocyte plasma membrane. TZPs link the surrounding follicle cells to the oocyte in most, and possibly all, mammalian species (Anderson and Albertini 1976; Hertig and Adams 1967), and similar structures have been observed in numerous nonmammalian species (Kessel et al. 1985; Schroeder 1981). Most TZPs contain an actin backbone, but a small fraction contains microtubules instead (Albertini and Rider 1994; De Smedt and Szollosi 1991). At the tips of the TZPs, where they contact the oocyte plasma membrane, are located gap junctions (which may also link TZPs to each other) (Anderson and Albertini 1976; Albertini and Rider 1994). These intercellular channels, composed of proteins known as connexins, permit the passage of molecules up to about 1 kDa in size (Kidder and Mhawi 2002). Connexin37 is the principal gap junction component in oocytes, and deletion of its encoding gene, *Gja4*, completely blocks gap junctional communication between the two cell types. As a result, oocytes of *Gja4*$^{-/-}$ mice fail to complete growth and do not become competent to develop as embryos (Simon et al. 1997). Thus, oocyte development depends crucially on gap junctional communication with the surrounding follicular granulosa cells.

On the other hand, oocyte-derived paracrine factors (ODPFs) regulate a wide range of granulosa cell activities (Emori and Sugiura 2014). For instance, in the absence of the growth-differentiation factor 9 (GDF9), a member of the transforming growth factor β (TGF β) family provided by the oocyte, the granulosa cells exhibit numerous abnormalities including a failure to proliferate to generate multilayered secondary follicles (Carabatsos et al. 1998; Dong et al. 1996). They also are unable to support normal oocyte development. These observations collectively underscore the essential nature of the bidirectional interactions between the somatic and germ-line compartments during folliculogenesis. By coordinating oocyte and follicular development from the time of growth initiation until the final stages of development, this communication network ensures that ovulation releases a developmentally competent egg. In the following, we discuss how these bidirectional interactions govern the various stages of follicle development.

8.4 Initiation of Oocyte Growth

At the time of primordial follicle formation, the association of the small oocytes with the somatic granulosa cells protects them from undergoing cell death (Pepling and Spradling 2001; Pesce et al. 1997; Rajah et al. 1992). Thus, interactions between the germ-line and somatic compartments of the follicle are indispensable even at the earliest stages of folliculogenesis, and indeed presage an ongoing interaction that is essential to yield a fertilizable egg. The first indication that a primordial follicle has been activated to begin growth is a morphological transition of the granulosa cells from a squamous to cuboidal morphology (Braw-Tal 2002; Hirshfield 1991). The external signals (if they exist) that initiate this transition remain unknown; however, it is associated with the increase in mitotic activity of the cells that is essential to enable them to fully enclose the growing oocyte (Da Silva-Buttkus et al. 2008). This morphological transition is apparently not indispensable for the initiation of oocyte growth though, because the oocytes of mice lacking forkhead box protein L2 (*Foxl2*) begin to grow even though the granulosa cells remain squamous (Schmidt et al. 2004; Uda et al. 2004). These oocytes grow only modestly before becoming arrested, however, suggesting that maintaining oocyte growth depends on this transition and/or continued proliferation of the granulosa cells.

The early morphological change observed in the granulosa cells of newly growing follicles suggests that follicular activation may initiate in this cellular

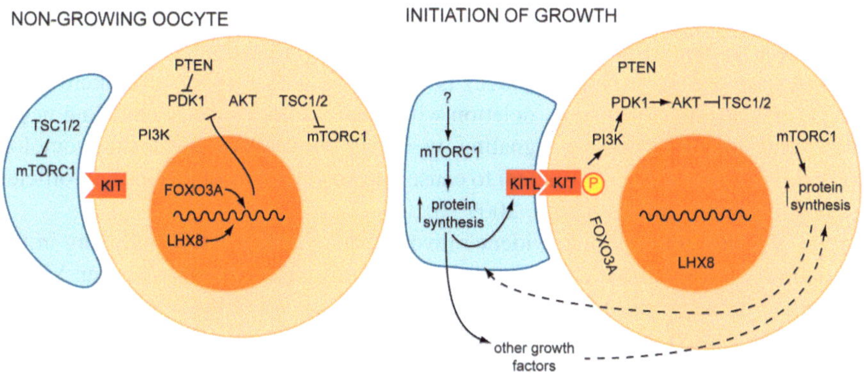

Fig. 8.2 Oocyte–granulosa communication and the initiation of growth. In a primordial follicle, TSC1/2 are active in granulosa cells, inhibiting mTORC1 activity and therefore limiting protein synthesis. In the oocyte, the transcription factors, FOXO3A and LHX8, prevent the oocytes from growing, possibly by inhibiting PI3-kinase activity. Signals that increase mTORC1 activity and therefore protein synthesis in granulosa cells are associated with a squamous-to-cuboidal morphological transition, mitotic proliferation, and increased production of KITL. This activates KIT on the oocyte plasma membrane, leading to increased protein synthesis. FOXO3A becomes excluded from the nucleus and activity of other transcription factors including LHX8 may be altered. Other growth factors from the granulosa cells may also be able to initiate oocyte growth, and triggering oocyte growth can also stimulate the morphological transition and mitotic proliferation of the granulosa cell (*dotted lines*)

compartment, and recent work supports this model. Mechanistic target of rapamycin complex-1 (mTORC1) regulates growth of many cell types, typically by increasing the rate of protein synthesis (Brunn et al. 1997). When Raptor (*Rptor*), which encodes an essential component of mTORC1, was selectively deleted in granulosa cells, the primordial follicles did not initiate growth (Zhang et al. 2014b). This was not due to cell death, as the follicles remained intact within the ovary for several weeks, although they subsequently degenerated. Conversely, when tuberous sclerosis 1 (*Tsc1*), which encodes an inhibitor of mTORC1 signaling, was deleted, the granulosa cells of most primordial follicles became cuboidal and the enclosed oocytes grew robustly (Zhang et al. 2014b). These results suggest that an increase in protein synthesis within the granulosa cells could be the signal that initiates oocyte growth (Fig. 8.2).

Several lines of evidence implicate KIT ligand (KITL) as a key growth-promoting signal for the oocyte. KITL is expressed by developing granulosa cells and binds to its receptor KIT on the oocyte, causing its phosphorylation (Manova et al. 1993) and activation of downstream signaling via the phosphatidylinositide (PI) 3-kinase pathway (Liu et al. 2006; Manova et al. 1993; Reddy et al. 2005, 2008). Early studies showed that adding KITL to cultures of ovarian fragments obtained from early postnatal animals induced follicle growth whereas an anti-KIT antibody known to block its function, or chemical inhibitors of KIT, reduced the number of growing follicles (Packer et al. 1994; Parrott and Skinner 1999; Yoshida et al. 1997). Similarly, fibroblasts expressing membrane-associated KITL induce growth of oocytes in the absence of granulosa cells (Thomas et al. 2008). Consistent with these results, deletion of *Tsc1* in granulosa cells of primordial follicles (which activates oocyte growth—see above) is associated with a substantial increase in KITL expression. Conversely, deletion within the oocyte of genes encoding negative regulators of PI3-kinase signaling, including phosphatase and tensin homolog (*Pten*), *Tsc1*, and *Tsc2*, is sufficient to cause the oocytes of most primordial follicles to begin growth (Adhikari et al. 2009, 2010).

Despite this considerable evidence favoring a key role for KIT signaling in the initiation of oocyte growth, the picture is not entirely clear. Mice carrying an Y719F mutation in KIT are unable to signal through the PI3-kinase pathway (John et al. 2009). Although females are subfertile, a small number of oocytes grow and give rise to apparently normal offspring (John et al. 2008; Kissel et al. 2000; Zhang et al. 2014b). This result seems to clearly indicate that oocyte growth does not absolutely require KIT activity or, by implication, KITL produced by the granulosa cells. Numerous other growth factors have been implicated in the activation of primordial follicles, including leukemia inhibiting factor (LIF), nerve growth factor (NGF), basic fibroblast growth factor (bFGF), keratinocyte growth factor (KGF), platelet-derived growth factor (PDGF), bone morphogenetic proteins (BMP), and connective tissue growth factor (CTGF) (Dissen et al. 2001; Kezele et al. 2005; Lee et al. 2001; Nilsson and Skinner 2001; Nilsson et al. 2006; Schindler et al. 2010). One possibility is that activation of KIT signaling by KITL plays an important role under normal conditions but that, when this pathway is disabled, other signals initiate oocyte growth. It is also worth noting that the relative importance of different growth factors may vary among species (Gilchrist et al. 2008).

Although the evidence described above favors granulosa cell-derived signals as initiators of primordial follicle activation and oocyte growth, several intriguing observations suggest that this is not an exclusive mechanism. Within the oocyte, the transcriptional regulator, forkhead box O 3A (FOXO3A), plays a central role in regulating the initiation of growth. FOXO3A is predominantly nuclear in oocytes within primordial follicles, but becomes translocated to the cytoplasm in growing oocytes (Li et al. 2010). As FOXO3A is thought to function primarily as a transcriptional repressor, this result suggests that it represses genes whose products promote oocyte growth (Brunet et al. 1999). In females lacking *Foxo3a*, the oocytes within virtually all primordial follicles begin to grow shortly after birth (Castrillon et al. 2003; Hosaka et al. 2004). In these follicles, the granulosa cells surrounding these oocytes become cuboidal and proliferate. Similarly, when the suppressors of the PI3-kinase pathway are genetically deleted in the oocyte, as described above, not only do the oocytes begin to grow, but the granulosa cells again become cuboidal and proliferate (Castrillon et al. 2003; Hosaka et al. 2004; John et al. 2008). Thus, it seems that activation of primordial follicles can be triggered by signals originating either in the granulosa cells or in the oocyte and that each cell type is capable of responding appropriately to signals received from the other so that oocyte and follicular growth are initiated in a coordinated manner (Fig. 8.2).

Other transcriptional regulators also play key roles in regulating the initiation of oocyte growth. LIM homeobox gene 8 (*Lhx8*) encodes a transcription factor that is abundantly expressed in oocytes in primordial follicles, as well as at later stages. When *Lhx8* was globally deleted, oocytes within primordial follicles failed to grow, suggesting an important role in oocyte growth (Choi et al. 2008a; Pangas et al. 2006). Recently, however, transgenic mice expressing Cre recombinase under the control of the promoter of *Gdf9* were used to selectively inactivate *Lhx8* in oocytes within primordial follicles (Ren et al. 2015). Many oocytes began to grow shortly after birth, accompanied by an increase in phosphorylated protein kinase B (AKT), indicating that PI3-kinase signaling was active and that FOXO3A translocated from the nucleus to the cytoplasm. Strikingly, however, the granulosa cells failed to undergo the squamous-cuboidal transition and the oocytes did not grow beyond ~30 μm in diameter (Ren et al. 2015). These results not only suggest that that LHX8 normally acts to maintain oocytes in a quiescent nongrowing condition, but may hint that signaling pathways that can relay the growth signal from the oocyte to the granulosa are dependent on LHX8. The spermatogenesis- and oogenesis-specific basic helix-loop-helix 1/2 (SOHLH1/2) proteins may play a similar role. These are abundant in oocytes of primordial follicles, but less detectable in growing oocytes. When either *Sohlh1* or *Sohlh2* is deleted, there is a rapid loss of oocytes and follicles. Some oocytes begin to grow, but the granulosa cells remain squamous (Choi et al. 2008b; Pangas et al. 2006), likely explaining why oocyte growth fails to progress. Taken together, the studies demonstrate that the entry of primordial follicles into the growth pool is an intricately coordinated process that can be regulated by multiple factors in both the oocyte and the granulosa cells.

8.5 Maintenance of Oocyte Growth and Follicle Development

8.5.1 Metabolic Support

The enormous increase in size of the growing oocyte makes it among the largest cells in the body. Growth is associated with an accumulation of a large stock of mRNAs and proteins, which direct late oocyte development and early embryonic development after fertilization, and of organelles such as mitochondria. This activity imposes a substantial metabolic burden on the oocyte (Collado-Fernandez et al. 2012). The oocyte does not shoulder this alone but outsources some essential tasks to the granulosa cells that surround it, allowing it to minimize its own energy expenditure (Fig. 8.3).

Follicular growth is associated with an increased consumption of glucose and oxygen, both in vivo (Harris et al. 2009) and in vitro (Boland et al. 1994; Harris et al. 2007). Many studies have shown that mammalian oocytes, including human (Tsutsumi et al. 1990) and mouse (Brinster 1971; Eppig 1976; Fagbohun and Downs 1992; Tsutsumi et al. 1992; Zuelke and Brackett 1992), cannot efficiently

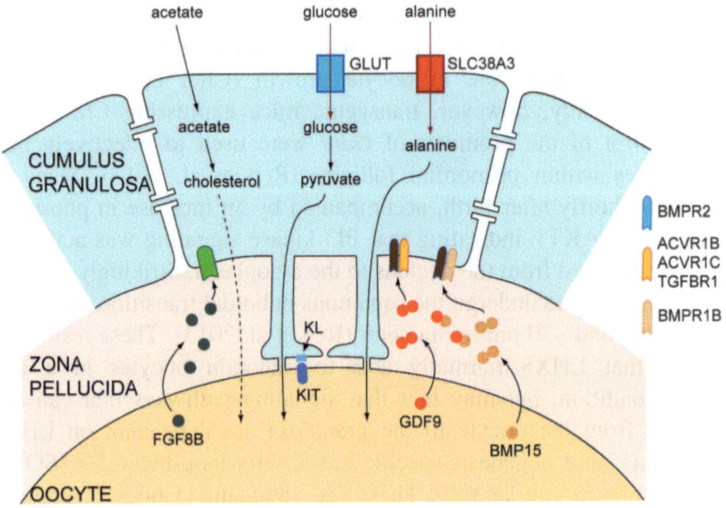

Fig. 8.3 Oocyte–granulosa cell communication during growth. The oocyte is unable to efficiently produce cholesterol or pyruvate or to take up certain amino acids such as alanine. The granulosa cells produce these essential factors and transfer them to the oocyte, via gap junctions in the case of pyruvate and amino acids. Conversely, the oocyte secretes several growth factors that are required for granulosa cell proliferation and/or differentiation. GDF9 and BMP15 may be presented to the granulosa cells as homo- or heterodimers

metabolize glucose to generate pyruvate. Indeed, oocytes are relatively deficient in expression of mRNAs encoding glycolytic enzymes including *Idoa*, *Enol*, *Ldha*, *Pfkp*, *Pkm2*, and *Tpi1*, whereas these genes are highly expressed by surrounding granulosa cells (Sugiura et al. 2005; Sugiura et al. 2007). Granulosa cells metabolize glucose and then transfer pyruvate to the oocyte (Brinster 1971; Tsutsumi et al. 1990). Nevertheless, one study traced the uptake of glucose by granulosa cells via glucose transporters and its transfer into the oocyte through gap junctions (Wang et al. 2012). Glucose transferred to the oocyte may be metabolized by non-glycolytic pathways such as the pentose phosphate or the hexosamine biosynthesis pathways (Collado-Fernandez et al. 2012). Hence, the oocyte relies on the somatic compartment to obtain its source of ATP. As reduced ATP content in the oocyte has been linked to defects of the meiotic spindle (Zhang et al. 2006), as well as diminished developmental competence (Igarashi et al. 2005; Van Blerkom et al. 1995), this cooperation between the oocyte and granulosa cells is essential for normal embryo development and fertility.

The protein synthetic activity of the oocyte evidently depends on a large supply of amino acids. But the oocyte is only poorly able to take up certain amino acids including alanine, glycine, and proline in vitro. Uptake significantly improves in the presence of surrounding granulosa cells, which transfer the amino acids through gap junctions (Colonna and Mangia 1983; Haghighat and Van Winkle 1990). Consistent with these observations, certain transporters including solute carrier (SLC) 38A3, whose substrates include histidine and alanine (Eppig et al. 2005), and SLC7A6, a cationic amino acid transporter (Corbett et al. 2014), are present only in granulosa cells. In addition, although other amino acid transport systems are active in denuded oocytes, they are functionally improved by the presence of surrounding granulosa cells (Colonna and Mangia 1983; Haghighat and Van Winkle 1990; Pelland et al. 2009). By supplying the oocyte with the amino acids for which it lacks transporters, granulosa cells thus greatly increase the efficiency of oocyte protein synthesis.

The oocyte also depends on the granulosa cells to obtain cholesterol. Many enzymes that act in the cholesterol biosynthetic pathway are expressed at relatively low levels in oocytes. Consistently, acetate is inefficiently converted to cholesterol in granulosa-free oocytes (Su et al. 2008). Oocytes also lack the ability to import cholesterol as they do not express receptors for HDL cholesterol and LDL cholesterol (Sato et al. 2003; Trigatti et al. 1999). Oocytes are thus deficient in both cholesterol synthesis and uptake of extracellular cholesterol. Instead, the surrounding granulosa cells, which express the necessary enzymes, synthesize cholesterol and transfer it to the oocyte (Su et al. 2008). Accumulation of cholesterol in oocytes is essential for normal embryo development (Comiskey and Warner 2007), as early preimplantation embryos are also unable to synthesize cholesterol (Pratt 1982). Thus, the metabolic cooperation between the granulosa cells and the oocyte, which ensures the deposition of cholesterol into the oocyte, is critical for the oocyte to acquire embryonic developmental competence.

The molecular deficiencies of the oocyte thus force it to be strictly dependent in several metabolic respects on the granulosa cells. However, the oocyte does not merely receive these metabolites from the granulosa cells. Rather, it instructs the granulosa cells to synthesize and/or supply the nutrients it needs to meet its metabolic requirements (Fig. 8.3). A clue to this was the observation that expression of several enzymes of the glycolytic pathway (Emori et al. 2013; Sugiura et al. 2005; Sugiura et al. 2007) and that of several amino acid transporters (Emori et al. 2013; Eppig et al. 2005) are all enriched in the cumulus cells adjacent to the oocyte as compared to mural granulosa on the inner wall of the follicle. It was then discovered that their expression, as well as that of several enzymes in the cholesterol biosynthetic pathway, is enhanced by oocytes or ODPFs, including GDF9 and BMP15 (Emori et al. 2013; Eppig et al. 2005; Nakamura et al. 2015; Sugiura et al. 2005, 2007; Su et al. 2008). These observations highlight the dynamic and bidirectional nature of the signaling between the oocyte and granulosa cells that allow the growing oocyte to meet its high metabolic demands.

8.5.2 Estradiol Synthesis

Steroidogenesis occurs in the adrenal gland, ovary, testis, and placenta. In the ovary, this process has long been considered as an example of intercellular cooperation, as exemplified by the "two-cell, two-gonadotropin" model describing cooperation between the granulosa and theca cells (Fortune and Armstrong

1977). Steroidogenesis involves the uptake and metabolism of cholesterol by steroidogenic enzymes and is driven by follicle-stimulating hormone (FSH) and LH (Conley et al. 1994; Parker and Schimmer 1995; Richards 1994; Richards et al. 2002; Tian et al. 1995). Because theca cells do not express the FSH receptor (Camp et al. 1991; Zeleznik et al. 1974), the theca–granulosa cooperation is necessary for estrogen synthesis.

Steroidogenesis is initiated by the import of cholesterol into mitochondria by steroidogenic acute regulatory protein (StAR) (Arakane et al. 1996; Clark et al. 1994; Stocco 2001). Cholesterol is then cleaved by the mitochondrial enzyme, cytochrome P450 (CYP) side-chain cleavage (CYP11A) producing pregnenolone. CYP17, 3-β-hydroxysteroid dehydrogenase (HSD), and 17β-HSD metabolize pregnenolone to produce androgens. These are then converted by aromatase (CYP19) to estradiol (Miller 1988; Peter and Dubuis 2000; Simpson et al. 1994). LH, whose receptor is on theca as well as granulosa cells (Camp et al. 1991; Erickson et al. 1979; Piquette et al. 1991; Segaloff et al. 1990; Zeleznik et al. 1974), drives androgen synthesis (Fortune and Armstrong 1977; Richards et al. 1987; Smyth et al. 1994), while FSH stimulates aromatase activity in granulosa cells (Dorrington et al. 1975; Erickson et al. 1979; Hickey et al. 1988; Fitzpatrick and Richards 1991; Whitelaw et al. 1992), enabling the production of estrogen from the theca-derived androgens. In combination with other follicular factors such as insulin-like growth factor 1 (IGF-1), LH (Devoto et al. 1999; McGee et al. 1996; Schoppee et al. 2002; Sekar et al. 2000; Simone et al. 1993; Willis et al. 1996; Zhang et al. 2000) and FSH (Adashi 1994; Balasubramanian et al. 1997; Glister et al. 2001; Hsu and Hammond 1987; Khalid et al. 2000; Monniaux and Pisselet 1992; Pescador et al. 1997; Silva and Price 2002; Spicer et al. 2002; Yoshimura 1998) increases expression of these steroidogenic enzymes in the theca and the granulosa cells, thus promoting androgen and estrogen synthesis, respectively, both in vivo and in vitro.

Hence, the combined actions of theca cells and granulosa cells enable steroidogenesis in the developing ovarian follicle. Recent findings suggest, however, that ovarian steroidogenesis is regulated by three, rather than two, follicular cell types. The removal of an oocyte from a granulosa–oocyte complex in vitro by oocytectomy leads to reduction in estradiol levels, suggesting that the oocyte promotes its production (Vanderhyden et al. 1993). Oocytes have been shown to enhance estradiol production in nonmammalian species as well (Sretarugsa and Wallace 1997; Yoshimura et al. 1994). In contrast, other studies have shown that oocyte signaling decreases the expression of steroidogenic enzymes (Diaz et al. 2007; Pangas et al. 2006; Wigglesworth et al. 2015) and suppresses FSH-stimulated estradiol synthesis (Glister et al. 2001; Vitt et al. 2000). These seemingly contradictory results may indicate that oocyte signaling promotes estradiol synthesis while also inhibiting the premature luteinization of the granulosa cells (Coskun et al. 1995; el-Fouly et al. 1970; Nekola and Nalbandov 1971) through independent pathways (Vanderhyden et al. 1993; Vanderhyden and Tonary 1995). The mechanism of how the oocyte achieves such a balanced effect remains to be elucidated.

The follicle itself is a target of the estradiol, as well as of the androgens, that it produces (Berisha et al. 2002; LaVoie et al. 2002; Rosenfeld et al. 2001; Walters et al. 2012). Estradiol drives folliculogenesis (Couse et al. 2005; Emmen et al. 2005; Krege et al. 1998) and prevents follicular atresia (Britt et al. 2000; Dupont et al. 2000; Richards 1980; Robker and Richards 1998; Rosenfeld et al. 2001). It promotes the survival of granulosa cells as well as the oocyte by inhibiting the expression of pro-apoptotic genes (Billig et al. 1993; Lund et al. 1999; Quirk et al. 2006; Toda et al. 2001). It also increases expression of IGF-1, thus promoting expression of FSH and LH targets within the follicle (Hsu and Hammond 1987; Khalid et al. 2000). Moreover, estradiol has also been shown to enhance the expression of natriuretic peptide receptor B (NPR2; see below), thus enabling the maintenance of meiotic arrest (Zhang et al. 2011). Hence, the cooperation between the various follicular compartments, resulting in the production of estradiol, is crucial for survival and development of the follicle as a whole.

8.6 Terminal Stages of Oocyte Growth and Development

8.6.1 Nuclear Differentiation of the Oocyte

Growing oocytes are highly active transcriptionally (Bachvarova 1985), likely reflecting that they must manufacture not only the RNA needed to sustain growth itself but also the RNA needed during the transcriptionally inactive periods of meiotic maturation and early embryonic development. In the mouse, the oocyte RNA content increases 300-fold during growth (Schultz et al. 1979; Sternlicht and Schultz 1981). Upon reaching full size, however, the oocyte dramatically reduces its transcriptional activity to an undetectable level (Bouniol-Baly et al. 1999; Christians et al. 1999; Worrad et al. 1994). Coincident with transcriptional arrest, the chromatin, which was previously dispersed throughout the nucleus in a configuration termed non-surrounded nucleolus (NSN), becomes partially condensed and prominently arrayed around the nucleolus in a configuration termed surrounded nucleolus (SN) (Mattson and Albertini 1990; Wickramasinghe et al. 1991). This conformational change, termed the NSN-SN transition, occurs in oocytes of many mammalian species (Fuhrer et al. 1989; Hinrichs et al. 1993; Parfenov et al. 1989). Although the events of chromatin remodeling and transcriptional silencing are temporally correlated, they do not strictly depend on each other (Abe et al. 2010; Andreu-Vieyra et al. 2010; De La Fuente 2006). For example, the dissociation of RNA polymerase II from DNA, an indication of transcriptional silencing, is not dependent on the NSN-SN transition (Abe et al. 2010).

The NSN-SN is associated with a number of events of late oocyte development, including a change in the nuclear concentration of the transcription factors, specificity protein 1 (SP1) and TATA box-binding protein (TBP), and a rearrangement of cytoplasmic components including the appearance of microtubule-organizing centers (Lodde et al. 2008; Mattson and Albertini 1990; Wickramasinghe et al. 1991). Hence, it is considered to be a reliable and consistent marker of oocyte developmental progression and generally strongly correlates with the ability of the oocyte to develop as an embryo following fertilization (Wickramasinghe et al. 1991; Zuccotti et al. 2002). Intriguingly, the granulosa cells play an important but not fully defined role in mediating the NSN-SN transition (De La Fuente 2006; Luciano et al. 2011). When the FSH analogue, equine chorionic gonadotropin, was injected into females and the cumulus cell–oocyte complexes recovered 2 days later from antral follicles, an increased number had undergone the NSN-SN transition and had ceased transcriptional activity as compared to oocytes obtained from non-injected females. This effect was not observed in oocytes that were not closely associated with cumulus cells at the time of recovery. Similarly, incubation of bovine cumulus–oocyte complexes in the presence of FSH promoted the NSN-SN transition together with increased gap junctional communication. As the FSH receptors are located on the granulosa cells, these observations imply that the granulosa send signals to the oocyte that promote its nuclear differentiation.

8.6.2 *Maintenance of Meiotic Arrest*

Throughout its growth, the oocyte remains arrested at late prophase I of the meiotic cell cycle. Upon the initiation of meiotic maturation, it completes the first meiotic division and then becomes arrested at metaphase of meiosis II. Physiologically, maturation is triggered by LH (Dekel et al. 1979; Eppig 1980), thereby coordinating it with ovulation. However, experiments in which oocytes of antral follicles are isolated and placed in culture showed that they acquire the ability to mature to metaphase II about midway through growth, which is well before the follicle acquires the ability to ovulate in response to LH (Sorensen and Wassarman 1976). This result established the concept that the follicular environment actively prevents oocytes from undergoing maturation until this inhibition is relieved by LH. The granulosa cells play a central role in mediating this inhibition.

Maturation is triggered by a protein complex, known as metaphase-promoting factor (MPF), that consists of a catalytic subunit, cyclin-dependent kinase 1 (CDK1), and a regulatory subunit, cyclin B1 (Brunet and Maro 2005). The activity of this complex is regulated by phosphorylation of specific sites on CDK1. The WEE1/MYT1 kinases phosphorylate two sites, which inhibits CDK1 activity, whereas the CDC25 phosphatases dephosphorylate these sites, thereby activating CDK1 (Lew and Kornbluth 1996; Lincoln et al. 2002). The relative activities of WEE/MYT1 and CDC25 thus control CDK1 activity. These two regulators are themselves controlled by phosphorylation. Phosphorylation by cyclic (c) AMP-dependent kinase A activates WEE1 and inhibits CDC25 (Han et al. 2005; Oh et al. 2010; Pirino et al. 2009; Zhang et al. 2008). Hence, high protein kinase A activity maintains CDK1 in an inactive state.

As would be anticipated based on the foregoing, cAMP levels have long been known to be high in follicle-enclosed oocytes and to drop rapidly when oocytes are removed from the follicle and placed in culture (Schultz et al. 1983). This suggests that the follicular environment prevents oocyte maturation by maintaining a high level of cAMP within the oocyte (Cho et al. 1974; Dekel and Beers 1978). Because the growth of antral follicles is promoted by FSH, whose receptors are located on the granulosa cells, and gonadotropins can increase intracellular cAMP, an attractive hypothesis was that the granulosa cells transferred cAMP to the oocyte via the communicating gap junctions. Subsequent work showed, however, that the oocyte is able to synthesize cAMP (Olsiewski and Beers 1983; Urner et al. 1983) through the activity of a G protein related receptor, GPR3 (Mehlmann et al. 2002, 2004). Importantly, when *Gpr3* was deleted from oocytes, they underwent maturation within the follicle in the absence of LH signaling (Mehlmann et al. 2004), providing convincing evidence that the oocyte-generated cAMP is needed to maintain high protein kinase A activity and inhibit maturation.

Although the granulosa cells may not be a primary source of cAMP for the oocyte, they nonetheless play a critical role in maintaining its meiotic arrest. cAMP in the oocyte is metabolized by phosphodiesterase (PDE)-3A (Bornslaeger et al. 1984; Dekel and Beers 1978, 1980; Jensen et al. 2002; Richard et al. 2001; Shitsukawa et al. 2001; Tsafriri et al. 1996; Vivarelli et al. 1983). Within the follicle, the granulosa cells transfer cyclic GMP, an inhibitor of PDE3A, to the oocyte, again via the gap junctions (Norris et al. 2009; Richard and Baltz 2014; Zhang et al. 2010). This helps to maintain a high steady-state level of cAMP within the oocyte. Intriguingly, the production of cGMP illustrates the cooperativity between the mural and cumulus granulosa cells. The mural cells secrete C-type natriuretic peptide C (CNP, encoded by *Nppc*), which binds to NPR2 located on the plasma membrane of the cumulus cells. NPR2 is associated with guanylate cyclase and thus upon its activation increases the production of cGMP (Fig. 8.4) (Franciosi et al. 2014; Kawamura et al. 2011; Zhang et al. 2010, 2011). The essential role of this cGMP is demonstrated by mice bearing mutations in either *Npr2* or *Nppc*. Their oocytes are unable to maintain meiotic arrest and thus undergo maturation within the ovarian follicle independently of LH signaling. Thus, by supplying an inhibitor of PDE3A activity, the granulosa cells maintain a high level of cAMP in the oocyte, which activates protein kinase A and keeps CDK1 in an inactive state.

Once again, the oocyte plays a key role in facilitating this granulosa cell function. GDF9 and BMP15 promote the expression of *Nppc* and *Npr2* (Lee et al. 2013; Wigglesworth et al. 2015; Zhang et al. 2010, 2011) and also enhance the expression of *Impdh*, an enzyme required for the production of cGMP (Wigglesworth et al. 2013). The oocyte thus plays a key role in maintaining its meiotic arrest. This cooperation between the two compartments, by preventing LH-independent meiotic maturation within the follicle, ensures that the production of a fertilizable egg is temporally coordinated with its ovulation.

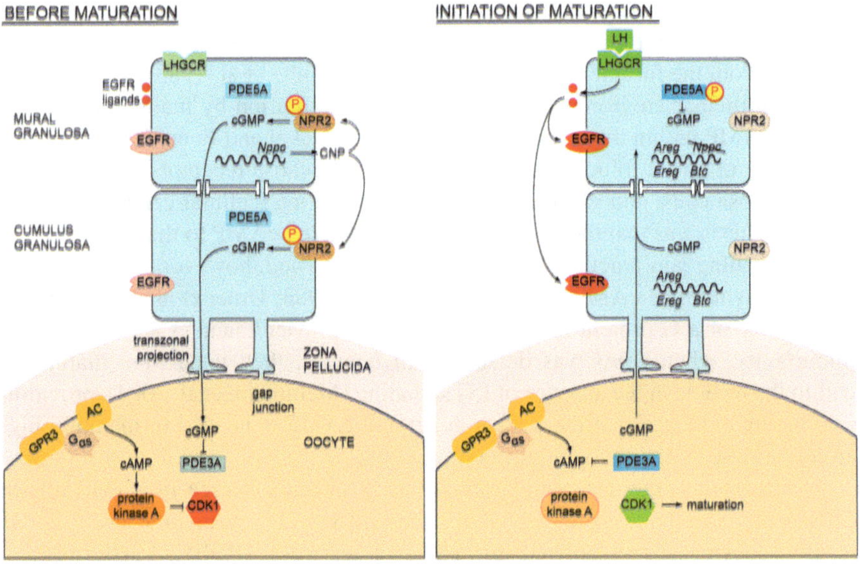

Fig. 8.4 Oocyte–granulosa communication and the regulation of meiotic maturation. Prior to maturation, CNP secreted by the granulosa cells activates NPR2 receptors that are coupled to guanylate cyclase, thereby activating cGMP synthesis. cGMP is transferred to the oocyte via gap junctions, where it inhibits PDE3A. This allows the level of cAMP in the oocyte, generated by the activity of Gs-coupled GPR3, to remain high. Protein kinase A is therefore active and CDK1 is inactive. Upon binding of LH to its receptor, LHGCR, on the mural granulosa cells, EGFR ligands (amphiregulin, beta-cellulin, and epiregulin) are released by the mural granulosa cells and activate EGFR on both mural and cumulus granulosa cells. Signaling through LHGCR and EGFR causes cGMP levels in the granulosa to fall via phosphorylation and activation of PDE5A, dephosphorylation of NPR2, and reduced *Nppc* transcription and CNP production. Transcription of *Areg*, *Btc*, and *Ereg* are also increased. The specific contributions of LHGCR and EGFR signaling to each of these events remain to be fully defined. The net result is a loss of cGMP from the oocyte, thus relieving the inhibition of PDE3A. Shortly after the initiation of maturation, gap junctional communication between granulosa cells and oocytes is disrupted

8.7 Ovulation-Associated Events

8.7.1 Meiotic Maturation

Meiotic maturation, the final stage of oocyte development, is triggered by the LH surge (Lei et al. 2001; Neal and Baker 1975). The LH receptors are expressed only on the mural granulosa cells within the follicle (as well as the thecal cells on the outside of the follicle). Therefore, this signal must be relayed to the oocyte. The discovery of the central role of granulosa cell-derived cGMP in maintaining meiotic arrest enabled the mechanism by which the LH signal is relayed to be deciphered. Less than 5 min after adding LH to antral follicles in vitro, the level of cGMP in the mural granulosa cells falls (Shuhaibar et al. 2015). This is followed by a decrease in cGMP in the cumulus granulosa cells and in the oocyte. Importantly, the rapid decrease in cumulus cell and oocyte cGMP is prevented by chemical inhibitors of gap junction permeability. cGMP decreases later in the cumulus cells and oocytes even when gap junctions are not functional, suggesting that multiple mechanisms may contribute to the decrease (Shuhaibar et al. 2015).

Several processes that could contribute to the fall in cGMP have been identified. Upon LH stimulation, NPR2 becomes rapidly dephosphorylated and inactivated, thus inhibiting the synthesis of new cGMP (Egbert et al. 2014; Robinson et al. 2012). In addition, PDE5A, which degrades cGMP, becomes phosphorylated and activated (Egbert et al. 2014). This likely increases the rate of degradation of existing cGMP. Following these early events, the amount of *Nppc* mRNA falls, which could lead to a reduction in the amount of CNP (Kawamura et al. 2011; Lee et al. 2013; Liu et al. 2014; Robinson et al. 2012). Finally, gap junctional communication becomes impaired both between granulosa cells and between the granulosa cells and oocyte, thus preventing the transfer of cGMP into the germ cell. It has been proposed that LH triggers a direct loss of cGMP in the mural granulosa cells that is followed by a flow of cGMP from the cumulus cells and oocyte to the mural cells to equalize the concentration of cGMP through the follicle (Shuhaibar et al. 2015). Thus, gap junctional communication is required for the initial reduction in cGMP within the cumulus cells and oocyte. Later, however, the mechanisms described above may directly reduce cGMP in cumulus cells, leading to a further fall in the oocyte cGMP.

The mechanism by which LH activates the multiple pathways that reduce cGMP remains to be fully elucidated. Upon binding of LH to its receptor, the mural granulosa cells release three members of the epidermal growth factor (EGF) family—amphiregulin (AREG), beta-cellulin (BTC), and epiregulin (EREG) (Conti et al. 2012; Park et al. 2004). As both the mural and cumulus granulosa cells express EGF receptors, this suggests that LH could act by transactivating EGF receptor (EGFR) signaling (Fig. 8.4). This provides a pathway by which the LH signal received by the mural granulosa cells could be relayed to the cumulus granulosa and ultimately the oocyte, and numerous observations support a central role for EGFR signaling in maturation. First, EGF is able to induce maturation of follicle-enclosed oocytes and expansion of the cumulus layer in the absence of LH (Ashkenazi et al. 2005; Downs and Chen 2008; Park et al. 2004). Second, the EGF-like ligands released by the granulosa cells in response to LH, as well as EGF, trigger the decrease in follicular cGMP independently of LH (Hsieh et al. 2011; Norris et al. 2010). Third, in mutants carrying the hypomorphic *Egfr*$^{wa2/wa2}$ allele or lacking *Areg* or *Ereg*, the oocytes fail to mature in response to LH (Hsieh et al. 2007, 2011).

Yet, other observations imply that LH may act independently of EGFR signaling. The extremely rapid drop in cGMP in the mural granulosa cells, which is detectable within a minute of adding LH in vitro (Shuhaibar et al. 2015), suggests that the gonadotropin acts directly on the granulosa cells. Consistent with this, LH induces a rapid drop in cGMP in the oocytes of *Egfr*$^{wa2/wa2}$ mice, although some EGFR activity remains in this mutant (Hsieh et al. 2011). In addition, the decrease in *Nppc* mRNA during maturation does not appear to require EGFR activity (Liu et al. 2014). It may be that LH acts through EGFR-dependent and -independent pathways to achieve a robust and stable decrease in cGMP throughout the preovulatory follicle, thus enabling activation of PDE3A in the oocyte and the consequent resumption of meiosis.

In addition to their role in the initiation of maturation, the granulosa cells also influence the maturation process itself. It is well established that cumulus cell-free oocytes can complete nuclear maturation to metaphase II in vitro, but develop poorly as embryos as compared to oocytes that mature within cumulus–oocyte complexes (Chang et al. 2005; De La Fuente et al. 1999). This suggests that cytoplasmic events of maturation may depend on or occur more efficiently in the presence of the cumulus cells. As noted earlier, a subset of mRNAs that had previously been synthesized and stored become translationally activated during maturation (Chen et al. 2011; Paynton and Bachvarova 1994; Su et al. 2007). Among these is *Tpx2*, encoding a protein required for proper assembly and function of the meiotic spindle (Chen et al. 2013). Although TPX2 increases when cumulus-free oocytes are matured in vitro, this increase is greater in cumulus-enclosed oocytes than in cumulus-free oocytes and is further enhanced in the presence of AREG. Translational activation is associated with an AREG-dependent increase in the activity of mTORC (mechanistic target of rapamycin complex) in the oocyte. Further evidence of the importance of signals from the cumulus cells comes from oocytes of mice deficient in either *Egfr* or *Areg* (Chen et al. 2013). These show reduced translational activation during maturation and impaired progression to the two-cell stage following fertilization and give rise to relatively small litters. These results indicate that the EGFR-mediated increase in translational activity during meiotic maturation is developmentally important and are consistent with previous reports that oocytes matured in vitro in the presence of EGF (De La Fuente et al. 1999; Richani et al. 2014) manifested increased developmental competence. Because gap junctional communication between the oocyte and granulosa cells is lost relatively early during the maturation process, it will be important to identify when and by what pathway the granulosa cells exert their beneficial effects on maturing oocytes.

8.7.2 Cumulus Expansion

As a result of LH-driven cumulus expansion, the egg is ovulated embedded in a mucified mass of cumulus granulosa cells. Cumulus expansion facilitates transport of the cumulus–oocyte complex through the oviduct (Chen et al. 1993) and is also essential for fertilization, likely by permitting the sperm to pass through the cumulus layers to reach the oocyte (Eisenbach 1999; Salustri et al. 2004; Van Soom et al. 2002). Cumulus expansion is characterized by the physical dispersal of the cumulus granulosa cells as an extracellular matrix becomes deposited around the individual cells (Fig. 8.5). This matrix consists mostly of the glycosaminoglycan, hyaluronan (HA), as well as HA-binding proteins (Chen et al. 1996; Fulop

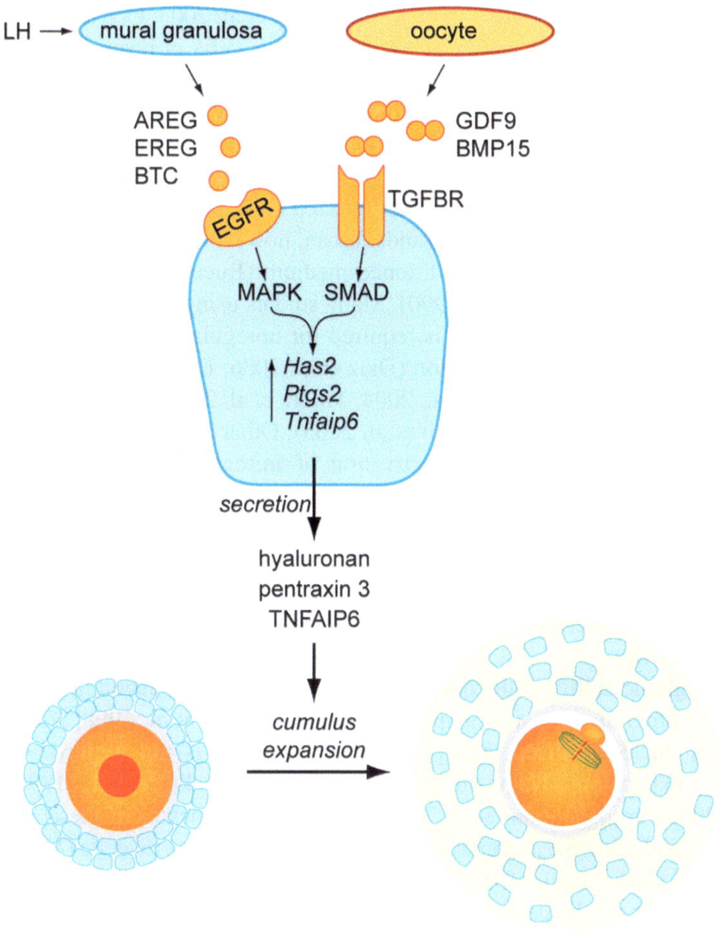

Fig. 8.5 Oocyte–granulosa communication and cumulus expansion. The combined influences of EGFR activation by its ligands and SMAD signaling activated by oocyte-derived factors such as GDF9 and BMP15 trigger the increased transcription in the cumulus cells of genes whose products are secreted and assembled into a matrix. This process, termed cumulus expansion, is required for efficient fertilization

et al. 1997a; Hess et al. 1998; Sato et al. 2001; Yoshioka et al. 2000) and proteoglycans (McArthur et al. 2000) that maintain the stability and structure of the matrix. The integrity of the mucified matrix also requires proteins such as pentraxin 3 (PTX3) and tumor necrosis factor-inducible gene 6 (TNFAIP6) (Baranova et al. 2014; Fulop et al. 2003; Ievoli et al. 2011; Salustri et al. 2004; Sanggaard et al. 2008; Scarchilli et al. 2007) which mediate the interactions between the components of the matrix. These factors, along with other proteins that are needed for cumulus expansion including *Ptgs2* (prostaglandin-endoperoxide synthase 2) (Davis et al. 1999; Ochsner et al. 2003b; Sirois et al. 1992), hyaluronan synthase

2 (*Has2*) (Sugiura et al. 2009), and *Tnfaip6* (Fulop et al. 2003; Ochsner et al. 2003a; Sugiura et al. 2009), are produced by the cumulus cells in response to LH surge (Fig. 8.5) (Eppig 1981; Fulop et al. 1997b; Salustri et al. 1999).

Although cumulus expansion occurs in response to the LH stimulus, which is transduced at least in part through the EGFR pathway, it also requires a signaling input from the oocyte. If the oocyte is microsurgically removed from a cumulus–oocyte complex, the remaining oocytectomized cumulus cell shell cannot undergo expansion in response to EGFR ligands. It can, however, expand if it is cocultured with oocytes or with oocyte-conditioned medium (Buccione et al. 1990; Salustri et al. 1990; Vanderhyden et al. 1990). Many studies using the mouse have demonstrated that signaling by ODPFs is required for upregulation of expansion-related mRNAs and for cumulus expansion (Diaz et al. 2006; Dragovic et al. 2005, 2007; Joyce et al. 2001; Mazerbourg et al. 2004; Moore et al. 2003; Sasseville et al. 2010; Su et al. 2003, 2010; Vanderhyden et al. 2003). Other studies have suggested that the ODPFs are needed for the activation of mitogen-activated protein kinase (MAPK) (Su et al. 2003) and for maintaining the activity of EGFR (Su et al. 2010) in the cumulus cells. The activity of the ODPFs appears to be potentiated by heparan sulfate proteoglycans (HSPG) acting as co-receptors (Watson et al. 2012). The oocyte thus enables cumulus expansion both indirectly through promoting EGFR activity and directly by increasing the expression of expansion-related transcripts.

Although the ability of the oocyte to induce cumulus expansion in vitro is also a feature of other species such as pig and rat (Prochazka et al. 1998; Vanderhyden 1993), in the cow, sheep, and pig, oocyte signaling is not required to induce cumulus expansion (Nagyova et al. 1999; Prochazka et al. 1991; Ralph et al. 1995; Varnosfaderani et al. 2013). Interestingly, oocytes of these species are capable of inducing the expansion of mouse cumulus granulosa cells (Vanderhyden 1993). Whether signaling from the oocyte is required during cumulus expansion in humans remains to be determined. In summary, in the mouse at least, the somatic and germ-line compartments of the follicle cooperate to induce the cumulus cells to expand and mucify in response to the LH signal, thus fulfilling an essential prerequisite for fertilization.

8.8 Conclusion

As in many tissues and organs, growth and differentiation of the ovarian follicle requires a precisely regulated program of intercellular communication and cooperation. We have described several important events during ovarian follicle growth that exemplify this cooperation, specifically between the oocyte and the granulosa cells. Although the gap junction-mediated support from the granulosa to the oocyte is best understood, it is plausible that the granulosa also support the growing oocyte through gap junction-independent pathways. For example, recent work suggests that bovine granulosa cells can transfer mRNAs and long noncoding RNAs to the oocyte (Macaulay et al. 2014). Yet uncharacterized secreted or membrane-associated molecules of the granulosa may also be important. In addition to the events we have discussed, other aspects of folliculogenesis also depend on intercellular cooperation, including the transition from a primary to a secondary follicle (Braw-Tal 2002; Dong et al. 1996; Galloway et al. 2000; Parrott and Skinner 1999; Yoshida et al. 1997), formation of the TZPs (Carabatsos et al. 1998), thecal cell recruitment (Dong et al. 1996; Liu et al. 2015; Spicer et al. 2008), differentiation of the granulosa into mural and cumulus cell compartments, and granulosa cell luteinization (Eppig et al. 1997; Glister et al. 2003; Li et al. 2000; Nekola and Nalbandov 1971; Vanderhyden 1993). All of these processes serve, directly or indirectly, the growth and development of the oocyte and thereby enable the ovulation of a mature and developmentally competent egg ready to give rise to a new organism.

Competing Interests The authors declare no competing interests.

Grant Support Supported by the Canadian Institutes for Health Research (H.J.C.). S.E.H. was supported by the CIHR Training Program in Reproduction, Early Development, and the Impact on Health, by the Réseau Québécois en Reproduction, and by the Research Institute of the McGill University Health Centre.

References

Abe K, Inoue A, Suzuki MG, Aoki F (2010) Global gene silencing is caused by the dissociation of RNA polymerase II from DNA in mouse oocytes. J Reprod Dev 56:502–507

Adashi EY (1994) Endocrinology of the ovary. Hum Reprod 9:815–827

Adhikari D, Flohr G, Gorre N et al (2009) Disruption of Tsc2 in oocytes leads to overactivation of the entire pool of primordial follicles. Mol Hum Reprod 15:765–770

Adhikari D, Zheng W, Shen Y et al (2010) Tsc/mTORC1 signaling in oocytes governs the quiescence and activation of primordial follicles. Hum Mol Genet 19:397–410

Albertini DF, Rider V (1994) Patterns of intercellular connectivity in the mammalian cumulus-oocyte complex. Microsc Res Tech 27:125–133

Anderson E, Albertini DF (1976) Gap junctions between the oocyte and companion follicle cells in the mammalian ovary. J Cell Biol 71:680–686

Andreu-Vieyra CV, Chen R, Agno JE et al (2010) MLL2 is required in oocytes for bulk histone 3 lysine 4 trimethylation and transcriptional silencing. PLoS Biol 8:e1000453

Arakane F, Sugawara T, Nishino H et al (1996) Steroidogenic acute regulatory protein (StAR) retains activity in the absence of its mitochondrial import sequence: implications for the mechanism of StAR action. Proc Natl Acad Sci U S A 93:13731–13736

Ashkenazi H, Cao X, Motola S et al (2005) Epidermal growth factor family members: endogenous mediators of the ovulatory response. Endocrinology 146:77–84

Bachvarova R (1985) Gene expression during oogenesis and oocyte development in mammals. Dev Biol (N Y 1985) 1:453–524

Balasubramanian K, Lavoie HA, Garmey JC et al (1997) Regulation of porcine granulosa cell steroidogenic acute regulatory protein (StAR) by insulin-like growth factor I: synergism with follicle-stimulating hormone or protein kinase A agonist. Endocrinology 138:433–439

Baranova NS, Inforzato A, Briggs DC et al (2014) Incorporation of pentraxin 3 into hyaluronan matrices is tightly regulated and promotes matrix cross-linking. J Biol Chem 289:30481–30498

Berisha B, Pfaffl MW, Schams D (2002) Expression of estrogen and progesterone receptors in the bovine ovary during estrous cycle and pregnancy. Endocrine 17:207–214

Billig H, Furuta I, Hsueh AJ (1993) Estrogens inhibit and androgens enhance ovarian granulosa cell apoptosis. Endocrinology 133:2204–2212

Boland NI, Humpherson PG, Leese HJ, Gosden RG (1994) The effect of glucose metabolism on murine follicle development and steroidogenesis in vitro. Hum Reprod 9:617–623

Bornslaeger EA, Wilde MW, Schultz RM (1984) Regulation of mouse oocyte maturation: involvement of cyclic AMP phosphodiesterase and calmodulin. Dev Biol 105:488–499

Borum K (1961) Oogenesis in the mouse. A study of the meiotic prophase. Exp Cell Res 24:495–507

Bouniol-Baly C, Hamraoui L, Guibert J et al (1999) Differential transcriptional activity associated with chromatin configuration in fully grown mouse germinal vesicle oocytes. Biol Reprod 60:580–587

Braw-Tal R (2002) The initiation of follicle growth: the oocyte or the somatic cells? Mol Cell Endocrinol 187:11–18

Brinster RL (1971) Oxidation of pyruvate and glucose by oocytes of the mouse and rhesus monkey. J Reprod Fertil 24:187–191

Britt KL, Drummond AE, Cox VA et al (2000) An age-related ovarian phenotype in mice with targeted disruption of the Cyp 19 (aromatase) gene. Endocrinology 141:2614–2623

Brunet S, Maro B (2005) Cytoskeleton and cell cycle control during meiotic maturation of the mouse oocyte: integrating time and space. Reproduction 130:801–811

Brunet A, Bonni A, Zigmond MJ et al (1999) Akt promotes cell survival by phosphorylating and inhibiting a Forkhead transcription factor. Cell 96:857–868

Brunn GJ, Hudson CC, Sekulic A et al (1997) Phosphorylation of the translational repressor PHAS-I by the mammalian target of rapamycin. Science 277:99–101

Buccione R, Vanderhyden BC, Caron PJ, Eppig JJ (1990) FSH-induced expansion of the mouse cumulus oophorus in vitro is dependent upon a specific factor(s) secreted by the oocyte. Dev Biol 138:16–25

Camp TA, Rahal JO, Mayo KE (1991) Cellular localization and hormonal regulation of follicle-stimulating hormone and luteinizing hormone receptor messenger RNAs in the rat ovary. Mol Endocrinol 5:1405–1417

Carabatsos MJ, Elvin J, Matzuk MM, Albertini DF (1998) Characterization of oocyte and follicle development in growth differentiation factor-9-deficient mice. Dev Biol 204:373–384

Castrillon DH, Miao L, Kollipara R et al (2003) Suppression of ovarian follicle activation in mice by the transcription factor Foxo3a. Science 301:215–218

Chang HC, Liu H, Zhang J et al (2005) Developmental incompetency of denuded mouse oocytes undergoing maturation in vitro is ooplasmic in nature and is associated with aberrant Oct-4 expression. Hum Reprod 20:1958–1968

Chen L, Russell PT, Larsen WJ (1993) Functional significance of cumulus expansion in the mouse: roles for the preovulatory synthesis of hyaluronic acid within the cumulus mass. Mol Reprod Dev 34:87–93

Chen L, Zhang H, Powers RW et al (1996) Covalent linkage between proteins of the inter-alpha-inhibitor family and hyaluronic acid is mediated by a factor produced by granulosa cells. J Biol Chem 271:19409–19414

Chen J, Melton C, Suh N et al (2011) Genome-wide analysis of translation reveals a critical role for deleted in azoospermia-like (Dazl) at the oocyte-to-zygote transition. Genes Dev 25:755–766

Chen J, Torcia S, Xie F et al (2013) Somatic cells regulate maternal mRNA translation and developmental competence of mouse oocytes. Nat Cell Biol 15:1415–1423

Cho WK, Stern S, Biggers JD (1974) Inhibitory effect of dibutyryl cAMP on mouse oocyte maturation in vitro. J Exp Zool 187:383–386

Choi Y, Ballow DJ, Xin Y, Rajkovic A (2008a) Lim homeobox gene, lhx8, is essential for mouse oocyte differentiation and survival. Biol Reprod 79:442–449

Choi Y, Yuan D, Rajkovic A (2008b) Germ cell-specific transcriptional regulator sohlh2 is essential for early mouse folliculogenesis and oocyte-specific gene expression. Biol Reprod 79:1176–1182
Christians E, Boiani M, Garagna S et al (1999) Gene expression and chromatin organization during mouse oocyte growth. Dev Biol 207:76–85
Clark BJ, Wells J, King SR, Stocco DM (1994) The purification, cloning, and expression of a novel luteinizing hormone-induced mitochondrial protein in MA-10 mouse Leydig tumor cells. Characterization of the steroidogenic acute regulatory protein (StAR). J Biol Chem 269:28314–28322
Clarke HJ (2012) Post-transcriptional control of gene expression during mouse oogenesis. Results Probl Cell Differ 55:1–21
Collado-Fernandez E, Picton HM, Dumollard R (2012) Metabolism throughout follicle and oocyte development in mammals. Int J Dev Biol 56:799–808
Colonna R, Mangia F (1983) Mechanisms of amino acid uptake in cumulus-enclosed mouse oocytes. Biol Reprod 28:797–803
Comiskey M, Warner CM (2007) Spatio-temporal localization of membrane lipid rafts in mouse oocytes and cleaving preimplantation embryos. Dev Biol 303:727–739
Conley AJ, Howard HJ, Slanger WD, Ford JJ (1994) Steroidogenesis in the preovulatory porcine follicle. Biol Reprod 51:655–661
Conti M, Hsieh M, Zamah AM, Oh JS (2012) Novel signaling mechanisms in the ovary during oocyte maturation and ovulation. Mol Cell Endocrinol 356:65–73
Corbett HE, Dube CD, Slow S et al (2014) Uptake of betaine into mouse cumulus-oocyte complexes via the SLC7A6 isoform of y+L transporter. Biol Reprod 90:81
Coskun S, Uzumcu M, Lin YC et al (1995) Regulation of cumulus cell steroidogenesis by the porcine oocyte and preliminary characterization of oocyte-produced factor(s). Biol Reprod 53:670–675
Couse JF, Yates MM, Deroo BJ, Korach KS (2005) Estrogen receptor-beta is critical to granulosa cell differentiation and the ovulatory response to gonadotropins. Endocrinology 146:3247–3262
Dalton CM, Carroll J (2013) Biased inheritance of mitochondria during asymmetric cell division in the mouse oocyte. J Cell Sci 126:2955–2964
Da Silva-Buttkus P, Jayasooriya GS, Mora JM et al (2008) Effect of cell shape and packing density on granulosa cell proliferation and formation of multiple layers during early follicle development in the ovary. J Cell Sci 121:3890–3900
Davis BJ, Lennard DE, Lee CA et al (1999) Anovulation in cyclooxygenase-2-deficient mice is restored by prostaglandin E2 and interleukin-1beta. Endocrinology 140:2685–2695
De La Fuente R (2006) Chromatin modifications in the germinal vesicle (GV) of mammalian oocytes. Dev Biol 292:1–12
De La Fuente R, O'Brien MJ, Eppig JJ (1999) Epidermal growth factor enhances preimplantation developmental competence of maturing mouse oocytes. Hum Reprod 14:3060–3068
De Smedt V, Szollosi D (1991) Cytochalasin D treatment induces meiotic resumption in follicular sheep oocytes. Mol Reprod Dev 29:163–171
Dekel N, Beers WH (1978) Rat oocyte maturation in vitro: relief of cyclic AMP inhibition by gonadotropins. Proc Natl Acad Sci U S A 75:4369–4373
Dekel N, Beers WH (1980) Development of the rat oocyte in vitro: inhibition and induction of maturation in the presence or absence of the cumulus oophorus. Dev Biol 75:247–254
Dekel N, Hillensjo T, Kraicer PF (1979) Maturational effects of gonadotropins on the cumulus-oocyte complex of the rat. Biol Reprod 20:191–197
Devoto L, Christenson LK, McAllister JM et al (1999) Insulin and insulin-like growth factor-I and -II modulate human granulosa-lutein cell steroidogenesis: enhancement of steroidogenic acute regulatory protein (StAR) expression. Mol Hum Reprod 5:1003–1010

Diaz FJ, O'Brien MJ, Wigglesworth K, Eppig JJ (2006) The preantral granulosa cell to cumulus cell transition in the mouse ovary: development of competence to undergo expansion. Dev Biol 299:91–104

Diaz FJ, Wigglesworth K, Eppig JJ (2007) Oocytes determine cumulus cell lineage in mouse ovarian follicles. J Cell Sci 120:1330–1340

Dissen GA, Romero C, Hirshfield AN, Ojeda SR (2001) Nerve growth factor is required for early follicular development in the mammalian ovary. Endocrinology 142:2078–2086

Dong J, Albertini DF, Nishimori K et al (1996) Growth differentiation factor-9 is required during early ovarian folliculogenesis. Nature 383:531–535

Dorrington JH, Moon YS, Armstrong DT (1975) Estradiol-17beta biosynthesis in cultured granulosa cells from hypophysectomized immature rats; stimulation by follicle-stimulating hormone. Endocrinology 97:1328–1331

Downs SM, Chen J (2008) EGF-like peptides mediate FSH-induced maturation of cumulus cell-enclosed mouse oocytes. Mol Reprod Dev 75:105–114

Dragovic RA, Ritter LJ, Schulz SJ et al (2005) Role of oocyte-secreted growth differentiation factor 9 in the regulation of mouse cumulus expansion. Endocrinology 146:2798–2806

Dragovic RA, Ritter LJ, Schulz SJ et al (2007) Oocyte-secreted factor activation of SMAD 2/3 signaling enables initiation of mouse cumulus cell expansion. Biol Reprod 76:848–857

Dupont S, Krust A, Gansmuller A et al (2000) Effect of single and compound knockouts of estrogen receptors alpha (ERalpha) and beta (ERbeta) on mouse reproductive phenotypes. Development 127:4277–4291

Egbert JR, Shuhaibar LC, Edmund AB et al (2014) Dephosphorylation and inactivation of NPR2 guanylyl cyclase in granulosa cells contributes to the LH-induced decrease in cGMP that causes resumption of meiosis in rat oocytes. Development 141:3594–3604

Eggan K, Jurga S, Gosden R et al (2006) Ovulated oocytes in adult mice derive from non-circulating germ cells. Nature 441:1109–1114

Eisenbach M (1999) Mammalian sperm chemotaxis and its association with capacitation. Dev Genet 25:87–94

el-Fouly MA, Cook B, Nekola M, Nalbandov AV (1970) Role of the ovum in follicular luteinization. Endocrinology 87:286–293

Emmen JM, Couse JF, Elmore SA et al (2005) In vitro growth and ovulation of follicles from ovaries of estrogen receptor (ER){alpha} and ER{beta} null mice indicate a role for ER{beta} in follicular maturation. Endocrinology 146:2817–2826

Emori C, Sugiura K (2014) Role of oocyte-derived paracrine factors in follicular development. Anim Sci J 85:627–633

Emori C, Wigglesworth K, Fujii W et al (2013) Cooperative effects of 17beta-estradiol and oocyte-derived paracrine factors on the transcriptome of mouse cumulus cells. Endocrinology 154:4859–4872

Ene AC, Park S, Edelmann W, Taketo T (2013) Caspase 9 is constitutively activated in mouse oocytes and plays a key role in oocyte elimination during meiotic prophase progression. Dev Biol 377:213–223

Eppig JJ (1976) Analysis of mouse oogenesis in vitro. Oocyte isolation and the utilization of exogenous energy sources by growing oocytes. J Exp Zool 198:375–382

Eppig JJ (1979) Gonadotropin stimulation of the expansion of cumulus oophori isolated from mice: general conditions for expansion in vitro. J Exp Zool 208:111–120

Eppig JJ (1980) Regulation of cumulus oophorus expansion by gonadotropins in vivo and in vitro. Biol Reprod 23:545–552

Eppig JJ (1981) Prostaglandin E2 stimulates cumulus expansion and hyaluronic acid synthesis by cumuli oophori isolated from mice. Biol Reprod 25:191–195

Eppig JJ, Schroeder AC (1989) Capacity of mouse oocytes from preantral follicles to undergo embryogenesis and development to live young after growth, maturation, and fertilization in vitro. Biol Reprod 41:268–276

Eppig JJ, Wigglesworth K, Pendola F, Hirao Y (1997) Murine oocytes suppress expression of luteinizing hormone receptor messenger ribonucleic acid by granulosa cells. Biol Reprod 56:976–984

Eppig JJ, Pendola FL, Wigglesworth K, Pendola JK (2005) Mouse oocytes regulate metabolic cooperativity between granulosa cells and oocytes: amino acid transport. Biol Reprod 73:351–357

Erickson GF, Wang C, Hsueh AJ (1979) FSH induction of functional LH receptors in granulosa cells cultured in a chemically defined medium. Nature 279:336–338

Espey LL (1980) Ovulation as an inflammatory reaction—a hypothesis. Biol Reprod 22:73–106

Fagbohun CF, Downs SM (1992) Requirement for glucose in ligand-stimulated meiotic maturation of cumulus cell-enclosed mouse oocytes. J Reprod Fertil 96:681–697

Findlay JK, Hutt KJ, Hickey M, Anderson RA (2015) How is the number of primordial follicles in the ovarian reserve established? Biol Reprod 93:111

Fitzpatrick SL, Richards JS (1991) Regulation of cytochrome P450 aromatase messenger ribonucleic acid and activity by steroids and gonadotropins in rat granulosa cells. Endocrinology 129:1452–1462

Fortune JE, Armstrong DT (1977) Androgen production by theca and granulosa isolated from proestrous rat follicles. Endocrinology 100:1341–1347

Franciosi F, Coticchio G, Lodde V et al (2014) Natriuretic peptide precursor C delays meiotic resumption and sustains gap junction-mediated communication in bovine cumulus-enclosed oocytes. Biol Reprod 91:61

Fuhrer F, Mayr B, Schellander K et al (1989) Maturation competence and chromatin behaviour in growing and fully grown cattle oocytes. Zentralbl Veterinarmed A 36:285–291

Fulop C, Kamath RV, Li Y et al (1997a) Coding sequence, exon-intron structure and chromosomal localization of murine TNF-stimulated gene 6 that is specifically expressed by expanding cumulus cell-oocyte complexes. Gene 202:95–102

Fulop C, Salustri A, Hascall VC (1997b) Coding sequence of a hyaluronan synthase homologue expressed during expansion of the mouse cumulus-oocyte complex. Arch Biochem Biophys 337:261–266

Fulop C, Szanto S, Mukhopadhyay D et al (2003) Impaired cumulus mucification and female sterility in tumor necrosis factor-induced protein-6 deficient mice. Development 130:2253–2261

Galloway SM, McNatty KP, Cambridge LM et al (2000) Mutations in an oocyte-derived growth factor gene (BMP15) cause increased ovulation rate and infertility in a dosage-sensitive manner. Nat Genet 25:279–283

Gilchrist RB, Lane M, Thompson JG (2008) Oocyte-secreted factors: regulators of cumulus cell function and oocyte quality. Hum Reprod Update 14:159–177

Glister C, Tannetta DS, Groome NP, Knight PG (2001) Interactions between follicle-stimulating hormone and growth factors in modulating secretion of steroids and inhibin-related peptides by nonluteinized bovine granulosa cells. Biol Reprod 65:1020–1028

Glister C, Groome NP, Knight PG (2003) Oocyte-mediated suppression of follicle-stimulating hormone- and insulin-like growth factor-induced secretion of steroids and inhibin-related proteins by bovine granulosa cells in vitro: possible role of transforming growth factor alpha. Biol Reprod 68:758–765

Grive KJ, Freiman RN (2015) The developmental origins of the mammalian ovarian reserve. Development 142:2554–2563

Haghighat N, Van Winkle LJ (1990) Developmental change in follicular cell-enhanced amino acid uptake into mouse oocytes that depends on intact gap junctions and transport system Gly. J Exp Zool 253:71–82

Han SJ, Chen R, Paronetto MP, Conti M (2005) Wee1B is an oocyte-specific kinase involved in the control of meiotic arrest in the mouse. Curr Biol 15:1670–1676

Harris SE, Adriaens I, Leese HJ, Gosden RG, Picton HM (2007) Carbohydrate metabolism by murine ovarian follicles and oocytes grown in vitro. Reproduction 134:415–424

Harris SE, Leese HJ, Gosden RG, Picton HM (2009) Pyruvate and oxygen consumption throughout the growth and development of murine oocytes. Mol Reprod Dev 76:231–238
Hasegawa A, Mochida N, Ogasawara T, Koyama K (2006) Pup birth from mouse oocytes in preantral follicles derived from vitrified and warmed ovaries followed by in vitro growth, in vitro maturation, and in vitro fertilization. Fertil Steril 86:1182–1192
Hertig AT, Adams EC (1967) Studies on the human oocyte and its follicle. I. Ultrastructural and histochemical observations on the primordial follicle stage. J Cell Biol 34:647–675
Hess KA, Chen L, Larsen WJ (1998) The ovarian blood follicle barrier is both charge- and size-selective in mice. Biol Reprod 58:705–711
Hickey GJ, Chen SA, Besman MJ et al (1988) Hormonal regulation, tissue distribution, and content of aromatase cytochrome P450 messenger ribonucleic acid and enzyme in rat ovarian follicles and corpora lutea: relationship to estradiol biosynthesis. Endocrinology 122:1426–1436
Hinrichs K, Schmidt AL, Friedman PP et al (1993) In vitro maturation of horse oocytes: characterization of chromatin configuration using fluorescence microscopy. Biol Reprod 48:363–370
Hirshfield AN (1991) Development of follicles in the mammalian ovary. Int Rev Cytol 124:43–101
Hosaka T, Biggs WH 3rd, Tieu D et al (2004) Disruption of forkhead transcription factor (FOXO) family members in mice reveals their functional diversification. Proc Natl Acad Sci U S A 101:2975–2980
Hsieh M, Lee D, Panigone S et al (2007) Luteinizing hormone-dependent activation of the epidermal growth factor network is essential for ovulation. Mol Cell Biol 27:1914–1924
Hsieh M, Thao K, Conti M (2011) Genetic dissection of epidermal growth factor receptor signaling during luteinizing hormone-induced oocyte maturation. PLoS One 6:e21574
Hsu CJ, Hammond JM (1987) Gonadotropins and estradiol stimulate immunoreactive insulin-like growth factor-I production by porcine granulosa cells in vitro. Endocrinology 120:198–207
Hutt KJ (2015) The role of BH3-only proteins in apoptosis within the ovary. Reproduction 149: R81–R89
Ievoli E, Lindstedt R, Inforzato A et al (2011) Implication of the oligomeric state of the N-terminal PTX3 domain in cumulus matrix assembly. Matrix Biol 30:330–337
Igarashi H, Takahashi T, Takahashi E et al (2005) Aged mouse oocytes fail to readjust intracellular adenosine triphosphates at fertilization. Biol Reprod 72:1256–1261
Jensen JT, Schwinof KM, Zelinski-Wooten MB et al (2002) Phosphodiesterase 3 inhibitors selectively block the spontaneous resumption of meiosis by macaque oocytes in vitro. Hum Reprod 17:2079–2084
John GB, Gallardo TD, Shirley LJ, Castrillon DH (2008) Foxo3 is a PI3K-dependent molecular switch controlling the initiation of oocyte growth. Dev Biol 321:197–204
John GB, Shidler MJ, Besmer P, Castrillon DH (2009) Kit signaling via PI3K promotes ovarian follicle maturation but is dispensable for primordial follicle activation. Dev Biol 331:292–299
Jorgensen JS (2013) Defining the neighborhoods that escort the oocyte through its early life events and into a functional follicle. Mol Reprod Dev 80:960–976
Joyce IM, Pendola FL, O'Brien M, Eppig JJ (2001) Regulation of prostaglandin-endoperoxide synthase 2 messenger ribonucleic acid expression in mouse granulosa cells during ovulation. Endocrinology 142:3187–3197
Kang MK, Han SJ (2011) Post-transcriptional and post-translational regulation during mouse oocyte maturation. BMB Rep 44:147–157
Kawamura K, Cheng Y, Kawamura N et al (2011) Pre-ovulatory LH/hCG surge decreases C-type natriuretic peptide secretion by ovarian granulosa cells to promote meiotic resumption of pre-ovulatory oocytes. Hum Reprod 26:3094–3101
Kessel RG, Tung HN, Roberts R, Beams HW (1985) The presence and distribution of gap junctions in the oocyte-follicle cell complex of the zebrafish, Brachydanio rerio. J Submicrosc Cytol 17:239–253

Kezele P, Nilsson EE, Skinner MK (2005) Keratinocyte growth factor acts as a mesenchymal factor that promotes ovarian primordial to primary follicle transition. Biol Reprod 73:967–973
Khalid M, Haresign W, Luck MR (2000) Secretion of IGF-1 by ovine granulosa cells: effects of growth hormone and follicle stimulating hormone. Anim Reprod Sci 58:261–272
Kidder GM, Mhawi AA (2002) Gap junctions and ovarian folliculogenesis. Reproduction 123:613–620
Kissel H, Timokhina I, Hardy MP et al (2000) Point mutation in kit receptor tyrosine kinase reveals essential roles for kit signaling in spermatogenesis and oogenesis without affecting other kit responses. EMBO J 19:1312–1326
Krege JH, Hodgin JB, Couse JF et al (1998) Generation and reproductive phenotypes of mice lacking estrogen receptor beta. Proc Natl Acad Sci U S A 95:15677–15682
LaVoie HA, DeSimone DC, Gillio-Meina C, Hui YY (2002) Cloning and characterization of porcine ovarian estrogen receptor beta isoforms. Biol Reprod 66:616–623
Lee WS, Otsuka F, Moore RK, Shimasaki S (2001) Effect of bone morphogenetic protein-7 on folliculogenesis and ovulation in the rat. Biol Reprod 65:994–999
Lee KB, Zhang M, Sugiura K et al (2013) Hormonal coordination of natriuretic peptide type C and natriuretic peptide receptor 3 expression in mouse granulosa cells. Biol Reprod 88:42
Lei L, Spradling AC (2013) Female mice lack adult germ-line stem cells but sustain oogenesis using stable primordial follicles. Proc Natl Acad Sci U S A 110:8585–8590
Lei ZM, Mishra S, Zou W et al (2001) Targeted disruption of luteinizing hormone/human chorionic gonadotropin receptor gene. Mol Endocrinol 15:184–200
Lew DJ, Kornbluth S (1996) Regulatory roles of cyclin dependent kinase phosphorylation in cell cycle control. Curr Opin Cell Biol 8:795–804
Li R, Norman RJ, Armstrong DT, Gilchrist RB (2000) Oocyte-secreted factor(s) determine functional differences between bovine mural granulosa cells and cumulus cells. Biol Reprod 63:839–845
Li J, Kawamura K, Cheng Y et al (2010) Activation of dormant ovarian follicles to generate mature eggs. Proc Natl Acad Sci U S A 107:10280–10284
Lincoln AJ, Wickramasinghe D, Stein P et al (2002) Cdc25b phosphatase is required for resumption of meiosis during oocyte maturation. Nat Genet 30:446–449
Liu M, Sims D, Calarco P, Talbot P (2003) Biochemical heterogeneity, migration, and pre-fertilization release of mouse oocyte cortical granules. Reprod Biol Endocrinol 1:77
Liu K, Rajareddy S, Liu L, Jagarlamudi K et al (2006) Control of mammalian oocyte growth and early follicular development by the oocyte PI3 kinase pathway: new roles for an old timer. Dev Biol 299:1–11
Liu X, Xie F, Zamah AM et al (2014) Multiple pathways mediate luteinizing hormone regulation of cGMP signaling in the mouse ovarian follicle. Biol Reprod 91:9
Liu C, Peng J, Matzuk MM, Yao HH (2015) Lineage specification of ovarian theca cells requires multicellular interactions via oocyte and granulosa cells. Nat Commun 6:6934
Lodde V, Modina S, Maddox-Hyttel P et al (2008) Oocyte morphology and transcriptional silencing in relation to chromatin remodeling during the final phases of bovine oocyte growth. Mol Reprod Dev 75:915–924
Luciano AM, Franciosi F, Modina SC, Lodde V (2011) Gap junction-mediated communications regulate chromatin remodeling during bovine oocyte growth and differentiation through cAMP-dependent mechanism(s). Biol Reprod 85:1252–1259
Lund SA, Murdoch J, Van Kirk EA, Murdoch WJ (1999) Mitogenic and antioxidant mechanisms of estradiol action in preovulatory ovine follicles: relevance to luteal function. Biol Reprod 61:388–392
Macaulay AD, Gilbert I, Caballero J et al (2014) The gametic synapse: RNA transfer to the bovine oocyte. Biol Reprod 91:90
Malki S, van der Heijden GW, O'Donnell KA et al (2014) A role for retrotransposon LINE-1 in fetal oocyte attrition in mice. Dev Cell 29:521–533

Manova K, Huang EJ, Angeles M et al (1993) The expression pattern of the c-kit ligand in gonads of mice supports a role for the c-kit receptor in oocyte growth and in proliferation of spermatogonia. Dev Biol 157:85–99

Mattson BA, Albertini DF (1990) Oogenesis: chromatin and microtubule dynamics during meiotic prophase. Mol Reprod Dev 25:374–383

Mazerbourg S, Klein C, Roh J et al (2004) Growth differentiation factor-9 signaling is mediated by the type I receptor, activin receptor-like kinase 5. Mol Endocrinol 18:653–665

McArthur ME, Irving-Rodgers HF, Byers S, Rodgers RJ (2000) Identification and immunolocalization of decorin, versican, perlecan, nidogen, and chondroitin sulfate proteoglycans in bovine small-antral ovarian follicles. Biol Reprod 63:913–924

McGee EA, Sawetawan C, Bird I et al (1996) The effect of insulin and insulin-like growth factors on the expression of steroidogenic enzymes in a human ovarian thecal-like tumor cell model. Fertil Steril 65:87–93

Mehlmann LM, Terasaki M, Jaffe LA, Kline D (1995) Reorganization of the endoplasmic reticulum during meiotic maturation of the mouse oocyte. Dev Biol 170:607–615

Mehlmann LM, Jones TL, Jaffe LA (2002) Meiotic arrest in the mouse follicle maintained by a Gs protein in the oocyte. Science 297:1343–1345

Mehlmann LM, Saeki Y, Tanaka S et al (2004) The Gs-linked receptor GPR3 maintains meiotic arrest in mammalian oocytes. Science 306:1947–1950

Meinecke B, Meinecke-Tillmann S (1979) Effects of gonadotropins on oocyte maturation and progesterone production by porcine ovarian follicles cultured in vitro. Theriogenology 11:351–365

Miller WL (1988) Molecular biology of steroid hormone synthesis. Endocr Rev 9:295–318

Monniaux D, Pisselet C (1992) Control of proliferation and differentiation of ovine granulosa cells by insulin-like growth factor-I and follicle-stimulating hormone in vitro. Biol Reprod 46:109–119

Moore RK, Otsuka F, Shimasaki S (2003) Molecular basis of bone morphogenetic protein-15 signaling in granulosa cells. J Biol Chem 278:304–310

Nagyova E, Prochazka R, Vanderhyden BC (1999) Oocytectomy does not influence synthesis of hyaluronic acid by pig cumulus cells: retention of hyaluronic acid after insulin-like growth factor-I treatment in serum-free medium. Biol Reprod 61:569–574

Nakamura T, Iwase A, Bayasula B et al (2015) CYP51A1 induced by growth differentiation factor 9 and follicle-stimulating hormone in granulosa cells is a possible predictor for unfertilization. Reprod Sci 22:377–384

Neal P, Baker TG (1975) Response of mouse graafian follicles in organ culture to varying doses of follicle-stimulating hormone and luteinizing hormone. J Endocrinol 65:27–32

Nekola MV, Nalbandov AV (1971) Morphological changes of rat follicular cells as influenced by oocytes. Biol Reprod 4:154–160

Nicosia SV, Wolf DP, Inoue M (1977) Cortical granule distribution and cell surface characteristics in mouse eggs. Dev Biol 57:56–74

Nilsson E, Skinner MK (2001) Cellular interactions that control primordial follicle development and folliculogenesis. J Soc Gynecol Investig 8:S17–S20

Nilsson EE, Detzel C, Skinner MK (2006) Platelet-derived growth factor modulates the primordial to primary follicle transition. Reproduction 131:1007–1015

Norris RP, Ratzan WJ, Freudzon M et al (2009) Cyclic GMP from the surrounding somatic cells regulates cyclic AMP and meiosis in the mouse oocyte. Development 136:1869–1878

Norris RP, Freudzon M, Nikolaev VO, Jaffe LA (2010) Epidermal growth factor receptor kinase activity is required for gap junction closure and for part of the decrease in ovarian follicle cGMP in response to LH. Reproduction 140:655–662

Oatley JM, Brinster RL (2012) The germline stem cell niche unit in mammalian testes. Physiol Rev 92:577–595

Ochsner SA, Day AJ, Rugg MS et al (2003a) Disrupted function of tumor necrosis factor-alpha-stimulated gene 6 blocks cumulus cell-oocyte complex expansion. Endocrinology 144:4376–4384
Ochsner SA, Russell DL, Day AJ et al (2003b) Decreased expression of tumor necrosis factor-alpha-stimulated gene 6 in cumulus cells of the cyclooxygenase-2 and EP2 null mice. Endocrinology 144:1008–1019
Oh JS, Han SJ, Conti M (2010) Wee1B, Myt1, and Cdc25 function in distinct compartments of the mouse oocyte to control meiotic resumption. J Cell Biol 188:199–207
Olsiewski PJ, Beers WH (1983) cAMP synthesis in the rat oocyte. Dev Biol 100:287–293
Packer AI, Hsu YC, Besmer P, Bachvarova RF (1994) The ligand of the c-kit receptor promotes oocyte growth. Dev Biol 161:194–205
Pangas SA, Choi Y, Ballow DJ et al (2006) Oogenesis requires germ cell-specific transcriptional regulators Sohlh1 and Lhx8. Proc Natl Acad Sci U S A 103:8090–8095
Parfenov V, Potchukalina G, Dudina L et al (1989) Human antral follicles: oocyte nucleus and the karyosphere formation (electron microscopic and autoradiographic data). Gamete Res 22:219–231
Park JY, Su YQ, Ariga M et al (2004) EGF-like growth factors as mediators of LH action in the ovulatory follicle. Science 303:682–684
Parker KL, Schimmer BP (1995) Transcriptional regulation of the genes encoding the cytochrome P-450 steroid hydroxylases. Vitam Horm 51:339–370
Parrott JA, Skinner MK (1999) Kit-ligand/stem cell factor induces primordial follicle development and initiates folliculogenesis. Endocrinology 140:4262–4271
Payne C, Schatten G (2003) Golgi dynamics during meiosis are distinct from mitosis and are coupled to endoplasmic reticulum dynamics until fertilization. Dev Biol 264:50–63
Paynton BV, Bachvarova R (1994) Polyadenylation and deadenylation of maternal mRNAs during oocyte growth and maturation in the mouse. Mol Reprod Dev 37:172–180
Pelland AM, Corbett HE, Baltz JM (2009) Amino Acid transport mechanisms in mouse oocytes during growth and meiotic maturation. Biol Reprod 81:1041–1054
Pepling ME (2012) Follicular assembly: mechanisms of action. Reproduction 143:139–149
Pepling ME, Spradling AC (2001) Mouse ovarian germ cell cysts undergo programmed breakdown to form primordial follicles. Dev Biol 234:339–351
Pescador N, Houde A, Stocco DM, Murphy BD (1997) Follicle-stimulating hormone and intracellular second messengers regulate steroidogenic acute regulatory protein messenger ribonucleic acid in luteinized porcine granulosa cells. Biol Reprod 57:660–668
Pesce M, Di Carlo A, De Felici M (1997) The c-kit receptor is involved in the adhesion of mouse primordial germ cells to somatic cells in culture. Mech Dev 68:37–44
Peter M, Dubuis JM (2000) Transcription factors as regulators of steroidogenic P-450 enzymes. Eur J Clin Invest 30(Suppl 3):14–20
Piquette GN, LaPolt PS, Oikawa M, Hsueh AJ (1991) Regulation of luteinizing hormone receptor messenger ribonucleic acid levels by gonadotropins, growth factors, and gonadotropin-releasing hormone in cultured rat granulosa cells. Endocrinology 128:2449–2456
Pirino G, Wescott MP, Donovan PJ (2009) Protein kinase A regulates resumption of meiosis by phosphorylation of Cdc25B in mammalian oocytes. Cell Cycle 8:665–670
Pratt HP (1982) Preimplantation mouse embryos synthesize membrane sterols. Dev Biol 89:101–110
Prochazka R, Nagyova E, Rimkevicova Z et al (1991) Lack of effect of oocytectomy on expansion of the porcine cumulus. J Reprod Fertil 93:569–576
Prochazka R, Nagyova E, Brem G et al (1998) Secretion of cumulus expansion-enabling factor (CEEF) in porcine follicles. Mol Reprod Dev 49:141–149
Quirk SM, Cowan RG, Harman RM (2006) The susceptibility of granulosa cells to apoptosis is influenced by oestradiol and the cell cycle. J Endocrinol 189:441–453
Rajah R, Glaser EM, Hirshfield AN (1992) The changing architecture of the neonatal rat ovary during histogenesis. Dev Dyn 194:177–192

Ralph JH, Telfer EE, Wilmut I (1995) Bovine cumulus cell expansion does not depend on the presence of an oocyte secreted factor. Mol Reprod Dev 42:248–253

Reddy P, Shen L, Ren C et al (2005) Activation of Akt (PKB) and suppression of FKHRL1 in mouse and rat oocytes by stem cell factor during follicular activation and development. Dev Biol 281:160–170

Reddy P, Liu L, Adhikari D et al (2008) Oocyte-specific deletion of Pten causes premature activation of the primordial follicle pool. Science 319:611–613

Ren Y, Suzuki H, Jagarlamudi K et al (2015) Lhx8 regulates primordial follicle activation and postnatal folliculogenesis. BMC Biol 13:39

Richani D, Wang X, Zeng HT et al (2014) Pre-maturation with cAMP modulators in conjunction with EGF-like peptides during in vitro maturation enhances mouse oocyte developmental competence. Mol Reprod Dev 81:422–435

Richard S, Baltz JM (2014) Prophase I arrest of mouse oocytes mediated by natriuretic peptide precursor C requires GJA1 (connexin-43) and GJA4 (connexin-37) gap junctions in the antral follicle and cumulus-oocyte complex. Biol Reprod 90:137

Richard FJ, Tsafriri A, Conti M (2001) Role of phosphodiesterase type 3A in rat oocyte maturation. Biol Reprod 65:1444–1451

Richards JS (1980) Maturation of ovarian follicles: actions and interactions of pituitary and ovarian hormones on follicular cell differentiation. Physiol Rev 60:51–89

Richards JS (1994) Hormonal control of gene expression in the ovary. Endocr Rev 15:725–751

Richards JS, Jahnsen T, Hedin L et al (1987) Ovarian follicular development: from physiology to molecular biology. Recent Prog Horm Res 43:231–276

Richards JS, Russell DL, Ochsner S et al (2002) Novel signaling pathways that control ovarian follicular development, ovulation, and luteinization. Recent Prog Horm Res 57:195–220

Robinson JW, Zhang M, Shuhaibar LC et al (2012) Luteinizing hormone reduces the activity of the NPR2 guanylyl cyclase in mouse ovarian follicles, contributing to the cyclic GMP decrease that promotes resumption of meiosis in oocytes. Dev Biol 366:308–316

Robker RL, Richards JS (1998) Hormone-induced proliferation and differentiation of granulosa cells: a coordinated balance of the cell cycle regulators cyclin D2 and p27Kip1. Mol Endocrinol 12:924–940

Rosenfeld CS, Wagner JS, Roberts RM, Lubahn DB (2001) Intraovarian actions of oestrogen. Reproduction 122:215–226

Salustri A, Ulisse S, Yanagishita M, Hascall VC (1990) Hyaluronic acid synthesis by mural granulosa cells and cumulus cells in vitro is selectively stimulated by a factor produced by oocytes and by transforming growth factor-beta. J Biol Chem 265:19517–19523

Salustri A, Camaioni A, Di Giacomo M et al (1999) Hyaluronan and proteoglycans in ovarian follicles. Hum Reprod Update 5:293–301

Salustri A, Garlanda C, Hirsch E et al (2004) PTX3 plays a key role in the organization of the cumulus oophorus extracellular matrix and in in vivo fertilization. Development 131:1577–1586

Sanggaard KW, Sonne-Schmidt CS, Krogager TP et al (2008) The transfer of heavy chains from bikunin proteins to hyaluronan requires both TSG-6 and HC2. J Biol Chem 283:18530–18537

Sasseville M, Ritter LJ, Nguyen TM et al (2010) Growth differentiation factor 9 signaling requires ERK1/2 activity in mouse granulosa and cumulus cells. J Cell Sci 123:3166–3176

Sato H, Kajikawa S, Kuroda S et al (2001) Impaired fertility in female mice lacking urinary trypsin inhibitor. Biochem Biophys Res Commun 281:1154–1160

Sato N, Kawamura K, Fukuda J et al (2003) Expression of LDL receptor and uptake of LDL in mouse preimplantation embryos. Mol Cell Endocrinol 202:191–194

Scarchilli L, Camaioni A, Bottazzi B et al (2007) PTX3 interacts with inter-alpha-trypsin inhibitor: implications for hyaluronan organization and cumulus oophorus expansion. J Biol Chem 282:30161–30170

Schindler R, Nilsson E, Skinner MK (2010) Induction of ovarian primordial follicle assembly by connective tissue growth factor CTGF. PLoS One 5:e12979

Schmidt D, Ovitt CE, Anlag K et al (2004) The murine winged-helix transcription factor Foxl2 is required for granulosa cell differentiation and ovary maintenance. Development 131:933–942
Schoppee PD, Garmey JC, Veldhuis JD (2002) Putative activation of the peroxisome proliferator-activated receptor gamma impairs androgen and enhances progesterone biosynthesis in primary cultures of porcine theca cells. Biol Reprod 66:190–198
Schroeder TE (1981) Microfilament-mediated surface change in starfish oocytes in response to 1-methyladenine: implications for identifying the pathway and receptor sites for maturation-inducing hormones. J Cell Biol 90:362–371
Schultz RM, Letourneau GE, Wassarman PM (1979) Program of early development in the mammal: changes in patterns and absolute rates of tubulin and total protein synthesis during oogenesis and early embryogenesis in the mouse. Dev Biol 68:341–359
Schultz RM, Montgomery RR, Belanoff JR (1983) Regulation of mouse oocyte meiotic maturation: implication of a decrease in oocyte cAMP and protein dephosphorylation in commitment to resume meiosis. Dev Biol 97:264–273
Segaloff DL, Wang HY, Richards JS (1990) Hormonal regulation of luteinizing hormone/chorionic gonadotropin receptor mRNA in rat ovarian cells during follicular development and luteinization. Mol Endocrinol 4:1856–1865
Sekar N, Garmey JC, Veldhuis JD (2000) Mechanisms underlying the steroidogenic synergy of insulin and luteinizing hormone in porcine granulosa cells: joint amplification of pivotal sterol-regulatory genes encoding the low-density lipoprotein (LDL) receptor, steroidogenic acute regulatory (stAR) protein and cytochrome P450 side-chain cleavage (P450scc) enzyme. Mol Cell Endocrinol 159:25–35
Shitsukawa K, Andersen CB, Richard FJ et al (2001) Cloning and characterization of the cyclic guanosine monophosphate-inhibited phosphodiesterase PDE3A expressed in mouse oocyte. Biol Reprod 65:188–196
Shuhaibar LC, Egbert JR, Norris RP et al (2015) Intercellular signaling via cyclic GMP diffusion through gap junctions restarts meiosis in mouse ovarian follicles. Proc Natl Acad Sci U S A 112:5527–5532
Silva JM, Price CA (2002) Insulin and IGF-I are necessary for FSH-induced cytochrome P450 aromatase but not cytochrome P450 side-chain cleavage gene expression in oestrogenic bovine granulosa cells in vitro. J Endocrinol 174:499–507
Simon AM, Goodenough DA, Li E, Paul DL (1997) Female infertility in mice lacking connexin 37. Nature 385:525–529
Simone DA, Chorich LP, Mahesh VB (1993) Mechanisms of action for an androgen-mediated autoregulatory process in rat thecal-interstitial cells. Biol Reprod 49:1190–1201
Simpson ER, Mahendroo MS, Means GD et al (1994) Aromatase cytochrome P450, the enzyme responsible for estrogen biosynthesis. Endocr Rev 15:342–355
Sirois J, Simmons DL, Richards JS (1992) Hormonal regulation of messenger ribonucleic acid encoding a novel isoform of prostaglandin endoperoxide H synthase in rat preovulatory follicles. Induction in vivo and in vitro. J Biol Chem 267:11586–11592
Smyth CD, Gosden RG, McNeilly AS, Hillier SG (1994) Effect of inhibin immunoneutralization on steroidogenesis in rat ovarian follicles in vitro. J Endocrinol 140:437–443
Sorensen RA, Wassarman PM (1976) Relationship between growth and meiotic maturation of the mouse oocyte. Dev Biol 50:531–536
Spicer LJ, Chamberlain CS, Maciel SM (2002) Influence of gonadotropins on insulin- and insulin-like growth factor-I (IGF-I)-induced steroid production by bovine granulosa cells. Domest Anim Endocrinol 22:237–254
Spicer LJ, Aad PY, Allen DT et al (2008) Growth differentiation factor 9 (GDF9) stimulates proliferation and inhibits steroidogenesis by bovine theca cells: influence of follicle size on responses to GDF9. Biol Reprod 78:243–253
Sretarugsa P, Wallace RA (1997) The developing Xenopus oocyte specifies the type of gonadotropin-stimulated steroidogenesis performed by its associated follicle cells. Dev Growth Differ 39:87–97

Sternlicht AL, Schultz RM (1981) Biochemical studies of mammalian oogenesis: kinetics of accumulation of total and poly(A)-containing RNA during growth of the mouse oocyte. J Exp Zool 215:191–200

Stocco DM (2001) Tracking the role of a star in the sky of the new millennium. Mol Endocrinol 15:1245–1254

Su YQ, Denegre JM, Wigglesworth K et al (2003) Oocyte-dependent activation of mitogen-activated protein kinase (ERK1/2) in cumulus cells is required for the maturation of the mouse oocyte-cumulus cell complex. Dev Biol 263:126–138

Su YQ, Sugiura K, Woo Y et al (2007) Selective degradation of transcripts during meiotic maturation of mouse oocytes. Dev Biol 302:104–117

Su YQ, Sugiura K, Wigglesworth K et al (2008) Oocyte regulation of metabolic cooperativity between mouse cumulus cells and oocytes: BMP15 and GDF9 control cholesterol biosynthesis in cumulus cells. Development 135:111–121

Su YQ, Sugiura K, Li Q et al (2010) Mouse oocytes enable LH-induced maturation of the cumulus-oocyte complex via promoting EGF receptor-dependent signaling. Mol Endocrinol 24:1230–1239

Sugiura K, Pendola FL, Eppig JJ (2005) Oocyte control of metabolic cooperativity between oocytes and companion granulosa cells: energy metabolism. Dev Biol 279:20–30

Sugiura K, Su YQ, Diaz FJ et al (2007) Oocyte-derived BMP15 and FGFs cooperate to promote glycolysis in cumulus cells. Development 134:2593–2603

Sugiura K, Su YQ, Eppig JJ (2009) Targeted suppression of Has2 mRNA in mouse cumulus cell-oocyte complexes by adenovirus-mediated short-hairpin RNA expression. Mol Reprod Dev 76:537–547

Suzuki T, Sasano H, Takaya R et al (1998) Cyclic changes of vasculature and vascular phenotypes in normal human ovaries. Hum Reprod 13:953–959

Thomas FH, Ismail RS, Jiang JY, Vanderhyden BC (2008) Kit ligand 2 promotes murine oocyte growth in vitro. Biol Reprod 78:167–175

Tian XC, Berndtson AK, Fortune JE (1995) Differentiation of bovine preovulatory follicles during the follicular phase is associated with increases in messenger ribonucleic acid for cytochrome P450 side-chain cleavage, 3 beta-hydroxysteroid dehydrogenase, and P450 17 alpha-hydroxylase, but not P450 aromatase. Endocrinology 136:5102–5110

Toda K, Takeda K, Okada T et al (2001) Targeted disruption of the aromatase P450 gene (Cyp19) in mice and their ovarian and uterine responses to 17beta-oestradiol. J Endocrinol 170:99–111

Trigatti B, Rayburn H, Vinals M et al (1999) Influence of the high density lipoprotein receptor SR-BI on reproductive and cardiovascular pathophysiology. Proc Natl Acad Sci U S A 96:9322–9327

Tsafriri A, Chun SY, Zhang R et al (1996) Oocyte maturation involves compartmentalization and opposing changes of cAMP levels in follicular somatic and germ cells: studies using selective phosphodiesterase inhibitors. Dev Biol 178:393–402

Tsutsumi O, Yano T, Satoh K et al (1990) Studies of hexokinase activity in human and mouse oocyte. Am J Obstet Gynecol 162:1301–1304

Tsutsumi O, Satoh K, Taketani Y, Kato T (1992) Determination of enzyme activities of energy metabolism in the maturing rat oocyte. Mol Reprod Dev 33:333–337

Uda M, Ottolenghi C, Crisponi L et al (2004) Foxl2 disruption causes mouse ovarian failure by pervasive blockage of follicle development. Hum Mol Genet 13:1171–1181

Urner F, Herrmann WL, Baulieu EE, Schorderet-Slatkine S (1983) Inhibition of denuded mouse oocyte meiotic maturation by forskolin, an activator of adenylate cyclase. Endocrinology 113:1170–1172

Van Blerkom J, Davis PW, Lee J (1995) ATP content of human oocytes and developmental potential and outcome after in-vitro fertilization and embryo transfer. Hum Reprod 10:415–424

Van Blerkom J, Davis P, Alexander S (2003) Inner mitochondrial membrane potential (DeltaPsim), cytoplasmic ATP content and free Ca2+ levels in metaphase II mouse oocytes. Hum Reprod 18:2429–2440

Vanderhyden BC (1993) Species differences in the regulation of cumulus expansion by an oocyte-secreted factor(s). J Reprod Fertil 98:219–227
Vanderhyden BC, Tonary AM (1995) Differential regulation of progesterone and estradiol production by mouse cumulus and mural granulosa cells by A factor(s) secreted by the oocyte. Biol Reprod 53:1243–1250
Vanderhyden BC, Caron PJ, Buccione R, Eppig JJ (1990) Developmental pattern of the secretion of cumulus expansion-enabling factor by mouse oocytes and the role of oocytes in promoting granulosa cell differentiation. Dev Biol 140:307–317
Vanderhyden BC, Cohen JN, Morley P (1993) Mouse oocytes regulate granulosa cell steroidogenesis. Endocrinology 133:423–426
Vanderhyden BC, Macdonald EA, Nagyova E, Dhawan A (2003) Evaluation of members of the TGFbeta superfamily as candidates for the oocyte factors that control mouse cumulus expansion and steroidogenesis. Reprod Suppl 61:55–70
Van Soom A, Tanghe S, De Pauw I et al (2002) Function of the cumulus oophorus before and during mammalian fertilization. Reprod Domest Anim 37:144–151
Varnosfaderani ShR, Ostadhosseini S, Hajian M et al (2013) Importance of the GDF9 signaling pathway on cumulus cell expansion and oocyte competency in sheep. Theriogenology 80:470–478
Vitt UA, McGee EA, Hayashi M, Hsueh AJ (2000) In vivo treatment with GDF-9 stimulates primordial and primary follicle progression and theca cell marker CYP17 in ovaries of immature rats. Endocrinology 141:3814–3820
Vivarelli E, Conti M, De Felici M, Siracusa G (1983) Meiotic resumption and intracellular cAMP levels in mouse oocytes treated with compounds which act on cAMP metabolism. Cell Differ 12:271–276
Wallace WH, Kelsey TW (2010) Human ovarian reserve from conception to the menopause. PLoS One 5:e8772
Walters KA, Middleton LJ, Joseph SR et al (2012) Targeted loss of androgen receptor signaling in murine granulosa cells of preantral and antral follicles causes female subfertility. Biol Reprod 87:151
Wang Q, Chi MM, Schedl T, Moley KH (2012) An intercellular pathway for glucose transport into mouse oocytes. Am J Physiol Endocrinol Metab 302:E1511–E1518
Watson LN, Mottershead DG, Dunning KR et al (2012) Heparan sulfate proteoglycans regulate responses to oocyte paracrine signals in ovarian follicle morphogenesis. Endocrinology 153:4544–4555
Whitelaw PF, Smyth CD, Howles CM, Hillier SG (1992) Cell-specific expression of aromatase and LH receptor mRNAs in rat ovary. J Mol Endocrinol 9:309–312
Wickramasinghe D, Ebert KM, Albertini DF (1991) Meiotic competence acquisition is associated with the appearance of M-phase characteristics in growing mouse oocytes. Dev Biol 143:162–172
Wigglesworth K, Lee KB, O'Brien MJ et al (2013) Bidirectional communication between oocytes and ovarian follicular somatic cells is required for meiotic arrest of mammalian oocytes. Proc Natl Acad Sci U S A 110:E3723–E3729
Wigglesworth K, Lee KB, Emori C et al (2015) Transcriptomic diversification of developing cumulus and mural granulosa cells in mouse ovarian follicles. Biol Reprod 92:23
Willis D, Mason H, Gilling-Smith C, Franks S (1996) Modulation by insulin of follicle-stimulating hormone and luteinizing hormone actions in human granulosa cells of normal and polycystic ovaries. J Clin Endocrinol Metab 81:302–309
Worrad DM, Ram PT, Schultz RM (1994) Regulation of gene expression in the mouse oocyte and early preimplantation embryo: developmental changes in Sp1 and TATA box-binding protein, TBP. Development 120:2347–2357
Yamamoto K, Otoi T, Koyama N et al (1999) Development to live young from bovine small oocytes after growth, maturation and fertilization in vitro. Theriogenology 52:81–89

Yoshida H, Takakura N, Kataoka H et al (1997) Stepwise requirement of c-kit tyrosine kinase in mouse ovarian follicle development. Dev Biol 184:122–137

Yoshimura Y (1998) Insulin-like growth factors and ovarian physiology. J Obstet Gynaecol Res 24:305–323

Yoshimura Y, Tischkau SA, Bahr JM (1994) Destruction of the germinal disc region of an immature preovulatory follicle suppresses follicular maturation and ovulation. Biol Reprod 51:229–233

Yoshioka S, Ochsner S, Russell DL et al (2000) Expression of tumor necrosis factor-stimulated gene-6 in the rat ovary in response to an ovulatory dose of gonadotropin. Endocrinology 141:4114–4119

Zeleznik AJ, Midgley AR Jr, Reichert LE Jr (1974) Granulosa cell maturation in the rat: increased binding of human chorionic gonadotropin following treatment with follicle-stimulating hormone in vivo. Endocrinology 95:818–825

Zhang G, Garmey JC, Veldhuis JD (2000) Interactive stimulation by luteinizing hormone and insulin of the steroidogenic acute regulatory (StAR) protein and 17alpha-hydroxylase/17,20-lyase (CYP17) genes in porcine theca cells. Endocrinology 141:2735–2742

Zhang X, Wu XQ, Lu S et al (2006) Deficit of mitochondria-derived ATP during oxidative stress impairs mouse MII oocyte spindles. Cell Res 16:841–850

Zhang Y, Zhang Z, Xu XY et al (2008) Protein kinase A modulates Cdc25B activity during meiotic resumption of mouse oocytes. Dev Dyn 237:3777–3786

Zhang M, Su YQ, Sugiura K, Xia G, Eppig JJ (2010) Granulosa cell ligand NPPC and its receptor NPR2 maintain meiotic arrest in mouse oocytes. Science 330:366–369

Zhang M, Su YQ, Sugiura K et al (2011) Estradiol promotes and maintains cumulus cell expression of natriuretic peptide receptor 2 (NPR2) and meiotic arrest in mouse oocytes in vitro. Endocrinology 152:4377–4385

Zhang H, Liu L, Li X et al (2014a) Life-long in vivo cell-lineage tracing shows that no oogenesis originates from putative germline stem cells in adult mice. Proc Natl Acad Sci U S A 111:17983–17988

Zhang H, Risal S, Gorre N et al (2014b) Somatic cells initiate primordial follicle activation and govern the development of dormant oocytes in mice. Curr Biol 24:2501–2508

Zhang H, Panula S, Petropoulos S et al (2015) Adult human and mouse ovaries lack DDX4-expressing functional oogonial stem cells. Nat Med 21:1116–1118

Zuccotti M, Ponce RH, Boiani M et al (2002) The analysis of chromatin organisation allows selection of mouse antral oocytes competent for development to blastocyst. Zygote 10:73–78

Zuelke KA, Brackett BG (1992) Effects of luteinizing hormone on glucose metabolism in cumulus-enclosed bovine oocytes matured in vitro. Endocrinology 131:2690–2696

Chapter 9
Biology of the Sertoli Cell in the Fetal, Pubertal, and Adult Mammalian Testis

Katarzyna Chojnacka, Marta Zarzycka, and Dolores D. Mruk

Abstract A healthy man typically produces between 50×10^6 and 200×10^6 spermatozoa per day by spermatogenesis; in the absence of Sertoli cells in the male gonad, this individual would be infertile. In the adult testis, Sertoli cells are sustentacular cells that support germ cell development by secreting proteins and other important biomolecules that are essential for germ cell survival and maturation, establishing the blood–testis barrier, and facilitating spermatozoa detachment at spermiation. In the fetal testis, on the other hand, pre-Sertoli cells form the testis cords, the future seminiferous tubules. However, the role of pre-Sertoli cells in this process is much less clear than the function of Sertoli cells in the adult testis. Within this framework, we provide an overview of the biology of the fetal, pubertal, and adult Sertoli cell, highlighting relevant cell biology studies that have expanded our understanding of mammalian spermatogenesis.

9.1 Introduction

Sperm are produced within seminiferous tubules, the functional unit of the testis, by spermatogenesis. Spermatogenesis, which is comprised of mitosis, meiosis, and spermiogenesis, is largely under the regulation of gonadotropins, androgens, and other important biomolecules that include gonadotropin-releasing hormone (Gn-RH), follicle-stimulating hormone (FSH), luteinizing hormone (LH), and testosterone (reviewed in Holdcraft and Braun 2004; Smith and Walker 2014; Walker and Cheng 2005) (Fig. 9.1). In the rat, spermatogenesis commences on postnatal day 5 (in the mouse, on postnatal day 3; in the human, between 9 and 14 years of age) when undifferentiated spermatogonial stem cells (SSCs; i.e., A_{single} spermatogonia) either self-renew into two SSCs or differentiate into two

K. Chojnacka • D.D. Mruk, Ph.D. (✉)
Center for Biomedical Research, Population Council, 1230 York Avenue, New York, NY 10065, USA
e-mail: mruk@popcbr.rockefeller.edu

M. Zarzycka
Department of Endocrinology, Institute of Zoology, Jagiellonian University, Krakow, Poland

R.P. Piprek (ed.), *Molecular Mechanisms of Cell Differentiation in Gonad Development*, Results and Problems in Cell Differentiation 58,
DOI 10.1007/978-3-319-31973-5_9

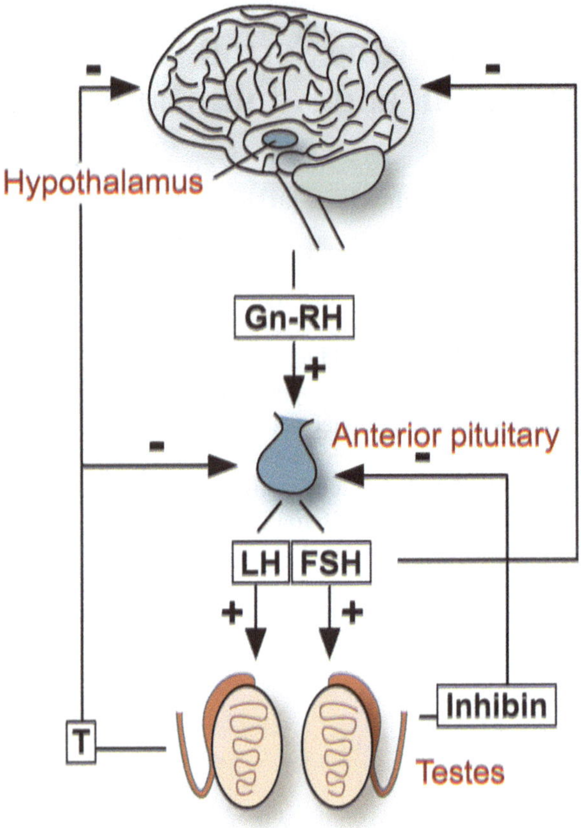

Fig. 9.1 The hypothalamic–pituitary–testicular axis. This axis, which is composed of several tightly regulated positive (+) and negative (–) feedback loops that span three organs (i.e., brain, pituitary, and testis), produces hormones, androgens, and other important biomolecules (*boxed*) that are essential for male reproductive development and function. *Gn-RH* gonadotropin-releasing hormone, *LH* luteinizing hormone, *FSH* follicle-stimulating hormone, *T* testosterone

spermatogonia connected by cytoplasmic bridges due to incomplete cytokinesis (i.e., A_{paired} spermatogonia) (reviewed in de Rooij and Russell 2000; de Rooij and Griswold 2012; Greenbaum et al. 2011; Oatley and Brinster 2008). Spermatogonia connected by cytoplasmic bridges undergo several mitotic divisions to give rise to 4-, 8-, or 16-cell chains of spermatogonia (i.e., $A_{aligned}$ spermatogonia), before proliferation significantly decreases. Thereafter, $A_{aligned}$ spermatogonia differentiate into differentiated spermatogonia by six mitotic divisions (i.e., type A1→A2→A3→A4→intermediate→type B spermatogonia), before giving rise to primary spermatocytes. This process is followed by meiosis I and meiosis II, which produce secondary spermatocytes and spermatids, respectively. In the last phase of spermatogenesis (i.e., spermiogenesis), spermatids undergo several changes that essentially transform round spermatids into elongated spermatids, the so-called spermatozoa. Thereafter, spermatozoa detach from the seminiferous epithelium, release into the seminiferous tubule lumen at spermiation, and enter the epididymis where they will mature and acquire the ability to fertilize ova (reviewed in Clermont 1972; Hermo et al. 2010). This entire process (not including transit through the epididymis) takes 51.6 days in the rat (reviewed in de Kretser and

Kerr 1988; Kerr et al. 2006; Schlatt and Ehmcke 2014). In the mouse and human, the total duration of spermatogenesis is 34.5 and 64 days, respectively (Amann 2008; Heller and Clermont 1963; Oakberg 1956).

Throughout spermatogenesis, Sertoli cells support developing germ cells as they traverse the seminiferous epithelium. Sertoli cells are nurse-like somatic cells that extend from the base to the lumen of the seminiferous tubule, and they are essential for spermatogenesis. For mature germ cells to reach the seminiferous tubule lumen, preleptotene/leptotene spermatocytes have to cross the blood–testis barrier, which is created by Sertoli cell junctions. The movement of preleptotene/leptotene spermatocytes across the blood–testis barrier involves coordinated restructuring of these cell junctions at specific stages of the seminiferous epithelial cycle. It is well established, however, that the blood–testis barrier is never completely disassembled or assembled but in a dynamic flux of the two states. Instead, spermatocytes cross the blood–testis barrier while enclosed within a sealed intermediate compartment (Dym and Cavicchia 1977; Russell 1978; Smith and Braun 2012; reviewed in Russell 1993a). As preleptotene/leptotene spermatocytes move toward the tubule lumen, junctions in front of germ cells disassemble, while those in back of germ cells assemble. This is just one example of the important function of Sertoli cells in spermatogenesis; certainly, there are others, and they are discussed herein.

There are several stages of testis development: (1) development of the genital ridge (i.e., bipotential gonad), (2) sex determination, (3) testis differentiation (i.e., testis cord formation), (4) testis development (i.e., elongation of testis cords and seminiferous tubule formation), and (5) testis maturation and restructuring. In this perspective, we discuss these stages of testis development, while highlighting the role of the Sertoli cell in the fetal, pubertal, and adult testis. We hope that the information in this review provides a strong framework for future studies in the field.

9.2 Sex Determination and Testis Cord Formation

9.2.1 Genital Ridge Formation and Sex Determination

Genetic sex is determined at fertilization with the acquisition of either an X or a Y chromosome from the father to produce an XX or XY individual. Gonad development begins with the migration of primordial germ cells (PGCs), which are specified extragonadally, to the bipotential genital ridge (otherwise known as the gonadal ridge), the site of the future ovary or testis. Structurally, there is no difference in the bipotential genital ridge from an XX and XY embryo. The gonadal ridge, which derives from the intermediate mesoderm, is comprised of somatic cells, as well as PGCs that migrate from the allantois through the hindgut mesentery and to the genital ridge. The fate of the bipotential gonads is largely determined by the genes expressed by the somatic cells of the genital ridges. If the somatic cells are XX, then an ovary will develop; if the somatic cells are XY, then a testis will form.

Testis differentiation is largely controlled by sex determination region on Y chromosome (*Sry*, previously termed testis-determining factor), a high-mobility group (HMG)-containing transcription factor that initiates the testis differentiation program (reviewed in Kashimada and Koopman 2010; Larney et al. 2014; Tanaka and Nishinakamura 2014; Wilhelm et al. 2007c). In the XY gonad, *Sry* is expressed by somatic cells (at this developmental stage, known as pre-Sertoli cells). Somatic cells, which derive from the coelomic epithelium that overlays the bipotential gonads, give rise to the Sertoli cells, the supporting cells in the adult testis. However, in the absence of *Sry* expression by the somatic cells the gonads differentiate into ovaries (Albrecht and Eicher 2001), the Wolffian ducts (the precursors to the epididymides, vas deferens, and secondary sex glands) regress and the Müllerian ducts develop into the fallopian tubes, uterus, cervix, and anterior vagina (Fig. 9.2). Aberrant *Sry* expression or regulation, however, impedes testis development in XY embryos, whereas a loss of or a mutation within *Sry* results in ovary development (Battiloro et al. 1997; Hawkins et al. 1992; Kato et al. 2013; McElreavey et al. 1996; Nagamine et al. 1999; reviewed in Larney et al. 2014; Ostrer 2014). Most mutations within *Sry* disrupt its ability to bind to a consensus-binding motif (A/TA/TCAAA/TG) and bend DNA into its proper conformation (Gubbay et al. 1990; Mitchell and Harley 2002; Schmitt-Ney et al. 1995).

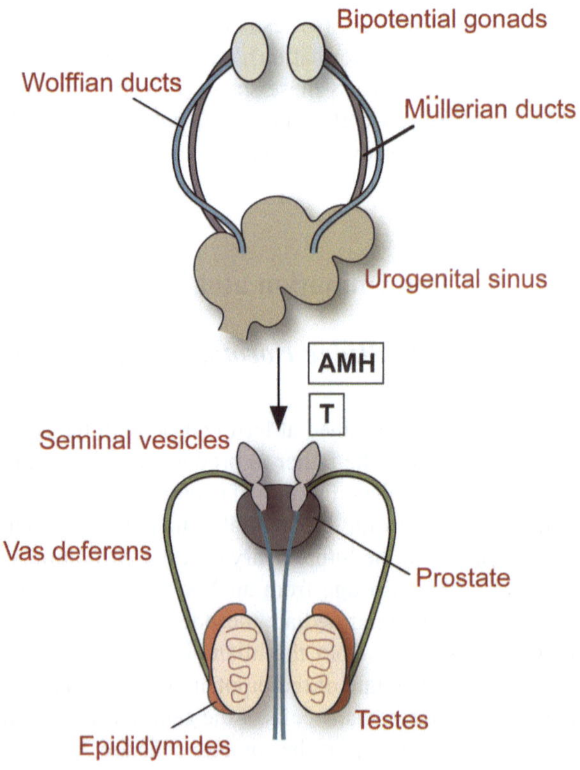

Fig. 9.2 Sex differentiation and testis development. Gonad development commences with the migration of primordial germ cells to the bipotential genital ridge, the site of the future ovary or testis. The fate of the primordial gonads is determined by the genes expressed by the somatic cells of the genital ridges. In the presence of anti-Müllerian hormone (AMH) and testosterone (T), the Müllerian ducts (the precursors to the fallopian tubes, uterus, cervix, and anterior vagina) regress and the Wolffian ducts (the precursors to the epididymides, vas deferens, and secondary sex glands) develop

Equally important, *Sry* expression, which is restricted to a very short period of testis development [*Sry* is expressed from 10.5 to 12.5 *dpc* (days *post coitum*) in the mouse (Hacker et al. 1995)], is strictly controlled by several genes (reviewed in Eggers et al. 2014; Larney et al. 2014). For instance, embryos homozygous for the boygirl (*byg*/*byg*) mutation, which renders MAPK kinase kinase (*Map3k4*, otherwise known as MAPK kinase 4) nonfunctional, exhibit sex reversal (i.e., XX embryos develop male organs and XY embryos develop female organs) due to a decrease in *Sry* expression (Bogani et al. 2009; Gierl et al. 2012; Warr et al. 2012; Wu et al. 2015). This effect is partly mediated by GADD45γ (growth arrest and DNA damage-inducible 45 gamma), a protein involved in cell differentiation, DNA repair, cell cycle control, apoptosis, and senescence (reviewed in Yang et al. 2009). Furthermore, the spatiotemporal pattern of *Gadd45γ* expression resembles that of *Sry* (Warr et al. 2012), suggesting that *Gadd45γ* and *Sry* function together in sex determination. Interestingly, C57BL/6J mice deficient for *Gadd45γ* exhibit sex reversal due to a decrease in *Sry* expression and p38 MAPK phosphorylation (Warr et al. 2012). GADD45γ, which is expressed by somatic cells in both XX and XY gonads, associates with MAP3K4 (Takekawa and Saito 1998). In addition, *Map3k4* overexpression rescues sex reversal in *Gadd45γ*-deficient embryos. Collectively, these studies demonstrate that *Map3k4* is critical for sex determination.

Other genes involved in the early stages of gonadogenesis are steroidogenic factor 1 (*Sf1*), GATA-binding protein 4 (*Gata4*), dosage-sensitive sex reversal, adrenal hypoplasia critical region, on chromosome X, gene 1 (*Dax-1*), Wilms' tumor suppressor 1 (*Wt1*), LIM homeobox 1 (*Lhx1*), *Lhx9*, and empty spiracles homeobox 2 (*Emx2*) (reviewed in Larney et al. 2014; Ludbrook and Harley 2004; Parker et al. 1999; Tanaka and Nishinakamura 2014; Wilhelm et al. 2013). For instance, mice null for *Sf1* show gonadal agenesis (i.e., the absence of one or both gonads), resulting from an arrest in genital ridge development (Luo et al. 1994). *Sf1* also regulates the transcription of genes that encode several hormones acting within the hypothalamic–pituitary–gonadal axis (reviewed in Kohler and Achermann 2010) (Fig. 9.1).

SRY binds SF1 to form a complex, which triggers the expression of *Sox9*, an HMG-containing transcription factor that specifies the Sertoli cell lineage (reviewed in Jakob and Lovell-Badge 2011; Jo et al. 2014; Koopman 1999). *Sox9* expression is upregulated at 11.5 *dpc* in the mouse when *Sry* expression is already highest at 10.5 *dpc*. Thereafter, *Sox9* downregulates *Sry* expression at 12.5 *dpc*, thereby leading to the cessation of its expression. To complete testis development, SOX9 recruits other proteins. Firstly, SOX9 activates lipocalin-type prostaglandin D synthase (PGDS), an enzyme that converts prostaglandin H2 into prostaglandin D2 (PGD2), and stimulates PGD2 secretion (Malki et al. 2005; Moniot et al. 2009; reviewed in Urade and Hayaishi 2000). The reverse is also true; PGDS2 promotes SOX9 activity by transforming pre-Sertoli cells and enhancing commitment to the male pathway (Wilhelm et al. 2005, 2007b). Interestingly, *Pgds* expression is induced immediately after *Sox9* expression (Adams and McLaren 2002; Wilhelm et al. 2007b). Secondly, *Sox9* activity promotes anti-Müllerian hormone (AMH), which inhibits Müllerian duct development (Behringer et al. 1994; reviewed in

Behringer 1995; Lee and Donahoe 1993) (Fig. 9.2). Collectively, these reports illustrate that SOX9 mediates pre-Sertoli cell differentiation. Similar to *Sry*, a loss of *Sox9* results in ovary development in XY embryos (Barrionuevo et al. 2006; Chaboissier et al. 2004). Sex determination is described in Chap. 3.

9.2.2 Testis Cord Formation

Testis cord formation, which occurs between 11.5 and 13.5 *dpc* in the mouse, involves the organization of pre-Sertoli cells and prospermatogonia (otherwise known as gonocytes) into testis cords, the future seminiferous tubules. Testis cord formation is composed of clustering (or coalescence), partitioning, and patterning (or remodeling) (Combes et al. 2009b). Initially pre-Sertoli cells and prospermatogonia are evenly distributed in the genital ridge, which is followed by the clustering of pre-Sertoli cells and the enclosure of germ cells by pre-Sertoli cells. During partitioning, endothelial cell precursors migrate from the mesonephros and invade the developing gonads. Thereafter, early testis cords grow, peritubular myoid cells (PMCs) surround testis cords and deposit the basement membrane together with pre-Sertoli cells, and testis cords and the vasculature mature and are remodeled (Combes et al. 2009a). At 13.5 *dpc* in the mouse, the number of testis cords that will become seminiferous tubules is established (reviewed in Cool et al. 2012).

The development of the vasculature is one of the most important processes to occur during testis cord formation. Endothelial cells detach from the arteries of the mesonephric plexus in the mesonephros, which is at the border with the gonad, and invade the XY gonad to create the distinct coelomic vessel on the surface of the testis (Brennan et al. 2002; Coveney et al. 2008; Martineau et al. 1997). Smaller vessels branch from the coelomic vessel and situate in the interstitium of the testis. Testis cord development is impeded in the absence of the vasculature with paths of migrating cells and branches of the coelomic vessel determining the shape and number of testis cords (Combes et al. 2009b; Cool et al. 2011). These results indicate that the vasculature is required for testis cord patterning. PMCs also drive testis cord development by contributing to the formation of the basement membrane, which initiates the polarization of pre-Sertoli cells (Combes et al. 2009b; Skinner et al. 1985; Tung and Fritz 1987).

Around birth, prospermatogonia migrate from the center of the testis cord to the basement membrane, which is critical for the establishment of the spermatogonial stem cell niche. Prospermatogonia, the precursors of spermatogonia, refer to germ cells from the time they inhabit the primordial gonads to the time they reach the basement membrane of the testis cords and differentiate into spermatogonia. Prospermatogonia are lost from the mouse testis a few days after birth (reviewed in Culty 2009). However, considering the molecular control of testis cord formation, testis cords do not form in the absence of *Sox9* expression (Barrionuevo et al. 2009). Other genes downstream of *Sry* and *Sox9* that are involved in testis development are fibroblast growth factor 9 (*Fgf9*), doublesex- and mab-3-related transcription factor 1 (*Dmrt1*), and *Dax-1* (reviewed in Brennan and Capel 2004; Eggers et al. 2014; Koopman 2001; Wilhelm et al. 2007c). For instance, *Sox9* upregulates *Fgf9* expression, and *Fgf9* upregulates *Sox9* expression, which are important for male sex determination (Kim et al. 2006). Furthermore, mutations within *Fgf9* result in the loss of *Sox9* expression and partial or complete sex reversal (Colvin et al. 2001). Thus, the strong upregulation of *Sox9* by several factors inhibits female sex-determining genes and canalizes the male pathway and fate (reviewed in Piprek 2009).

9.3 Differentiation of Sertoli Cells in Fetal, Neonatal, and Prepubertal Testes

Vast changes in pre-Sertoli cell structure underlie the formation and expansion of testis cords. At 11.5–12.5 *dpc* in the mouse, pre-Sertoli cells undergo a mesenchymal-to-epithelial transition and then enclose either a single prospermatogonium or a group of prospermatogonia, resulting in the formation of testis cords (reviewed in Cool et al. 2012). The testis cords are the precursors of the seminiferous tubules that segregate germ cell development from androgen production in the adult testis. At 12.5 *dpc* in the mouse, a basement membrane surrounds testis cords. At 13.5 *dpc*, Sertoli cells position their nuclei near the basement membrane and extend their body into the center of the testis cords, which is indicative of cell polarization. Cell polarization also associates with dynamic changes in the cytoskeleton with actin filaments accumulating at the basal domain of Sertoli cells (Kanai et al. 1992). During fetal development, Sertoli cells proliferate (Orth 1982, 1984; reviewed in Sharpe et al. 2003), which is controlled by several factors that include FSH, thyroid hormone (T_3), and activin A, as well as the insulin-like growth factor 1 (IGF1) pathway (reviewed in Lucas et al. 2014; Sharpe et al. 2003). For instance, Sertoli cell proliferation increases after the administration of recombinant FSH to neonatal rats (Meachem et al. 1996). It deserves emphasis that the final number of Sertoli cells determines the number of germ cells they will support in adulthood (i.e., spermatogenic capacity increases with the number of Sertoli cells) so that these factors are crucial for testis development (Orth 1982; Orth et al. 1988; reviewed in Griswold 1995). During early postnatal development, Sertoli cells continue to proliferate, differentiate, and mature, which marks the transition of testis cords into seminiferous tubules. At the onset of puberty, however, they cease to proliferate and undergo additional changes in structure and function (Orth 1982; reviewed in Sharpe et al. 2003). While adult Sertoli cells are generally considered to be terminally differentiated, Sertoli cells in the adult human and hamster can proliferate under certain hormonal conditions (Tarulli et al. 2006, 2013; reviewed in Tarulli et al. 2012). In the human, mouse, and rat, each adult

Sertoli cell supports approximately 11, 35, and 20 developing germ cells, respectively (reviewed in Kerr et al. 2006).

At the onset of spermatogenesis, there is extensive proliferation of germ cells, which coincides with an increase in Sertoli cell size and the development of extensive Sertoli cell cytoplasmic processes that establish contact with germ cells. Sertoli cells establish the blood–testis barrier, which creates a specialized microenvironment for meiotic spermatocytes and post-meiotic spermatids throughout adulthood (reviewed in Fawcett et al. 1970; Mruk and Cheng 2015; Setchell and Waites 1975). Successful spermatogenesis relies on these early events in Sertoli cells, because abnormal Sertoli cell proliferation or differentiation can subsequently disrupt fertility (Orth et al. 1988; reviewed in Sharpe et al. 2003). Upon the appearance of spermatids in the seminiferous epithelium, Sertoli cell differentiation is complete. Interestingly, in the adult mouse testis, there are only 15–20 seminiferous tubules that add up to 2 m in length; in the human, there are 250–1000 seminiferous tubules that add up to 300–900 m (Bascom and Osterud 1925). Collectively, these studies demonstrate that Sertoli cells not only establish sex determination; they also orchestrate testis development throughout fetal, neonatal, and prepubertal development.

9.4 Cellular Organization of the Adult Mammalian Testis

9.4.1 The Seminiferous Tubule

Germ cell development occurs within seminiferous tubules under the regulation of gonadotropins, androgens, and other important factors that include Gn-RH, FSH, LH, and testosterone (reviewed in Holdcraft and Braun 2004; Sharpe 1994; Smith and Walker 2014; Walker and Cheng 2005). The basic organization of the rat testis, which is visible by light microscopy when adult testis cross sections are stained with either periodic acid–Schiff or hematoxylin and eosin, is comprised of seminiferous tubules that contain Sertoli and germ cells organized into 14 discrete associations or stages (denoted as stages I–XIV). Each stage is comprised of four or five generations of germ cells that are concentrically arranged within the seminiferous tubule. These stages run in succession so that stage I will eventually develop into stage II, and so on, thereby constituting a 12.9-day cycle in the rat. The major difference in the seminiferous epithelial cycle across different mammals is in the length of the cycle. The cycle then repeats along the entire length of the seminiferous tubule (Hess 1990; LeBlond and Clermont 1952; reviewed in Hess and Renato de Franca 2008; Russell et al. 1990).

The different stages of the seminiferous epithelial cycle are discerned by examining the relative position of the oldest generation of spermatids within each seminiferous tubule cross section. For instance, the heads of step 17 spermatids, which are narrow and defined, deeply embed within Sertoli cell crypts at stage V of

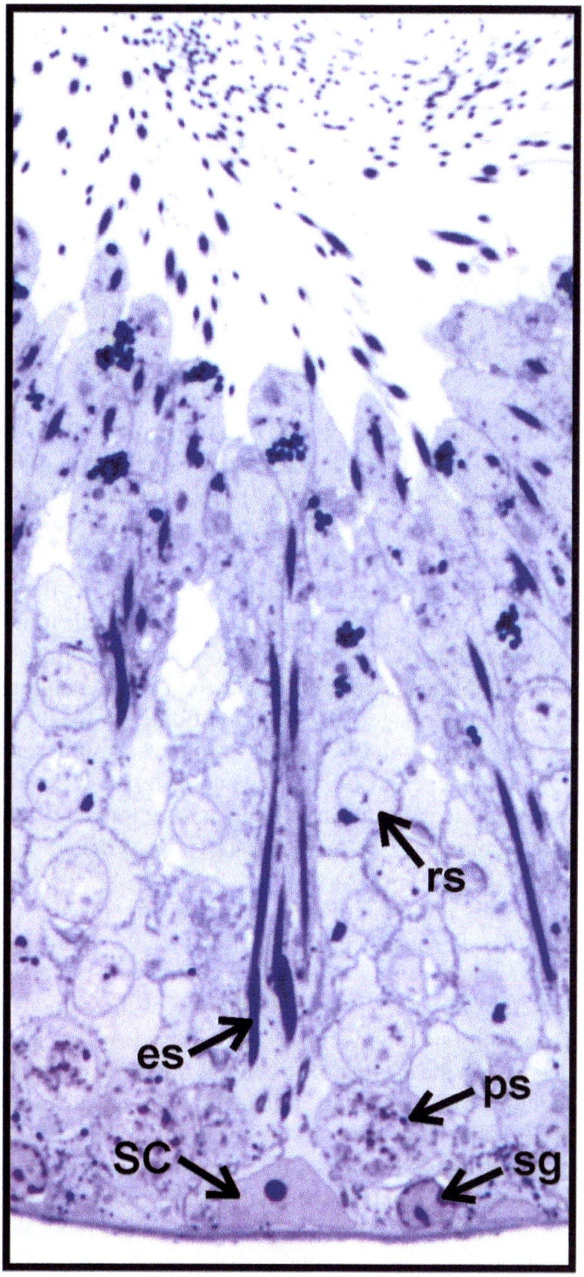

Fig. 9.3 A highly magnified image that shows a portion of a stage V seminiferous tubule from the adult rat testis. In the rat, each adult Sertoli cell (SC) supports approximately 20 developing germ cells. *sg* spermatogonium, *ps* pachytene spermatocyte, *rs* round spermatid, *es* elongated spermatid

the seminiferous epithelial cycle in the rat (Hess 1990; LeBlond and Clermont 1952; reviewed in Hess and Renato de Franca 2008; Russell et al. 1990) (Fig. 9.3). Sertoli cells in which elongated spermatids are embedded within crypts are defined as type B; Sertoli cells without crypts are defined as type A. Additional criteria such as the presence of type B spermatogonia at stages IV–VI, which stain for dense chromatin, can be used when a stage is ambiguous (e.g., to discern stage III from stage IV). However, correct staging using this and other similar criteria calls for excellent preservation of testis morphology, which can be difficult to achieve (Russell et al. 1990).

9.4.2 The Interstitium

As previously discussed, Sertoli and germ cells reside in the seminiferous tubule, which is surrounded by PMCs. In the adult testis, PMCs mediate the contraction of seminiferous tubules, which sends spermatozoa to the epididymis (reviewed in Maekawa et al. 1996). In the rodent, a single layer of PMCs surrounds seminiferous tubules (Dym and Fawcett 1970; Gardner and Holyoke 1964; Regaud 1901). In the human, however, they are circumscribed by three or four layers of PMCs (Ross and Long 1966). In the rat (but not in other species), PMCs restrict the entry of lanthanum nitrate, a small electron-dense tracer, into approximately 85 % of seminiferous tubules (Dym and Fawcett 1970), indicating these cells partly contribute to barrier function. During development and throughout adulthood, PMCs work with Sertoli cells to deposit the basement membrane [it is largely composed of laminin, collagen IV, heparin sulfate proteoglycan, entactin, and fibronectin (Hadley and Dym 1987)] that surrounds seminiferous tubules (Skinner and Fritz 1985; Skinner et al. 1985; Tung and Fritz 1987; reviewed in Dym 1994) (Fig. 9.4), indicating that PMC–Sertoli cell interactions are critical for seminiferous tubule architecture and spermatogenesis (reviewed in Skinner et al. 1991; Verhoeven et al. 2000).

Equally important, PMCs control spermatogenesis through the androgen receptor (AR). For instance, Welsh et al. (2009) report that PMC-specific *Ar* knockout mice, which were created by crossing mice heterozygous for Cre recombinase driven by the PMC-specific promoter of the smooth muscle myosin heavy chain gene with mice homozygous for a floxed *Ar*, are azoospermic. Interestingly, this decrease in the spermatozoa count is not the result of a defect in SSC function, even though PMCs associate intimately with SSCs residing in the niche (Shinohara et al. 2001; Tegelenbosch and de Rooij 1993). SSCs, which either self-renew to replenish the SSC pool or differentiate into spermatogonia connected by cytoplasmic bridges, maintain spermatogenesis (reviewed in de Rooij and Russell 2000; Greenbaum et al. 2011; Oatley and Brinster 2008). While the significance of these interactions is not yet clear, these results support the importance of PMCs in spermatogenesis. Additional studies are needed to determine where PMCs derive from.

In the interstitium of the adult testis, Leydig cells are the major component. There are two populations of Leydig cells: fetal (FLCs) and adult Leydig cells (ALCs) (reviewed in Griswold and Behringer 2009; Habert et al. 2001; Haider

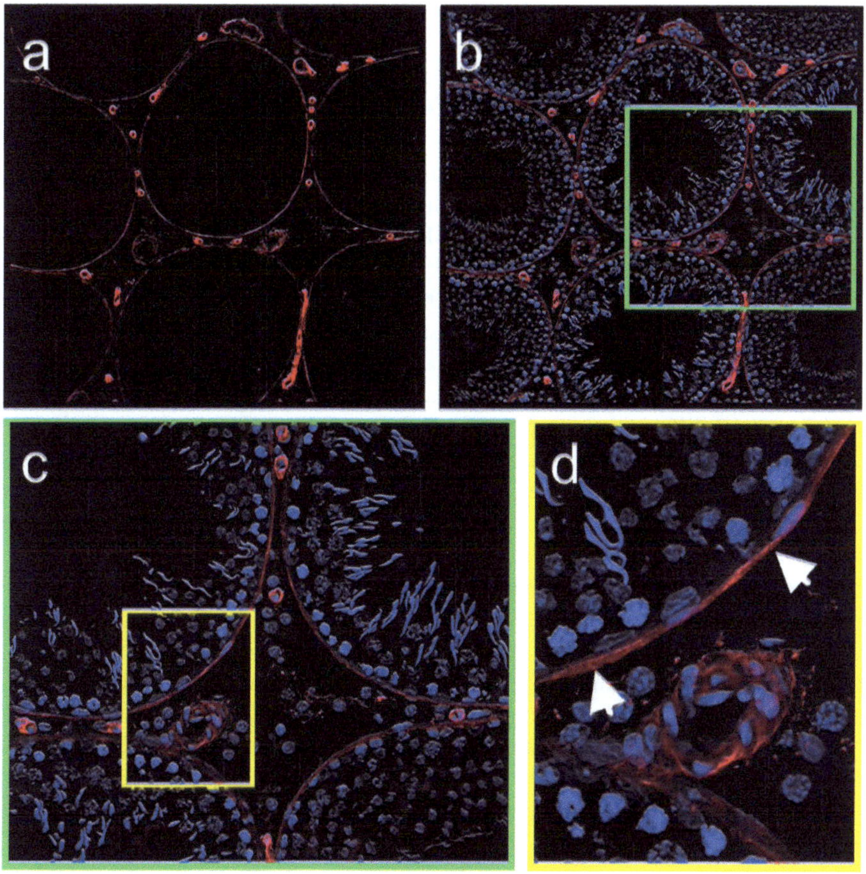

Fig. 9.4 The immunolocalization of collagen IV in the adult rat testis. Frozen testis cross sections were immunostained for collagen IV (*red* fluorescence, **a–d**), a protein of the basement membrane that is largely produced and deposited by peritubular myoid cells. The *green boxed area* (**b**) is magnified in the neighboring image (**c**); and the *yellow boxed area* (**c**) is magnified in the neighboring image (**d**). *Arrows* (**d**) point to collagen IV. The vasculature was also immunoreactive for collagen IV. Cell nuclei were stained with 4′,6-diamidino-2-phenylindole (DAPI, *blue* fluorescence)

2004; Mendis-Handagama and Ariyaratne 2001; O'Shaughnessy et al. 2006; Tremblay 2015). FLCs, which do not express *Sry* (Zwingman et al. 1994), differentiate in response to Sertoli cell factors that include platelet-derived growth factor A (PDGFA) and in response to Dessert Hedgehog (DHH) and Notch signaling (Tang et al. 2008; reviewed in Barsoum and Yao 2006; O'Shaughnessy et al. 2006; Svingen and Koopman 2013). FLCs are critical for the elongation of testis cords. FLCs masculinize the reproductive tract through testosterone and induce testis descent via insulin-like growth factor 3 (IGF3), a secreted Leydig cell protein (Wilhelm et al. 2007a; reviewed in Bay et al. 2011; Hutson et al. 2013). FLCs initially synthesize testosterone in the absence of LH. However, they subsequently express luteinizing hormone receptors (LHR) and respond to LH (O'Shaughnessy et al. 1998). Shortly after birth, FLCs undergo apoptosis. Thus, FLCs do not appear to give rise to ALCs (Ariyaratne et al. 2000). The origin and differentiation of FLCs and ALCs are described in Chap. 5.

ALCs, on the other hand, derive from Leydig stem cells, which are capable of self-renewal. Leydig stem cells develop into progenitor Leydig cells, which express 3β-hydroxysteroid dehydrogenase (3βHSD) and LHR (Dong et al. 2007), which is followed by differentiation into immature and adult cells that no longer proliferate (reviewed in Griswold and Behringer 2009; Haider 2004). ALCs produce testosterone in the presence of LH, which is essential for the establishment and maintenance of secondary sex characteristics and the continuation of spermatogenesis (reviewed in Ge et al. 2008; Smith and Walker 2014; Walker 2011). Sertoli cells are critical for the differentiation of FLCs and the preservation of ALCs. For instance, the ablation of Sertoli cells in a transgenic adult mouse model results in a marked reduction in Leydig cell number, as well as the loss of all germ cells (Rebourcet et al. 2014a, b). PMC function is also affected in these mice. Collectively, these reports illustrate that Sertoli cells are critical for Leydig cell function.

Immune cells (e.g., macrophages, T-cells, mast cells, natural killer cells, and dendritic cells) are also present in the interstitium of the adult testis where they function in innate and adaptive immune responses (reviewed in Perez et al. 2013). Of these immune cells, macrophages are the most abundant, constituting approximately 25 % of the interstitial cells in the adult rodent testis (Giannessi et al. 2005; Niemi et al. 1986). This population of macrophages associates with Leydig cells and the vasculature (Hume et al. 1984; Hutson 2006). In a recent study, DeFalco et al. (2015) describe a second population of macrophages that associates with PMCs and the vasculature, residing within the spermatogonial stem cell niche. Interestingly, the number of $A_{aligned}$, but not A_{single}, spermatogonia declines after the ablation of both populations of macrophages, indicating there is cross talk between immune cells and spermatogonia. Macrophages also establish cell junctions with Leydig cells, which facilitate cross talk between cells (reviewed in Christensen and Gillim 1969; Meinhardt and Hedger 2011; Perez et al. 2013). For example, macrophages secrete 25-hydroxycholesterol, which is used by ALCs for testosterone synthesis (Lukyanenko et al. 2001; Nes et al. 2000). Cytokines produced by activated macrophages can also modulate Leydig cell steroidogenesis (Hales et al. 1992; reviewed in Bornstein et al. 2004; Hales 2002). Collectively, these studies demonstrate that the function of macrophages in the testis goes beyond that of the immune response.

9.5 Establishment and Maintenance of the Sertoli Cell Barrier

The blood–testis barrier, a physical barrier that isolates meiotic and post-meiotic germ cells from immune and lymphatic systems, is one of the tightest tissue barriers based on studies that revealed the testis to be impenetrable to intravenously injected dyes (Ribbert 1904). It is constituted by Sertoli cell junctions that divide the seminiferous epithelium into a basal compartment, where spermatogonia and early primary spermatocytes (i.e., preleptotene/leptotene spermatocytes) reside, and an adluminal compartment, where more-mature primary spermatocytes (i.e., zygotene, pachytene, and diplotene spermatocytes), secondary spermatocytes, and spermatids dwell (Cavicchia and Dym 1977; Dym and Fawcett 1970; reviewed in Cheng and Mruk 2012; Setchell and Waites 1975). Strictly speaking, germ cells that occupy the basal compartment are not isolated from immune and lymphatic systems. To reflect its location within the seminiferous epithelium, the blood–testis barrier is more appropriately defined as the Sertoli cell barrier instead of the blood–testis barrier. However, the term "blood–testis barrier" is more commonly used in the literature. The blood–testis barrier has three main functions: it prevents the entry of unwanted substances into the adluminal compartment, regulates the passage of substances into/out of the same compartment, and sequesters meiotic and post-meiotic germ cells from immune and lymphatic systems. Thus, the blood–testis barrier is marked by anatomical, physiological, and immunological features.

There are several important differences between the blood–testis barrier and other blood–tissue barriers that include the blood–brain, blood–retinal, and blood–epididymal barriers. Firstly, the blood–testis barrier is constituted by tight junctions, basal ectoplasmic specializations, desmosomes, and gap junctions that localize to the basal domain of Sertoli cells (Dym and Fawcett 1970; reviewed in Cheng and Mruk 2012). This is different from other blood–tissue barriers that are largely comprised of tight junctions that are restricted to the apical domains of epithelial and endothelial cells (reviewed in Matter and Balda 2003). Furthermore, the localization of cell junctions at the blood–testis barrier is highly variable with desmosomes and gap junctions coexisting with tight junctions and basal ectoplasmic specializations. Secondly, the blood–testis barrier does not assemble until puberty unlike most other blood–tissue barriers that form in utero or early postnatal development (reviewed in Rizzolo 2007; Yao et al. 2014). In the rat, interstitially injected tracers permeate seminiferous tubules until postnatal day 16 (Vitale et al. 1973), indicating the blood–testis barrier assembles on postnatal day 16. A delay in blood–testis barrier formation halts meiosis (Chihara et al. 2013b; Hosoi et al. 2002; Toyama et al. 2001). Likewise, the assembly of the blood–epididymis barrier completes on postnatal day 21 in the rat (Agarwal and Hoffer 1989). Thirdly, the blood–testis barrier restructures to accommodate the passage of preleptotene/leptotene spermatocytes at stages VIII–XI (Dym and Cavicchia 1977; Russell 1978; Smith and Braun 2012; reviewed in Russell 1993a). Previous studies show that preleptotene/leptotene spermatocytes cross the blood–testis barrier while enclosed within an intermediate compartment that is created by Sertoli cell junctions. The entry of spermatocytes into the adluminal compartment initiates when cell junctions ahead of spermatocytes disassemble, while new cell junctions assemble behind them.

The blood–testis barrier is largely comprised of tight junctions and basal ectoplasmic specializations. Basal ectoplasmic specializations are testis-specific anchoring junctions whose component proteins directly or indirectly attach to the actinomyosin cytoskeleton (reviewed in Fawcett et al. 1970; Mruk and Cheng 2004b, 2010; Vogl et al. 2008). Desmosomes and gap junctions, which coexist with tight junctions and basal ectoplasmic specializations, also contribute to blood–testis barrier function (reviewed in Lie et al. 2011a; Pointis et al. 2010). Continuous cross talk among the component proteins of these cell junctions is essential for spermatogenesis, and blood–testis barrier function is affected if cross talk is perturbed. Tight junctions are highly complex regions of close apposition between cells that divide the plasma membrane of polarized epithelial and endothelial cells into apical and basal domains (reviewed in Anderson and Van Itallie 2008; Furuse 2010). Tight junctions regulate the passage of molecules through the paracellular pathway (otherwise known as the gate function) and restrict the movement of proteins and lipids between apical and basal domains (otherwise known as the fence function) (reviewed in Madara 1998; Shin et al. 2006). There are two major types of tight junctions: bicellular tight junctions form between two cells, while tricellular tight junctions form where three cells meet. While claudins are the main structural and functional components of the bicellular tight junction, tricellulin is that for the tricellular tight junction (reviewed in Furuse et al. 2014; Tsukita and Furuse 2000). Occludin is another example of a transmembrane tight junction protein. It binds zona occludens-1 (ZO-1), a cytoplasmic protein (Fig. 9.5).

In the rodent testis, claudins (*Cldn*) 3 and 11 are best studied. In the adult mouse testis, claudin 3 expression is highest at stages VI–IX. However, it is not expressed by the rat testis (Kaitu'u-Lino et al. 2007). Claudin 3 localizes to newly assembled tight junctions behind migrating spermatocytes (Chihara et al. 2013a; Meng et al. 2005; Smith and Braun 2012), indicating that it reseals the blood–testis barrier after spermatocyte movement. Furthermore, claudin 3 expression is significantly reduced in $Ar^{\mathrm{invflox(exl\text{-}neo)Y}}$;Tg(*Amh-Cre*) mice, which were created by crossing mice with a hypomorphic inverted floxed *Ar* with mice expressing Cre recombinase driven by the Sertoli cell-specific promoter of the *Amh* gene. These mice present for conditional androgen sensitivity and azoospermia (Meng et al. 2005), which indicates that claudin 3 is regulated by androgens. The permeability of the blood–testis barrier is also affected in these mice (Meng et al. 2005). Claudin 11 is also critical for blood–testis barrier integrity, because tight junction function is disrupted in mice in which *Cldn11* is constitutively deleted. These mice are infertile due to the inability of germ cells to differentiate beyond the spermatocyte stage (Gow et al. 1999). In addition, Sertoli cells proliferate in *Cldn11*-deficient mice (Mazaud-Guittot et al. 2010), illustrating that claudin 11 contributes to the terminal differentiation of Sertoli cells.

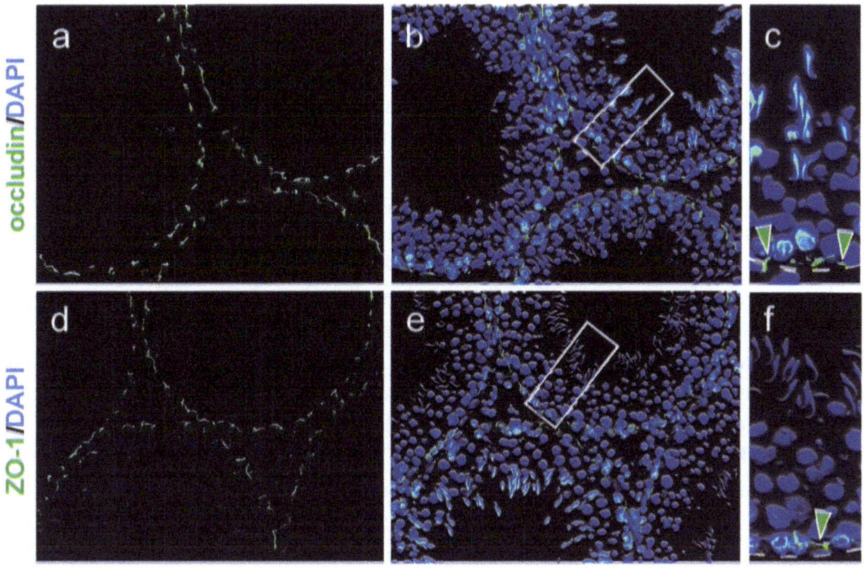

Fig. 9.5 The immunolocalization of occludin and zona occludens-1 (ZO-1) in the adult rat testis. Frozen testis cross sections were immunostained for occludin (*green* fluorescence, **a–c**) or ZO-1 (*green* fluorescence, **d–f**), component proteins of the tight junction. The *white boxed areas* (**b**, **e**) are magnified in the adjacent images (**c**, **f**), respectively. *Arrowheads* point to occludin (**c**) and ZO-1 (**f**) at the blood–testis barrier. Cell nuclei were stained with DAPI (*blue* fluorescence)

9.6 Sertoli Cells in the Adult Testis

In the adult mammalian testis, terminally differentiated Sertoli cells are characterized by the ability to (1) support the structure of the seminiferous epithelium, (2) assemble the blood–testis barrier, (3) secrete proteins and other biomolecules that are needed by developing germ cells, (4) facilitate spermatozoa detachment at spermiation, and (5) phagocytose germ cell residual bodies (reviewed in Bardin et al. 1988; Griswold 1998; Mruk and Cheng 2004b; Russell 1993b). For instance, Sertoli cells secrete approximately 15 % of the proteins that they produce, many of which are secreted stage specifically (reviewed in Bardin et al. 1988; Djakiew and Onoda 1993; Griswold 1993; Skinner 1993). Activin A (activates FSH secretion, thereby stimulating Sertoli cell proliferation) and inhibins A and B (inhibit FSH secretion) are possibly the best studied Sertoli cell proteins (reviewed in de Kretser 1990; de Kretser et al. 2004). During testis development, the production of activin A significantly decreases, which contributes to the differentiation of Sertoli cells (Barakat et al. 2008; Mithraprabhu et al. 2010).

Throughout spermatogenesis, germ cells attach to Sertoli cells via adhesion junctions, which enable spermatogonia to develop into spermatozoa with the support of somatic cells. The developmental stage of the germ cell determines whether adhesion is mediated by intermediate filament-based desmosomes or actin-based apical ectoplasmic specializations. In the adult testis, desmosomes are found between Sertoli cells and all germ cells up to, but not including, elongating and elongated spermatids (i.e., step 8 and beyond). They are also found at the blood–testis barrier where they contribute to barrier integrity (Lie et al. 2010; Russell 1977; Russell et al. 1983; reviewed in Lie et al. 2011a; Mruk and Cheng 2011). Desmosomes are comprised of transmembrane proteins of the desmosomal cadherin family (i.e., desmoglein and desmocollin) that connect to intermediate filaments through cytoplasmic proteins of armadillo (e.g., plakoglobin, plakophilin) and plakin (e.g., desmoplakin) families (reviewed in Harmon and Green 2013; Nekrasova and Green 2013) (Fig. 9.6). Apical ectoplasmic specializations, on the other hand, are found between Sertoli cells and elongating and elongated spermatids (reviewed in Mruk and Cheng 2004a; Vogl et al. 1993, 2008). Adhesion mediated by the apical ectoplasmic specialization is robust and dynamic, and it involves cross talk between several multiprotein complexes. When the strength needed to detach germ cells from Sertoli cells in vitro is measured with a micropipette pressure transducing system, adhesion is strongest between step 8 spermatids and Sertoli cells (Wolski et al. 2005). Of the multiprotein complexes present at this structure, nectin (*Pvrl*), a Ca^{2+}-independent integral membrane protein, is critical for apical ectoplasmic specialization function, because mice null for *Pvrl2* or *Pvrl3* show defects in the apical ectoplasmic specialization, actin distribution, and sperm morphology, rendering mice infertile (Bouchard et al. 2000; Inagaki et al. 2006; Ozaki-Kuroda et al. 2002).

It is well established that both desmosomes and apical ectoplasmic specializations undergo restructuring, which facilitates germ cell movement across the seminiferous epithelium. However, it is not known how these cell junctions disassemble and reassemble throughout spermatogenesis. While component proteins of the tight junction and basal ectoplasmic specialization internalize after Sertoli cells are treated with cytokines (Lie et al. 2011b; Xia et al. 2009; Yan et al. 2008), it is not clear whether desmosomes are regulated by a similar mechanism. Keratinocytes, which use desmosomes for adhesion, maintain two distinct cell adhesion states: stable hyper-adhesion (i.e., Ca^{2+}-independent) and dynamic weak adhesion (Ca^{2+}-dependent), which are prompted by wound healing or experimental Ca^{2+} switch. At the molecular level, protein kinases such as protein kinase C, proto-oncogene tyrosine-protein kinase SRC, and epidermal growth factor receptor induce weak adhesion, resulting in the internalization of desmoglein (Aoyama and Kitajima 1999; Calkins, et al. 2006). In the testis, desmosomes lack the electron-dense midline, which is characteristic of desmosomes in keratinocytes, and they likely exhibit weak adhesion (Russell 1977). Further studies are needed in the testis, because there are apparent differences in the regulation of desmosomes across different epithelia.

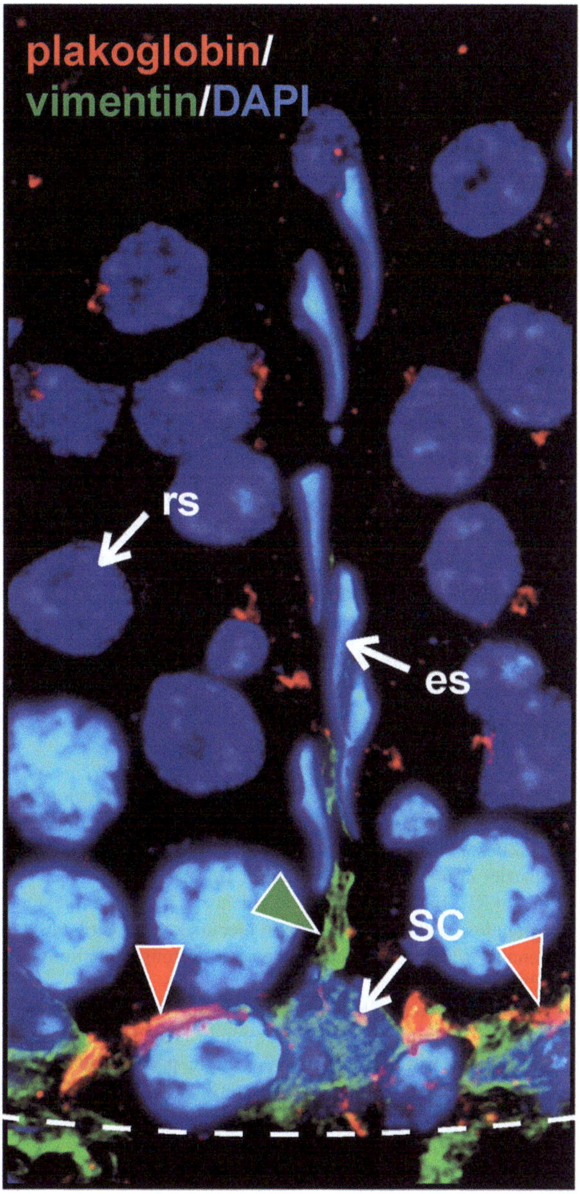

Fig. 9.6 The immunolocalization of plakoglobin and vimentin in the adult rat testis. Frozen testis cross sections were immunostained for plakoglobin (*red* fluorescence) and vimentin (*green* fluorescence), component proteins of the desmosome. *Red* and *green arrowheads* point to plakoglobin and vimentin, respectively, in a stage V seminiferous tubule. Cell nuclei were stained with DAPI (*blue* fluorescence). *SC* Sertoli cell, *rs* round spermatid, *es* elongated spermatid

9.7 Conclusion

In this perspective, we have highlighted the importance of Sertoli cells in the development of the fetal, pubertal, and adult testis. Collectively, these reports illustrate that Sertoli cells are critical for the initiation of spermatogenesis, as well as for its maintenance throughout adulthood. Our knowledge on the Sertoli cell has mostly come from in vitro studies, because these cells are relatively easy to isolate from the testes of early pubertal rats (i.e., 18–20 days old). For instance, Sertoli cells initiate polarization, establish cell junctions, and secrete proteins into the apical and/or basal compartment when they are cultured at high density on Matrigel-coated bicameral units, similar to Sertoli cells in vivo (reviewed in Djakiew and Onoda 1993; Steinberger and Jakubowiak 1993). Thus, Sertoli cell cultures are a good system for present and future studies. Recent studies show that adult human Sertoli cells can be cultured and expanded in vitro while maintaining their primary characteristics (Chui et al. 2011; Guo et al. 2015), indicating that they may have use in reproductive medicine. Future studies should continue to investigate the proliferative ability of human Sertoli cells in vivo.

The Sertoli cell—its structure, function, and regulation—has intrigued reproductive cell biologists for more than a century, resulting in thousands of publications. The goal of this review was to briefly present the biology of the Sertoli cell, highlighting specific milestones and major achievements made by several investigators. Interested readers are strongly encouraged to refer to the references cited herein. We hope that the information in this review provides a strong framework for future studies in the field.

References

Adams IR, McLaren A (2002) Sexually dimorphic development of mouse primordial germ cells: switching from oogenesis to spermatogenesis. Development 129:1155–1164

Agarwal A, Hoffer AP (1989) Ultrastructural studies on the development of the blood–epididymis barrier in immature rats. J Androl 10:425–431

Albrecht KH, Eicher EM (2001) Evidence that Sry is expressed in pre-Sertoli cells and Sertoli and granulosa cells have a common precursor. Dev Biol 240:92–107

Amann RP (2008) The cycle of the seminiferous epithelium in human: a need to revisit? J Androl 29:469–487

Anderson JM, Van Itallie CM (2008) Tight junctions. Curr Biol 18:R941–R943

Aoyama Y, Kitajima Y (1999) Pemphigus vulgaris IgG causes a rapid depletion of desmoglein 3 (Dsg 3) from Triton X-100 soluble pools, leading to the formation of Dsg 3-depleted desmosomes in a human squamous carcinoma cell line, DJM-1 cells. J Invest Dermatol 112:67–71

Ariyaratne HB, Mendis-Handagama SM, Hales BD, Mason JI (2000) Studies on the onset of Leydig precursor cell differentiation in the prepubertal rat testis. Biol Reprod 63:165–171

Barakat B, O'Connor AE, Gold E, de Kretser DM, Loveland KL (2008) Inhibin, activin, follistatin and FSH serum levels and testicular production are highly modulated during the first spermatogenic wave in mice. Reproduction 136:345–359

Bardin CW, Cheng CY, Musto NA, Gunsalus GL (1988) The Sertoli cell. In: Knobil E, Neill JD, Ewing LL, Greenwald GS, Markert CL, Pfaff DW (eds) The Physiology of Reproduction. Raven, New York, pp 933–974

Barrionuevo F, Bagheri-Fam S, Klattig J, Kist R, Taketo MM, Englert C, Scherer G (2006) Homozygous inactivation of Sox9 causes complete XY sex reversal in mice. Biol Reprod 74:195–201

Barrionuevo F, Georg I, Scherthan H, Lecureuil C, Guillou F, Wegner M, Scherer G (2009) Testis cord differentiation after the sex determination stage is independent of *Sox9* but fails in the combined absence of *Sox9* and *Sox8*. Dev Biol 327:301–312
Barsoum I, Yao HH (2006) The road to maleness: from testis to Wolffian duct. Trends Endocrinol Metab 17:223–228
Bascom KF, Osterud HL (1925) Quantitative studies of the testicle. II. Pattern and total tubule length in the testicles of certain common mammals. Anat Rec 31:159–169
Battiloro E, Angeletti B, Tozzi MC, Bruni L, Tondini S, Vignetti P, Verna R, D'Ambrosio E (1997) A novel double nucleotide substitution in the HMG box of the SRY gene associated with Swyer syndrome. Hum Genet 100:585–587
Bay K, Main KM, Toppari J, Skakkebaek NE (2011) Testicular descent: INSL3, testosterone, genes and the intrauterine milieu. Nat Rev Urol 8:187–196
Behringer RR (1995) The Müllerian inhibitor and mammalian sexual development. Philos Trans R Soc Lond B Biol Sci 350:285–288
Behringer RR, Finegold MJ, Cate RL (1994) Müllerian-inhibiting substance function during mammalian sexual development. Cell 79:415–425
Bogani D, Siggers P, Brixey R, Warr N, Beddow S, Edwards J (2009) Loss of mitogen-activated protein kinase kinase kinase 4 (MAP3K4) reveals a requirement for MAPK signalling in mouse sex determination. PLoS Biol 7:e1000196
Bornstein SR, Rutkowski H, Vrezas I (2004) Cytokines and steroidogenesis. Mol Cell Endocrinol 215:135–141
Bouchard MJ, Dong Y, McDermott BM Jr, Lam DH, Brown KR, Shelanski M, Bellve AR, Racaniello VR (2000) Defects in nuclear and cytoskeletal morphology and mitochondrial localization in spermatozoa of mice lacking nectin-2, a component of cell-cell adherens junctions. Mol Cell Biol 20:2865–2873
Brennan J, Capel B (2004) One tissue, two fates: molecular genetic events that underlie testis versus ovary development. Nat Rev Genet 5:509–521
Brennan J, Karl J, Capel B (2002) Divergent vascular mechanisms downstream of Sry establish the arterial system in the XY gonad. Dev Biol 244:418–428
Calkins CC, Setzer SV, Jennings JM, Summers S, Tsunoda K, Amagai M, Kowalczyk A (2006) Desmoglein endocytosis and desmosome disassembly are coordinated responses to pemphigus autoantibodies. J Biol Chem 281:7623–7634
Cavicchia JC, Dym M (1977) Relative volume of Sertoli cells in monkey seminiferous epithelium. Am J Anat 150:501–503
Chaboissier MC, Kobayashi A, Vidal VI, Lutzkendorf S, van de Kant HJ, Wegner M, de Rooij DG, Behringer RR, Schedl A (2004) Functional analysis of Sox8 and Sox9 during sex determination in the mouse. Development 131:1891–1901
Cheng CY, Mruk DD (2012) The blood–testis barrier and its implication in male contraception. Pharmacol Rev 64:16–64
Chihara M, Ikebuchi R, Otsuka S, Ichii O, Hashimoto Y, Suzuki A, Saga Y, Kon Y (2013a) Mice stage-specific claudin 3 expression regulates progression of meiosis in early stage spermatocytes. Biol Reprod 89:1–12
Chihara M, Otsuka S, Ichii O, Kon Y (2013b) Vitamin A deprivation affects the progression of the spermatogenic wave and initial formation of the blood–testis barrier, resulting in irreversible testicular degeneration in mice. J Reprod Dev 59:525–535
Christensen AK, Gillim SW (1969) The correlation of fine structure and function in steroid-secreting cells, with emphasis on those of the gonads. In: McKerns KW (ed) The Gonads. Appleton-Century-Crofts, New York, pp 415–488
Chui K, Trivedi A, Cheng CY, Cherbavaz DB, Dazin PF, Huynh AL, Mitchell JB, Rabinovich GA, Noble-Haeusslein LJ, John CM (2011) Characterization and functionality of proliferative human Sertoli cells. Cell Transplant 20:619–635
Clermont Y (1972) Kinetics of spermatogenesis in mammals: seminiferous epithelium cycle and spermatogonial renewal. Physiol Rev 52:198–235

Colvin JS, Green RP, Schmahl J, Capel B, Ornitz DM (2001) Male-to-female sex reversal in mice lacking fibroblast growth factor 9. Cell 104:875–889
Combes AN, Lesieur E, Harley VR, Sinclair AH, Little MH, Wilhelm D, Koopman P (2009a) Three-dimensional visualization of testis cord morphogenesis, a novel tubulogenic mechanism in development. Dev Dyn 238:1033–1041
Combes AN, Wilhelm D, Davidson T, Dejana E, Harley V, Sinclair A, Koopman P (2009b) Endothelial cell migration directs testis cord formation. Dev Biol 326:112–120
Cool J, DeFalco TJ, Capel B (2011) Vascular-mesenchymal cross-talk through Vegf and Pdgf drives organ patterning. Proc Natl Acad Sci U S A 108:167–172
Cool J, DeFalco T, Capel B (2012) Testis formation in the fetal mouse: dynamic and complex de novo tubulogenesis. Wiley Interdiscip Rev Dev Biol 1:847–859
Coveney D, Cool J, Oliver T, Capel B (2008) Four-dimensional analysis of vascularization during primary development of an organ, the gonad. Proc Natl Acad Sci U S A 105:7212–7217
Culty M (2009) Gonocytes, the forgotten cells of the germ cell lineage. Birth Defects Res C Embryo Today 87:1–26
de Kretser DM (1990) Inhibin. Mol Cell Endocrinol 69:C17–C20
de Kretser DM, Kerr JB (1988) The cytology of the testis. In: Knobil E, Neill JB, Ewing LL, Greenwald GS, Markert CL, Pfaff DW (eds) The Physiology of Reproduction. Raven, New York, pp 837–932
de Kretser DM, Buzzard JJ, Okuma Y, O'Connor AE, Hayashi T, Lin SY, Morrison JR, Loveland KL (2004) The role of activin, follistatin and inhibin in testicular physiology. Mol Cell Endocrinol 225:57–64
de Rooij DG, Griswold MD (2012) Questions about spermatogonia posed and answered since 2000. J Androl 33:1085–1095
de Rooij DG, Russell LD (2000) All you wanted to know about spermatogonia but were afraid to ask. J Androl 21:776–798
DeFalco T, Potter SJ, Williams AV, Waller B, Kan MJ, Capel B (2015) Macrophages contribute to the spermatogonial niche in the adult testis. Cell Rep 12:1107–1119
Djakiew D, Onoda M (1993) Multichamber cell culture and directional secretion. In: Russell LD, Griswold MD (eds) The Sertoli Cell. Cache River, Clearwater, pp 181–194
Dong L, Jelinsky SA, Finger JN, Johnston DS, Kopf GS, Sottas CM, Hardy MP, Ge RS (2007) Gene expression during development of fetal and adult Leydig cells. Ann NY Acad Sci 1120:16–35
Dym M (1994) Basement membrane regulation of Sertoli cells. Endocr Rev 15:102–115
Dym M, Cavicchia JC (1977) Further observations on the blood–testis barrier in monkeys. Biol Reprod 17:390–403
Dym M, Fawcett DW (1970) The blood–testis barrier in the rat and the physiological compartmentation of the seminiferous epithelium. Biol Reprod 3:308–326
Eggers S, Ohnesorg T, Sinclair A (2014) Genetic regulation of mammalian gonad development. Nat Rev Endocrinol 10:673–683
Fawcett DW, Leak LV, Heidger PM (1970) Electron microscopic observations on the structural components of the blood–testis barrier. J Reprod Fertil 10:105–122
Furuse M (2010) Molecular basis of the core structure of tight junctions. Cold Spring Harb Perspect Biol 2:a002907
Furuse M, Izumi Y, Oda Y, Higashi T, Iwamoto N (2014) Molecular organization of tricellular tight junctions. Tissue Barriers 2:e28960
Gardner PJ, Holyoke EA (1964) Fine structure of the seminiferous tubule of the Swiss mouse. I. The limiting membrane, Sertoli cell, spermatogonia, and spermatocytes. Anat Rec 150:391–404
Ge RS, Chen G, Hardy MP (2008) The role of the Leydig cell in spermatogenic function. Adv Exp Med Biol 636:255–269
Giannessi F, Giambelluca MA, Scavuzzo MC, Ruffoli R (2005) Ultrastructure of testicular macrophages in aging mice. J Morphol 263:39–46

Gierl MS, Gruhn WH, von Seggern A, Maltry N, Niehrs C (2012) GADD45G functions in male sex determination by promoting p38 signaling and Sry expression. Dev Cell 23:1032–1042
Gow A, Southwood CM, Li JS, Pariali M, Riordan GP, Brodie SE, Danias J, Bronstein JM, Kachar B, Lazzarini RA (1999) CNS myelin and Sertoli cell tight junction strands are absent in Osp/claudin-11 null mice. Cell 99:649–659
Greenbaum MP, Iwamori T, Buchold GM, Matzuk MM (2011) Germ cell intercellular bridges. Cold Spring Harb Perspect Biol 3:a005850
Griswold MD (1993) Protein secretion by Sertoli cells: general considerations. In: Russell LD, Griswold MD (eds) The Sertoli Cell. Cache River, Clearwater, pp 195–200
Griswold MD (1995) Interactions between germ cells and Sertoli cells in the testis. Biol Reprod 52:211–216
Griswold MD (1998) The central role of Sertoli cells in spermatogenesis. Semin Cell Dev Biol 9:411–416
Griswold SL, Behringer RR (2009) Fetal Leydig cell origin and development. Sex Dev 3:1–15
Gubbay J, Koopman P, Collignon J, Burgoyne P, Lovell-Badge R (1990) Normal structure and expression of *Zfy* genes in XY female mice mutant in *Tdy*. Development 109:647–653
Guo Y, Hai Y, Yao C, Chen Z, Hou J, Li Z, He Z (2015) Long-term culture and significant expansion of human Sertoli cells whilst maintaining stable global phenotype and AKT and SMAD1/5 activation. Cell Commun Signal 13:20
Habert R, Lejeune H, Saez JM (2001) Origin, differentiation and regulation of fetal and adult Leydig cells. Mol Cell Endocrinol 179:47–74
Hacker A, Capel B, Goodfellow P, Lovell-Badge R (1995) Expression of Sry, the mouse sex determining gene. Development 121:1603–1614
Hadley MA, Dym M (1987) Immunocytochemistry of extracellular matrix in the lamina propria of the rat testis: electron microscopic localization. Biol Reprod 37:1283–1289
Haider SG (2004) Cell biology of Leydig cells in the testis. Int Rev Cytol 233:181–241
Hales DB (2002) Testicular macrophage modulation of Leydig cell steroidogenesis. J Reprod Immunol 57:3–18
Hales DB, Xiong Y, Tur-Kaspa I (1992) The role of cytokines in the regulation of Leydig cell P450c17 gene expression. J Steroid Biochem Mol Biol 43:907–914
Harmon RM, Green KJ (2013) Structural and functional diversity of desmosomes. Cell Commun Adhes 20:171–187
Hawkins JR, Taylor A, Berta P, Levilliers J, Van der Auwera B, Goodfellow PN (1992) Mutational analysis of SRY: nonsense and missense mutations in XY sex reversal. Hum Genet 88:471–474
Heller CG, Clermont Y (1963) Spermatogenesis in man: an estimation of its duration. Science 140:184–186
Hermo L, Pelletier RM, Cyr DG, Smith CE (2010) Surfing the wave, cycle, life history, and genes/proteins expressed by testicular germ cells. Part 2: Changes in spermatid organelles associated with development of spermatozoa. Microsc Res Tech 73:279–319
Hess RA (1990) Quantitative and qualitative characteristics of the stages and transitions in the cycle of the rat seminiferous epithelium: light microscopic observations of perfusion-fixed and plastic-embedded testes. Biol Reprod 43:525–542
Hess RA, Renato de Franca L (2008) Spermatogenesis and cycle of the seminiferous epithelium. Adv Exp Med Biol 636:1–15
Holdcraft RW, Braun RE (2004) Hormonal regulation of spermatogenesis. Int J Androl 27:335–342
Hosoi I, Toyama Y, Maekawa M, Ito H, Yuasa S (2002) Development of the blood–testis barrier in the mouse is delayed by neonatally administered diethylstilbestrol but not by β-estradiol 3-benzoate. Andrologia 34:255–262
Hume DA, Halpin D, Charlton HM, Gordon S (1984) The mononuclear phagocyte system of the mouse defined by immunohistochemical localization of antigen F4/80: macrophages of endocrine organs. Proc Natl Acad Sci U S A 81:4174–4177

Hutson JC (2006) Physiologic interactions between macrophages and Leydig cells. Exp Biol Med (Maywood) 231:1–7

Hutson JM, Southwell BR, Li R, Lie G, Ismail K, Harisis G, Chen N (2013) The regulation of testicular descent and the effects of cryptorchidism. Endocr Rev 34:725–752

Inagaki M, Irie K, Ishizaki H, Tanaka-Okamoto M, Miyoshi J, Takai Y (2006) Role of cell adhesion molecule nectin-3 in spermatid development. Genes Cells 11:1125–1132

Jakob S, Lovell-Badge R (2011) Sex determination and the control of Sox9 expression in mammals. FEBS J 278:1002–1009

Jo A, Denduluri S, Zhang B, Wang Z, Yin L, Yan Z, Kang R, Shi LL, Mok J, Lee MJ et al (2014) The versatile functions of Sox9 in development, stem cells, and human diseases. Genes Dis 1:149–161

Kaitu'u-Lino TJ, Sluka P, Foo CF, Stanton PG (2007) Claudin-11 expression and localisation is regulated by androgens in rat Sertoli cells *in vitro*. Reproduction 133:1169–1179

Kanai Y, Kawakami H, Takata K, Kurohmaru M, Hirano H, Hayashi Y (1992) Involvement of actin filaments in mouse testicular cord organization *in vivo* and *in vitro*. Biol Reprod 46:233–245

Kashimada K, Koopman P (2010) Sry: the master switch in mammalian sex determination. Development 137:3921–3930

Kato T, Miyata K, Sonobe M, Yamashita S, Tamano M, Miura K, Kanai Y, Miyamoto S, Sakuma T, Yamamoto T et al (2013) Production of Sry knockout mouse using TALEN via oocyte injection. Sci Rep 3:3136

Kerr JB, Loveland KL, O'Bryan MK, de Kretser DM (2006) Cytology of the testis and intrinsic control mechanisms. In: Neill JD (ed) Knobil and Neill's Physiology of Reproduction. Elsevier, New York, pp 827–947

Kim Y, Kobayashi A, Sekido R, DiNapoli L, Brennan J, Chaboissier MC, Poulat F, Behringer RR, Lovell-Badge R, Capel B (2006) Fgf9 and Wnt4 act as antagonistic signals to regulate mammalian sex determination. PLoS Biol 4:e187

Kohler B, Achermann JC (2010) Update–steroidogenic factor 1 (SF-1, NRA1). Minerva Endocrinol 35:73–86

Koopman P (1999) Sry and Sox9: mammalian testis-determining genes. Cell Mol Life Sci 55:839–856

Koopman P (2001) The genetics and biology of vertebrate sex determination. Cell 105:843–847

Larney C, Bailey TL, Koopman P (2014) Switching on sex: transcriptional regulation of the testis-determining gene Sry. Development 141:2195–2205

LeBlond CP, Clermont Y (1952) Definition of the stages of the cycle of the seminiferous epithelium in the rat. Ann NY Acad Sci 55:548–573

Lee MM, Donahoe PK (1993) Müllerian inhibiting substance: a gonadal hormone with multiple functions. Endocr Rev 14:152–164

Lie PPY, Cheng CY, Mruk DD (2010) The desmoglein-2/desmocollin-2/Src kinase protein complex regulates blood–testis barrier dynamics. Int J Biochem Cell Biol 42:975–986

Lie PPY, Cheng CY, Mruk DD (2011a) The biology of the desmosome-like junction: a versatile anchoring junction and signal transducer in the seminiferous epithelium. Int Rev Cell Mol Biol 286:223–269

Lie PPY, Cheng CY, Mruk DD (2011b) Interleukin-1α is a regulator of the blood–testis barrier. FASEB J 25:1244–1253

Lucas TFG, Nascimento AR, Pisolato R, Pimenta MT, Lazari MFM, Porto CS (2014) Receptors and signaling pathways involved in proliferation and differentiations of Sertoli cells. Spermatogenesis 4:e28138

Ludbrook LM, Harley VR (2004) Sex determination: a 'window' of Dax1 activity. Trends Endocrinol Metab 15:116–121

Lukyanenko YO, Chen JJ, Hutson JC (2001) Production of 25-hydroxycholesterol by testicular macrophages and its effects on Leydig cells. Biol Reprod 64:790–796

Luo X, Ikeda Y, Parker KL (1994) A cell-specific nuclear receptor is essential for adrenal and gonadal development and sexual differentiation. Cell 77:481–490
Madara JL (1998) Regulation of the movement of solutes across tight junctions. Annu Rev Physiol 60:143–159
Maekawa M, Kamimura K, Nagano T (1996) Peritubular myoid cells in the testis: their structure and function. Arch Histol Cytol 59:1–13
Malki S, Nef S, Notarnicola C, Thevenet L, Gasca S, Mejean C, Berta P, Poulat F, Boizet-Bonhoure B (2005) Prostaglandin D2 induces nuclear import of the sex-determining factor SOX9 via its cAMP-PKA phosphorylation. EMBO J 24:1798–1809
Martineau J, Nordqvist K, Tilmann C, Lovell-Badge R, Capel B (1997) Male-specific cell migration into the developing gonad. Curr Biol 7:958–968
Matter K, Balda MS (2003) Signaling to and from tight junctions. Nat Rev Mol Cell Biol 4:225–236
Mazaud-Guittot S, Meugnier E, Pesenti S, Wu XJ, Vidal H, Gow A, Le Magueresse-Battistoni B (2010) Claudin 11 deficiency in mice results in loss of the Sertoli cell epithelial phenotype in the testis. Biol Reprod 82:202–213
McElreavey K, Vilain E, Barbaux S, Fuqua JS, Fechner PY, Souleyreau N, Doco-Fenzy M, Gabriel R, Quereux C, Fellous M et al (1996) Loss of sequences 3′ to the testis-determining gene, SRY, including the Y pseudoautosomal boundary associated with partial testicular determination. Proc Natl Acad Sci U S A 93:8590–8594
Meachem SJ, McLachlan RI, de Kretser DM, Robertson DM, Wreford NG (1996) Neonatal exposure of rats to recombinant follicle stimulating hormone increases adult Sertoli and spermatogenic cell numbers. Biol Reprod 54:36–44
Meinhardt A, Hedger MP (2011) Immunological, paracrine and endocrine aspects of testicular immune privilege. Mol Cell Endocrinol 335:60–68
Mendis-Handagama SM, Ariyaratne HB (2001) Differentiation of the adult Leydig cell population in the postnatal testis. Biol Reprod 65:60–671
Meng J, Holdcraft RW, Shima JE, Griswold MD, Braun RE (2005) Androgens regulate the permeability of the blood–testis barrier. Proc Natl Acad Sci U S A 102:16696–16700
Mitchell CL, Harley VR (2002) Biochemical defects in eight SRY missense mutations causing XY gonadal dysgenesis. Mol Genet Metab 77:217–225
Mithraprabhu S, Mendis S, Meachen SJ, Tubino L, Matzuk MM, Brown CW, Loveland KL (2010) Activin bioactivity affects germ cell differentiation in the postnatal mouse testis *in vivo*. Biol Reprod 82:980–990
Moniot B, Declosmenil F, Barrionuevo F, Scherer G, Aritake K, Malki S, Marzi L, Cohen-Solal A, Georg I, Klattig J et al (2009) The PGD2 pathway, independently of FGF9, amplifies SOX9 activity in Sertoli cells during male sexual differentiation. Development 136:1813–1821
Mruk DD, Cheng CY (2004a) Cell-cell interactions at the ectoplasmic specialization in the testis. Trends Endocrinol Metab 15:439–447
Mruk DD, Cheng CY (2004b) Sertoli-Sertoli and Sertoli-germ cell interactions and their significance in germ cell movement in the seminiferous epithelium during spermatogenesis. Endocr Rev 25:747–806
Mruk DD, Cheng CY (2010) Tight junctions in the testis: new perspectives. Philos Trans R Soc Lond B Biol Sci 365:1621–1635
Mruk DD, Cheng CY (2011) Desmosomes in the testis: moving into an unchartered territory. Spermatogenesis 1:47–51
Mruk DD, Cheng CY (2015) The mammalian blood–testis barrier: its biology and regulation. Endocr Rev 36:564–591
Nagamine CM, Morohashi K, Carlisle C, Chang DK (1999) Sex reversal caused by *Mus musculus domesticus* Y chromosomes linked to variant expression of the testis-determining gene *Sry*. Dev Biol 216:182–194
Nekrasova OE, Green KJ (2013) Desmosome assembly and dynamics. Trends Cell Biol 23:537–546

Nes WD, Lukyanenko YO, Jia ZH, Quideau S, Howald WN, Pratum TK, West RR, Hutson JC (2000) Identification of the lipophilic factor produced by macrophages that stimulates steroidogenesis. Endocrinology 141:953–958

Niemi M, Sharpe RM, Brown WR (1986) Macrophages in the interstitial tissue of the rat testis. Cell Tissue Res 243:337–344

Oakberg EF (1956) Duration of spermatogenesis in the mouse and timing of stages of the cycle of the seminiferous epithelium. Am J Anat 99:507–516

Oatley JM, Brinster RL (2008) Regulation of spermatogonial stem cell self-renewal in mammals. Annu Rev Cell Dev Biol 24:263–286

Orth JM (1982) Proliferation of Sertoli cells in fetal and postnatal rats: a quantitative autoradiographic study. Anat Rec 203:485–492

Orth JM (1984) The role of follicle-stimulating hormone in controlling Sertoli cell proliferation in testes of fetal rats. Endocrinology 115:1248–1255

Orth JM, Gunsalus GM, Lamperti AA (1988) Evidence from Sertoli cell-depleted rats indicates that spermatid numbers in adults depends on numbers of Sertoli cells produced during perinatal development. Endocrinology 122:787–794

O'Shaughnessy PJ, Baker PJ, Sohnius U, Haavisto AM, Charlton HM, Huhtaniemi I (1998) Fetal development of Leydig cell activity in the mouse is independent of pituitary gonadotroph function. Endocrinology 139:1141–1146

O'Shaughnessy PJ, Baker PJ, Johnston H (2006) The foetal Leydig cell—differentiation, function and regulation. Int J Androl 29:90–95

Ostrer H (2014) Disorders of sex development (DSDs): an update. J Clin Endocrinol Metab 99:1503–1509

Ozaki-Kuroda K, Nakanishi H, Ohta H, Tanaka H, Kurihara H, Mueller S, Irie K, Ikeda W, Sakai T, Wimmer E et al (2002) Nectin couples cell-cell adhesion and the actin scaffold at heterotypic testicular junctions. Curr Biol 12:1145–1150

Parker KL, Schimmer BP, Schedl A (1999) Genes essential for early events in gonadal development. Cell Mol Life Sci 55:831–838

Perez CV, Theas MS, Jacobo PV, Jarazo-Dietrich S, Guazzone VA, Lustig L (2013) Dual role of immune cells in the testis. Spermatogenesis 3:e23870

Piprek R (2009) Genetic mechanisms underlying male sex determination in mammals. J Appl Genet 50:347–360

Pointis G, Gilleron J, Carette D, Segretain D (2010) Physiological and physiopathological aspects of connexins and communicating gap junctions in spermatogenesis. Philos Trans R Soc Lond B Biol Sci 365:1607–1620

Rebourcet D, O'Shaughnessy PJ, Monteiro A, Milne L, Cruichshanks L, Jeffrey N (2014a) Sertoli cells maintain Leydig cell number and peritubular myoid cell activity in the adult mouse testis. PLoS One 9:e105687

Rebourcet D, O'Shaughnessy PJ, Pitetti JL, Monteiro A, O'Hara L, Milne L, Tsai YT, Cruickshanks L, Riethmacher D, Guillou F et al (2014b) Sertoli cells control peritubular myoid cell fate and support adult Leydig cell development in the prepubertal testis. Development 141:2139–2149

Regaud C (1901) Études sur la structure des tubes séminifères et sur la spermatogénèse chez les mammifères. Arch Anat Microsc Morphol Exp 4:101–156, 231–380

Ribbert H (1904) Die abscheidung intravenos injizierten gelosten Karmins in den Geweben. Z Allg Physiol 4:201–214

Rizzolo LJ (2007) Development and role of tight junctions in the retinal pigment epithelium. Int Rev Cytol 258:195–234

Ross MH, Long IR (1966) Contractile cells in human seminiferous tubules. Science 153:1271–1273

Russell LD (1977) Desmosome-like junctions between Sertoli and germ cells in the rat testis. Am J Anat 148:301–312

Russell LD (1978) The blood–testis barrier and its formation relative to spermatocyte maturation in the adult rat: a lanthanum tracer study. Anat Rec 190:99–111

Russell LD (1993a) Morphological and functional evidence for Sertoli-germ cell relationships. In: Russell LD, Griswold MD (eds) The Sertoli Cell. Cache River, Clearwater, pp 365–390

Russell LD (1993b) Role in spermiation. In: Russell LD, Griswold MD (eds) The Sertoli Cell. Cache River, Clearwater, pp 269–303

Russell LD, Tallon-Doran M, Weber JE, Wong V, Peterson RN (1983) Three-dimensional reconstruction of a rat stage V Sertoli cell: III. A study of specific cellular relationships. Am J Anat 167:181–192

Russell LD, Ettlin RA, Hikim APS, Clegg ED (eds) (1990) Histological and Histopathological Evaluation of the Testis. Cache River, St. Louis

Schlatt S, Ehmcke J (2014) Regulation of spermatogenesis: an evolutionary biologist's perspective. Semin Cell Dev Biol 29:2–16

Schmitt-Ney M, Thiele H, Kaltwasser P, Bardoni B, Cisternino M, Scherer G (1995) Two novel SRY missense mutations reducing DNA binding identified in XY females and their mosaic fathers. Am J Hum Genet 56:862–869

Setchell BP, Waites GMB (1975) The blood–testis barrier. In: Hamilton DW, Greep RO (eds) The Handbook of Physiology. Williams and Wilkens, Baltimore, pp 143–172

Sharpe RM (1994) Regulation of spermatogenesis. In: Knobil E, Neill JD (eds) The Physiology of Reproduction. Raven, New York, pp 1363–1434

Sharpe RM, McKinnell C, Kivlin C, Fisher JS (2003) Proliferation and functional maturation of Sertoli cells, and their relevance to disorders of testis function in adulthood. Reproduction 125:769–784

Shin K, Fogg VC, Margolis B (2006) Tight junctions and cell polarity. Annu Rev Cell Dev Biol 22:207–235

Shinohara T, Orwig KE, Avarbock MR, Brinster RL (2001) Remodeling of the postnatal mouse testis is accompanied by dramatic changes in stem cell number and niche accessibility. Proc Natl Acad Sci U S A 98:6186–6191

Skinner MK (1993) Secretion of growth factors and other regulatory factors. In: Russell LD, Griswold MD (eds) The Sertoli Cell. Cache River, Clearwater, pp 237–247

Skinner MK, Fritz IB (1985) Structural characterization of proteoglycans produced by testicular peritubular myoid cells *in vitro* in the rat. J Biol Chem 260:11874–11883

Skinner MK, Tung PS, Fritz IB (1985) Cooperativity between Sertoli cells and testicular peritubular cells in the production and deposition of extracellular matrix components. J Cell Biol 100:1941–1947

Skinner MK, Norton JN, Mullaney BP, Rosselli M, Whaley PD, Anthony CT (1991) Cell-cell interactions and the regulation of testis function. Ann NY Acad Sci 637:354–363

Smith BE, Braun RE (2012) Germ cell migration across Sertoli cell tight junctions. Science 338:798–802

Smith LB, Walker WH (2014) The regulation of spermatogenesis by androgens. Semin Cell Dev Biol 30:2–13

Steinberger A, Jakubowiak A (1993) Sertoli cell culture: historical perspective and review of methods. In: Russell LD, Griswold MD (eds) The Sertoli Cell. Cache River, Clearwater, pp 155–180

Svingen T, Koopman P (2013) Building the mammalian testis: origins, differentiation, and assembly of the component populations. Genes Dev 27:2409–2426

Takekawa M, Saito H (1998) A family of stress-inducible GADD45-like proteins mediate activation of the stress-responsive MTK1/MEKK4 MAPKKK. Cell 95:521–530

Tanaka SS, Nishinakamura R (2014) Regulation of male sex determination: genital ridge formation of *Sry* activation in mice. Cell Mol Life Sci 71:4781–4802

Tang H, Brennan J, Karl J, Hamada Y, Raetzman L, Capel B (2008) Notch signaling maintains Leydig cell progenitor cells in the mouse testis. Development 135:3745–3753

Tarulli GA, Stanton PG, Lerchl A, Meachem SJ (2006) Adult Sertoli cells are not terminally differentiated in the Djungarian hamster: effect of FSH on proliferation and junction protein organization. Biol Reprod 74:798–806

Tarulli GA, Stanton PG, Meachem SJ (2012) Is the adult Sertoli cell terminally differentiated? Biol Reprod 87:1–11

Tarulli GA, Stanton PG, Loveland KL, Rajpert-De Meyts E, McLachlan RI, Meachem SJ (2013) A survey of Sertoli cell differentiation in men after gonadotropin suppression and in testicular cancer. Spermatogenesis 3:e24014

Tegelenbosch RA, de Rooij DG (1993) A quantitative study of spermatogonial multiplication and stem cell renewal in the C3H/101 F1 hybrid mouse. Mutat Res 290:193–200

Toyama Y, Ohkawa M, Oku R, Maekawa M, Yuasa S (2001) Neonatally administered diethylstilbestrol retards the development of the blood–testis barrier in the rat. J Androl 22:413–423

Tremblay JJ (2015) Molecular regulation of steroidogenesis in endocrine Leydig cells. Steroids 103:3–10

Tsukita S, Furuse M (2000) Pores in the wall: claudins constitute tight junction strands containing aqueous pores. J Cell Biol 149:13–16

Tung PS, Fritz IB (1987) Morphogenetic restructuring and formation of basement membranes by Sertoli cells and testis peritubular cells in co-culture: inhibition of the morphogenetic cascade by cyclic AMP derivatives and by blocking direct cell contact. Dev Biol 120:139–153

Urade Y, Hayaishi O (2000) Prostaglandin D synthase: structure and function. Vitam Horm 58:89–120

Verhoeven G, Hoeben E, De Gendt K (2000) Peritubular cell–Sertoli cell interactions: factors involved in PmodS activity. Andrologia 43:42–45

Vitale R, Fawcett DW, Dym M (1973) The normal development of the blood–testis barrier and the effects of clomiphene and estrogen treatment. Anat Rec 176:333–344

Vogl AW, Pfeiffer DC, Redenbach DM, Grove BD (1993) Sertoli cell cytoskeleton. In: Russell LD, Griswold MD (eds) The Sertoli Cell. Cache River, Clearwater, pp 39–86

Vogl AW, Vaid KS, Guttman JA (2008) The Sertoli cell cytoskeleton. In: Cheng CY (ed) Molecular Mechanisms in Spermatogenesis. Landes and Springer, Austin, pp 186–211

Walker WH (2011) Testosterone signaling and the regulation of spermatogenesis. Spermatogenesis 1:116–120

Walker WH, Cheng J (2005) FSH and testosterone signaling in Sertoli cells. Reproduction 130:15–28

Warr N, Carre GA, Siggers P, Faleato JV, Brixey R, Pope M, Bogani D, Childers M, Wells S, Scudamore CL et al (2012) Gadd45γ and Map3k4 interactions regulate mouse testis determination via p38 MAPK-mediated control of Sry expression. Dev Cell 23:1020–1031

Welsh M, Saunders PT, Atanassova N, Sharpe RM, Smith LB (2009) Androgen action via testicular peritubular myoid cells is essential for male fertility. FASEB J 23:4218–4230

Wilhelm D, Martinson F, Bradford S, Wilson MJ, Combes AN, Beverdam A, Bowles J, Mizusaki H, Koopman P (2005) Sertoli cell differentiation is induced both cell-autonomously and through prostaglandin signaling during mammalian sex determination. Dev Biol 287:111–124

Wilhelm D, Hiramatsu R, Mizusaki H, Widjaja L, Combes AM, Kanai Y, Koopman P (2007a) Leydig cell-derived factors (testosterone and insulin-like growth factor 3) play important roles in masculinization and testis descent. J Biol Chem 282:10553–10560

Wilhelm D, Hiramatsu R, Mizusaki H, Widjaja L, Combes AN, Kanai Y, Koopman P (2007b) SOX9 regulates prostaglandin D synthase gene transcription *in vivo* to ensure testis development. J Biol Chem 282:10553–10560

Wilhelm D, Palmer S, Koopman P (2007c) Sex determination and gonadal development in mammals. Physiol Rev 87:1–28

Wilhelm D, Yang JX, Tomas P (2013) Mammalian sex determination and gonad development. Curr Top Dev Biol 106:89–121

Wolski KM, Perrault C, Tran-Son-Tay R, Cameron DF (2005) Strength measurement of the Sertoli–spermatid junctional complex. J Androl 26:354–359
Wu Q, Fukuda K, Weinstein M, Graff JM, Saga Y (2015) SMAD2 and p38 signaling pathways act in concert to determine XY primordial germ cell fate in mice. Development 142:575–586
Xia W, Wong EWP, Mruk DD, Cheng CY (2009) TGF-β3 and TNFα perturb blood–testis barrier (BTB) dynamics by accelerating the clathrin-mediated endocytosis of integral membrane proteins: a new concept of BTB regulation during spermatogenesis. Dev Biol 327:48–61
Yan HHN, Mruk DD, Lee WM, Cheng CY (2008) Blood–testis barrier dynamics are regulated by testosterone and cytokines via their differential effects on the kinetics of protein endocytosis and recycling in Sertoli cells. FASEB J 22:1945–1959
Yang Z, Song L, Huang C (2009) Gadd45 proteins as critical signal transducers linking NF-κB to MAPK cascades. Curr Cancer Drug Targets 9:915–930
Yao HHC, Wang T, Deng J, Liu D, Li X, Deng J (2014) The development of blood–retinal barrier during the interaction of astrocytes with vascular wall cells. Neural Regen Res 9:1047–1054
Zwingman T, Fujimoto H, Lai LW, Boyer T, Ao A, Stalvey JR, Blecher SR, Erickson RP (1994) Transcription of circular and noncircular forms of Sry in mouse testes. Mol Reprod Dev 37:370–381

Chapter 10
Mechanisms Regulating Spermatogonial Differentiation

Jennifer M. Mecklenburg and Brian P. Hermann

Abstract Mammalian spermatogenesis is a complex and highly ordered process by which male germ cells proceed through a series of differentiation steps to produce haploid flagellated spermatozoa. Underlying this process is a pool of adult stem cells, the spermatogonial stem cells (SSCs), which commence the spermatogenic lineage by undertaking a differentiation fate decision to become progenitor spermatogonia. Subsequently, progenitors acquire a differentiating spermatogonia phenotype and undergo a series of amplifying mitoses while becoming competent to enter meiosis. After spermatocytes complete meiosis, post-meiotic spermatids must then undergo a remarkable transformation from small round spermatids to a flagellated spermatozoa with extremely compacted nuclei. This chapter reviews the current literature pertaining to spermatogonial differentiation with an emphasis on the mechanisms controlling stem cell fate decisions and early differentiation events in the life of a spermatogonium.

10.1 Introduction

Spermatogenesis is the process by which a diploid population of germline stem cells propagate and differentiate to give rise to millions of haploid flagellated spermatozoa every day throughout the life span of adult male mammals. Sustaining this tremendous productivity to maintain male fertility requires that germ cells undergo a series of programmed differentiation steps instructed by their microenvironment. Spermatogenic differentiation begins with commitment of spermatogonial stem cells (SSCs) to this differentiation pathway (versus the alternate cell fate, self-renewal) followed by a series of additional hallmark events, including spermatogonial differentiation concurrent with mitotic amplification, entry into meiosis, and post-meiotic spermiogenesis before ultimately culminating in sperm release. Intensive research efforts over the past several decades have revealed

J.M. Mecklenburg • B.P. Hermann, Ph.D. (✉)
Department of Biology, The University of Texas at San Antonio, One UTSA Circle, San Antonio, TX 78249, USA
e-mail: brian.hermann@utsa.edu

R.P. Piprek (ed.), *Molecular Mechanisms of Cell Differentiation in Gonad Development*, Results and Problems in Cell Differentiation 58,
DOI 10.1007/978-3-319-31973-5_10

many of the molecular and cellular cues, both germ cell intrinsic and arising from the somatic niche, which instruct these differentiation events and permit male fertility. Likewise, investigation of a variety of animal models and humans has revealed conservation of many, but not all, of the regulatory mechanisms controlling spermatogenic differentiation.

This chapter begins with a brief overview of the organization of spermatogenesis to frame an in-depth discussion of the key early events in which spermatogonia become committed to the differentiation pathway, initiate that differentiation, and known mechanisms of their control. To facilitate the translatability of this information, this chapter also highlights which processes are conserved or divergent among mammalian species, bringing to bear knowledge from rodent and nonhuman primate animal models, as well as data from humans.

10.2 Organization of Spermatogenesis

10.2.1 Testicular Anatomy and Cellular Components

Within the mammalian testis, there are two distinct partitions which physically separate the major biological functions of the testis. Seminiferous tubules comprise the gametogenic compartment in which spermatogenesis takes place, and the interstitial compartment between seminiferous tubules is the major site of testicular steroidogenesis (Russell et al. 1990). While the architecture of seminiferous tubules is highly organized, consisting of an epithelium of polar Sertoli cells and multiple layers of germ cells at different stages of spermatogenic development, the interstitial space is relatively disorganized and consists of numerous cell types and structures, including steroidogenic Leydig cells, blood and lymphatic vessels, macrophages, and connective tissue (Fawcett et al. 1973). Within the seminiferous epithelium, Sertoli cells envelop developing germ cells and provide substantial trophic support (Russell et al. 1990). Neighboring Sertoli cells are connected by tight junctions which further divide the seminiferous epithelium into adluminal and basal compartments and thereby separate meiotic and post-meiotic germ cells (leptotene spermatocytes and later) from the blood supply and prevent their immune recognition (Fawcett et al. 1973; Russell 1977). Beyond providing nutritional and structural support, Sertoli cells are considered to play a major role in regulating germ cell differentiation by providing microenvironmental cues for both stem cell maintenance and each of the major spermatogenic differentiation events. Beneath the Sertoli cell epithelium, the seminiferous tubule basement membrane is surrounded by a layer of peritubular myoid cells (PMCs) which may evoke peristaltic fluid movement in the tubule lumen to expel spermatozoa (Leeson and Forman 1981; Maekawa et al. 1996; Virtanen et al. 1986), but which have also recently been implicated in control of germ cell differentiation during spermatogenesis (Chen et al. 2014, 2016). Leydig cells in the interstitial space between tubules are steroidogenic, producing testosterone which is required for spermatogonial proliferation, meiotic progression, and spermiation (reviewed by O'Shaughnessy 2014). Multiple distinct populations of macrophages are located in the interstitial compartment (DeFalco et al. 2015; Oatley et al. 2009). One macrophage population in particular is located on the outer surface of seminiferous tubules, immediately adjacent to the peritubular myoid cells, and are enriched at segments of seminiferous tubules that contain more undifferentiated spermatogonia (DeFalco et al. 2015). These macrophages have been implicated as regulators of spermatogonial differentiation, although the mechanisms of their involvement remain unclear.

10.2.2 *Spermatogenic Lineage*

Spermatogenesis occurs within the seminiferous epithelium and is divisible into three distinct developmental phases which occur sequentially in distinct cell types: (1) mitotic proliferation/clonal amplification, which takes place in spermatogonia, (2) meiosis in spermatocytes (primary and secondary), and (3) post-meiotic spermiogenesis in spermatids (round and elongating; Fig. 10.1). This organization is highly conserved among mammalian species (Fig. 10.1b), although the number of amplifying divisions of mitotic spermatogonia can vary quite substantially (Fig. 10.2).

At the foundation of the spermatogenic lineage is a population of spermatogonia that are adult stem cells, termed spermatogonial stem cells (SSCs). SSCs are responsible for maintaining spermatogenesis throughout the life span of adult mammals (Griswold and Oatley 2013; Oatley and Brinster 2008; Yang and Oatley 2014). As a population, SSCs must balance self-renewal to maintain the stem cell pool and differentiation to produce committed progenitor spermatogonia to meet the biological demand for sperm production and sustain the stem cell population (Jaenisch and Young 2008; Oatley and Brinster 2008, 2012; Yang and Oatley 2014) (Fig. 10.2). Indeed, excessive SSC differentiation or self-renewal impedes spermatogenesis and leads to male infertility by either depleting the stem cell pool or failing to produce differentiating germ cells to support spermatogenesis, respectively (de Rooij and Grootegoed 1998). At differentiating divisions of SSCs and at all subsequent divisions among differentiating male germ cells, division (mitotic or meiotic) is accompanied by incomplete cytokinesis which maintains daughter cells of the cell division as clones connected by intercellular cytoplasmic bridges (Fawcett 1959; Weber and Russell 1987; Greenbaum et al. 2006). Consequentially, spermatogenic cells develop in synchrony as syncytial clones which exchange transcripts, proteins, and organelles (Braun et al. 1989).

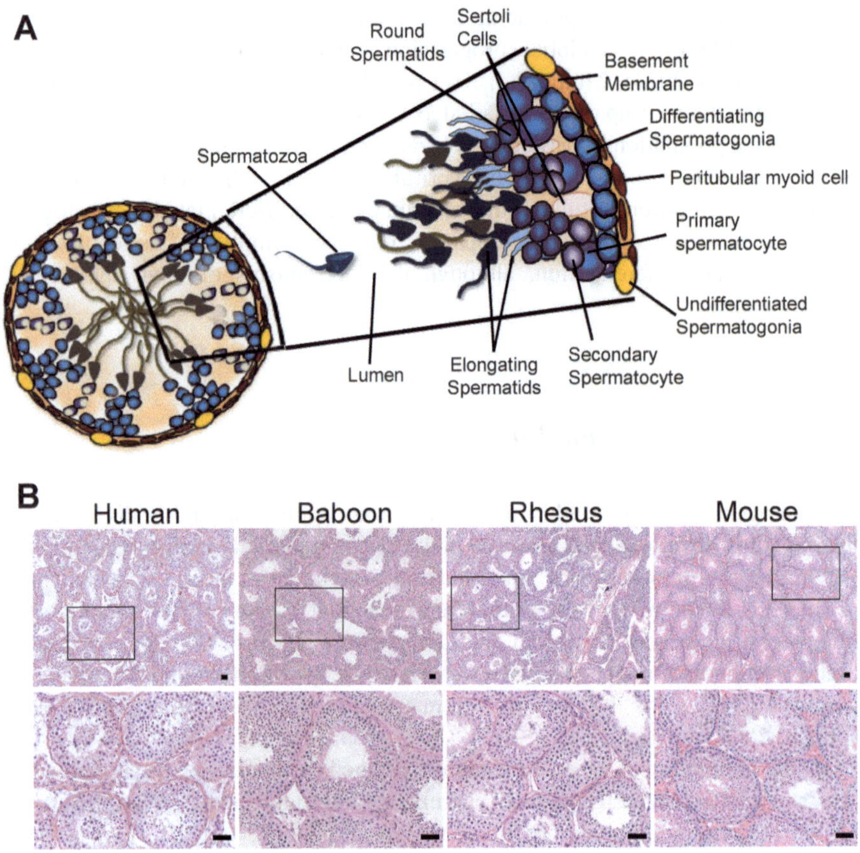

Fig. 10.1 Cellular organization of seminiferous tubules. (**a**) Seminiferous tubules are comprised of multiple layers of germ cells at distinct phases of spermatogenic development interdigitating between Sertoli cells. All spermatogonia, undifferentiated and differentiating, reside on the basement membrane. A layer of peritubular myoid cells is located on the outside of the seminiferous tubule basement membrane. (**b**) Histological sections through adult human, baboon, rhesus macaque, and mouse testes demonstrate extraordinary conservation in the organization of spermatogenesis among mammalian species. *Bar* = 50 μm

10.2.3 Undifferentiated and Differentiating Spermatogonia

Spermatogonia have historically been broadly classified as undifferentiated and differentiating, initially based upon histological appearance of nuclear morphology (Chiarini-Garcia et al. 2001; Russell et al. 1990; de Rooij and Grootegoed 1998). Undifferentiated spermatogonia generally lack heterochromatic nuclear architecture and, instead, have nuclei which appear more euchromatic (de Rooij and Russell 2000). In rodents, the multiple subtypes of spermatogonia are nearly all classified as differentiating: Type A_1, A_2, A_3, A_4, Intermediate (In), and B (Clermont et al. 1959; Roosen-Runge and Giesel 1950). Among Type A spermatogonia, the first five

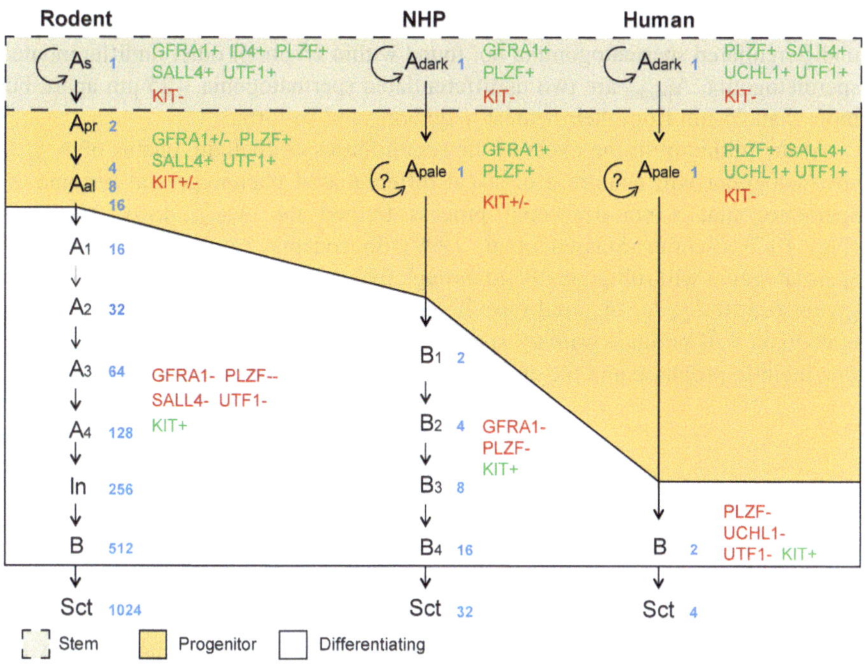

Fig. 10.2 The spermatogenic lineage. This schematic representation of the spermatogenic lineage depicts spermatogonial clonal amplification up to primary spermatocytes in rodents, nonhuman primates (NHP), and humans. Highlighting behind spermatogonial subtypes indicates whether cells are considered to be stem cells (*tan*), progenitor spermatogonia (*beige*), or differentiating spermatogonia (*white*). These subgroups also represent the major differentiation transitions among spermatogonia, initiation of differentiation among SSCs, and the conversion of progenitors to a differentiating spermatogonia phenotype (consistent with A_{al}–A_1 transition in rodents). Expression of select molecular markers is noted in *green* (expressed/detected) or red (not expressed/undetectable) and noted next to the respective lineages. *Blue* numbers indicate the degree of clonal amplification (cell numbers). For primate lineages, the precise number of transient amplifying divisions among A_{pale} progenitors is unknown (noted with *curved arrow* with *question mark*), but the absolute number of amplifying divisions from stem cells to sperm is considered to be fewer among primate species compared with rodents

generations of spermatogonia which precede differentiating Type A_1 are classified as undifferentiated, including A_{single}, A_{paired}, and $A_{aligned}$ clones of 4–16 spermatogonia based on the number of spermatogonia with similar nuclear morphology in a given clone (Fig. 10.2) (Huckins 1971b; Oakberg 1971; Oatley and Brinster 2012). Collectively, A_{single}, A_{paired}, and $A_{aligned}$ spermatogonia constitute the undifferentiated spermatogonial pool in rodent testes, which can all be distinguished from differentiated spermatogonia histologically in tissue section, but can only be further subdivided based on clone size using intact (whole-mounted) seminiferous tubules (Phillips et al. 2010). Spermatogonia with similar nuclear morphology that have an inter-nuclear distance $\leq$25 μm are considered to be within a single clone (de Rooij and Russell 2000; Huckins 1971b), although exceptions to this 25 μm rule have been noted (Tokuda et al. 2007). Thus, an A_{single} is an undifferentiated spermatogonium not found within 25 μm of other undifferentiated spermatogonia, A_{paired} are two undifferentiated spermatogonia ≤25 μm apart, but more than 25 μm from their nearest neighbors, and so forth.

At the midpoint in the cycle of the seminiferous epithelium, chains of $A_{aligned}$ spermatogonia will acquire a classical differentiated phenotype and become A_1 spermatogonia, a non-divisional process termed the $A_{aligned}$-to-A_1 transition (Fig. 10.2) (Schrans-Stassen et al. 1999). In rodents, resulting differentiating spermatogonia will subsequently go through five more clonal amplification mitoses giving rise to A_2, A_3, A_4, and type B spermatogonia, respectively. Type B spermatogonia will produce primary spermatocytes representing the beginning of the first meiotic prophase and the end of mitotic amplification (Oakberg 1956).

10.2.4 Stages of the Cycle of the Seminiferous Epithelium (Spermatogenic Stages)

During spermatogenesis, the cellular associations between different spermatogenic cell types are maintained in a repeated, cyclic fashion termed the cycle of the seminiferous epithelium (Clermont 1972; de Rooij and Russell 2000). For the sake of description, the cycle of the seminiferous epithelium cycle can be divided into a series of species-specific "stages," from as few as 6 in humans (Clermont 1963) to as many as 14 in rats (Leblond and Clermont 1952a, b), which correspond to segments of seminiferous tubules (or cross sections, when determined histologically) that always have the same complement of mitotic, meiotic, and post-meiotic germ cells. These spermatogenic stages occur sequentially along the length of the tubule, a characteristic referred to as the spermatogenic wave (Perey et al. 1961). This allows one to predict the subsequent or previous stages of the spermatogenic cycle based on a stage within one tubule cross section (de Rooij and Russell 2000). As a result, it is possible to follow the spermatogenic differentiation process over time along the length of a given seminiferous tubule. In both rodents and macaques, spermatogenic stages are arranged longitudinally, where each particular segment of a seminiferous tubule would have no more than one spermatogenic stage (Ehmcke and Schlatt 2006). However, in baboons and humans, multiple stages of spermatogenesis can be observed in each tubule cross section (Amann 2008; Chowdhury and Steinberger 1976; Chowdhury and Marshall 1980; Ehmcke et al. 2005a), likely the result of fewer amplifying mitoses (Hermann et al. 2010). The duration of one cycle of the seminiferous epithelium (the amount of time for a segment of seminiferous tubule to proceed from stage I through the final stage), which is also variable between species (Hermann et al. 2010), is a useful benchmark for the various hallmark differentiation events in the spermatogenic lineage because nearly all of these events happen in a repeated fashion in concert with the cycle. The duration of one cycle of the seminiferous epithelium is distinct from the complete duration of spermatogenesis, from SSC to spermatozoa, which occurs over the span of several cycles of the seminiferous epithelium.

10.3 Hallmark Differentiation Events in the Life Span of Spermatogonia

After formation of the spermatogonial stem cell pool, spermatogonial differentiation can be simplified into two major transitions/events in rodents: (1) SSC fate determination (self-renewal or initiation of differentiation) and (2) the A_{al}-to-A_1 transition. Subsequently, spermatogonia undergo a programmed clonal amplification (A_1–A_4, In., B), enter and proceed through meiosis, and undertake spermiogenesis, the programmed post-meiotic differentiation and physiological maturation that produces flagellated spermatozoa. In the human testis, the terminology employed to describe pre-meiotic spermatogenic cell types is not nearly as refined as rodents, which makes direct interspecies comparison of the differentiation events challenging. Adding to the confusion, dogma holds that the stem cell system functions differently in primates than rodents. This section of the chapter will expound on the relevant mechanisms regulating the two hallmark spermatogonial differentiation events noted above and highlight key differences between rodents and primates.

10.3.1 Formation of the Foundational SSC Pool

In mice, after arrival of primordial germ cells (PGCs) to the developing testis at mid-gestation [~12.5 *dpc* (days *post coitum*); Fig 10.3] (McLaren 2003), M-prospermatogonia (mitotic-prospermatogonia) are formed and proliferate for a short time before becoming mitotically quiescent (designated T1-prospermatogonia; transitional-prospermatogonia) until birth (Hilscher et al. 1974; McCarrey 2013). Between postnatal days 0–3 (0–3 *dpp*, days *postpartum*) in mice, these prospermatogonia reenter the cell cycle (and are renamed T2-prospermatogonia) and proliferate in the middle of the seminiferous cords (Hilscher et al. 1974; McCarrey 2013). The terminology describing prospermatogonial types has been the subject of recent debate with some preferring a simplified description whereby all descendants of male PGCs are termed "gonocytes" until relocating to the basement membrane beginning at about 3 *dpp* (Culty 2013; McCarrey 2013). Between 3 and 6 *dpp*, nascent spermatogonia migrate to the basement membrane of seminiferous cords asynchronously and some will directly differentiate to A_2 spermatogonia and produce the first wave of spermatogenesis (Kluin and de Rooij 1981; Yoshida et al. 2006), while the remainder will be specified as foundational SSCs that will maintain spermatogenesis. There are two

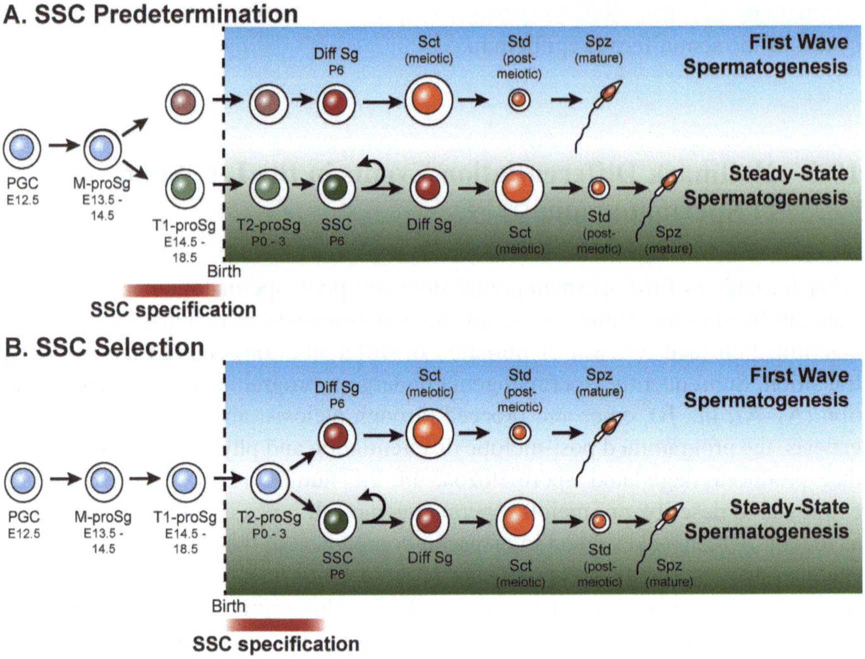

Fig. 10.3 Hypothetical modes of mouse SSC specification. Primordial germ cells (PGCs) arrive at the testis at mid-gestation (~embryonic day 12.5; E12.5), after which time they are considered to be prospermatogonia. These M-prospermatogonia proliferate for a short time before becoming mitotically quiescent (T1-prospermatogonia) until birth. Subsequently, in the first few days after birth (P0–P3), T2-prospermatogonia reenter the cell cycle and proliferate in the middle of the seminiferous cords. During the next 3 days (P3–P6), T2-prospermatogonia become spermatogonia and migrate to the basement membrane of seminiferous cords asynchronously. Some of the resulting spermatogonia will produce the first differentiated spermatogonia and produce the first wave of spermatogenesis, while the remainder will form the pool of foundational SSCs that will maintain spermatogenesis. There are two alternate theoretical models for how the SSCs are specified from prospermatogonia. (**a**) The predetermination model holds that there are subpopulations of prospermatogonia that are more or less likely to be specified as SSCs. (**b**) The selection model holds that prospermatogonia are equipotent and SSCs are selected stochastically. Modified from Hermann et al. (2015)

competing theories that describe potential mechanisms driving specification of the foundational SSCs from prospermatogonia—predetermination and selection—and we will discuss the evidence for each mechanism (Fig. 10.3).

The predetermination theory is based on the concept that there are actually yet-to-be-described subpopulations of prospermatogonia that are more or less likely to become SSCs (Fig. 10.3a). Alternately, a subpopulation of prospermatogonia could be selected stochastically to form the SSC pool (Fig. 10.3b). The predetermination theory was based on histological evaluation of fetal testes performed by Kluin and de Rooij, who suggested that a substantial proportion of mouse late fetal prospermatogonia exhibit a nuclear morphology similar to adult Type A_1

differentiating spermatogonia (Kluin and de Rooij 1981). Consequentially, those cells directly give rise to Type A_2 differentiating spermatogonia in the neonatal period, producing the first wave of spermatogenesis (Kluin and de Rooij 1981). Subsequently, Yoshida and colleagues suggested that the rodent first wave of spermatogenesis, which is not thought to have a human analog, initiates directly from differentiating prospermatogonia based on *Neurog3* and KIT marker expression (Yoshida et al. 2006). It follows that SSC specification may occur as a result of molecular divergence among prospermatogonia during an earlier developmental window that establishes subpopulations with higher or lower likelihoods of producing foundational SSCs (Fig. 10.3a). Results of analyses of spontaneous mutation frequency among prospermatogonia, spermatogonia, and later spermatogenic cells support this mechanism. Specifically, there appears to be a winnowing of mutation-bearing male germ cells between the prospermatogonia and SSC stage in which only those cells that bear a low mutation load produce the SSC pool, thereby suggesting therefore that these cells are predetermined to this fate (Walter et al. 1998; Murphey et al. 2013). We recently published the results of a single-cell gene expression study in which we defined the extent of molecular heterogeneity among neonatal mouse spermatogonia (Hermann et al. 2015). Single-cell qRT-PCR was done on a panel of 172 genes using enriched populations of spermatogonia from postnatal day 6 (6 *dpp*) testes, including cells from *Id4*-eGFP transgenic mice that express eGFP in a small fraction of undifferentiated spermatogonia (Chan et al. 2014). These analyses separated P6 testis cells into four major clusters based on distinct gene expression signatures, consisting of contaminating somatic cells and three groups of spermatogonia (Fig. 10.4). Thus, we found subpopulations of neonatal undifferentiated spermatogonia with discrete mRNA abundance signatures that may correlate with specific subtypes that differ in their functional capacities. In unpublished studies, we have since expanded our assessment of spermatogonial heterogeneity transcriptome wide and we have probed the functional implications of this heterogeneity with transplant analysis, confirming that these discrete transcriptomes separate cells with distinct function. Thus, the gene expression heterogeneity we observed at P6 spermatogonia supports the existence of multiple subtypes of undifferentiated spermatogonia at this stage (Fig. 10.4). Additional studies are needed to expand these results to earlier developmental time-points to determine if these discrete cell types emerge relatively earlier or later in development, which would tend to support the predetermination or selection theories, respectively. Direct evidence for the selection theory is largely lacking, although through examination of one or a few markers among fetal and neonatal germ cells has revealed much homogeneity in those markers (Busada et al. 2014), leading many to gravitate toward the selection hypothesis. Clearly, though, a more thorough examination of mRNA levels at the single-cell level indicates more substantial heterogeneity among spermatogonia and potentially their precursors.

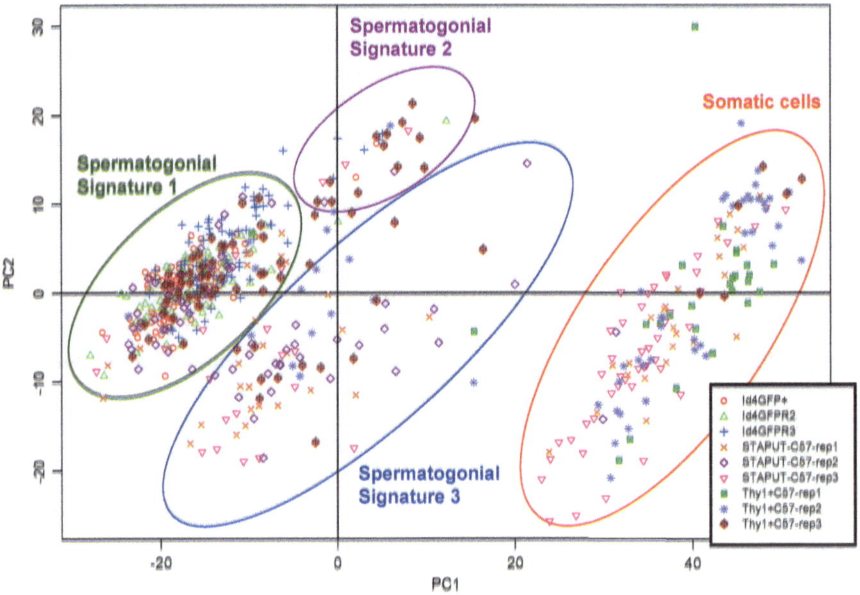

Fig. 10.4 Distinct clusters of P6 spermatogonia (signatures) revealed by mRNA abundance in single cells. Principal component analysis comparison of mRNA levels for 172 genes in 584 individual P6 testis cells (each point is an individual cell). Contaminating somatic cells (*red ellipse*) and three distinct spermatogonial signatures were noted—spermatogonial signatures 1, 2, and 3 (*green*, *violet*, and *blue ellipses*). Legend shows sample sources. Reprinted from Hermann et al. (2015)

10.3.2 Models of Spermatogonial Stem Cell Renewal and Differentiation

The first differentiation event in spermatogenesis results from SSC commitment to the differentiation pathway. Since this fate decision necessarily results from the mechanisms that balance stem cell self-renewal and initiation of differentiation, it is useful to review the kinetic models that have been proposed for this process. The undifferentiated spermatogonia discussed above are further divisible into subpopulations that differ in their biological capacities, including some cells which are considered to be SSCs that exhibit the ability to perpetually self-renew and initiate differentiation by producing committed progenitors (Fig. 10.2). Progenitors arising from SSCs have distinct developmental potential in that they have committed to the differentiation pathway and exhibit a finite transient-amplifying replicative capacity (Fig. 10.2; Hermann et al. 2015; Seaberg and van der Kooy 2003). The prevailing model for renewal and differentiation of rodent SSCs holds that the stem cells are A_{single} spermatogonia that symmetrically self-renew by completing cytokinesis to produce two new A_{single} spermatogonia or commit to differentiate and remain as an A_{paired} clone which will produce a chain of four progenitor cells

($A_{aligned\text{-}4}$) at the next mitosis (Fig. 10.2; Huckins 1971b; Oakberg 1971). This A_{single} model has gained wide (but not universal) acceptance in the field and is consistent with the basic tenets of stem cell biology (Potten 1992).

An alternative model, which we will term the "clone fragmentation" model has been advanced recently based on results from live imaging of transgenic mouse models bearing fluorescent reporters (Nakagawa et al. 2007; Yoshida et al. 2007a). In this model, renewal of spermatogonial stem cells is considered to arise from fragmentation of clones of $A_{aligned}$ spermatogonia. Moreover, while the vast majority of A_{paired} and $A_{aligned}$ typically have committed to differentiation in steady state, and thus, would normally not self-renew, this model holds that these progenitors still possess some latent stemness that can be mobilized under abnormal conditions, such as transplantation (Morimoto et al. 2009; Orwig et al. 2008; Yoshida et al. 2007a). This clone fragmentation model shares considerable similarities with an early alternative model of SSC self-renewal and differentiation put forth by Clermont and Bustos-Obregon (1968), termed the A_0/A_1 model and is also known as the "reserve stem cell" theory. Rather than a single pool of SSCs, this model describes an active stem cell pool that maintains steady-state spermatogenesis and a separate reserve stem cell pool that remains largely quiescent unless the testis is faced with a significant toxic insult. The A_0/A_1 model held that an "active" stem cell population (A_1) was renewed by fragmentation of differentiating Type A_1–A_4 spermatogonia, while a population of quiescent A_0 spermatogonia were "reserve" stem cells. Among the five generations of Type A spermatogonia (A_0, A_1, A_2, A_3, A_4) recognized by Clermont in this model, A_1 were arranged in clones of 4, 8, or 16 cells and most abundant in stages II–VII, much like $A_{aligned}$ spermatogonia from the A_{single} model, and the A_0 spermatogonia appear to reflect A_{single} and A_{paired} spermatogonia. The A_0/A_1 model was supplanted by the A_{single} model because the compendium of data were more consistent with the progressive, clonal amplification and renewal kinetics posited by the A_{single} model (Clermont and Bustos-Obregon 1968; de Rooij 1973; de Rooij and Russell 2000; Huckins 1971a, c; Huckins and Oakberg 1978; Oakberg 1971).

10.3.3 Models of Primate SSC Renewal and Differentiation

Historically, the identity of the SSC pool in primate testes and their mode of renewal has been distinct from the rodent A_{single} model. In the 1950s, Clermont's group initially described two morphologically distinct types of undifferentiated spermatogonia in the testes of rhesus macaques (Clermont and Leblond 1959) and designated these cells A_1 and A_2 (later renamed A_{dark} and A_{pale}). Initially, Clermont proposed that A_{dark} were bona fide SSCs, which undergo self-renewing divisions to maintain the stem cell pool and give rise to A_{pale} that subsequently generate differentiating type B spermatogonia (Clermont and Leblond 1959), much like the rodent model. Yet, a decade later, this model was revised based on observations in the vervet monkey (*Cercopithecus aethiops*) that A_{dark} failed to label with ^{3}H-thymidine (i.e., they did not appear to proceed through S-phase and self-renew) (Clermont 1969). Rather, Clermont proposed based on these new results that A_{dark} and A_{pale} represented distinct "reserve" and "active" stem cells, respectively. Subsequently, three studies have reported results from S-phase pulse-labeling experiments in various nonhuman primate species and there is a lack of consensus about whether any labeling is observed in A_{dark} within a few hours or several days after the pulse (Clermont and Antar 1973; Ehmcke et al. 2005b; Fouquet and Dadoune 1986). According to this "reserve stem cell" model, ongoing spermatogenesis is maintained by the "active" pool of A_{pale} SSCs under normal circumstances, while the quiescent pool of A_{dark} is only mobilized to regenerate spermatogenesis when spermatogenesis (including A_{pale}) is destroyed by a gonadotoxic insult (e.g., radiation). This A_{dark}/A_{pale} "reserve stem cell" model of primate SSCs shares similarities with the A_0/A_1 model that Clermont advanced for rodents (Bartmanska and Clermont 1983; Clermont and Bustos-Obregon 1968; Clermont and Hermo 1975; Dym and Clermont 1970). As discussed above, the A_0/A_1 model fell out of favor when the alternative A_{single} model, in which there is a single population of stem cells (A_{single} spermatogonia) that divides regularly, but infrequently, was put forth for rodent (Huckins 1971b, c; Oakberg 1971). More recently, the numbers of A_{pale} divisions, which likely reflect progenitor amplification analogous to A_{pr}–A_{al} chains in rodents, have been subtly revised based on contemporary whole-mount immunofluorescent techniques (Ehmcke et al. 2006; Ehmcke and Schlatt 2006), yet the A_{dark}/A_{pale} "reserve stem cell" model has yet to be significantly revisited. Marker analyses (PLZF, GFRA1, NEUROG3, KIT) in adult rhesus monkey testes suggested that A_{dark} spermatogonia bear closer resemblance to rodent A_{single} spermatogonia, while A_{pale} spermatogonia, some of which were KIT+, mimicked A_{paired}–$A_{aligned}$ progenitors (Hermann et al. 2009, 2010). Additional studies will be required to more thoroughly characterize the A_{dark} and A_{pale} spermatogonia populations at the whole transcriptome levels in nonhuman primate and human testes to connect these populations with stem and progenitor phenotypes. Ultimately, though, a functional test (e.g., transplantation) may be needed to resolve uncertainty in the hierarchy among primate undifferentiated spermatogonia.

10.3.4 Identifying Spermatogonial Stem Cells

To further our understanding of the process by which SSCs undertake either self-renewal or differentiation, it is essential to identify the distinguishing features of renewing and differentiating progeny of SSCs. However, this has proven difficult because these cells are extremely rare (~3000 per adult testis based on transplantation; Nagano 2003), and there is currently no method to prospectively identify SSCs in any species (Valli et al. 2015). Conventional SSC identification has been accomplished retrospectively using a functional transplantation assay that measures the ability to produce and maintain spermatogenesis (Table 10.1; Aloisio et al. 2014; Avarbock et al. 1996; Brinster and Zimmermann 1994; Buaas et al. 2004; Buageaw et al. 2005; Chan et al. 2014; Costoya et al. 2004; Kanatsu-Shinohara et al. 2004, 2014; Kubota et al. 2003; Oatley et al. 2011; Shinohara et al. 1999, 2000; Tokuda et al. 2007; Yang et al. 2013a, c). Knockout approaches have also been used quite extensively to confirm that loss of spermatogenesis follows loss of specific gene products, which, when coupled with lineage tracing, allows demonstration of gene expression that is required for SSC function, in vivo (Table 10.1; Agbor et al. 2013; Aloisio et al. 2014; Ballow et al. 2006; Buaas et al. 2004; Costoya et al. 2004; Falender et al. 2005; Goertz et al. 2011; Greenbaum et al. 2006; Hobbs et al. 2012; Hu et al. 2013; Kanatsu-Shinohara et al. 2014; Lovasco et al. 2015; Meng et al. 2000; Nakagawa et al. 2007; Oatley et al. 2011; Raverot et al. 2005; Schlesser et al. 2008; Suzuki et al. 2012; Yang et al. 2013a; Yoshida et al. 2004, 2007b). Furthermore, whole-mount immunostaining analyses have provided an avenue to define gene products which have testicular expression patterns that are limited, at least primarily, to undifferentiated spermatogonia. A number of such markers have been identified, including (but not limited to) glial cell line-derived neurotrophic factor receptor 1 alpha (GFRA1), inhibitor of DNA binding 4 (ID4), Neurogenin 3 (NEUROG3), Paired box 7 (PAX7), POU domain, class 5, transcription factor 1 (POU5F1; aka: OCT4), Sal-like 4 (SALL4, aka: Tex20), spermatogenesis- and oogenesis-specific basic helix-loop-helix transcription factor (SOHLH1), SRY-box containing gene 3 (SOX3), undifferentiated embryonic cell transcription factor 1 (UTF1), and zinc finger and BTB domain containing 16 (ZBTB16; aka: PLZF) (Table 10.1; Aloisio et al. 2014; Ballow et al. 2006; Buaas et al. 2004; Hobbs et al. 2012; Gassei and Orwig 2013; Greenbaum et al. 2006; Nakagawa et al. 2007; Oatley et al. 2011; Raverot et al. 2005; Schlesser et al. 2008; Tokuda et al. 2007; van Bragt et al. 2008; Wang et al. 2001; Yoshida et al. 2004, 2007b). Expression of many of these genes has been confirmed among spermatogonial types in nonhuman primate and human testes (Fig. 10.2; Altman et al. 2014; Dovey et al. 2013; Hermann et al. 2007, 2009, 2010, 2011; Izadyar et al. 2011; Valli et al. 2014). Yet, connecting these gene expression patterns to regulation of SSC fate decisions in any species has proven difficult because defects in either SSC renewal or differentiation produce an essentially indistinguishable block in spermatogenesis.

Emerging out of these studies are examples of just a couple of genes which exhibit expression profiles that are restricted to only A_{single} spermatogonia, which contain the entire presumptive SSC pool. One such gene product, Inhibitor of DNA binding 4 (ID4), is a helix-loop-helix factor lacking a basic region which has an expression pattern restricted to a subpopulation of A_{single} spermatogonia among mitotic germ cells in the testis and is required for SSC maintenance and spermatogenesis (Oatley et al. 2011). The LT-11 *Id4*-eGFP BAC transgenic mouse exhibits eGFP+ expression only in a fraction of $A_{singles}$, and in cultures of THY1+ spermatogonia eGFP is observed in about 10 % of spermatogonia (Chan et al. 2014). Importantly, all stem cell activity from cultured spermatogonia was observed in the eGFP+ fraction (Chan et al. 2014). Thus, by transitive logic, if *Id4*-eGFP is restricted to A_{single} spermatogonia and only eGFP+ spermatogonia exhibit

Table 10.1 Key genes expressed by spermatogenic cell types involved in differentiation

Marker/ gene (alias)	Expression profile (method)	Mutant phenotype	Citation(s)
Atm	Undifferentiated spermatogonia (FACS, IHC, BrdU, TP, RT-PCR)	$Atm^{-/-}$ depletion of undifferentiated spermatogonia followed by cell cycle arrest as well as activation of pathways associated with DNA damage	Shiloh (2003); Takubo et al. (2008)
Bcl6b	Undifferentiated spermatogonia (siRNA, TP)	Sertoli-only tubules and complete loss of SSCs in Bcl6b-deficient mice	Oatley et al. (2006)
Cd24a	SSCs (FACS, TG, THY1, Tr)	N/A	Kubota et al. (2003)
Cd81	Germline stem cells (FACS)	CD81 is a cell surface marker for SSCs highly expressed on GS cells	Kanatsu-Shinohara et al. (2012a)
Cd9	SSCs (Tr, FACS)	Selecting cells positive for CD9 show enrichment of SSCs	Kanatsu-Shinohara et al. (2004)
Cdh1	Undifferentiated spermatogonia (WM, IHC, WB, BrdU, Tr)	CDH1-positive cells formed clusters of one, two, four, and eight cells and survived after busulfan treatment colonizing recipient testes	Tokuda et al. (2007)
Cldn3	SSCs, Sertoli cells (RT-PCR, WB, IHC, siRNA)	Rac mutant cells show reduction in expression of *Cldn3*—role in the testis–blood barrier	Takashima et al. (2011)
Csf1	Leydig cells, peritubular myoid cells (IHC)	Addition of CSF1 to media increased the self-renewal capacity of SSCs in THY1 cultures without affecting germ cell expansion	Oatley et al. (2009)
Csf1r	Undifferentiated spermatogonia (IHC, ICC THY1, FACS)	N/A	Oatley et al. (2009)
Cxcr4	Undifferentiated spermatogonia (IHC, THY1, RT-PCR, Tr, FACS, shRNA, WB)	CXCL12/CXCR4 signaling necessary for transplant colonization, inhibition—SSC loss and buildup of progenitor spermatogonia	Yang et al. (2013b)
Dazl	Type A spermatogonia, intermediate, and type B spermatogonia (WM, IHC)	*Dazl*$^{-/-}$; blocked transition from A_{al} to A_1 spermatogonia	Schrans-Stassen et al. (2001)

(continued)

Table 10.1 (continued)

Marker/ gene (alias)	Expression profile (method)	Mutant phenotype	Citation(s)
Dmrtl	Spermatogonia and Sertoli cells (cKO, TG, BrdU, ChIP, IHC, WM, RT-PCR)	*Dmrtl*$^{-/-}$; premature meiotic entry, increased RA responsiveness	Matson et al. (2010); Agbor et al. (2013)
Dnmt3l	Prospermatogonia, Spermatocytes (Mu, KO, THY1)	*Dnmt3l*$^{-/-}$; meiotic-entry defect preceded by a progenitor differentiation defect (PLZF)	Webster et al. (2005); Liao et al. (2014)
Egr2	Undifferentiated spermatogonia (RT-PCR, Tr)	*Egr3* is GDNF responsive	Oatley et al. (2006)
Egr3	Undifferentiated spermatogonia (RT-PCR, Tr)	*Egr3* is GDNF responsive	Oatley et al. (2006)
Epcam	SSCs (FACS, siRNA, Tr, TG, Tu, RT-PCR, shRNA)	N/A	Anderson et al. (1999); Kanatsu-Shinohara et al. (2011)
Epha2	SSCs (FACS), subpopulation of *Id4*-eGFP+ (single-cell qRT-PCR)	N/A	Kanatsu-Shinohara et al. (2012a); Hermann et al. (2015)
Etv5	Sertoli cells, germ cells (Mu, TP)	*Etv5*$^{-/-}$; gradual germ cell depletion consistent with SSC renewal defect	Chen et al. (2005); Morrow et al. (2007)
Foxol	Prospermatogonia, undifferentiated spermatogonia (KO, IHC, RT-PCR)	Deficiency of *Foxol*, *Foxo3*, and *Foxo4* resulted in impairment of SSCs and block of differentiation	Goertz et al. (2011)
Gndf	Sertoli cells (TG, NB, WB, ISH, BrdU)	*Gdnf*$^{+/-}$; SSC depletion, overexpression leads to accumulation of undifferentiated spermatogonia	Meng et al. (2000); He et al. (2008); Johnston et al. (2011); Chen et al. (2014, 2016); Takashima et al. (2015)
Gfral	Undifferentiated spermatogonia and some SSCs (WMIHC, Tr, siRNA)	*Gfral* silencing leads to spermatogonial differentiation	Buageaw et al. (2005); He et al. (2007); Grasso et al. (2012); Hara et al. (2014)
Gpr125	Undifferentiated spermatogonia (IHC, WM, Tr, FACS)	N/A	Seandel et al. (2007)
Id4	Exclusive to A_{single} in adults (TG, IHC, FACS, TP, siRNA, WM, THY1, WB)	*Id4*$^{-/-}$ progressive loss of undifferentiated spermatogonia—essential for SSC self-renewal. *Id4*-eGFP BAC transgenic confirmed SSC activity in ID4-expressing cells	Oatley et al. (2011); Chan et al. (2014)

(continued)

Table 10.1 (continued)

Marker/ gene (alias)	Expression profile (method)	Mutant phenotype	Citation(s)
Itga6	SSCs (FACS, Tr), spermatogonia (IHC)	N/A	Shinohara et al. (1999, 2000)
Itgb1	SSCs (FACS, Tr), spermatogonia (IHC)	N/A	Shinohara et al. (1999, 2000)
Kdr (*Vegfr2*, *Flk1*)	Prospermatogonia, spermatogonia, Sertoli cells (IHC, RT-PCR, Tr)	Conditional *Vegfa*$^{-/-}$ with *Dmrt1*-Cre—defects in spermatogonial differentiation	Caires et al. (2012); Lu et al. (2013); Sargent et al. (2016)
Lin28a/b	PGCs, undifferentiated spermatogonia (WB, TG, IHC, WM, siRNA)	*Lin28a* needed for normal PGC expansion	Zheng et al. (2009); Shinoda et al. (2013); Chakraborty et al. (2014)
Lhx1	Undifferentiated spermatogonia (IHC, TP, siRNA)	*Lhx1* expression is upregulated by GDNF. *Lhx1* knockdown (in vitro) impaired SSC maintenance	Oatley et al. (2007)
Mcam	SSCs (siRNA, FACS/Tr)	*Mcam* knockdown (in vitro) reduced transplant colonization	Kanatsu-Shinohara et al. (2012b)
Mir221/ *Mir222*	Undifferentiated spermatogonia (THY1, IHC, RT-PCR, FACS, shRNA, ISH, WB, NB, Tr)	Overexpression prevents RA-dependent differentiation	Yang et al. (2013c)
Mycn	Undifferentiated spermatogonia (THY1, RT-PCR)	*N/A*	Braydich-Stolle et al. (2007)
Nanos2	A_{single} and A_{paired} (TG, WB, RT-PCR, IHC, cKO, KO, WM)	Conditional *Nanos2*$^{-/-}$; rapid progressive loss of SSCs. Overexpression—accumulation of undifferentiated spermatogonia	Suzuki et al. (2007, 2009); Sada et al. (2009); Zhou et al. (2015)
Nanos3	Undifferentiated spermatogonia (KO, NB, WB, IHC, FACS, IF)	*Nanos3* overexpression causes accumulation of cells in the G1 phase, blocks differentiation, and causes failure of meiotic entry and progression	Lolicato et al. (2008); Julaton and Reijo Pera (2011)
Neurog3 (*Ngn3*)	Undifferentiated spermatogonia (FACS, ISH, RT-PCR, TG, cKO, WM, THY1, TP, ChIP, shRNA, siRNA, WB)	Neurog3 marks early differentiating SSCs (perhaps coincident with entry into cell cycle)	Yoshida et al. (2004, 2006); Nakagawa et al. (2007); Ikami et al. (2015)

(continued)

Table 10.1 (continued)

Marker/ gene (alias)	Expression profile (method)	Mutant phenotype	Citation(s)
Notch1	Sertoli cells (Notch signaling reporter mice—ICC/IHC)	Conditional *Rbpj*$^{-/-}$ in Sertoli cells with *Amh*-Cre (mediator of NOTCH signaling) increases SSC number. Constitutive NOTCH activation leads to premature prospermatogonia differentiation	Dirami et al. (2001); Garcia and Hofmann (2013); Garcia et al. (2013); Garcia et al. (2014)
Nox1	Undifferentiated spermatogonia (IHC, KO, Tr)	Loss of *Nox1* results in suppression of spermatogonial proliferation (in vitro)	Morimoto et al. (2013)
Nox2 (*Cybb*)	Undifferentiated spermatogonia (FACS, RT-PCR, Tr)	*Nox2* initiates the AKT pathway inducing cellular proliferation and is important in NADPH oxidation	Morimoto et al. (2013)
Nox3	Undifferentiated spermatogonia (RT-PCR, shRNA, Tr)	shRNA-mediated knockdown prevents GDNF-stimulated ROS and SSC renewal in cultured and fresh spermatogonia	Morimoto et al. (2013, 2015)
Nox4	Undifferentiated spermatogonia (RT-PCR, shRNA)	Knockdown of *Nox4* (in vitro) suppresses spermatogonial proliferation	Morimoto et al. (2013)
Pax7	Subpopulation of A_{single} spermatogonia (IHC, LT, WM, Tr, cKO, WB)	Dispensable for spermatogenesis (*Pax7*$^{-/-}$)	Aloisio et al. (2014)
Pou3f1 (*Oct6*)	Undifferentiated spermatogonia (THY1, siRNA, IHC, RT-PCR, WB, TP)	Knockdown of *Pou3f1* in cultured spermatogonia induced apoptosis and impaired SSC maintenance	Wu et al. (2010)
Pvr	SSCs (FACS), subpopulation of *Id4*-eGFP (single-cell qRT-PCR)	N/A	Kanatsu-Shinohara et al. (2012a)
Rarg	$A_{aligned}$ spermatogonia and later spermatogonia	Spermatogonial differentiation incomplete in conditional *Rarγ*$^{-/-}$ (block A_{al}–A_1 transition)	Gely-Pernot et al. (2012)

(continued)

Table 10.1 (continued)

Marker/ gene (alias)	Expression profile (method)	Mutant phenotype	Citation(s)
Rb1	Prospermatogonia, undifferentiated spermatogonia, Sertoli cells (cKO, IHC, WB, siRNA, TP, RT-PCR)	Conditional *Rb1*$^{-/-}$ in prospermatogonia (*Ddx4*-Cre) have normal first wave, but are SSCs lost. Conditional *Rb1*$^{-/-}$ in progenitors (*Neurog3*-Cre) normal spermatogenesis	Yang et al. (2013a); Hu et al. (2013)
Sall4	Undifferentiated spermatogonia (IHC, WM, WB)	Conditional *Sall4*$^{-/-}$ (*Stra8*-Cre or *Ddx4*-Cre) loss of spermatogenesis consistent with defects in both SSC renewal and differentiation	Hobbs et al. (2012); Gassei and Orwig (2013)
Sohlh1	Type A1-B spermatogonia, spermatocytes (IHC, RT-PCR, Mu, BrdU)	*Sohlh1*$^{-/-}$; increased spermatogonial apoptosis, few spermatocytes in some tubules	Ballow et al. (2006)
Sohlh2	Undifferentiated spermatogonia (Mu, RT-PCR, WB, IHC, WM, ChIp)	*Sohlh2*$^{-/-}$; reduced numbers of A and B spermatogonia	Ballow et al. (2006); Toyoda et al. (2009)
Stat3	SSCs (siRNA, THY1, ChIP, IHC, shRNA, RT-PCR, WB, TP)	*Stat3*$^{-/-}$; block to *Neurog3* expression and differentiation in cultured A_{undiff}	Kaucher et al. (2012)
Sox3	Undifferentiated spermatogonia (Mu, IHC, RT-PCR)	SOX3 co-localizes with *Neurog3*; *Sox3*$^{-/-}$ mice have blocked spermatogonial differentiation beyond Type A spermatogonia (P10)	Raverot et al. (2005)
Stra8	A_{al}-B spermatogonia, spermatocytes (BrdU, IHC, RT-PCR, Mu)	*Stra8* mRNA and translation induced by RA and required for meiotic initiation and A_{al}–A_1 transition	Anderson et al. (2008); Snyder et al. (2011); Endo et al. (2015)
T (*Brachyury*)	Expressed by subpopulation of cultured spermatogonia (*Id4*-eGFP+)	*Etv5*$^{-/-}$ have decreased T activation	Wu et al. (2011); Chan et al. (2014)
Taf4b	Prospermatogonia, spermatocytes, Sertoli cells (Mu, IHC, TP, WB, RT-PCR, IF)	*Taf4b*$^{-/-}$; reduced prospermatogonia numbers at E18-P1, progressive loss of spermatogenesis consistent with SSC renewal defect	Falender et al. (2005); Lovasco et al. (2015)

(continued)

Table 10.1 (continued)

Marker/ gene (alias)	Expression profile (method)	Mutant phenotype	Citation(s)
Tspan8	Undifferentiated spermatogonia, SSCs (RT-PCR, Tr), subpopulation of *Id4*-eGFP (single-cell qRT-PCR)	*Tspan8* downregulated in cultured THY1+ spermatogonia following 1 h GDNF withdrawal	Oatley et al. (2006); Hermann et al. (2015)
Thy1 (*Cd90*)	SSCs (transplant), but also widely expressed by other somatic cells and progenitor spermatogonia	N/A	Kubota et al. (2003); Reding et al. (2010)
Uchl1 (*Pgp9.5*)	Undifferentiated spermatogonia	*Uchl1*-deficient gracile axonal dystrophy (GAD) mutant mice; seminiferous tubule atrophy, decreased germ cell numbers	Kwon et al. (2003, 2005); Luo et al. (2006)
Utf1	Prospermatogonia and early type A undifferentiated spermatogonia (RT-PCR, IHC)	UTF1 distribution suggests that it plays a role in maintaining the undifferentiated state	van Bragt et al. (2008)
Utp14b (*Jsd*)	Spermatogonia, higher in spermatocytes to round spermatids (RACE-PCR, TG, RT-PCR)	*Utp14b* mutants are sterile due to spermatogonial depletion	Rohozinski and Bishop (2004); Zhao et al. (2007)
Zbtb16 (*Plzf*)	Undifferentiated spermatogonia (Mu, ISH, RT-PCR, KO, TP, WM, IHC, FACS)	*Plzf* mutants (Luxoid or KO) exhibit progressive SSC loss due to compromised SSC renewal	Costoya et al. (2004); Buaas et al. (2004); Filipponi et al. (2007); Hobbs et al. (2010, 2012)

BrdU 5-bromo-2′-deoxyuridine DNA synthesis, *ChIP* chromatin-immunoprecipitation assay, *cKO* conditional knockout, *FACS* fluorescence-activated cell sorting, *IF* immunofluorescent staining, *IHC* immunohistochemistry, *ISH* in situ hybridization, *KO* knockout *LT* lineage tracing, *Mu* mutant animals, *RACE-PCR* rapid amplification of cDNA ends, *RT-PCR* real-time PCR, *TP* transplantation, *WB* Western Blot analysis, *WM* whole-mount, *THY1* THY1+ spermatogonia culture, *siRNA* in vitro knockdown experiment using siRNA, *shRNA* lentivirus vector transduction, *TG* transgenic animals, *Tu* TUNEL cell apoptosis assay

transplantable stem cell activity, then one could conclude that all SSCs are A_{single} spermatogonia. These results provide perhaps the most compelling argument in favor of the A_{single} model of SSC renewal and differentiation (Chan et al. 2014; Oatley et al. 2011). However, it does not appear that all *Id4*-eGFP+ spermatogonia exhibit stem cell activity (unpublished results and Chan et al. 2014; Oatley et al. 2011). Likewise, both endogenous *Id4* mRNA and *Id4*-eGFP are detectable in pachytene spermatocytes, and while the significance of *Id4* expression to spermatocyte function is not clear, this reinforces the concept that *Id4*-eGFP expression alone is not sufficient to delineate stem cells in the testis. Indeed, there is currently no method, including using of molecular markers, that can distinguish SSCs from their differentiating progeny in the testis from any species.

Another candidate SSC marker, PAX7, was reported recently in the literature (Aloisio et al. 2014), which, like ID4, was observed exclusively among A_s spermatogonia in the neonatal and adult testis. While lineage-tracing experiments suggest that progeny of *Pax7*+ cells can produce complete spermatogenesis and have regenerative capacity after cytotoxic insult, this gene is also known to be dispensable for spermatogenesis. Examination of the transcriptomes of *Id4*-eGFP+ and *Id4*-eGFP– cultured spermatogonia and unpublished results from single-cell gene expression studies failed to identify any *Pax7* transcripts (Chan et al. 2014), raising the concern that PAX7 is not an effective marker of SSCs .

10.3.5 Regulation of SSC Self-Renewal: Instruction from the Niche

The fate decision of dividing SSCs to either self-renew or initiate differentiation represents the first differentiation step in the spermatogenic lineage. It is well recognized that cells comprising the SSC niche produce signals (e.g., GDNF, KITL, RA, FGF2, CSF1, NOTCH2) which help instruct this balance by promoting either self-renewal or differentiation (Busada et al. 2015a; Chen et al. 2014, 2016; Dann et al. 2008; Garcia et al. 2014; He et al. 2008; Ishii et al. 2012; Kubota et al. 2004; Meng et al. 2000; Oatley et al. 2006; Oatley et al. 2007, 2009; Ohta et al. 2000; Schrans-Stassen et al. 1999; Takashima et al. 2015). Molecules that promote SSC renewal (e.g., GDNF, FGF2, CSF1) appear to play a predominant role in this process since SSCs do not appear to be responsive to differentiation signals (e.g., RA, KITL), but rather, their capacity to respond to differentiation signals is acquired after commitment to differentiation by progenitors.

10.3.5.1 GDNF and GFRA1/RET

Among the best studied niche-derived factor is glial cell line-derived neurotrophic factor (GDNF), which was initially described in 2000 as a key regulator of this critical balance (Meng et al. 2000). GDNF is a paracrine factor produced by testicular somatic cells in the niche (e.g., Sertoli cells, peritubular myoid cells) which must be produced in the correct amount to sustain spermatogenesis (Chen et al. 2014, 2016; Meng et al. 2000). Animals bearing one null allele of *Gdnf* produce roughly half the normal GDNF leads and exhibit reduced SSC self-renewal and exhaustion of the stem cell pool (Meng et al. 2000). Reciprocally, transgenic overexpression of *Gdnf* leads to excessive SSC self-renewal and failure to produce differentiating spermatogenic cells (Meng et al. 2000). This concept that GDNF levels must be exquisitely regulated to remain in a narrow range to promote balanced SSC self-renewal and differentiation has since been supported by the results of numerous in vitro and in vivo studies (Grasso et al. 2012; He et al. 2007, 2008; Jain et al. 2004; Jijiwa et al. 2008; Kubota et al. 2004, 2011; Lee et al. 2007; Naughton et al. 2006; Oatley et al. 2006, 2007; Parker et al. 2014; Ryu et al. 2005; Savitt et al. 2012; Tadokoro et al. 2002; Wu et al. 2010). GDNF acts by binding glial cell line-derived neurotrophic factor family receptor alpha 1 (GFRA1), which induces signaling through the RET tyrosine kinase receptor, both of which are expressed by undifferentiated type A spermatogonia (Jijiwa et al. 2008; Naughton et al. 2006). GDNF-dependent signaling through GFRA1/RET (Arighi et al. 2005) involves activation of at least three cascades in SSCs including PI3kinase/AKT (Lee et al. 2007; Oatley et al. 2007), RAS/ERK1/2 (He et al. 2008), and SRC family kinases (SFKs; Oatley et al. 2007). Manipulation of GDNF levels in the medium of THY1+ spermatogonia cultured leads to changes in the expression of at least 269 genes, including several known to be involved in SSC self-renewal and differentiation, such as B cell CLL/lymphoma 6 member b (*Bcl6b*), Ets variant gene 5 (*Etv5*), Forkhead box protein O1 (*Foxo1*), and Lim homeobox protein 1 (*Lhx1*) (Goertz et al. 2011; Morrow et al. 2007; Oatley et al. 2006, 2007; Schlesser et al. 2008; Wu et al. 2011). More recently, the cellular source of GDNF required for spermatogenesis has come into question. Initially, it was presumed that Sertoli cells were the primary source of GDNF required for spermatogenesis, based on expression studies (Fouchecourt et al. 2006; Johnston et al. 2011; Katoh-Semba et al. 2007). However, it now appears that peritubular myoid cells also produce GDNF in a manner dependent on testosterone (Chen et al. 2014) and that GDNF production by peritubular myoid cells is required for normal spermatogenesis (Chen et al. 2016). It remains to be seen whether Sertoli cell-derived GDNF is also necessary for sustaining SSC self-renewal and spermatogenesis.

10.3.5.2 Colony-Stimulating Factor 1

Colony-stimulating factor 1 (CSF1) is a cytokine that is produced by cells in the testicular interstitium (near clusters of Leydig cells) and some peritubular myoid cells which is known to promote SSC self-renewal (Oatley et al. 2009). Colony-stimulating factor 1 receptor (*Csf1r*) was found to be highly expressed in THY1+ germ cells, indicating that CSF1 acts directly on SSCs. Exposure of cultures of THY1+ spermatogonia to CSF1 in the presence of GDNF enhances mouse SSC self-renewal in vitro (Oatley et al. 2009). Interestingly, exposure of cultured mouse THY1+ spermatogonia to CSF1 alone (i.e., in the absence of GDNF) does not expand SSCs (Oatley et al. 2009), suggesting a requisite cooperation between both growth factors that likely involves signal transduction cross talk. Furthermore, testicular macrophages were recently shown to participate in regulation of SSC or spermatogonial fate, perhaps by secreting CSF1 or another related factor (DeFalco et al. 2015). A related cytokine, granulocyte colony-stimulating factor (G-CSF or CSF3), was recently identified as a potential somatic-derived self-renewal factor based on the potential of exogenous CSF3 to protect spermatogenesis from alkylating chemotherapy (Benavides-Garcia et al. 2015). The CSF3 receptor (CSF3R) is present at the mRNA and protein level in undifferentiated spermatogonia, but the cellular source of CSF3 in normal testes is not clear (Benavides-Garcia et al. 2015). These data support the concept that other testicular somatic cell types, beyond Sertoli cells, may play critical roles in driving SSC fate during normal steady-state spermatogenesis.

10.3.5.3 FGF2 (Basic FGF)

Like CSF1, fibroblast growth factor 2 (FGF2) is known to enhance mouse, rat, and hamster SSC self-renewal in vitro, but also like CSF1, this factor is insufficient to support SSC renewal and expansion alone (Kanatsu-Shinohara et al. 2008; Kubota et al. 2004; Ryu et al. 2005). In these studies, addition of FGF2 to the medium for cultured THY1+ spermatogonia is able to promote SSC expansion only in the presence of GDNF. Like GDNF, though, it appears that the primary testicular source of FGF2 is Sertoli cells (Mullaney and Skinner 1991). FGF2 promotes self-renewal of SSCs by enhancing expression of GDNF-regulated genes and acting as a mitogen (Ishii et al. 2012; Takashima et al. 2015).

10.3.6 Intrinsic Control of SSC Fate

In order to regulate the alternative fates among progeny of SSCs, niche-derived signals (e.g., GDNF) must be mediated by alternative mechanisms within SSCs that translate these signals into distinct gene expression patterns. Intrinsic control of SSC fate decisions, however, has remained poorly understood due to the extraordinary technical challenge of studying a cell population that comprises roughly 0.003 % of the germ cells in the adult testis (Nagano 2003; Tegelenbosch and de Rooij 1993). Thus, the field has focused on understanding the roles of gene products which exhibit an expression pattern that is restricted (or primarily restricted) to undifferentiated A_{single}, A_{paired}, and $A_{aligned}$ spermatogonia in early spermatogenesis. As a result, while numerous gene products with this restricted expression profile and which are necessary for spermatogenesis have been identified (Table 10.1), for the most part, it is not clear at which point in the spermatogenic lineage (i.e., in which male germ cells) such genes are essential, and thus, their precise role(s) in SSC fate is not clear. Despite this uncertainty, it is clear from such studies that spermatogenesis can be disrupted at the very earliest stages by deletion or perturbation of a variety of genes and gene products (Table 10.1).

Curiously, a majority of gene products examined which are required for undifferentiated spermatogonial function are involved in transcriptional regulation (e.g., ID4, PLZF, and SALL4). Thus, as is the case in other developmental programs, such genes might play roles as master regulators of cell fate decisions by executing essential gene expression programs. Here we will focus on three examples of potential master transcriptional regulators of SSC fate, ID4, PLZF, and SALL4, for which functional evidence is available. Knockouts for all three factors have confirmed that they are essential for proper control of SSC fate (Buaas et al. 2004; Costoya et al. 2004; Hobbs et al. 2012; Oatley et al. 2011). Two different PLZF mutants exhibit the same phenotype, a progressive depletion of the SSC pool after the first wave of spermatogenesis leading to azoospermia (Buaas et al. 2004; Costoya et al. 2004). This defect appears to result from a shift in the balance in SSC fate away from self-renewal and toward differentiation, indicating that PLZF either inhibits genes involved in SSC differentiation or activates SSC self-renewal genes. In support of this, we recently reported the full regulatory repertoire for PLZF in undifferentiated spermatogonia using ChIP-seq and found that both renewal and differentiation genes were bound by and required PLZF for expression (Lovelace in press). While whole-animal knockouts for *Sall4* are embryonic lethal (Sakaki-Yumoto et al. 2006), conditional *Sall4* knockouts generated using the *Vasa*-Cre transgene exhibit a progressive spermatogenic deficiency reminiscent of the PLZF mutants, suggesting that both maintenance and differentiation of SSCs are perturbed (Hobbs et al. 2012). Similar ChIP-seq results for SALL4 in undifferentiated spermatogonia demonstrate it is targeted to the genome by PLZF and the differentiation factor DMRT1 and appears to be required for expression of both self-renewal and differentiation genes (Lovelace in press). Most recently, *Id4*-null animals also exhibited a progressive spermatogenic loss defect consistent with failure of SSC renewal (Oatley et al. 2011). ID4 is a helix-loop-helix protein that lacks a basic region, meaning that it can dimerize with other HLH factors and act as a dominant-negative protein to prevent their DNA binding. It is not clear which proteins complex with ID4 in A_{single} spermatogonia, but two possibilities include E2F and NEUROG3, which may have implications for spermatogonial proliferation and differentiation, respectively. Thus, ID4, PLZF, and SALL4, which have expression patterns restricted to subpopulations of undifferentiated spermatogonia, including SSCs and progenitor spermatogonia, and which are known to be functionally required for ongoing spermatogenesis, likely are involved in creating the favorable transcriptomes for SSC self-renewal and/or differentiation. Future studies further examining the cistromes for these and other transcription factors with similar expression profiles will undoubtedly reveal key regulatory networks that predispose the ability of SSCs to renew or differentiate.

10.3.7 Spermatogonial Progenitor Response to Differentiation Signals

The reciprocal SSC fate, differentiation, is only discernable once progenitor spermatogonia acquire the capacity to respond to niche-derived differentiation factors (e.g., RA, KITL). At postnatal day 6 (6 *dpp*), nearly all *Id4*-eGFP+ spermatogonia in the mouse testis express mRNAs for *Stra8* and *Kit*, including presumptive SSCs and progenitor spermatogonia (Hermann et al. 2015), which would be indicative of an intact retinoid receptor response and be indicative of spermatogonial differentiation. Yet, it is also clear that neither STRA8 nor KIT protein was detectable in the vast majority of *Id4*-eGFP+ spermatogonia (Hermann et al. 2015). Indeed, in that study, KIT and STRA8 protein were only localized to a small percentage of spermatogonia with weak eGFP fluorescence intensity which is thought to be indicative of spermatogonia that have transitioned out of an SSC state to become progenitor spermatogonia (Chan et al. 2014; Hermann et al. 2015). These results raise an important concept of early spermatogonial differentiation in which regulation of cell fate appears to occur predominantly at the level of regulation of mRNA utilization (i.e., translational control), rather than transcriptional regulation (Busada et al. 2014; Chappell et al. 2013). Acquisition of RA responsiveness and KIT protein expression appears to be the rate-limiting step in poising progenitor spermatogonia for competency to undergo the A_{al}–A_1 transition (Ikami et al. 2015).

10.3.7.1 Retinoic Acid

Retinoic acid (RA) is a biologically active metabolite of vitamin A (retinol) and is essential for male fertility (Anderson et al. 2008; Bowles et al. 2006). The requirement for RA in spermatogenesis is easily recognized using the vitamin A-deficient (VAD) model in which dietary-derived RA is essentially absent. VAD mice or rats exhibit a block in spermatogenesis at the undifferentiated spermatogonia stage (Morales and Griswold 1987). VAD followed by release of the RA deficiency also synchronizes spermatogenesis and 24–48 h after injection with either RA or retinol (vitamin A); previously arrested A_{al} reenter the cell cycle and differentiate into A_1 spermatogonia (Morales and Griswold 1987). Likewise, inhibition of ALDH1A2, the rate-limiting enzyme responsible for RA biosynthesis from retinaldehyde, using the bisdichloroacetyldiamine (BDAD) WIN 18,446, mimics VAD and induces spermatogenic arrest by blocking A_{al}–A_1 transition (Heller et al. 1961). It is not clear which retinoid receptors (RARs or RXRs) are involved in cellular changes induced by RA in progenitor spermatogonia, although one publication suggests that RARγ is required for the A_{al}–A_1 transition (Gely-Pernot et al. 2012). Independent of the nuclear RAR/RXR response, it appears that a primary consequence of RA stimulation of progenitor spermatogonia is to induce translation of mRNAs encoding the KIT tyrosine kinase and STRA8, via a mechanism involving P13K/AKT/mTORC1 signaling (Busada et al. 2015a, b).

During the cycle of the seminiferous epithelium, a mid-cycle pulse of RA (centered on stages VI–VII) drives transition of $A_{aligned}$ spermatogonia to A_1 spermatogonia (Endo et al. 2015; Hogarth et al. 2015a), although not all $A_{aligned}$ appear to differentiate. It is not clear why only some undifferentiated spermatogonia respond to RA, although it is possible that expression of CYP26A1 and CYP26B1 enzymes, which catalyze the degradation of RA, may restrict RA action to subpopulations of progenitor spermatogonia. Recent conditional mutants of CYP26A1 and CYP26B1 within either Sertoli cells (*Amh*-Cre) or differentiating spermatogonia (*Stra8*-iCre) suggest that neither enzyme is essential for spermatogenesis, since these animals were fertile (Hogarth et al. 2015b). Subtle spermatogenic defects were observed in these animals, most markedly, an increase in the number of STRA8+ differentiating spermatogonia, but overall suggesting that redundancy might preclude complete differentiation of the undifferentiated spermatogonial pool when RA levels cannot be reduced.

10.3.7.2 KIT and KIT-Ligand

Acquisition of KIT expression among A_{al} spermatogonia is another crucial step promoting differentiation to Type A_1 spermatogonia. KIT is type III receptor tyrosine kinase which is considered to be required for normal spermatogonial differentiation (Besmer et al. 1993; Dym et al. 1995; Koshimizu et al. 1992; Manova and Bachvarova 1991; Sorrentino et al. 1991; Yoshinaga et al. 1991; Zhang et al. 2011). KIT is the membrane receptor for stem cell factor (SCF, aka: KIT ligand) and is expressed in some A_{al} progenitor spermatogonia (chains of 8–16 cells), differentiating spermatogonia, and up through preleptotene spermatocytes (Schrans-Stassen et al. 1999; Yoshinaga et al. 1991). Male mice bearing mutant alleles of *Kit* (W/W^v, dominant-white spotting locus) or *Kitl* (Steele) are largely sterile (Besmer et al. 1993; Coulombre and Russell 1954; Geissler et al. 1981; Koshimizu et al. 1992). Absence of KIT or SCF expression does not affect proliferation of progenitor spermatogonia, demonstrated by both transplantation and in vitro THY1+ culture experiments (Kubota et al. 2009; Ohta et al. 2003). However, under certain transplant conditions (W/W^v pup recipients), KIT-mutant germ cells are still able to differentiate to produce complete spermatogenesis (Kubota et al. 2009).

The *Kit* gene is also perhaps the best characterized putative PLZF target (Filipponi et al. 2007; Hobbs et al. 2012; Puszyk et al. 2013). PLZF binding to the *Kit* promoter in spermatogonia occurred through a "consensus" PLZF binding motif (5′-ATACAGT-3′) which was identified by chromatography with a GST fusion to the seven most carboxy-terminal Zn fingers found in human PLZF (Li et al. 1997). Unpublished ChIP-seq results for PLZF from the Hermann lab demonstrate modest binding to a site in first intron of *Kit*, but no evidence of PLZF binding to the *Kit* promoter in THY1+ spermatogonia. Moreover, the in vitro-selected putative PLZF binding motif (5′-A-T/G-G/C-T-A/C-A/C-A-G-T-3′) was not among the top ten motifs from PLZF ChIP-seq in undifferentiated spermatogonia and was not significantly represented among PLZF binding sites. Regardless, it is hard to conceive how PLZF directly represses transcription of *Kit* in undifferentiated spermatogonia given that PLZF and *Kit* mRNA are co-expressed, as confirmed recently by single-cell gene expression studies in undifferentiated spermatogonia (Hermann et al. 2015). That is, expression of both PLZF and *Kit* in the vast majority of *Id4*-eGFP+ spermatogonia makes it unlikely that PLZF directly represses *Kit* transcription. Still, *Kit* mRNA and KIT translation are both induced by RA and represent among the first cellular consequences of RA-induced spermatogonial differentiation (Busada et al. 2014, 2015a; Dann et al. 2008).

10.4 Conclusions

Spermatogonial differentiation is essential for spermatogenesis and male fertility, but is also a highly complex process requiring exquistite extrinsic and intrinsic instruction to properly control cell fate. The mechanisms responsible for executing the earliest phase of this differentiation cascade, stem cell commitment to a progenitor phenotype, have been the most elusive. That is, the mechanisms which cause SSC progeny to differentially respond to niche-derived signals and undertake alternate self-renewal or differentiation fate decisions are not known because it has been impossible to precisely identify and selectively recover these cell types from among the heterogeneous pool of undifferentiated spermatogonia in the testis. Consequently, new studies that can separate cells undertaking these alternate fates will help resolve long-standing questions in the field—*how is the balance between SSC renewal and differentiation controlled?* Ultimately, knowledge of such processes will permit insight into the etiology of certain types of male infertility (e.g., non-obstructive azoospermia), as well as revealing new therapeutic avenues to address naturally occurring defects in sperm production, mitigate iatrogenic male infertility (e.g., arising from cancer therapy), or intervene for male contraception.

References

Agbor VA, Tao S, Lei N et al (2013) A Wt1-Dmrt1 transgene restores DMRT1 to Sertoli cells of Dmrt1(-/-) testes: a novel model of DMRT1-deficient germ cells. Biol Reprod 88:51

Aloisio GM, Nakada Y, Saatcioglu HD et al (2014) PAX7 expression defines germline stem cells in the adult testis. J Clin Invest 124:3929–3944

Altman E, Yango P, Moustafa R et al (2014) Characterization of human spermatogonial stem cell markers in fetal, pediatric, and adult testicular tissues. Reproduction 148:417–427

Amann RP (2008) The cycle of the seminiferous epithelium in humans: a need to revisit? J Androl 29:469–487

Anderson R, Schaible K, Heasman J et al (1999) Expression of the homophilic adhesion molecule, Ep-CAM, in the mammalian germ line. J Reprod Fertil 116:379–384

Anderson EL, Baltus AE, Roepers-Gajadien HL et al (2008) Stra8 and its inducer, retinoic acid, regulate meiotic initiation in both spermatogenesis and oogenesis in mice. Proc Natl Acad Sci U S A 105:14976–14980

Arighi E, Borrello MG, Sariola H (2005) RET tyrosine kinase signaling in development and cancer. Cytokine Growth Factor Rev 16:441–467

Avarbock MR, Brinster CJ, Brinster RL (1996) Reconstitution of spermatogenesis from frozen spermatogonial stem cells. Nat Med 2:693–696

Ballow D, Meistrich ML, Matzuk M et al (2006) Sohlh1 is essential for spermatogonial differentiation. Dev Biol 294:161–167

Bartmanska J, Clermont Y (1983) Renewal of type A spermatogonia in adult rats. Cell Tissue Kinet 16:135–143

Benavides-Garcia R, Joachim R, Pina NA et al (2015) Granulocyte colony-stimulating factor prevents loss of spermatogenesis after sterilizing busulfan chemotherapy. Fertil Steril 103:270–280

Besmer P, Manova K, Duttlinger R et al (1993) The kit-ligand (steel factor) and its receptor c-kit/W: pleiotropic roles in gametogenesis and melanogenesis. Dev Suppl 125–137
Bowles J, Knight D, Smith C et al (2006) Retinoid signaling determines germ cell fate in mice. Science 312:596–600
Braun RE, Behringer RR, Peschon JJ et al (1989) Genetically haploid spermatids are phenotypically diploid. Nature 337:373–376
Braydich-Stolle L, Kostereva N, Dym M et al (2007) Role of Src family kinases and N-Myc in spermatogonial stem cell proliferation. Dev Biol 304:34–45
Brinster RL, Zimmermann JW (1994) Spermatogenesis following male germ-cell transplantation. Proc Natl Acad Sci U S A 91:11298–11302
Buaas FW, Kirsh AL, Sharma M et al (2004) Plzf is required in adult male germ cells for stem cell self-renewal. Nat Genet 36:647–652
Buageaw A, Sukhwani M, Ben-Yehudah A et al (2005) GDNF family receptor alpha1 phenotype of spermatogonial stem cells in immature mouse testes. Biol Reprod 73:1011–1016
Busada JT, Kaye EP, Renegar RH et al (2014) Retinoic acid induces multiple hallmarks of the prospermatogonia-to-spermatogonia transition in the neonatal mouse. Biol Reprod 90:64
Busada JT, Chappell VA, Niedenberger BA et al (2015a) Retinoic acid regulates Kit translation during spermatogonial differentiation in the mouse. Dev Biol 397:140–149
Busada JT, Niedenberger BA, Velte EK et al (2015b) Mammalian target of rapamycin complex 1 (mTORC1) is required for mouse spermatogonial differentiation in vivo. Dev Biol 407:90–102
Caires KC, de Avila JM, Cupp AS et al (2012) VEGFA family isoforms regulate spermatogonial stem cell homeostasis in vivo. Endocrinology 153:887–900
Chakraborty P, Buaas FW, Sharma M et al (2014) LIN28A marks the spermatogonial progenitor population and regulates its cyclic expansion. Stem Cells 32:860–873
Chan F, Oatley MJ, Kaucher AV et al (2014) Functional and molecular features of the Id4+ germline stem cell population in mouse testes. Genes Dev 28:1351–1362
Chappell VA, Busada JT, Keiper BD et al (2013) Translational activation of developmental messenger RNAs during neonatal mouse testis development. Biol Reprod 89:61
Chen C, Ouyang W, Grigura V et al (2005) ERM is required for transcriptional control of the spermatogonial stem cell niche. Nature 436:1030–1034
Chen LY, Brown PR, Willis WB et al (2014) Peritubular myoid cells participate in male mouse spermatogonial stem cell maintenance. Endocrinology 155:4964–4974
Chen LY, Willis WD, Eddy EM (2016) Targeting the Gdnf Gene in peritubular myoid cells disrupts undifferentiated spermatogonial cell development. Proc Natl Acad Sci U S A 113:1829–1834
Chiarini-Garcia H, Hornick JR, Griswold MD et al (2001) Distribution of type A spermatogonia in the mouse is not random. Biol Reprod 65:1179–1185
Chowdhury AK, Marshall GR (1980) Irregular pattern of spermatogenesis in the baboon (Papio anubis) and its possible mechanism. In: Steinberger A, Steinberger E (eds) Testicular development, structure and function. Raven, New York, pp 129–137
Chowdhury AK, Steinberger E (1976) A study of germ cell morphology and duration of spermatogenic cycle in the baboon, Papio anubis. Anat Rec 185:155–169
Clermont Y (1963) The cycle of the seminiferous epithelium in man. Am J Anat 112:35–51
Clermont Y (1969) Two classes of spermatogonial stem cells in the monkey (Cercopithecus aethiops). Am J Anat 126:57–71
Clermont Y (1972) Kinetics of spermatogenesis in mammals: seminiferous epithelium cycle and spermatogonial renewal. Physiol Rev 52:198–236
Clermont Y, Antar M (1973) Duration of the cycle of the seminiferous epithelium and the spermatogonial renewal in the monkey Macaca arctoides. Am J Anat 136:153–165
Clermont Y, Bustos-Obregon E (1968) Re-examination of spermatogonial renewal in the rat by means of seminiferous tubules mounted "in toto". Am J Anat 122:237–247
Clermont Y, Hermo L (1975) Spermatogonial stem cells in the albino rat. Am J Anat 142:159–175

Clermont Y, Leblond CP (1959) Differentiation and renewal of spermatogonia in the monkey, Macacus rhesus. Am J Anat 104:237–271

Clermont Y, Leblond CP, Messier B (1959) Duration of the cycle of the seminal epithelium of the rat. Arch Anat Microsc Morphol Exp 48(Suppl):37–55

Costoya JA, Hobbs RM, Barna M et al (2004) Essential role of Plzf in maintenance of spermatogonial stem cells. Nat Genet 36:653–659

Coulombre J, Russell ES (1954) Analysis of the pleiotropism at the W-locus in the mouse. The effects of W and Wv substitution upon postnatal development of germ cells. J Exp Zool 126:277–295

Culty M (2013) Gonocytes, from the fifties to the present: is there a reason to change the name? Biol Reprod 89:46

Dann CT, Alvarado AL, Molyneux LA et al (2008) Spermatogonial stem cell self-renewal requires OCT4, a factor downregulated during retinoic acid-induced differentiation. Stem Cells 26:2928–2937

DeFalco T, Potter SJ, Williams AV et al (2015) Macrophages contribute to the spermatogonial niche in the adult testis. Cell Rep 12:1107–1119

de Rooij DG (1973) Spermatogonial stem cell renewal in the mouse. I. Normal situation. Cell Tissue Kinet 6:281–287

de Rooij DG, Grootegoed JA (1998) Spermatogonial stem cells. Curr Opin Cell Biol 10:694–701

de Rooij DG, Russell LD (2000) All you wanted to know about spermatogonia but were afraid to ask. J Androl 21:776–798

Dirami G, Ravindranath N, Achi MV et al (2001) Expression of Notch pathway components in spermatogonia and Sertoli cells of neonatal mice. J Androl 22:944–952

Dovey SL, Valli H, Hermann BP et al (2013) Eliminating malignant potential from therapeutic human spermatogonial stem cells. J Clin Invest 123:1833–1843

Dym M, Clermont Y (1970) Role of spermatogonia in the repair of the seminiferous epithelium following x-irradiation of the rat testis. Am J Anat 128:265–282

Dym M, Jia MC, Dirami G et al (1995) Expression of c-kit receptor and its autophosphorylation in immature rat type A spermatogonia. Biol Reprod 52:8–19

Ehmcke J, Schlatt S (2006) A revised model for spermatogonial expansion in man: lessons from non-human primates. Reproduction 132:673–680

Ehmcke J, Luetjens CM, Schlatt S (2005a) Clonal organization of proliferating spermatogonial stem cells in adult males of two species of non-human primates, Macaca mulatta and Callithrix jacchus. Biol Reprod 72:293–300

Ehmcke J, Simorangkir DR, Schlatt S (2005b) Identification of the starting point for spermatogenesis and characterization of the testicular stem cell in adult male rhesus monkeys. Hum Reprod 20:1185–1193

Ehmcke J, Wistuba J, Schlatt S (2006) Spermatogonial stem cells: questions, models and perspectives. Hum Reprod Update 12:275–282

Endo T, Romer KA, Anderson EL et al (2015) Periodic retinoic acid-STRA8 signaling intersects with periodic germ-cell competencies to regulate spermatogenesis. Proc Natl Acad Sci U S A 112:E2347–E2356

Falender AE, Freiman RN, Geles KG et al (2005) Maintenance of spermatogenesis requires TAF4b, a gonad-specific subunit of TFIID. Genes Dev 19:794–803

Fawcett DW (1959) The occurrence of intercellular bridges in groups of cells exhibiting synchronous differentiation. J Biophys Biochem Cytol 3:453–460

Fawcett DW, Neaves WB, Flores MN (1973) Comparative observations on intertubular lymphatics and the organization of the interstitial tissue of the mammalian testis. Biol Reprod 9:500–532

Filipponi D, Hobbs RM, Ottolenghi S et al (2007) Repression of kit expression by Plzf in germ cells. Mol Cell Biol 27:6770–6781

Fouchecourt S, Godet M, Sabido O et al (2006) Glial cell-line-derived neurotropic factor and its receptors are expressed by germinal and somatic cells of the rat testis. J Endocrinol 190:59–71

Fouquet JP, Dadoune JP (1986) Renewal of spermatogonia in the monkey (Macaca fascicularis). Biol Reprod 35:199–207
Garcia TX, Hofmann MC (2013) NOTCH signaling in Sertoli cells regulates gonocyte fate. Cell Cycle 12:2538–2545
Garcia TX, DeFalco T, Capel B et al (2013) Constitutive activation of NOTCH1 signaling in Sertoli cells causes gonocyte exit from quiescence. Dev Biol 377:188–201
Garcia TX, Farmaha JK, Kow S et al (2014) RBPJ in mouse Sertoli cells is required for proper regulation of the testis stem cell niche. Development 141:4468–4478
Gassei K, Orwig KE (2013) SALL4 expression in gonocytes and spermatogonial clones of postnatal mouse testes. PLoS One 8:e53976
Geissler EN, McFarland EC, Russell ES (1981) Analysis of pleiotropism at the dominant white-spotting (W) locus of the house mouse: a description of ten new W alleles. Genetics 97:337–361
Gely-Pernot A, Raverdeau M, Celebi C et al (2012) Spermatogonia differentiation requires retinoic acid receptor gamma. Endocrinology 153:438–449
Goertz MJ, Wu Z, Gallardo TD et al (2011) Foxo1 is required in mouse spermatogonial stem cells for their maintenance and the initiation of spermatogenesis. J Clin Invest 121:3456–3466
Grasso M, Fuso A, Dovere L et al (2012) Distribution of GFRA1-expressing spermatogonia in adult mouse testis. Reproduction 143:325–332
Greenbaum MP, Yan W, Wu MH et al (2006) TEX14 is essential for intercellular bridges and fertility in male mice. Proc Natl Acad Sci U S A 103:4982–4987
Griswold MD, Oatley JM (2013) Concise review: defining characteristics of mammalian spermatogenic stem cells. Stem Cells 31:8–11
Hara K, Nakagawa T, Enomoto H et al (2014) Mouse spermatogenic stem cells continually interconvert between equipotent singly isolated and syncytial states. Cell Stem Cell 14:658–672
He Z, Jiang J, Hofmann MC et al (2007) Gfra1 silencing in mouse spermatogonial stem cells results in their differentiation via the inactivation of RET tyrosine kinase. Biol Reprod 77:723–733
He Z, Jiang J, Kokkinaki M et al (2008) Gdnf upregulates c-Fos transcription via the Ras/Erk1/2 pathway to promote mouse spermatogonial stem cell proliferation. Stem Cells 26:266–278
Heller CG, Moore DJ, Paulsen CA (1961) Suppression of spermatogenesis and chronic toxicity in men by a new series of bis(dichloroacetyl) diamines. Toxicol Appl Pharmacol 3:1–11
Hermann BP, Sukhwani M, Lin CC et al (2007) Characterization, cryopreservation and ablation of spermatogonial stem cells In adult rhesus macaques. Stem Cells 25:2330–2338
Hermann BP, Sukhwani M, Simorangkir DR et al (2009) Molecular dissection of the male germ cell lineage identifies putative spermatogonial stem cells in rhesus macaques. Hum Reprod 24:1704–1716
Hermann BP, Sukhwani M, Hansel MC, Orwig KE (2010) Spermatogonial stem cells in higher primates: are there differences to those in rodents? Reproduction 139:479–493
Hermann BP, Sukhwani M, Salati J et al (2011) Separating spermatogonia from cancer cells in contaminated prepubertal primate testis cell suspensions. Hum Reprod 26:3222–3231
Hermann BP, Mutoji KN, Velte EK et al (2015) Transcriptional and Translational Heterogeneity among Neonatal Mouse Spermatogonia. Biol Reprod 92:54
Hilscher B, Hilscher W, Bulthoff-Ohnolz B et al (1974) Kinetics of gametogenesis. I. Comparative histological and autoradiographic studies of oocytes and transitional prospermatogonia during oogenesis and prespermatogenesis. Cell Tissue Res 154:443–470
Hobbs RM, Seandel M, Falciatori I et al (2010) Plzf regulates germline progenitor self-renewal by opposing mTORC1. Cell 142:468–479
Hobbs RM, Fagoonee S, Papa A et al (2012) Functional antagonism between Sall4 and Plzf defines germline progenitors. Cell Stem Cell 10:284–298
Hogarth CA, Arnold S, Kent T et al (2015a) Processive pulses of retinoic acid propel asynchronous and continuous murine sperm production. Biol Reprod 92:37

Hogarth CA, Evans E, Onken J et al (2015b) CYP26 enzymes are necessary within the postnatal seminiferous epithelium for normal murine spermatogenesis. Biol Reprod 93:19

Hu YC, de Rooij DG, Page DC (2013) Tumor suppressor gene Rb is required for self-renewal of spermatogonial stem cells in mice. Proc Natl Acad Sci U S A 110:12685–12690

Huckins C (1971a) Cell cycle properties of differentiating spermatogonia in adult Sprague-Dawley rats. Cell Tissue Kinet 4:139–154

Huckins C (1971b) The spermatogonial stem cell population in adult rats. I. Their morphology, proliferation and maturation. Anat Rec 169:533–557

Huckins C (1971c) The spermatogonial stem cell population in adult rats. II. A radioautographic analysis of their cell cycle properties. Cell Tissue Kinet 4:313–334

Huckins C, Oakberg EF (1978) Morphological and quantitative analysis of spermatogonia in mouse testes using whole mounted seminiferous tubules, I. The normal testes. Anat Rec 192:519–528

Ikami K, Tokue M, Sugimoto R et al (2015) Hierarchical differentiation competence in response to retinoic acid ensures stem cell maintenance during mouse spermatogenesis. Development 142:1582–1592

Ishii K, Kanatsu-Shinohara M, Toyokuni S et al (2012) FGF2 mediates mouse spermatogonial stem cell self-renewal via upregulation of Etv5 and Bcl6b through MAP2K1 activation. Development 139:1734–1743

Izadyar F, Wong J, Maki C et al (2011) Identification and characterization of repopulating spermatogonial stem cells from the adult human testis. Hum Reprod 26:1296–1306

Jaenisch R, Young R (2008) Stem cells, the molecular circuitry of pluripotency and nuclear reprogramming. Cell 132:567–582

Jain S, Naughton CK, Yang M et al (2004) Mice expressing a dominant-negative Ret mutation phenocopy human Hirschsprung disease and delineate a direct role of Ret in spermatogenesis. Development 131:5503–5513

Jijiwa M, Kawai K, Fukihara J et al (2008) GDNF-mediated signaling via RET tyrosine 1062 is essential for maintenance of spermatogonial stem cells. Genes Cells 13:365–374

Johnston DS, Olivas E, DiCandeloro P et al (2011) Stage-specific changes in GDNF expression by rat Sertoli cells: a possible regulator of the replication and differentiation of stem spermatogonia. Biol Reprod 85:763–769

Julaton VT, Reijo Pera RA (2011) NANOS3 function in human germ cell development. Hum Mol Genet 20:2238–2250

Kanatsu-Shinohara M, Toyokuni S, Shinohara T (2004) CD9 is a surface marker on mouse and rat male germline stem cells. Biol Reprod 70:70–75

Kanatsu-Shinohara M, Muneto T, Lee J et al (2008) Long-term culture of male germline stem cells from hamster testes. Biol Reprod 78:611–617

Kanatsu-Shinohara M, Takashima S, Ishii K et al (2011) Dynamic changes in EPCAM expression during spermatogonial stem cell differentiation in the mouse testis. PLoS One 6:e23663

Kanatsu-Shinohara M, Inoue K, Takashima S et al (2012a) Reconstitution of mouse spermatogonial stem cell niches in culture. Cell Stem Cell 11:567–578

Kanatsu-Shinohara M, Morimoto H, Shinohara T (2012b) Enrichment of mouse spermatogonial stem cells by melanoma cell adhesion molecule expression. Biol Reprod 87:139

Kanatsu-Shinohara M, Onoyama I, Nakayama KI et al (2014) Skp1-Cullin-F-box (SCF)-type ubiquitin ligase FBXW7 negatively regulates spermatogonial stem cell self-renewal. Proc Natl Acad Sci U S A 111:8826–8831

Katoh-Semba R, Tsuzuki M, Miyazaki N et al (2007) Distribution and immunohistochemical localization of GDNF protein in selected neural and non-neural tissues of rats during development and changes in unilateral 6-hydroxydopamine lesions. Neurosci Res 59:277–287

Kaucher AV, Oatley MJ, Oatley JM (2012) NEUROG3 is a critical downstream effector for STAT3-regulated differentiation of mammalian stem and progenitor spermatogonia. Biol Reprod 86:164, 1–11

Kluin PM, de Rooij DG (1981) A comparison between the morphology and cell kinetics of gonocytes and adult type undifferentiated spermatogonia in the mouse. Int J Androl 4:475–493
Koshimizu U, Watanabe D, Tajima Y et al (1992) Effects of W (c-kit) gene mutation on gametogenesis in male mice: agametic tubular segments in Wf/Wf testes. Development 114:861–867
Kubota H, Avarbock MR, Brinster RL (2003) Spermatogonial stem cells share some, but not all, phenotypic and functional characteristics with other stem cells. Proc Natl Acad Sci U S A 100:6487–6492
Kubota H, Avarbock MR, Brinster RL (2004) Growth factors essential for self-renewal and expansion of mouse spermatogonial stem cells. Proc Natl Acad Sci U S A 101:16489–16494
Kubota H, Avarbock MR, Schmidt JA et al (2009) Spermatogonial stem cells derived from infertile Wv/Wv mice self-renew in vitro and generate progeny following transplantation. Biol Reprod 81:293–301
Kubota H, Wu X, Goodyear SM et al (2011) Glial cell line-derived neurotrophic factor and endothelial cells promote self-renewal of rabbit germ cells with spermatogonial stem cell properties. FASEB J 25:2604–2614
Kwon J, Kikuchi T, Setsuie R et al (2003) Characterization of the testis in congenitally ubiquitin carboxy-terminal hydrolase-1 (Uch-L1) defective (gad) mice. Exp Anim 52:1–9
Kwon J, Mochida K, Wang YL et al (2005) Ubiquitin C-terminal hydrolase L-1 is essential for the early apoptotic wave of germinal cells and for sperm quality control during spermatogenesis. Biol Reprod 73:29–35
Leblond CP, Clermont Y (1952a) Definition of the stages of the cycle of the seminiferous epithelium in the rat. Ann N Y Acad Sci 55:548–573
Leblond CP, Clermont Y (1952b) Spermiogenesis of rat, mouse, hamster and guinea pig as revealed by the periodic acid-fuchsin sulfurous acid technique. Am J Anat 90:167–215
Lee J, Kanatsu-Shinohara M, Inoue K et al (2007) Akt mediates self-renewal division of mouse spermatogonial stem cells. Development 134:1853–1859
Leeson CR, Forman DE (1981) Postnatal development and differentiation of contractile cells within the rabbit testis. J Anat 132:491–511
Li JY, English MA, Ball HJ et al (1997) Sequence-specific DNA binding and transcriptional regulation by the promyelocytic leukemia zinc finger protein. J Biol Chem 272:22447–22455
Liao HF, Chen WS, Chen YH et al (2014) DNMT3L promotes quiescence in postnatal spermatogonial progenitor cells. Development 141:2402–2413
Lolicato F, Marino R, Paronetto MP et al (2008) Potential role of Nanos3 in maintaining the undifferentiated spermatogonia population. Dev Biol 313:725–738
Lovasco LA, Gustafson EA, Seymour KA et al (2015) TAF4b is required for mouse spermatogonial stem cell development. Stem Cells 33:1267–1276
Lovelace DL, Gao Z, Mutoji K et al. The regulatory repertoire of PLZF and SALL4 in undifferentiated spermatogonia. Development (in press)
Lu N, Sargent KM, Clopton DT et al (2013) Loss of vascular endothelial growth factor A (VEGFA) isoforms in the testes of male mice causes subfertility, reduces sperm numbers, and alters expression of genes that regulate undifferentiated spermatogonia. Endocrinology 154:4790–4802
Luo J, Megee S, Rathi R et al (2006) Protein gene product 9.5 is a spermatogonia-specific marker in the pig testis: application to enrichment and culture of porcine spermatogonia. Mol Reprod Dev 73:1531–1540
Maekawa M, Kamimura K, Nagano T (1996) Peritubular myoid cells in the testis: their structure and function. Arch Histol Cytol 59:1–13
Manova K, Bachvarova RF (1991) Expression of c-kit encoded at the W locus of mice in developing embryonic germ cells and presumptive melanoblasts. Dev Biol 146:312–324
Matson CK, Murphy MW, Griswold MD et al (2010) The mammalian doublesex homolog DMRT1 is a transcriptional gatekeeper that controls the mitosis versus meiosis decision in male germ cells. Dev Cell 19:612–624

McCarrey JR (2013) Toward a more precise and informative nomenclature describing fetal and neonatal male germ cells in rodents. Biol Reprod 89:47
McLaren A (2003) Primordial germ cells in the mouse. Dev Biol 262:1–15
Meng X, Lindahl M, Hyvonen ME et al (2000) Regulation of cell fate decision of undifferentiated spermatogonia by GDNF. Science 287:1489–1493
Morales C, Griswold MD (1987) Retinol-induced stage synchronization in seminiferous tubules of the rat. Endocrinology 121:432–434
Morimoto H, Kanatsu-Shinohara M, Takashima S et al (2009) Phenotypic plasticity of mouse spermatogonial stem cells. PLoS One 4:e7909
Morimoto H, Iwata K, Ogonuki N et al (2013) ROS are required for mouse spermatogonial stem cell self-renewal. Cell Stem Cell 12:774–786
Morimoto H, Kanatsu-Shinohara M, Shinohara T (2015) ROS-generating oxidase Nox3 regulates the self-renewal of mouse spermatogonial stem cells. Biol Reprod 92:147
Morrow CM, Hostetler CE, Griswold MD et al (2007) ETV5 is required for continuous spermatogenesis in adult mice and may mediate blood testes barrier function and testicular immune privilege. Ann N Y Acad Sci 1120:144–151
Mullaney BP, Skinner MK (1991) Growth factors as mediators of testicular cell-cell interactions. Baillieres Clin Endocrinol Metab 5:771–790
Murphey P, McLean DJ, McMahan CA et al (2013) Enhanced genetic integrity in mouse germ cells. Biol Reprod 88:6
Nagano MC (2003) Homing efficiency and proliferation kinetics of male germ line stem cells following transplantation in mice. Biol Reprod 69:701–707
Nakagawa T, Nabeshima Y, Yoshida S (2007) Functional identification of the actual and potential stem cell compartments in mouse spermatogenesis. Dev Cell 12:195–206
Naughton CK, Jain S, Strickland AM et al (2006) Glial cell-line derived neurotrophic factor-mediated RET signaling regulates spermatogonial stem cell fate. Biol Reprod 74:314–321
Oakberg EF (1956) A description of spermiogenesis in the mouse and its use in analysis of the cycle of the seminiferous epithelium and germ cell renewal. Am J Anat 99:391–413
Oakberg EF (1971) Spermatogonial stem-cell renewal in the mouse. Anat Rec 169:515–531
Oatley JM, Brinster RL (2008) Regulation of spermatogonial stem cell self-renewal in mammals. Annu Rev Cell Dev Biol 24:263–286
Oatley JM, Brinster RL (2012) The germline stem cell niche unit in mammalian testes. Physiol Rev 92:577–595
Oatley JM, Avarbock MR, Telaranta AI et al (2006) Identifying genes important for spermatogonial stem cell self-renewal and survival. Proc Natl Acad Sci U S A 103:9524–9529
Oatley JM, Avarbock MR, Brinster RL (2007) Glial cell line-derived neurotrophic factor regulation of genes essential for self-renewal of mouse spermatogonial stem cells is dependent on Src family kinase signaling. J Biol Chem 282:25842–25851
Oatley JM, Oatley MJ, Avarbock MR et al (2009) Colony stimulating factor 1 is an extrinsic stimulator of mouse spermatogonial stem cell self-renewal. Development 136:1191–1199
Oatley MJ, Kaucher AV, Racicot KE et al (2011) Inhibitor of DNA binding 4 is expressed selectively by single spermatogonia in the male germline and regulates the self-renewal of spermatogonial stem cells in mice. Biol Reprod 85:347–356
Ohta H, Yomogida K, Dohmae K et al (2000) Regulation of proliferation and differentiation in spermatogonial stem cells: the role of c-kit and its ligand SCF. Development 127:2125–2131
Ohta H, Tohda A, Nishimune Y (2003) Proliferation and differentiation of spermatogonial stem cells in the w/wv mutant mouse testis. Biol Reprod 69:1815–1821
Orwig KE, Ryu BY, Master SR et al (2008) Genes involved in post-transcriptional regulation are overrepresented in stem/progenitor spermatogonia of cryptorchid mouse testes. Stem Cells 26:927–938
O'Shaughnessy PJ (2014) Hormonal control of germ cell development and spermatogenesis. Semin Cell Dev Biol 29:55–65

Parker N, Falk H, Singh D et al (2014) Responses to glial cell line-derived neurotrophic factor change in mice as spermatogonial stem cells form progenitor spermatogonia which replicate and give rise to more differentiated progeny. Biol Reprod 91:92
Perey B, Clermont Y, Leblond CP (1961) The wave of the seminiferous epithelium of the rat. Am J Anat 108:47–77
Phillips BT, Gassei K, Orwig KE (2010) Spermatogonial stem cell regulation and spermatogenesis. Philos Trans R Soc Lond B Biol Sci 365:1663–1678
Potten CS (1992) Cell lineages. In: McGee JO, Isaacson PG, Wright NA (eds) Oxford textbook of pathology. Oxford University Press, Oxford, pp 43–52
Puszyk W, Down T, Grimwade D et al (2013) The epigenetic regulator PLZF represses L1 retrotransposition in germ and progenitor cells. EMBO J 32:1941–1952
Raverot G, Weiss J, Park SY et al (2005) Sox3 expression in undifferentiated spermatogonia is required for the progression of spermatogenesis. Dev Biol 283:215–225
Reding SC, Stepnoski AL, Cloninger EW et al (2010) THY1 is a conserved marker of undifferentiated spermatogonia in the pre-pubertal bull testis. Reproduction 139:893–903
Rohozinski J, Bishop CE (2004) The mouse juvenile spermatogonial depletion (jsd) phenotype is due to a mutation in the X-derived retrogene, mUtp14b. Proc Natl Acad Sci U S A 101:11695–11700
Roosen-Runge EC, Giesel LO Jr (1950) Quantitative studies on spermatogenesis in the albino rat. Am J Anat 87:1–30
Russell L (1977) Desmosome-like junctions between Sertoli and germ cells in the rat testis. Am J Anat 148:301–312
Russell LD, Ettlin RA, Sinha-Hikim AP, Clegg ED (1990) Histological and histopathological evaluation of the testis. Cache River, Clearwater, FL
Ryu BY, Kubota H, Avarbock MR et al (2005) Conservation of spermatogonial stem cell self-renewal signaling between mouse and rat. Proc Natl Acad Sci U S A 102:14302–14307
Sada A, Suzuki A, Suzuki H et al (2009) The RNA-binding protein NANOS2 is required to maintain murine spermatogonial stem cells. Science 325:1394–1398
Sakaki-Yumoto M, Kobayashi C, Sato A et al (2006) The murine homolog of SALL4, a causative gene in Okihiro syndrome, is essential for embryonic stem cell proliferation, and cooperates with Sall1 in anorectal, heart, brain and kidney development. Development 133:3005–3013
Sargent KM, Clopton DT, Lu N et al (2016) VEGFA splicing: divergent isoforms regulate spermatogonial stem cell maintenance. Cell Tissue Res 363(1):31–45
Savitt J, Singh D, Zhang C et al (2012) The in vivo response of stem and other undifferentiated spermatogonia to the reversible inhibition of glial cell line-derived neurotrophic factor signaling in the adult. Stem Cells 30:732–740
Schlesser HN, Simon L, Hofmann MC et al (2008) Effects of ETV5 (ets variant gene 5) on testis and body growth, time course of spermatogonial stem cell loss, and fertility in mice. Biol Reprod 78:483–489
Schrans-Stassen BH, van de Kant HJ, de Rooij DG et al (1999) Differential expression of c-kit in mouse undifferentiated and differentiating type A spermatogonia. Endocrinology 140:5894–5900
Schrans-Stassen BH, Saunders PT, Cooke HJ et al (2001) Nature of the spermatogenic arrest in Dazl -/- mice. Biol Reprod 65:771–776
Seaberg RM, van der Kooy D (2003) Stem and progenitor cells: the premature desertion of rigorous definitions. Trends Neurosci 26:125–131
Seandel M, James D, Shmelkov SV et al (2007) Generation of functional multipotent adult stem cells from GPR125+ germline progenitors. Nature 449:346–350
Shiloh Y (2003) ATM and related protein kinases: safeguarding genome integrity. Nat Rev Cancer 3:155–168
Shinoda G, de Soysa TY, Seligson MT et al (2013) Lin28a regulates germ cell pool size and fertility. Stem Cells 31:1001–1009

Shinohara T, Avarbock MR, Brinster RL (1999) β1- and α6-integrin are surface markers on mouse spermatogonial stem cells. Proc Natl Acad Sci U S A 96:5504–5509
Shinohara T, Orwig KE, Avarbock MR et al (2000) Spermatogonial stem cell enrichment by multiparameter selection of mouse testis cells. Proc Natl Acad Sci U S A 97:8346–8351
Snyder EM, Davis JC, Zhou Q et al (2011) Exposure to retinoic acid in the neonatal but not adult mouse results in synchronous spermatogenesis. Biol Reprod 84:886–893
Sorrentino V, Giorgi M, Geremia R et al (1991) Expression of the c-kit proto-oncogene in the murine male germ cells. Oncogene 6:149–151
Suzuki A, Tsuda M, Saga Y (2007) Functional redundancy among Nanos proteins and a distinct role of Nanos2 during male germ cell development. Development 134:77–83
Suzuki H, Sada A, Yoshida S et al (2009) The heterogeneity of spermatogonia is revealed by their topology and expression of marker proteins including the germ cell-specific proteins Nanos2 and Nanos3. Dev Biol 336:222–231
Suzuki H, Ahn HW, Chu T et al (2012) SOHLH1 and SOHLH2 coordinate spermatogonial differentiation. Dev Biol 361:301–312
Tadokoro Y, Yomogida K, Ohta H et al (2002) Homeostatic regulation of germinal stem cell proliferation by the GDNF/FSH pathway. Mech Dev 113:29–39
Takashima S, Kanatsu-Shinohara M, Tanaka T et al (2011) Rac mediates mouse spermatogonial stem cell homing to germline niches by regulating transmigration through the blood-testis barrier. Cell Stem Cell 9:463–475
Takashima S, Kanatsu-Shinohara M, Tanaka T et al (2015) Functional differences between GDNF-dependent and FGF2-dependent mouse spermatogonial stem cell self-renewal. Stem Cell Rep 4:489–502
Takubo K, Ohmura M, Azuma M et al (2008) Stem cell defects in ATM-deficient undifferentiated spermatogonia through DNA damage-induced cell-cycle arrest. Cell Stem Cell 2:170–182
Tegelenbosch RAJ, de Rooij DG (1993) A quantitative study of spermatogonial multiplication and stem cell renewal in the C3H/101 F1 hybrid mouse. Mutat Res 290:193–200
Tokuda M, Kadokawa Y, Kurahashi H et al (2007) CDH1 is a specific marker for undifferentiated spermatogonia in mouse testes. Biol Reprod 76:130–141
Toyoda S, Miyzaki T, Miyazaki S et al (2009) Sohlh2 affects differentiation of KIT positive oocytes and spermatogonia. Dev Biol 325:238–248
Valli H, Sukhwani M, Dovey SL et al (2014) Fluorescence- and magnetic-activated cell sorting strategies to isolate and enrich human spermatogonial stem cells. Fertil Steril 102:566–580
Valli H, Phillips BT, Orwig KE, Gassei K, Nagano MC (2015) Spermatogonial stem cells and spermatogenesis. In: Plant TM, Zeleznik AJ (eds) Knobil and Neill's physiology of reproduction, 4th edn. Academic, New York, pp 595–635
van Bragt MP, Roepers-Gajadien HL, Korver CM et al (2008) Expression of the pluripotency marker UTF1 is restricted to a subpopulation of early A spermatogonia in rat testis. Reproduction 136:33–40
Virtanen I, Kallajoki M, Narvanen O et al (1986) Peritubular myoid cells of human and rat testis are smooth muscle cells that contain desmin-type intermediate filaments. Anat Rec 215:10–20
Walter CA, Intano GW, McCarrey JR et al (1998) Mutation frequency declines during spermatogenesis in young mice but increases in old mice. Proc Natl Acad Sci U S A 95:10015–10019
Wang PJ, McCarrey JR, Yang F et al (2001) An abundance of X-linked genes expressed in spermatogonia. Nat Genet 27:422–426
Weber JE, Russell LD (1987) A study of intercellular bridges during spermatogenesis in the rat. Am J Anat 180:1–24
Webster KE, O'Bryan MK, Fletcher S et al (2005) Meiotic and epigenetic defects in Dnmt3L-knockout mouse spermatogenesis. Proc Natl Acad Sci U S A 102:4068–4073
Wu X, Oatley JM, Oatley MJ et al (2010) The POU domain transcription factor POU3F1 is an important intrinsic regulator of GDNF-induced survival and self-renewal of mouse spermatogonial stem cells. Biol Reprod 82:1103–1111

Wu X, Goodyear SM, Tobias JW et al (2011) Spermatogonial stem cell self-renewal requires ETV5-mediated downstream activation of Brachyury in mice. Biol Reprod 85:1114–1123
Yang QE, Oatley JM (2014) Spermatogonial stem cell functions in physiological and pathological conditions. Curr Top Dev Biol 107:235–267
Yang QE, Gwost I, Oatley MJ et al (2013a) Retinoblastoma protein (RB1) controls fate determination in stem cells and progenitors of the mouse male germline. Biol Reprod 89:113
Yang QE, Kim D, Kaucher A et al (2013b) CXCL12-CXCR4 signaling is required for the maintenance of mouse spermatogonial stem cells. J Cell Sci 126:1009–1020
Yang QE, Racicot KE, Kaucher AV et al (2013c) MicroRNAs 221 and 222 regulate the undifferentiated state in mammalian male germ cells. Development 140:280–290
Yoshida S, Takakura A, Ohbo K et al (2004) Neurogenin3 delineates the earliest stages of spermatogenesis in the mouse testis. Dev Biol 269:447–458
Yoshida S, Sukeno M, Nakagawa T et al (2006) The first round of mouse spermatogenesis is a distinctive program that lacks the self-renewing spermatogonia stage. Development 133:1495–1505
Yoshida S, Nabeshima Y, Nakagawa T (2007a) Stem cell heterogeneity: actual and potential stem cell compartments in mouse spermatogenesis. Ann N Y Acad Sci 1120:47–58
Yoshida S, Sukeno M, Nabeshima Y (2007b) A vasculature-associated niche for undifferentiated spermatogonia in the mouse testis. Science 317:1722–1726
Yoshinaga K, Nishikawa S, Ogawa M et al (1991) Role of c-kit in mouse spermatogenesis: identification of spermatogonia as a specific site of c-kit expression and function. Development 113:689–699
Zhang L, Tang J, Haines CJ et al (2011) c-kit and its related genes in spermatogonial differentiation. Spermatogenesis 1:186–194
Zhao M, Rohozinski J, Sharmac M et al (2007) Utp14b: a unique retrogene within a gene that has acquired multiple promoters and a specific function in spermatogenesis. Dev Biol 304:848–859
Zheng K, Wu X, Kaestner KH et al (2009) The pluripotency factor LIN28 marks undifferentiated spermatogonia in mouse. BMC Dev Biol 9:38
Zhou Z, Shirakawa T, Ohbo K et al (2015) RNA binding protein Nanos2 organizes post-transcriptional buffering system to retain primitive state of mouse spermatogonial stem cells. Dev Cell 34:96–107

Chapter 11
Stem Cells in Mammalian Gonads

Ji Wu, Xinbao Ding, and Jian Wang

Abstract Stem cells have great value in clinical application because of their ability to self-renew and their potential to differentiate into many different cell types. Mammalian gonads, including testes for males and ovaries for females, are composed of germline and somatic cells. In male mammals, spermatogonial stem cells maintain spermatogenesis which occurs continuously in adult testis. Likewise, a growing body of evidence demonstrated that female germline stem cells could be found in mammalian ovaries. Meanwhile, prior studies have shown that somatic stem cells exist in both testes and ovaries. In this chapter, we focus on mammalian gonad stem cells and discuss their characteristics as well as differentiation potentials.

Abbreviations

3β-HSD	3β-hydroxysteroid dehydrogenase
Aldh1	Aldehyde dehydrogenase 1
Blimp-1	Blymphocyte-induced maturation protein
BMP4	Bone morphogenetic protein 4
c-Kit	Kit oncogene
CSF1	Colony-stimulating factor-1

J. Wu (✉)
Key Laboratory for the Genetics of Developmental & Neuropsychiatric Disorders (Ministry of Education), Bio-X Institutes, Shanghai Jiao Tong University, No. 800. Dongchuan Road, Minhang District, Shanghai 200240, China

Key Laboratory of Fertility Preservation and Maintenance of Ministry of Education, Ningxia Medical University, Yinchuan 750004, China

Shanghai Key Laboratory of Reproductive Medicine, Shanghai 200025, China
e-mail: jiwu@sjtu.edu.cn

X. Ding • J. Wang
Key Laboratory for the Genetics of Developmental & Neuropsychiatric Disorders (Ministry of Education), Bio-X Institutes, Shanghai Jiao Tong University, No. 800. Dongchuan Road, Minhang District, Shanghai 200240, China

R.P. Piprek (ed.), *Molecular Mechanisms of Cell Differentiation in Gonad Development*, Results and Problems in Cell Differentiation 58,
DOI 10.1007/978-3-319-31973-5_11

CXCR4	C-X-C chemokine receptor type 4
Dazl	Deleted in azoospermia-like
DMC1	DNA meiotic recombinase 1
Dnaic2	Dynein axonemal intermediate chain 2
Figla	Factor in the germline alpha
Fragilis	Also commonly referred to as Ifitm3, interferon-induced transmembrane protein 3
Fshr	Follicle-stimulating hormone receptor
GCNA	Germ cell nuclear antigen
GDF9	Growth differentiation factor 9
GFP	Green fluorescent protein
GFRA1	GDNF family receptor alpha-1
Gli2	GLI family zinc finger 2
IGF1	Insulin-like growth factor I
Ihh	Indian hedgehog
Lef1	Lymphoid enhancer-binding factor 1
Lgr5	Leucine-rich repeat-containing G-protein coupled receptor 5
LHR	Luteinizing hormone receptor
LHX8	LIM homeobox 8
LIFR	Leukemia inhibitory factor receptor alpha
Mvh	Mouse vasa homolog also termed DEAD box polypeptide 4, DDX4
Nanog	Nanog homeobox
NOBOX	NOBOX oogenesis homeobox
OCT4	Octamer-binding transcription factor 4
PDGFRα	Platelet-derived growth factor receptor α
Prdm14	PR domain containing 14 with ZNF domain
Ptch2	Patched homolog 2
Rex-1	Also commonly referred to as ZFP42, zinc finger protein 42
Sca-1	Sarco-endoplasmic reticulum calcium ATPase
SCF	Stem cell factor
SCP1–3	Synaptonemal complex protein 1–3
Sox-2	SRY-box containing gene 2
SSEA-1	Stage-specific embryonic antigen-1
SSEA-4	Stage-specific embryonic antigen-4
StAR	Steroidogenic acute regulatory protein
Stella	Also commonly referred to as DPPA3, developmental pluripotency-associated 3
Stpb-c	Short-type pituitary gland and brain-cadherin
Stra8	Stimulated by retinoic acid gene 8
TEK	Endothelial-specific receptor tyrosine kinase
TIE	Tyrosine kinase with immunoglobulin-like and EGF-like domains
UCHL1	Ubiquitin carboxy-terminal hydrolase L1
VEGFR1	Vascular endothelial growth factor receptor 1
VEGFR2	Vascular endothelial growth factor receptor 2

vWF	von Willebrand factor
YBX2	Y-box protein 2
ZP1–3	Zona pellucida glycoprotein 1–3

11.1 Introduction

The precursor of the gonads is the genital ridge, which is identical in male and female individuals. The genital ridge is an undifferentiated primordium composed of bipotential precursor cells, mainly including supporting and steroid-secreting cells, which can follow one of two possible fates. Supporting precursor cells can develop into either testis-specific Sertoli cells or ovary-specific granulosa cells. Steroid-secreting precursor cells might be the common precursor of Leydig cells in the testis and theca cells in the ovary (Wilhelm et al. 2007).

Germ cells are responsible for the transmission of genetic information to subsequent generations. In mammals, primordial germ cells (PGCs) are the common precursors of oocytes and spermatozoa in the ovaries and testes, respectively. They differentiate from epiblast cells in response to bone morphogenic protein (BMP) signals from extra-embryonic tissues at 6.25 *dpc* (days *post coitum*) in mice. It has been demonstrated that BMP signals could activate *Blimp-1* and *Prdm14*, which leads to PGC differentiation (Saitou and Yamaji 2010). In mice, PGCs begin to migrate from the primitive streak to the endoderm at 7.5 *dpc*, through the hindgut endoderm at 8 *dpc*, toward the dorsal body wall at 9.5 *dpc*, and finally colonize the gonadal ridges at 10.5 *dpc* (Richardson and Lehmann 2010).

Stem cells are unique cell types that can differentiate into specialized cells and produce more stem cells via self-renewal. In mammals, they mainly include (1) embryonic stem cells (ESCs) (Evans and Kaufman 1981; Martin 1981; Thomson et al. 1998) derived from the inner cell mass of blastocysts, (2) embryonic germ cells (EGCs) (Matsui et al. 1992; Resnick et al. 1992; Shamblott et al. 1998) derived from PGCs, (3) induced pluripotent stem cells (iPSCs) (Yu et al. 2007; Takahashi et al. 2007; Takahashi and Yamanaka 2006) reprogrammed from adult cells, and (4) adult stem cells (ASCs) found in various tissues or organs. ESCs, EGCs, and iPSCs are pluripotent stem cells with the ability to differentiate into any type of cell in the body. Very recently, a new stem cell type, termed region-selective pluripotent stem cells (rsPSCs), was established from mouse embryos and primate pluripotent stem cells, including humans (Wu et al. 2015). ASCs can differentiate into specialized cell types of the tissues. Here, we summarize the ASCs in gonads including germline stem cells (GSCs) and somatic stem cells (non-germline ASCs) (Fig. 11.1).

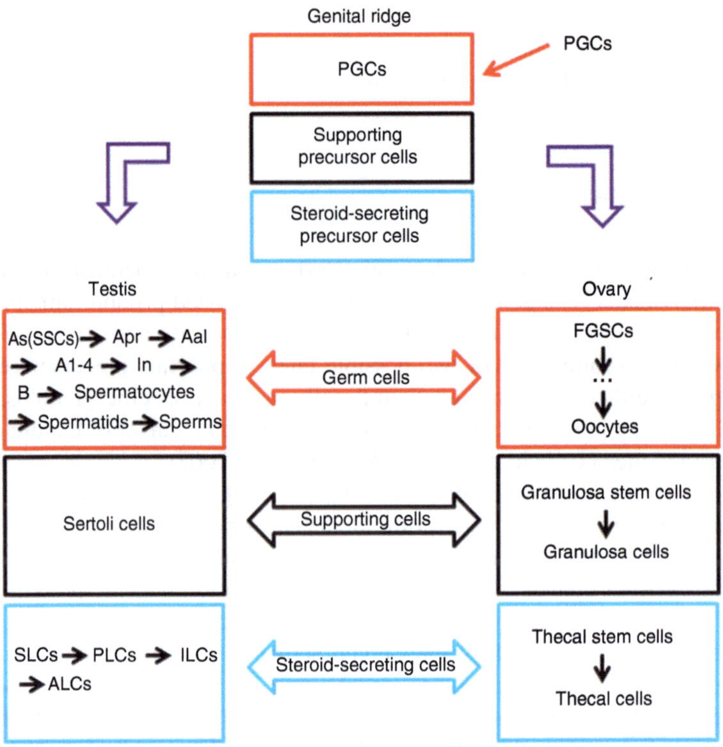

Fig. 11.1 Stem cells and their in vivo physiological status in mammalian gonads. *PGCs* primordial germ cells, *SSCs* spermatogonial stem cells, A_s A-single spermatogonia, A_{pr} A-paired spermatogonia, A_{al} A-aligned spermatogonia, *A1–4* types A1–A4 spermatogonia, *In* intermediate spermatogonia, *B* type B spermatogonia, *SLCs* stem Leydig cells, *PLCs* progenitor Leydig cells, *ILCs* immature Leydig cells, *ALCs* adult Leydig cells, *FGSCs* female germline stem cells

11.2 Germline Stem Cells

GSCs belong to ASCs and maintain their balance between self-renewal and differentiation in their niche. One of the characteristics of GSCs is that they are responsible for gametogenesis and transmitting genetic information from one generation to the next. In this section, we summarize the spermatogonial stem cells (SSCs) in the testes and the newly found female germline stem cells (FGSCs) in the ovaries.

11.2.1 Spermatogonial Stem Cells

Spermatogonial stem cells (SSCs, also known as male germline stem cells) are located in the seminiferous tubules of the testes and have the capacity to differentiate into sperm continuously for adult mammals.

11.2.1.1 The Origin and Location of SSCs

Gonocytes, generated by PGCs after migrating into the gonad (McLaren 2003), are the sole source of a functional reservoir of SSCs in the postnatal testis. Yasuda et al. (1986) used a succession of terms of mitotic- (M), transitional 1- (T1), and transitional 2- (T2) prospermatogonia to describe the successive phases of gonocyte development based on the cells being proliferative or quiescent. M-prospermatogonia, mitotic fetal germ cells, will become T1-prospermatogonia arrested at about 16.5 *dpc* of mouse development and subsequently transform to T2-prospermatogonia and resume proliferation after birth. M-prospermatogonia are far away from the basal membrane and are located in the center of the testicular cords, whereas T2-prospermatogonia are already migrating into the seminiferous tubule basement membrane and form undifferentiated type A spermatogonia (Culty 2013; Yoshida et al. 2007). In rodents, undifferentiated type A spermatogonia are classified as A-single (A_s), A-paired (A_{pr}), and A-aligned (A_{al}, groups of 4, 8, or 16 cells) spermatogonia according to their morphological arrangement. It is generally believed that A_s cells are SSCs, while A_{pr} and A_{al} cells represent the progenitor spermatogonial populations committed to proliferate and eventually produce haploid cells. However, in humans, A_{dark} and the A_{pale} spermatogonia were identified. The A_{dark} spermatogonia are considered to be the reserve stem cells, whereas the A_{pale} spermatogonia are the renewing stem cells (Clermont 1963, 1966). SSCs reside at the basement membrane of seminiferous tubules in the testis and are almost completely surrounded by somatic Sertoli cells, resembling a microenvironment, or niche, to regulate the self-renewal and differentiation of SSCs (Brinster 2002). SSCs are rare and occupy only about 0.02–0.03 % of the total testis cell population (Tagelenbosch and de Rooij 1993).

11.2.1.2 Differentiation and Trans-differentiation of SSCs

The mechanisms of regulating SSC differentiation have been well summarized previously (Mei et al. 2015). In this section, we focus on the differentiation and trans-differentiation potentials of SSCs.

The interaction of stem cell factor (SCF; also known as KITL), which is secreted by Sertoli cells, and C-KIT receptor, which is expressed in differentiated type A and B spermatogonia, is critical for spermatogenesis (Vincent et al. 1998). Feng et al. (2002) created telomerase-immortalized, mouse type A spermatogonia cell lines. In the presence of SCF, these cells could differentiate into spermatocytes and spermatids. Carlomagno et al. (2010) found that BMP4 (bone morphogenetic protein 4) upregulates the expression of *c-Kit* and affects cell adhesion pathways in a rat SSC line. Sertoli cells secrete cytokines, such as BMP4 and Activin A, which are involved in regulating SSC differentiation (De Rooij 2009). Accordingly, Riboldi et al. (2012) cultured Sertoli cells in medium with addition of $CD49f^+$, a marker of SSCs, cells from testicular sperm extraction samples of azoospermic patients for generating haploid cells in vitro. It is generally accepted that all-trans retinoic acid (atRA) induces the expression of a meiotic gatekeeper gene, *Stra8* (stimulated by retinoic acid gene 8), leading to meiotic initiation (Anderson et al. 2008; Bowles et al. 2006; Koubova et al. 2006; Mark et al. 2008). Yang et al. (2014) reported that RA and SCF together could efficiently differentiate human SSCs (positive for UCHL1 and GFRA1) from cryptorchid patients into functional haploid cells, demonstrated by fertilization after round spermatid microinjection into mouse oocytes. Lee et al. (2006) isolated SSC-like cells (positive for OCT4 and integrin β1) from testicular tissue of non-obstructive azoospermic patients, including maturation arrest and Sertoli cell-only syndrome. They differentiated these SSC-like cells into haploid germ cells by encapsulation in differentiation medium-containing calcium alginate. The haploid germ cells were demonstrated by fluorescence in situ hybridization and activation of human oocytes after the injection of round spermatid-like cells in vitro. Organ culture is another strategy to differentiate SSCs into sperm in vitro (Gohbara et al. 2010; Sato et al. 2011a, b, 2012). In 1937, Martinovitch (1937) performed the first organ culture of newborn mouse testes. After 11 days of culture in vitro, the early meiotic prophase was observed. However, complete spermatogenesis and production of haploid cells was not observed. Sato et al. (2011a) demonstrated that addition of knockout serum replacement, instead of fetal bovine serum, to the organ culture medium could increase the efficiency of meiosis completion and the formation of spermatids and sperms. The same group subsequently injected SSCs into the seminiferous tubules within recipient testes and applied organ culture in vitro to form haploid gametes. These cells are fertile and give rise to healthy offspring through micro-insemination (Sato et al. 2011b). The authors also demonstrated that their system can be used to correct spermatogenic defects in vitro. Colony-stimulating factor-1 (CSF1) and recombinant KITL in the medium of organ cultures could induce spermatogenesis to proceed through meiosis in Sl/Sl^{d} mouse testes that contain only primitive spermatogonia as germ cells and lack spermatogenesis (Sato et al. 2012).

Apart from being able to differentiate into sperm, SSCs can directly trans-differentiate into a wide range of other type cells. Boulanger et al. (2007) mixed adult testicular cells with dispersed mammary epithelial cells and subsequently injected them into epithelium-divested mammary fat pads. Several weeks later, they found that the cells from adult seminiferous tubules achieved mammary epithelial progenitor cell properties and differentiated into functional mammary epithelial cells. Simon et al. (2009) demonstrated that stem/progenitor spermatogonia could directly trans-differentiate into tissues of all germ layers, including prostatic, uterine, and skin epithelium, when recombined with the appropriate mesenchyme and transplanted in vivo. Ning et al. (2010) injected GFP^{+} (green fluorescent protein) SSCs ($CD49f^{+}$) into the bone marrow (BM) of busulfan-treated GFP^{-} female mice. Twelve weeks after transplantation, they detected GFP^{+} and Y^{+} chromosome cells in BM, peripheral blood, and the spleen and concluded that SSCs have the potential to trans-differentiate into hematopoietic cells in vivo. Zhang et al. (2013) developed an in vitro protocol to differentiate SSCs into hepatic stem-like cells, which subsequently differentiate into small hepatocytes and mature hepatocyte-like cells.

11.2.2 *Female Germline Stem Cells*

11.2.2.1 The Development of Germ Cells in Postnatal Mammalian Ovaries

GSCs in male mammals have been well studied. However, in most female mammals, there has been a long-persisting belief that the number of oocytes available throughout life is determined during the perinatal period (Anderson and Hirshfield 1992; Borum 1961; McLaren 1984; Zuckerman 1951) even though there are different voices (Butler and Juma 1970; Generoso et al. 1971a, b; Ioannou 1967). But recent evidence over the past decade has challenged this belief. For example, in 2004, Johnson et al. (2004) indicated that the female gonad may have regenerative activity in postnatal mice, including juvenile and adult mice in vivo. One year later, the same research group showed the expression of germline markers in BM and thus performed BM transplantation. Subsequently, they indicated that BM could restore oocyte production in wild-type mice after transplantation and concluded that the cells from BM and peripheral blood may serve as a source of germ cells that could sustain oocyte production in adulthood (Johnson et al. 2005). However, this conclusion has been received with controversy. Eggan et al. (2006) established transplantation and parabiotic mouse models in which the vasculature of wild-type mice was surgically connected to that of transgenic mice expressing GFP to assess this novel view. They have shown that bone marrow cells, or other normally circulating cells, are not involved in the formation of mature ovulated oocytes. Veitia et al. (2007) examined a rare clinical situation of a woman who gave birth to a child after allogeneic BM transplant (BMT). This woman suffers from Fanconi anemia, a genetic condition caused by mutations in DNA repair genes. After studying the polymorphic microsatellites from the patient (mother), the daughter, and the donor, they demonstrated the genetic relationship between the mother and the daughter. Thus, the recovery of fertility after BMT can result from incomplete depletion of follicle reserve, not from its replacement by donor BM-derived GSCs. Similarly, Nabhan et al. (2010) found that recovery of normal ovarian function and a viable pregnancy is a realistic but relatively rare possibility in Fanconi anemia patients following allogeneic hematopoietic stem cell transplantation. They did not detect germ cell transmission from the donor. Santiquet et al. (2012) transplanted BM cells in PU.1 mice and in SCID mice after treatment with chemotherapeutic agents. They found no evidence that transplanted BM cells provide new fertilizable oocytes in PU.1 mice or SCID mice.

In 2009, our group established the first lines of FGSCs from postnatal mouse ovaries with the capacity to produce fully functional oocytes and fertile offspring after transplantation into ovaries (Zou et al. 2009). More recently, Reizel et al. (2012) studied female germline dynamics in mice using reconstructed cell lineage trees from somatic mutations and found that female germline is embryonically separate from hematopoietic and mesenchymal stem cells, indicating that these populations belong to distinct lineages. Therefore, the conclusion is that the putative GSCs that may contribute to postnatal oocyte renewal are progenies of the primordial germline, but not cells from BM and peripheral blood.

11.2.2.2 Isolation of FGSCs from Postnatal Mammalian Ovaries

We first isolated and sorted female germ cells from neonatal and adult mouse ovaries by magnetic-activated cell sorting with an antibody against the C-terminus of mouse vasa homolog (Mvh, also termed DEAD box polypeptide 4, Ddx4, a germ cell-specific marker) (Castrillon et al. 2000; Tanaka et al. 2000). After in vitro culture, some isolated cells were able to grow on an inactive STO (SIM thioguanine/ouabain-resistant mouse fibroblast cell line) feeder layer with a normal karyotype. These long-term cultured FGSCs consistently expressed germline-specific markers, such as *Oct4*, *Mvh*, *Fragilis*, *Stella*, *Rex1*, *Dazl*, and *Blimp-1*, and had the capacity to produce fully functional oocytes and fertile offspring after transplantation into chemotherapy-damaged mouse ovaries (Zou et al. 2009). Pacchiarotti et al. (2010) isolated GFP^+ cells from ovaries of neonatal and adult transgenic mice in which GFP is expressed under a germ cell-specific *Oct4* promoter. The GFP^+ cells were able to be cultured up to 1 year and expressed germ cell markers such as germ cell nuclear antigen (GCNA), *c-Kit*, and *Mvh*. They also could form follicle-like structures with granulosa cells after suspension culture. We screened three candidate markers comprising short-type pituitary gland and brain-cadherin (Stpb-c), CD9, a marker used for purifying SSCs (Kanatsu-Shinohara et al. 2004b), and Fragilis (also termed Ifitm3) that are expressed in germ cells. We found that the efficiency of FGSCs' purification was remarkably enhanced using Fragilis, compared with that of using Ddx4 (Zou et al. 2011). Subsequently, White et al. (2012) successfully isolated and purified oogonial stem cells (OSCs; also termed FGSCs) from healthy reproductive-age women using fluorescence-activated cell sorting (FACS) with the same antibody reported by us. They compared the antibodies against C-terminal and N-terminal DDX4 for sorting and confirmed the former antibody was suitable for isolating FGSCs. $DDX4^+$ cells propagated in vitro could spontaneously generate into oocyte-like cells and form chimeric follicles in human ovarian cortical pieces when subcutaneously xenografted into adult SCID female mice.

Recently, we have isolated and purified FGSCs from postnatal female rats using the antibody against Fragilis. Rat FGSCs have similar characteristics to those of mouse FGSCs. *Fat-1*, which encodes an n−3 fatty acid desaturase that converts n-6 to n-3 fatty acids (Kang et al. 2004), transgenic rats were also generated using rat FGSC lines (Zhou et al. 2014).

11.2.2.3 Characteristics of FGSCs

In our studies, FGSCs show large cell bodies with little cytoplasm and spherical nuclei with slight staining, and the FGSCs have similar morphology and growth patterns to those of SSCs (Xie et al. 2014). The FGSCs were positive for BrdU and Mvh expression. The cells also expressed *Oct4*, *Dazl*, *Blimp-1*, *Fragilis*, *Stella*, and *Rex-1*. In contrast, the cells did not express *c-Kit*, *Figla*, *Sox2*, *Nanog*, *Scp1-3*, or *Zp3* (Zou et al. 2009). We performed a microarray hybridization for detecting global gene expression profiles of FGSCs and SSCs and showed similar signatures at the transcriptome level (Xie et al. 2014). The FGSCs also showed stem cell characteristics, including high telomerase activity. Cytogenetic analysis by DAPI immunofluorescence showed that the FGSCs had a normal karyotype (40, XX). In addition, the FGSCs were positive for alkaline phosphatase staining. However, the intensity seemed weaker than that of ESCs. Combined bisulfite restriction analysis (COBRA) indicated that the maternally imprinted regions were partially methylated and the paternally imprinted regions were demethylated in FGSCs, indicative of a female imprinting pattern (Zou et al. 2009). When the FGSCs were transplanted into the ovaries of sterile female recipients, the cells could differentiate into functional oocytes and fertile offspring (Zou et al. 2009). Mouse and human FGSCs (or OSCs) could spontaneously generate oocyte-like cells expressing oocyte- and meiosis-specific markers, such as *SYCP3*, *DMC1*, *NOBOX*, *YBX2*, *LHX8*, *GDF9*, and *ZP1/2/3*. In addition, DNA ploidy analysis detected haploid cells (White et al. 2012). We developed a three-step system to differentiate FGSCs into GV-like oocytes in vitro, as judged by morphological criteria and molecular markers (Zhou et al. 2014). With germ cell characteristics for FGSCs, we demonstrated that FGSCs could serve as an alternative approach to generate transgenic and gene knock-down mice. We successfully produced *Oocyte-G1*, which is involved in ovarian follicular development (Zhang and Wu 2009), and dynein axonemal intermediate chain 2 (*Dnaic2*) (Yang and Wu 2008) transgenic mice, as well as *Oocyte-G1* knock-down mice (Zhang et al. 2011). It was previously reported that SSCs can be converted into pluripotent stem cells without gene manipulation under certain culture conditions (Guan et al. 2006; Izadyar et al. 2008; Kanatsu-Shinohara et al. 2004a; Seandel et al. 2007). Similarly, we demonstrated that stably proliferating mouse FGSCs could be converted to female embryonic stem-like cells under certain culture conditions and exhibit properties similar to those of ESCs, including genomic imprinting pattern, potential to differentiate into the three germ layers, and chimera production capacity (Wang et al. 2014).

11.3 Somatic Stem Cells

Apart from GSCs in gonads, somatic stem cells have also been reported in the gonads of both sexes. In this section, we have summarized the somatic stem cells, including stem Leydig cells (SLCs) located in the testes (the presence of stem cells for Sertoli cells and peritubular myoid cells has not been found until now), and the very small embryonic-like stem cells (VSELs), $Lgr5^+$ cells, ranulose stem cells, and thecal stem cells located in the ovaries.

11.3.1 Stem Leydig Cells

Leydig cells have distinct populations during testis development, including fetal Leydig cells (FLCs) and adult Leydig cells (ALCs). FLCs are present in the embryonic testis from shortly after sex determination to birth. Subsequently, the ALC population replaces the FLC population and expands at puberty (Griswold and Behringer 2009). ALCs are the testosterone-producing cells of the adult testis. They arise from SLCs and undergo four distinct developmental stages, including SLCs, progenitor Leydig cells (PLCs), immature Leydig cells (ILCs), and ALCs, judged by morphological and biochemical criteria (Chen et al. 2009). Ge et al. (2006) were the first to isolate putative SLCs, including FLCs, from rat testes at postnatal day 7 by Percoll gradient centrifugation. The FLCs were removed by immunoselection for luteinizing hormone receptor (LHR)-negative cells and the putative SLCs were further purified by immunoselection for the stem cell marker platelet-derived growth factor receptor α (PDGFRα) from the remaining cells. The cultured SLCs expressed PDGFRα, C-KIT, and LIFR, but not 3β-hydroxysteroid dehydrogenase (3β-HSD) or LHR. These SLCs have the potential to differentiate into Leydig cells in vitro and in vivo. In the presence of PDGFβ homodimer (PDGF-ββ), LH, thyroid hormone, and IGF-1, SLCs could differentiate into Leydig cells expressing P450scc, 3β-HSD, P450c17, LHR, and steroidogenic acute regulatory protein (StAR), as well as producing testosterone. After transplanting into adult rat testes depleted of Leydig cells, SLCs were found in the interstitial compartment. Stanley et al. (2012) isolated $PDGFR\alpha^+/3\beta\text{-}HSD^-$ cells from ethane dimethanesulfonate-(EDS, an alkylating agent for depleting ALCs) treated adult rat testes. These cells could proliferate indefinitely and differentiate in vitro, as well as produce testosterone.

11.3.2 Stem Cells in OSE

The ovarian surface epithelium (OSE) covers the ovary as a layer of mostly squamous or cuboidal cells. The OSE undergoes damage and repair with each ovulatory cycle and there are different theories about their origins (Hummitzsch et al. 2015). Here, we have summarized the VSELs and Lgr5$^+$ cells in OSE.

Kucia et al. (2006) first identified a population of CXCR4$^+$/OCT4$^+$/SSEA-1$^+$/Sca-1$^+$/Lin$^-$/CD45$^-$ small cells (approximately 2–4 μm) from adult murine bone marrow. They are positive for pluripotent markers such as SSEA-1, OCT4, and NANOG; thus, they were named very small embryonic-like stem cells (VSELs). Subsequently, the same group found that VSELs widely exist in many organs, including the brain, kidneys, muscles, and pancreas (Zuba-Surma et al. 2008). Likewise, VSELs were also found in adult gonads. VSELs have been proposed to be present in the OSE of various mammalian species, including mice, rabbits, monkeys, sheep, and humans (Parte et al. 2011, 2013; Bhartiya et al. 2012; Patel et al. 2013). Virant-Klun et al. (2008) isolated small round cells of 2–4 μm in diameter from OSEs scraped from postmenopausal and premature ovarian failure (POF) women. These small cells expressed early embryonic developmental markers, including SSEA-4, OCT4, NANOG, SOX2, and C-KIT. They could develop into oocyte-like cells in vitro, as judged by morphological criteria and oocyte markers, but lack meiotic markers.

LGR5 (Leucine-rich repeat-containing G-protein coupled receptor 5, also termed GPR49 or GPR67) is known to be a stem cell marker in the murine small intestine and colon (Barker et al. 2007). Szotek et al. (2008) used BrdU (5-bromodeoxyuridine) to define a slow cycling cell population in the OSE with the enhanced growth characteristics expected of somatic stem cells. Flesken-Nikitin et al. (2013) identified the cells in hilum OSE of postnatal mice and found that these cells expressed stem/progenitor cell markers such as Aldh1 (aldehyde dehydrogenase 1), Lgr5, Lef1 (lymphoid enhancer-binding factor 1), CD133 (cluster of differentiation 133), and cytokeratin 6B, as well as proliferated long-term in vitro. They also found that the cells in hilum OSE were prone to malignant transformation. Subsequently, Ng et al. (2014) identified Lgr5$^+$ cells at E13.5 through the ovary surface and subsurface and subsequently restricted to the OSE monolayer at postnatal day 7 and adult mice. Using in vivo lineage tracing, they found that embryonic and neonatal Lgr5$^+$ cells as stem/progenitor cells contributed to OSE development whereas the adult Lgr5$^+$ cells maintained OSE homeostasis. However, the differentiation potentials of Lgr5$^+$ cells in OSE need to be studied in further research.

11.3.3 Granulosa Stem Cells

In 1994, Lavranos et al. proposed that some granulosa cells have stem cell properties (Lavranos et al. 1994). Using the antibody against FSHR, luteinizing granulosa cells from follicular aspirates produced in in vitro fertilization centers were sorted by FACS. Cultured cells exhibited two distinct morphologies, epithelial and fibroblastic structures, and the latter morphology was consistently observed in prolonged culture. Furthermore, these cells expressed markers positive for mesenchymal stem cells (MSCs), such as CD29, CD44, CD105, CD117, and CD166, but not CD73 (Kossowska-Tomaszczuk et al. 2009). Under certain culture conditions, the cells have the potential to differentiate into neurons, chondrocytes, and osteoblasts in vitro and generate tissues of mesenchymal origin after being transplanted into the back of nude mice (Kossowska-Tomaszczuk et al. 2009). Dzafic et al. (2014) compared the gene expression profile of aspirated follicular cells (AFCs) from follicular aspirates of infertile women produced in in vitro fertilization with bone marrow-derived mesenchymal stem cells (BM-MSCs) and dermal fibroblasts and revealed their specific stemness was different from the BM-MSCs and the fibroblasts. Similar to luteinizing granulosa cells (Kossowska-Tomaszczuk et al. 2009), under certain conditions AFCs have the potential to differentiate into adipogenic-, osteogenic-, and pancreatic-like cells in vitro (Dzafic et al. 2014). Mattioli et al. (2012) compared the osteogenic potential of growing and luteinizing granulosa cells (GGCs and LGCs). They found that granulosa cell characteristics disappeared in both cell types after in vitro expansion. The cells incorporated in poly lactic-co-glycolic acid (PLGA) scaffolds were cultured in osteogenic medium and subsequently implanted in the dorsal region of SCID mice to assess their osteogenic potential. They demonstrated that they both have osteogenic potential, while LGCs have more efficiency. Similar with the results from Mattioli et al. (2012), Oki et al. (2012) showed that under certain culture conditions porcine granulosa cells could dedifferentiate into fibroblast-like cells (dedifferentiated follicular granulosa, DFOG) which lost the characteristics of granulosa cells. Furthermore, DFOG cells could differentiate into osteoblasts in vitro and in vivo.

However, a subpopulation of endothelial-like cells was also identified in granulosa cells. Antczak and Van Blerkom (2000) demonstrated that mural and cumulus subpopulations of human and mouse follicular granulosa cells expressed endothelial-like markers, such as tyrosine kinase with immunoglobulin-like and EGF-like domains (TIE), endothelial-specific receptor tyrosine kinase (TEK), C-KIT, vascular endothelial growth factor receptor 1 (VEGFR1), CD31, and von Willebrand factor (vWF), as well as rapidly internalized acetylated low-density lipoprotein. Additionally, they could engage in tube-forming activity in vitro. Merkwitz et al. (2010) found that colonies developed in post-confluent bovine granulosa cell cultures from antral follicles expressed progenitor cell markers, such as SOX2, OCT3/4, C-KIT, and alkaline phosphatase. The specific subpopulation of progenitor cells was identified by co-expression of C-KIT and CD14, CD45, CD133, or VEGFR2. Double-positive cells were purified by magnetic-activated cell sorting and differentiated into endothelial cells by the hanging drop technique using hematopoietic differentiation medium.

11.3.4 Thecal Stem Cells

Honda et al. (2007) originally claimed that they isolated putative thecal stem cells from newborn mouse ovaries, according to the methods for isolating male SSCs without sorting. The colonies mixed with oocytes are weakly positive for alkaline phosphatase staining. RT-PCR revealed that markers of thecal cells (*Gli2* and *Ptch2*), but not granulosa cells (*Fshr* and *Ihh*), were detected after differentiation. Intraovarian transplantation of thecal stem cells demonstrated that they distributed in the theca cell layers and the ovarian interstitium. The same research group subsequently used this system to culture primordial oocytes, which incorporated into these thecal cell colonies. After in vitro culture, the oocytes could form a zona pellucida structure and fuse with spermatozoa (Honda et al. 2009). Another study in pigs found ovarian theca-derived multipotent stem cells expressed mesenchymal surface markers, including CD29, CD44, and CD90, and a pluripotency marker SOX2. After being induced using specific culture conditions, they could differentiate into osteogenic, adipogenic, and oocyte-like cells, demonstrating their multipotent differentiation potential (Lee et al. 2013).

11.4 Conclusion

GSCs and somatic stem cells have been found in mammalian gonads. GSCs, including SSCs and FGSCs, are unique cells that can transmit genetic information to the next generation. In vivo, they maintain an undifferentiated state by self-renewing, while simultaneously differentiating into gametes. Functional haploid gametes have been differentiated from SSCs in vitro. This is of great significance for the clinical treatment of male infertility caused by azoospermia. In addition, SSCs can be induced to trans-differentiate into different cell types under specific culture conditions or after transplantation into a different niche microenvironment. Thus, SSCs could be an alternative source of cells for regenerative medicine. Likewise, the differentiation potential is an important characteristic of FGSCs. FGSCs can generate functional oocytes when transplanted back into chemotherapy-damaged adult mouse ovaries. GV-like oocytes can be differentiated from rat FGSCs in vitro. There is no doubt that the clinical application value will be great if FGSCs can differentiate into functional mature and genomic imprinting oocytes in vitro. However, such evidence will take a long time and will undoubtedly be restricted by the current methods, as well as ethical concerns.

There are three major somatic cell types in mammalian testis, including Sertoli cells, Leydig cells, and peritubular myoid cells. Sertoli cells and peritubular myoid cells are considered as terminally differentiated cells. The presence of specific stem cells for Sertoli cells and peritubular myoid cells has not been found until now. However, SLCs have previously been identified in the testes. It has been shown that SLCs have the potential to differentiate into Leydig cells in vitro and in vivo. The potential for SLCs to differentiate into other cell types has yet to be proven.

Granulosa stem cells and thecal stem cells are the main somatic stem cells existing in the mammalian ovary. They are able to give rise to granulosa cells and thecal cells, respectively, but their potential to differentiate into other cell types remains to be determined. In addition, VSELs, which have “stemness” qualities, have been found in OSE. Previous research showed that VSELs can develop into oocyte-like cells in in vitro culture.

References

Anderson LD, Hirshfield AN (1992) An overview of follicular development in the ovary: from embryo to the fertilized ovum in vitro. Md Med J 41(7):614–620

Anderson EL, Baltus AE, Roepers-Gajadien HL, Hassold TJ, de Rooij DG, van Pelt AMM, Page DC (2008) Stra8 and its inducer, retinoic acid, regulate meiotic initiation in both spermatogenesis and oogenesis in mice. Proc Natl Acad Sci U S A 105(39):14976–14980

Antczak M, Van Blerkom J (2000) The vascular character of ovarian follicular granulosa cells: phenotypic and functional evidence for an endothelial-like cell population. Hum Reprod 15 (11):2306–2318

Barker N, van Es JH, Kuipers J, Kujala P, van den Born M, Cozijnsen M, Haegebarth A, Korving J, Begthel H, Peters PJ, Clevers H (2007) Identification of stem cells in small intestine and colon by marker gene Lgr5. Nature 449(7165):1003–1007

Bhartiya D, Sriraman K, Gunjal P, Modak H (2012) Gonadotropin treatment augments postnatal oogenesis and primordial follicle assembly in adult mouse ovaries? J Ovarian Res 5:32

Borum K (1961) Oogenesis in the mouse: a study of the meiotic prophase. Exp Cell Res 24 (3):495–507

Boulanger CA, Mack DL, Booth BW, Smith GH (2007) Interaction with the mammary microenvironment redirects spermatogenic cell fate in vivo. Proc Natl Acad Sci U S A 104 (10):3871–3876

Bowles J, Knight D, Smith C, Wilhelm D, Richman J, Mamiya S, Yashiro K, Chawengsaksophak K, Wilson MJ, Rossant J, Hamada H, Koopman P (2006) Retinoid signaling determines germ cell fate in mice. Science 312(5773):596–600

Brinster RL (2002) Germline stem cell transplantation and transgenesis. Science 296 (5576):2174–2176

Butler H, Juma MB (1970) Oogenesis in an adult prosimian. Nature 226(5245):552–553

Carlomagno G, van Bragt MPA, Korver CM, Repping S, de Rooij DG, van Pelt AMM (2010) BMP4-induced differentiation of a rat spermatogonial stem cell line causes changes in its cell adhesion properties. Biol Reprod 83(5):742–749

Castrillon DH, Quade BJ, Wang TY, Quigley C, Crum CP (2000) The human VASA gene is specifically expressed in the germ cell lineage. Proc Natl Acad Sci U S A 97(17):9585–9590

Chen H, Ge R-S, Zirkin BR (2009) Leydig cells: from stem cells to aging. Mol Cell Endocrinol 306(1–2):9–16

Clermont Y (1963) The cycle of the seminiferous epithelium in man. Am J Anat 112(1):35–51

Clermont Y (1966) Renewal of spermatogonia in man. Am J Anat 118(2):509–524

Culty M (2013) Gonocytes, from the fifties to the present: is there a reason to change the name? Biol Reprod 89(2):41–46

De Rooij DG (2009) The spermatogonial stem cell niche. Microsc Res Tech 72(8):580–585

Dzafic E, Stimpfel M, Novakovic S, Cerkovnik P, Virant-Klun I (2014) Expression of mesenchymal stem cells-related genes and plasticity of aspirated follicular cells obtained from infertile women. BioMed Res Int 2014. doi:10.1155/2014/508216

Eggan K, Jurga S, Gosden R, Min IM, Wagers AJ (2006) Ovulated oocytes in adult mice derive from non-circulating germ cells. Nature 441(7097):1109–1114

Evans MJ, Kaufman MH (1981) Establishment in culture of pluripotential cells from mouse embryos. Nature 292(5819):154–156

Feng LX, Chen Y, Dettin L, Reijo Pera RA, Herr JC, Goldberg E, Dym M (2002) Generation and in vitro differentiation of a spermatogonial cell line. Science 297(5580):392–395

Flesken-Nikitin A, Hwang CI, Cheng CY, Michurina TV, Enikolopov G, Nikitin AY (2013) Ovarian surface epithelium at the junction area contains a cancer-prone stem cell niche. Nature 495(7440):241–245

Ge RS, Dong Q, Sottas CM, Papadopoulos V, Zirkin BR, Hardy MP (2006) In search of rat stem Leydig cells: identification, isolation, and lineage-specific development. Proc Natl Acad Sci U S A 103(8):2719–2724

Generoso WM, Huff SW, Stout SK (1971a) Chemically induced dominant-lethal mutations and cell killing mouse oocytes in the advanced stages of follicular development. Mutat Res 11 (4):411–420

Generoso WM, Stout SK, Huff SW (1971b) Effects of alkylating chemicals on reproductive capacity of adult female mice. Mutat Res 13(2):171–184

Gohbara A, Katagiri K, Sato T, Kubota Y, Kagechika H, Araki Y, Araki Y, Ogawa T (2010) In vitro murine spermatogenesis in an organ culture system. Biol Reprod 83(2):261–267

Griswold SL, Behringer RR (2009) Fetal Leydig cell origin and development. Sex Dev 3(1):1–15

Guan K, Nayernia K, Maier LS, Wagner S, Dressel R, Lee JH, Nolte J, Wolf F, Li M, Engel W, Hasenfuss G (2006) Pluripotency of spermatogonial stem cells from adult mouse testis. Nature 440(7088):1199–1203

Honda A, Hirose M, Hara K, Matoba S, Inoue K, Miki H, Hiura H, Kanatsu-Shinohara M, Kanai Y, Kono T, Shinohara T, Ogura A (2007) Isolation, characterization, and in vitro and in vivo differentiation of putative thecal stem cells. Proc Natl Acad Sci U S A 104 (30):12389–12394

Honda A, Hirose M, Inoue K, Hiura H, Miki H, Ogonuki N, Sugimoto M, Abe K, Kanatsu-Shinohara M, Kono T, Shinohara T, Ogura A (2009) Large-scale production of growing oocytes in vitro from neonatal mouse ovaries. Int J Dev Biol 53(4):605–613

Hummitzsch K, Anderson RA, Wilhelm D, Wu J, Telfer EE, Russell DL, Robertson SA, Rodgers RJ (2015) Stem cells, progenitor cells, and lineage decisions in the ovary. Endocr Rev 36 (1):65–91

Ioannou JM (1967) Oogenesis in adult prosimians. J Embryol Exp Morphol 17(1):139–145

Izadyar F, Pau F, Marh J, Slepko N, Wang T, Gonzalez R, Ramos T, Howerton K, Sayre C, Silva F (2008) Generation of multipotent cell lines from a distinct population of male germ line stem cells. Reproduction 135(6):771–784

Johnson J, Canning J, Kaneko T, Pru JK, Tilly JL (2004) Germline stem cells and follicular renewal in the postnatal mammalian ovary. Nature 428(6979):145–150

Johnson J, Bagley J, Skaznik-Wikiel M, Lee H-J, Adams GB, Niikura Y, Tschudy KS, Tilly JC, Cortes ML, Forkert R, Spitzer T, Iacomini J, Scadden DT, Tilly JL (2005) Oocyte generation in adult mammalian ovaries by putative germ cells in bone marrow and peripheral blood. Cell 122 (2):303–315

Kanatsu-Shinohara M, Inoue K, Lee J, Yoshimoto M, Ogonuki N, Miki H, Baba S, Kato T, Kazuki Y, Toyokuni S, Toyoshima M, Niwa O, Oshimura M, Heike T, Nakahata T, Ishino F, Ogura A, Shinohara T (2004a) Generation of pluripotent stem cells from neonatal mouse testis. Cell 119(7):1001–1012

Kanatsu-Shinohara M, Toyokuni S, Shinohara T (2004b) CD9 is a surface marker on mouse and rat male germline stem cells. Biol Reprod 70(1):70–75

Kang JX, Wang J, Wu L, Kang ZB (2004) Transgenic mice: fat-1 mice convert n-6 to n-3 fatty acids. Nature 427(6974):504

Kossowska-Tomaszczuk K, De Geyter C, De Geyter M, Martin I, Holzgreve W, Scherberich A, Zhang H (2009) The multipotency of luteinizing granulosa cells collected from mature ovarian follicles. Stem Cells 27(1):210–219

Koubova J, Menke DB, Zhou Q, Capel B, Griswold MD, Page DC (2006) Retinoic acid regulates sex-specific timing of meiotic initiation in mice. Proc Natl Acad Sci U S A 103(8):2474–2479

Kucia M, Reca R, Campbell FR, Zuba-Surma E, Majka M, Ratajczak J, Ratajczak MZ (2006) A population of very small embryonic-like (VSEL) CXCR4+SSEA-1+Oct-4+ stem cells identified in adult bone marrow. Leukemia 20(5):857–869

Lavranos TC, Rodgers HF, Bertoncello I, Rodgers RJ (1994) Anchorage-independent culture of bovine granulosa cells: the effects of basic fibroblast growth factor and dibutyryl cAMP on cell division and differentiation. Exp Cell Res 211(2):245–251

Lee DR, Kim K-S, Yang YH, Oh HS, Lee SH, Chung TG, Cho JH, Kim HJ, Yoon TK, Cha KY (2006) Isolation of male germ stem cell-like cells from testicular tissue of non-obstructive azoospermic patients and differentiation into haploid male germ cells in vitro. Hum Reprod 21 (2):471–476

Lee YM, Kumar BM, Lee JH, Lee WJ, Kim TH, Lee SL, Ock SA, Jeon BG, Park BW, Rho GJ (2013) Characterisation and differentiation of porcine ovarian theca-derived multipotent stem cells. Vet J 197(3):761–768

Mark M, Jacobs H, Oulad-Abdelghani M, Dennefeld C, Féret B, Vernet N, Codreanu C-A, Chambon P, Ghyselinck NB (2008) STRA8-deficient spermatocytes initiate, but fail to complete, meiosis and undergo premature chromosome condensation. J Cell Sci 121 (19):3233–3242

Martin GR (1981) Isolation of a pluripotent cell line from early mouse embryos cultured in medium conditioned by teratocarcinoma stem cells. Proc Natl Acad Sci U S A 78 (12):7634–7638

Martinovitch PN (1937) Development in vitro of the mammalian gonad. Nature 139(3514):413

Matsui Y, Zsebo K, Hogan BLM (1992) Derivation of pluripotential embryonic stem cells from murine primordial germ cells in culture. Cell 70(5):841–847

Mattioli M, Gloria A, Turriani M, Berardinelli P, Russo V, Nardinocchi D, Curini V, Baratta M, Martignani E, Barboni B (2012) Osteo-regenerative potential of ovarian granulosa cells: an in vitro and in vivo study. Theriogenology 77(7):1425–1437

McLaren A (1984) Meiosis and differentiation of mouse germ cells. Symp Soc Exp Biol 38:7–23

McLaren A (2003) Primordial germ cells in the mouse. Dev Biol 262(1):1–15

Mei XX, Wang J, Wu J (2015) Extrinsic and intrinsic factors controlling spermatogonial stem cell self-renewal and differentiation. Asian J Androl 17(3):347–354

Merkwitz C, Ricken AM, Lösche A, Sakurai M, Spanel-Borowski K (2010) Progenitor cells harvested from bovine follicles become endothelial cells. Differentiation 79(4–5):203–210

Nabhan SK, Bitencourt MA, Duval M, Abecasis M, Dufour C, Boudjedir K, Rocha V, Socié G, Passweg J, Goi K, Sanders J, Snowden J, Yabe H, Pasquini R, Gluckman E (2010) Fertility recovery and pregnancy after allogeneic hematopoietic stem cell transplantation in Fanconi anemia patients. Haematologica 95(10):1783–1787

Ng A, Tan S, Singh G, Rizk P, Swathi Y, Tan TZ, Huang RY-J, Leushacke M, Barker N (2014) Lgr5 marks stem/progenitor cells in ovary and tubal epithelia. Nat Cell Biol 16(8):745–757

Ning L, Goossens E, Geens M, Van Saen D, Van Riet I, He D, Tournaye H (2010) Mouse spermatogonial stem cells obtain morphologic and functional characteristics of hematopoietic cells in vivo. Hum Reprod 25(12):3101–3109

Oki Y, Ono H, Motohashi T, Sugiura N, Nobusue H, Kano K (2012) Dedifferentiated follicular granulosa cells derived from pig ovary can transdifferentiate into osteoblasts. Biochem J 447 (2):239–248

Pacchiarotti J, Maki C, Ramos T, Marh J, Howerton K, Wong J, Pham J, Anorve S, Chow Y-C, Izadyar F (2010) Differentiation potential of germ line stem cells derived from the postnatal mouse ovary. Differentiation 79(3):159–170

Parte S, Bhartiya D, Telang J, Daithankar V, Salvi V, Zaveri K, Hinduja I (2011) Detection, characterization, and spontaneous differentiation in vitro of very small embryonic-like putative stem cells in adult mammalian ovary. Stem Cells Dev 20(8):1451–1464

Parte S, Bhartiya D, Manjramkar DD, Chauhan A, Joshi A (2013) Stimulation of ovarian stem cells by follicle stimulating hormone and basic fibroblast growth factor during cortical tissue culture. J Ovarian Res 6(1):20

Patel H, Bhartiya D, Parte S, Gunjal P, Yedurkar S, Bhatt M (2013) Follicle stimulating hormone modulates ovarian stem cells through alternately spliced receptor variant FSH-R3. J Ovarian Res 6(1):52

Reizel Y, Itzkovitz S, Adar R, Elbaz J, Jinich A, Chapal-Ilani N, Maruvka YE, Nevo N, Marx Z, Horovitz I, Wasserstrom A, Mayo A, Shur I, Benayahu D, Skorecki K, Segal E, Dekel N, Shapiro E (2012) Cell lineage analysis of the mammalian female germline. PLoS Genet 8(2): e1002477

Resnick JL, Bixler LS, Cheng L, Donovan PJ (1992) Long-term proliferation of mouse primordial germ cells in culture. Nature 359(6395):550–551

Riboldi M, Rubio C, Pellicer A, Gil-Salom M, Simón C (2012) In vitro production of haploid cells after coculture of CD49f+ with Sertoli cells from testicular sperm extraction in nonobstructive azoospermic patients. Fertil Steril 98(3):580–590.e584

Richardson BE, Lehmann R (2010) Mechanisms guiding primordial germ cell migration: strategies from different organisms. Nat Rev Mol Cell Biol 11(1):37–49

Saitou M, Yamaji M (2010) Germ cell specification in mice: signaling, transcription regulation, and epigenetic consequences. Reproduction 139(6):931–942

Santiquet N, Vallières L, Pothier F, Sirard MA, Robert C, Richard F (2012) Transplanted bone marrow cells do not provide new oocytes but rescue fertility in female mice following treatment with chemotherapeutic agents. Cell Reprogram 14(2):123–129

Sato T, Katagiri K, Gohbara A, Inoue K, Ogonuki N, Ogura A, Kubota Y, Ogawa T (2011a) In vitro production of functional sperm in cultured neonatal mouse testes. Nature 471 (7339):504–507

Sato T, Katagiri K, Yokonishi T, Kubota Y, Inoue K, Ogonuki N, Matoba S, Ogura A, Ogawa T (2011b) In vitro production of fertile sperm from murine spermatogonial stem cell lines. Nat Commun 2:472

Sato T, Yokonishi T, Komeya M, Katagiri K, Kubota Y, Matoba S, Ogonuki N, Ogura A, Yoshida S, Ogawa T (2012) Testis tissue explantation cures spermatogenic failure in c-Kit ligand mutant mice. Proc Natl Acad Sci U S A 109(42):16934–16938

Seandel M, James D, Shmelkov SV, Falciatori I, Kim J, Chavala S, Scherr DS, Zhang F, Torres R, Gale NW, Yancopoulos GD, Murphy A, Valenzuela DM, Hobbs RM, Pandolfi PP, Rafii S (2007) Generation of functional multipotent adult stem cells from GPR125 + germline progenitors. Nature 449(7160):346–350

Shamblott MJ, Axelman J, Wang S, Bugg EM, Littlefield JW, Donovan PJ, Blumenthal PD, Huggins GR, Gearhart JD (1998) Derivation of pluripotent stem cells from cultured human primordial germ cells. Proc Natl Acad Sci U S A 95(23):13726–13731

Simon L, Ekman GC, Kostereva N, Zhang Z, Hess RA, Hofmann M-C, Cooke PS (2009) Direct transdifferentiation of stem/progenitor spermatogonia into reproductive and nonreproductive tissues of all germ layers. Stem Cells 27(7):1666–1675

Stanley E, Lin C-Y, Jin S, Liu J, Sottas CM, Ge R, Zirkin BR, Chen H (2012) Identification, Proliferation, and Differentiation of Adult Leydig Stem Cells. Endocrinology 153 (10):5002–5010

Szotek PP, Chang HL, Brennand K, Fujino A, Pieretti-Vanmarcke R, Lo Celso C, Dombkowski D, Preffer F, Cohen KS, Teixeira J, Donahoe PK (2008) Normal ovarian surface epithelial label-retaining cells exhibit stem/progenitor cell characteristics. Proc Natl Acad Sci U S A 105 (34):12469–12473

Tagelenbosch RAJ, de Rooij DG (1993) A quantitative study of spermatogonial multiplication and stem cell renewal in the C3H/101 F1 hybrid mouse. Mutat Res 290(2):193–200

Takahashi K, Yamanaka S (2006) Induction of pluripotent stem cells from mouse embryonic and adult fibroblast cultures by defined factors. Cell 126(4):663–676

Takahashi K, Tanabe K, Ohnuki M, Narita M, Ichisaka T, Tomoda K, Yamanaka S (2007) Induction of pluripotent stem cells from adult human fibroblasts by defined factors. Cell 131 (5):861–872

Tanaka SS, Toyooka Y, Akasu R, Katoh-Fukui Y, Nakahara Y, Suzuki R, Yokoyama M, Noce T (2000) The mouse homolog of Drosophila Vasa is required for the development of male germ cells. Genes Dev 14(7):841–853

Thomson JA, Itskovitz-Eldor J, Shapiro SS, Waknitz MA, Swiergiel JJ, Marshall VS, Jones JM (1998) Embryonic stem cell lines derived from human blastocysts. Science 282 (5391):1145–1147

Veitia RA, Gluckman E, Fellous M, Soulier J (2007) Recovery of female fertility after chemotherapy, irradiation, and bone marrow allograft: further evidence against massive oocyte regeneration by bone marrow-derived germline stem cells. Stem Cells 25(5):1334–1335

Vincent S, Segretain D, Nishikawa S, Nishikawa SI, Sage J, Cuzin F, Rassoulzadegan M (1998) Stage-specific expression of the Kit receptor and its ligand (KL) during male gametogenesis in the mouse: a Kit-KL interaction critical for meiosis. Development 125(22):4585–4593

Virant-Klun I, Zech N, Rožman P, Vogler A, Cvjetičanin B, Klemenc P, Maličev E, Meden-Vrtovec H (2008) Putative stem cells with an embryonic character isolated from the ovarian surface epithelium of women with no naturally present follicles and oocytes. Differentiation 76 (8):843–856

Wang H, Jiang M, Bi H, Chen X, He L, Li X, Wu J (2014) Conversion of female germline stem cells from neonatal and prepubertal mice into pluripotent stem cells. J Mol Cell Biol 6 (2):164–171

White YAR, Woods DC, Takai Y, Ishihara O, Seki H, Tilly JL (2012) Oocyte formation by mitotically active germ cells purified from ovaries of reproductive-age women. Nat Med 18 (3):413–421

Wilhelm D, Palmer S, Koopman P (2007) Sex determination and gonadal development in mammals. Physiol Rev 87(1):1–28

Wu J, Okamura D, Li M, Suzuki K, Luo C, Ma L, He Y, Li Z, Benner C, Tamura I, Krause MN, Nery JR, Du T, Zhang Z, Hishida T, Takahashi Y, Aizawa E, Kim NY, Lajara J, Guillen P, Campistol JM, Esteban CR, Ross PJ, Saghatelian A, Ren B, Ecker JR, Belmonte JCI (2015) An alternative pluripotent state confers interspecies chimaeric competency. Nature 521 (7552):316–321, advance online publication

Xie W, Wang H, Wu J (2014) Similar morphological and molecular signatures shared by female and male germline stem cells. Sci Rep 4. doi:10.1038/srep05580

Yang Z, Wu J (2008) Mouse dynein axonemal intermediate chain 2: cloning and expression. DNA Cell Biol 27(9):479–488

Yang S, Ping P, Ma M, Li P, Tian R, Yang H, Liu Y, Gong Y, Zhang Z, Li Z, He Z (2014) Generation of haploid spermatids with fertilization and development capacity from human spermatogonial stem cells of cryptorchid patients. Stem Cell Rep 3(4):663–675

Yasuda Y, Konishi H, Matuso T, Tanimura T (1986) Accelerated differentiation in seminiferous tubules of fetal mice prenatally exposed to ethinyl estradiol. Anat Embryol 174(3):289–299

Yoshida S, Sukeno M, Nabeshima Y-i (2007) A vasculature-associated niche for undifferentiated spermatogonia in the mouse testis. Science 317 (5845):1722–1726

Yu J, Vodyanik MA, Smuga-Otto K, Antosiewicz-Bourget J, Frane JL, Tian S, Nie J, Jonsdottir GA, Ruotti V, Stewart R, Slukvin IL, Thomson JA (2007) Induced pluripotent stem cell lines derived from human somatic cells. Science 318:1917–1920

Zhang Y, Wu J (2009) Molecular cloning and characterization of a new gene, Oocyte-G1. J Cell Physiol 218(1):75–83

Zhang Y, Yang Z, Yang Y, Wang S, Shi L, Xie W, Sun K, Zou K, Wang L, Xiong J, Xiang J, Wu J (2011) Production of transgenic mice by random recombination of targeted genes in female germline stem cells. J Mol Cell Biol 3(2):132–141

Zhang Z, Gong Y, Guo Y, Hai Y, Yang H, Yang S, Liu Y, Ma M, Liu L, Li Z, Gao WQ, He Z (2013) Direct transdifferentiation of spermatogonial stem cells to morphological, phenotypic and functional hepatocyte-like cells via the ERK1/2 and Smad2/3 signaling pathways and the inactivation of cyclin A, cyclin B and cyclin E. Cell Commun Signal 11:67

Zhou L, Wang L, Kang JX, Xie W, Li X, Wu C, Xu B, Wu J (2014) Production of fat-1 transgenic rats using a post-natal female germline stem cell line. Mol Hum Reprod 20(3):271–281

Zou K, Yuan Z, Yang Z, Luo H, Sun K, Zhou L, Xiang J, Shi L, Yu Q, Zhang Y, Hou R, Wu J (2009) Production of offspring from a germline stem cell line derived from neonatal ovaries. Nat Cell Biol 11(5):631–636

Zou K, Hou L, Sun K, Xie W, Wu J (2011) Improved efficiency of female germline stem cell purification using fragilis-based magnetic bead sorting. Stem Cells Dev 20(12):2197–2204

Zuba-Surma EK, Kucia M, Wu W, Klich I, Lillard JW, Ratajczak J, Ratajczak MZ (2008) Very small embryonic-like stem cells are present in adult murine organs: ImageStream-based morphological analysis and distribution studies. Cytometry A 73A(12):1116–1127

Zuckerman S (1951) The number of oocytes in the mature ovary. Recent Prog Horm Res 6:63–109

Chapter 12
A Role of MicroRNAs in Cell Differentiation During Gonad Development

Hadas Grossman and Ruth Shalgi

Abstract MicroRNAs (miRNAs) are a group of small noncoding RNA molecules that play a major role in posttranscriptional regulation of gene expression and are expressed in an organ-specific manner. One miRNA can potentially regulate the expression of several genes, depending on cell type and differentiation stage. miRNAs are differentially expressed in the male and female gonads and have an organ-specific reproductive function. Exerting their affect through germ cells and gonadal somatic cells, miRNAs regulate key proteins necessary for gonad development. The role of miRNAs in the testes is only starting to emerge though they have been shown to be required for adequate spermatogenesis. Widely explored in the ovary, miRNAs were suggested to play a fundamental role in follicles' assembly, growth, differentiation, and ovulation. In this chapter, we focus on data obtained from mice in which distinct proteins that participate in the biosynthesis of miRNAs were conditionally knocked out from germ cells (spermatogonial cells or oocytes) or gonadal somatic cells (Sertoli or granulosa cells). We detail recent advances in identification of particular miRNAs and their significance in the development and function of male and female gonads. miRNAs can serve as biomarkers and therapeutic agents of pathological conditions; thus, elucidating the branched and complex network of reproduction-related miRNAs will aid understanding of gonads' physiology and managing reproduction disorders.

12.1 miRNAs Biogenesis

MicroRNAs (miRNAs) are a group of small noncoding RNA sequences, ranging in size from 18 to 25 nucleotides (nt) and playing a major role in posttranscriptional gene expression regulation by several mechanisms. There are two distinct subclasses of miRNAs, which differ by their biogenesis: the canonical and the noncanonical pathways. Canonical miRNAs are initially transcribed as long primary RNA transcripts (pri-miRNAs) that fold to form hairpin structure (Quick-

H. Grossman • R. Shalgi (✉)
Department of Cell Biology and Development, Tel Aviv University, Ramat Aviv, Israel
e-mail: shalgir@tauex.tau.ac.il

R.P. Piprek (ed.), *Molecular Mechanisms of Cell Differentiation in Gonad Development*, Results and Problems in Cell Differentiation 58,
DOI 10.1007/978-3-319-31973-5_12

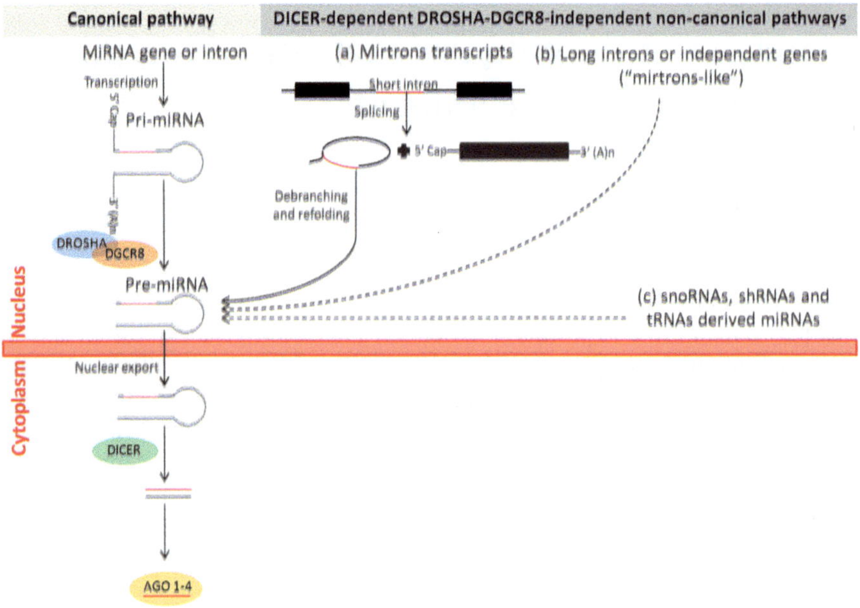

Fig. 12.1 Biogenesis of miRNAs. *Canonical miRNAs pathway*: canonical miRNAs are transcribed in the nucleus from a miRNA gene or intron to long primary transcripts (Pri-miRNAs) that form a hairpin structure. Pri-miRNAs are processed to 60–70 nt precursor miRNAs (pre-miRNAs) by the DROSHA-DGCR8 complex. The Pre-miRNAs are exported to the cytoplasm where they are cleaved by DICER to double-stranded mature miRNAs, which are then loaded onto Argonaute proteins (AGO1–4). *Noncanonical miRNAs pathways*: these include DICER-independent and DROSHA-DGCR8-dependent as well as DICER-dependent and DROSHA-DGCR8-independent pathways. The latter constitute the majority of the noncanonical miRNAs and include several distinct pathways: (a) Mirtrons are transcribed from short introns, spliced and debranched into a product that is recognized by Dicer, hence bypassing Drosha cleavage. (b) Mirtron-like miRNAs are transcribed from long introns or independent genes. (c) snoRNAs, shRNAs, and tRNAs-derived miRNAs require the activity of Dicer but are independent of DROSHA-DGCR8

Cleveland et al. 2014). Pri-miRNAs are recognized by DGCR8 (DiGeorge Syndrome Critical Region 8), a double-strand RNA binding protein that directs DROSHA, an RNase III enzyme, to cleave the pri-miRNA; the resulting product is a short stem-loop precursor miRNA (pre-miRNA; Nguyen et al. 2015). The pre-miRNA is then transported to the cytoplasm where it is cleaved by the RNase III enzyme DICER to a double-stranded mature miRNA (Fig. 12.1; reviewed by Ha and Kim 2014). Biogenesis of noncanonical miRNAs includes several pathways that bypass components of the machinery used for the synthesis of canonical miRNAs (Fig. 12.1; Yang and Lai 2011). Whereas most of the noncanonical miRNAs are DICER dependent and DROSHA or DGCR8 independent (Chong et al. 2010; Ender et al. 2008; Ruby et al. 2007), some are DICER independent (Cheloufi et al. 2010; Havens et al. 2012).

Generally, once a mature miRNA duplex is formed, one strand is selected and incorporated into one of the four Argonaute (AGO) proteins, forming the core of the RNA-induced silencing complex (RISC; Gregory et al. 2005; Hutvagner and Simard 2008). Within the RISC complex, the miRNA binds to its target mRNA by sequence complementarity between the seed sequence of the miRNA (nt 2–8) and the 3′UTR of the target mRNA. Inhibition of target mRNA expression can be executed by deadenylation of mRNA leading to its degradation, by mRNA cleavage, by translation inhibition, or by localization to sub-cytoplasmic compartments, known as P-bodies, where the mRNA can be either stored or degraded (reviewed by Stefani and Slack 2008; Inui et al. 2010). Based on current database information, there are more than 3000 miRNAs in the mouse genome; many of them are expressed in the gonadal somatic and germ cells and are assumed to have key roles in the development and physiological function of both female and male gonads.

12.2 miRNAs and Sex Determination

Expression of miRNAs is known to be diversely regulated in different organs during various stages of differentiation. Mishima et al. (2008) used cloning analysis to demonstrate different expression profiles of miRNAs in adult ovaries and testes, indicating that miRNAs have organ-specific reproductive functions. Comparison between prenatal XY and XX gonads, during the period of sex determination, revealed characteristic miRNA signatures in the developing ovaries and testes, implying that miRNAs play a role in the development of female and male gonads, already at early stages (Rakoczy et al. 2013).

The development of a gonad into an ovary or testis is dependent on the presence or absence of the *Sry* gene (located on Y chromosome), which is expressed in pre-Sertoli cells of 10.5–12.5 days *post coitum* (*dpc*) mouse embryos (Ewen and Koopman 2010). A critical target of SRY is the SRY-related HMG-box gene 9 transcription factor, known as SOX9 (reviewed by Ungewitter and Yao 2012), which is expressed at low levels in the developing testes and ovaries. Once the expression of SRY is upregulated, the expression of SOX9 in the testes increases, and in the ovaries it becomes undetectable (Kobayashi et al. 2005).

Ectopic expression of SOX9 in mice XX gonads leads to the development of XX testes (Vidal et al. 2001), whereas lack of SOX9 in XY mice leads to the development of XY ovaries (Barrionuevo et al. 2006). Recently, SOX9 was shown to regulate the expression of miR-202-5p and miR-202-3p, two sexually dimorphic miRNAs that are predominantly expressed in the somatic cells of XY gonads at the time of sex determination (Wainwright et al. 2013). The expression of miR-124 is similar in 11.5 *dpc* (prior to sexual differentiation) XY and XX gonads and increases dramatically only in XX gonads at 13.5 *dpc* (post sex determination). Knockdown of miR-124 in 13.5 *dpc* embryonic XX gonads resulted in elevated

SOX9 expression, implying that miR-124 have a role in downregulation of SOX9 in XX gonad from the stage of sex determination onward (Real et al. 2013).

12.3 miRNAs in the Testis

Of the many miRNAs present in mouse testes, some are preferential or are exclusively expressed in the testes (Ro et al. 2007). Furthermore, testicular miRNAs are differentially expressed during mouse spermatogenesis (Buchold et al. 2010; Ro et al. 2007; Sree et al. 2014; Yan et al. 2007). The following chapter details data obtained from experiments in models of knockout (KO) mice in which components of the miRNA biogenesis machinery have been conditionally knocked out (cKO) from either Sertoli cells or spermatogonial cells. The role of specific miRNAs in the testes function will be briefly summarized.

12.3.1 miRNAs in Sertoli Cell

MiRNAs hold a great potential in regulating physiological and pathological processes in the male reproductive system (Kotaja 2014) and are highly abundant in Sertoli cells, compared to other noncoding RNAs (Tan et al. 2014). Nevertheless, their role in Sertoli cells is only starting to emerge. The expression of many miRNAs in adult Sertoli cells is testosterone dependent; the expression of some is differentially regulated during the first wave of spermatogenesis (Panneerdoss et al. 2012), indicating that androgens regulate spermatogenesis through a miRNAs-dependent mechanism.

KO mouse model was used to gain insight into the involvement of miRNAs in the function of Sertoli cells and their contribution to spermatogenesis. According to this model, *Dicer1* is conditionally deleted from Sertoli cells using the Cre-loxP recombination system (Schwenk et al. 1995) under regulation of the *Anti-Müllerian hormone* (*Amh*) promoter (Kim et al. 2010a; Papaioannou et al. 2009). Because DICER is essential for the biogenesis of canonical and most noncanonical miRNAs, *Dicer1* cKO cells exhibit dramatic changes in their miRNAs' profile (reviewed by Abdelfattah and Choi 2015). Papaioannou et al. (2009) found that deletion of *Dicer1* from Sertoli cells caused a drastic reduction in mice testes size already 15 days *postpartum* (*dpp*), leading to infertility. In testes of 15 *dpp* mice, a significant proportion of both Sertoli and spermatogonial cells have undergone apoptosis, inferring that miRNAs of Sertoli cells are required for spermatogenesis (Papaioannou et al. 2009). These findings are in agreement with another study that showed an increased rate of apoptosis in Sertoli cells and spermatogonial cells, in the testes of *Dicer1* cKO mice, already at 3 *dpp* (Kim et al. 2010a). Whereas morphological changes caused by deletion of *Dicer1* were observed in testes of 3 *dpp* mice, transcriptome analysis of testes from *Dicer1* cKO mice indicated alterations in gene expression already at 0 *dpp* (Kim et al. 2010a). The proteomic profile of *Dicer1* cKO mice was also dramatically altered already at the time of birth (Papaioannou et al. 2011).

12.3.2 miRNAs in Spermatogonial Cells

As mentioned, miRNAs are differentially expressed during mouse spermatogenesis (Buchold et al. 2010; Ro et al. 2007; Sree et al. 2014; Yan et al. 2007). Interestingly, miRNAs that are differentially expressed during spermatogenesis have a nonrandomized chromosomal distribution; an enrichment of miRNAs derived from chromosome 12 was found in testes of 7 *dpp* old mice and from chromosomes 2 and X in testes of 14 *dpp* old mice (Buchold et al. 2010). This was further supported by Sree *et al.* who found dramatic changes in the expression of miRNAs in testes of 8 and 16 *dpp* old mice, with many of the differentially expressed miRNAs mapped to X chromosome (Sree et al. 2014). This chromosomal localization of many of the miRNAs that are preferentially expressed in testes at the time of meiotic sex chromosome inactivation (MSCI) is somewhat surprising. However, a significant percentage of the X-linked miRNAs is transcribed in pachytene spermatocytes (12–14 *dpp*), in which MSCI is ongoing. Of these, some miRNAs remained upregulated in round spermatids, indicating that they also escape post-meiotic sex chromatin (PMSC), thus implying that X-linked miRNAs can escape MSCI and PMSC and may have a functional role in spermatogenesis (Song et al. 2009).

As in Sertoli cells, the general role of miRNAs in the spermatogonial cells was initially evaluated using cKO mice in which *Dicer1* was deleted from spermatogonial cells using the Cre-loxP recombination system (Schwenk et al. 1995). However, as opposed to Sertoli cells, several models have been employed to conditionally delete *Dicer1* from spermatogonial cells, by utilizing different promoters to regulate the expression of Cre recombinase (reviewed by Kotaja 2014). Accordingly, deletion of *Dicer1* was achieved at different stages of spermatogenesis, yielding various phenotypic results. Using the *Tissue nonspecific alkaline phosphatase* (*Tnap*) promoter, Cre recombinase is preferentially expressed in primordial germ cells (PGCs) from 10 *dpc* onward (Lomelí et al. 2000). Testes of 12.5 *dpc Dicer1* cKO mice had a decreased number of germ cells due to attenuated PGCs proliferation (Hayashi et al. 2008); their differentiated elongated spermatids exhibited abnormal morphology (Maatouk et al. 2008). When *Vasa/Ddx4* promoter is used to drive the expression of Cre recombinase, deletion of *Dicer1* begins at 12.5 *dpc* and reaches ~90 % by 18 *dpc* (Gallardo et al. 2007). These *Dicer1* cKO mice are non-fertile and exhibit decreased-size testes and abnormal structure of the seminiferous tubules, due to increased apoptosis that started around 15 *dpp* and impaired spermiogenesis (Liu et al. 2012; Romero et al. 2011; Zimmermann et al. 2014). To evaluate whether the time of *Dicer1* deletion has an impact on mice fertility, Liu et al. used two promoters: *Stimulated by retinoic acid 8* (*Stra8*) and *Phosphoglycerate kinase 2* (*Pgk2*) to drive the expression of Cre recombinase. Deletion of *Dicer1* using *Stra8* promoter occurred at 3 *dpp*; the cKO male mice were sub-fertile and exhibited phenotypes similar to *Vasa/Ddx4*-mediated *Dicer1*-deleted mice, though to a lesser degree (Liu et al. 2012). However, using a similar method (*Stra8* promoter), other male cKO mice were completely non-fertile (Greenlee et al. 2012; Wu et al. 2012). On the contrary to the *Stra8* promoter, deletion of *Dicer1* is achieved only in spermatocytes and spermatids using the *Pgk2* promoter and these cKO mice had no prominent phenotype (Liu et al. 2012). Selective deletion of *Dicer1* using *neurogenin 3* (*Ngn3*) promoter takes place in spermatogonial cell only from 5 *dpp* onward, causing defects in relatively progressive stages of spermatogenesis, indicated by normal appearance of early meiotic cells, defects in differentiation of haploid cells, and increased apoptosis (Korhonen et al. 2011). Chang et al. used the *protamine 1* (*Prm1*) promoter to drive the expression of Cre recombinase (O'Gorman et al. 1997), causing deletion of *Dicer1* at 18 *dpp* when the first wave of round spermatids appears (Bellvé et al. 1977). According to this model, male cKO mice exhibited increased apoptosis of round spermatids, various sperms morphological abnormalities, and improper chromatin compaction and spermiation failure. Surprisingly, the fertility of cKO males was not completely abolished; they sired fewer litters with reduced litter sizes (Chang et al. 2012). Taken together, it appears that DICER is required throughout male germ cell differentiation. The diverse severity of the different cKO mice model phenotypes may be attributed both by the time in which deletion of *Dicer1* is achieved and by incomplete deletion of *Dicer1*.

DICER is required for the biogenesis of miRNAs and of small-interference RNAs (siRNAs; reviewed by Carthew and Sontheimer 2009). To distinguish between the contribution of miRNAs and siRNAs to the phenotype of *Dicer1* cKO mice, several models have been employed in which different proteins, specific for the biogenesis of miRNAs, have been conditionally knocked out in spermatogonial cells (reviewed by Kotaja 2014). Deletion of *Drosha* interfered with processing of pri-miRNAs into mature miRNAs, resulting in a decreased expression of canonical miRNAs but retaining an intact siRNA pathway (Abdelfattah and Choi 2015). Specific deletion of *Drosha* in spermatogonial cells, using *Stra8* promoter to drive the expression of Cre recombinase, resulted in male sterility due to disrupted spermatogenesis (Wu et al. 2012). Interestingly, *Drosha* cKO mice exhibited more severe spermatogenic disruptions than *Stra8*-mediated *Dicer1* cKO mice. miRNAs sequencing in pachytene spermatocytes and round spermatids identified several upregulated miRNAs in *Drosha* cKO cells that were downregulated or absent in *Dicer1* cKO cells, and vice versa, representing distinct subgroups of noncanonical miRNAs (Wu et al. 2012). DROSHA needs to form a complex with DGCR8 for successfully processing pri-miRNAs into mature miRNAs (Nguyen et al. 2015). Zimmermann et al. compared the phenotype of *Dicer1* cKO mice to the phenotype of *Dgcr8* cKO mice, using *Vasa/Ddx4* promoter to drive the expression of Cre recombinanse. cKO mice of both phenotypes displayed similar defects, though the *Dgcr8* phenotype was milder (Zimmermann et al. 2014). These findings reinforce the importance of miRNAs in the developing testes and emphasize the role of siRNAs. When *Dicer1* or *Dgcr8* were conditionally deleted by the *Vasa/Ddx4* promoter, cKO spermatocytes exhibited drastic alterations in the morphology of sex chromosomes and failed to progress through meiosis (Modzelewski et al. 2015). This phenotype is likely to be a result of dysregulation of Ataxia telangiectasia mutated (ATM) kinase that possesses many potential sites for miRNA binding. Of these, miR-18a/b, miR-183, and miR-16 are all expressed in spermatocytes and are depleted in *Dgcr8* and *Dicer* cKO cells (Modzelewski et al. 2015). miR-18a/b, miR-183, and miR-16 are differentially expressed during the first wave of spermatogenesis, though they constitute only a small percentage of the miRNAs expressed in spermatogonial cells during this period (Buchold et al. 2010). The chromosomal localization of these miRNAs is heterogenic and only miR-18b is mapped to X chromosome (Buchold et al. 2010). Furthermore, miR-18a and miR-18b are part of the miR-17/92 and miR-106a/363 clusters, respectively (Mogilyansky and Rigoutsos 2013). The first cluster is highly expressed in PGCs and spermatogonia cells (Hayashi et al. 2008) and is downregulated during retinoic acid (RA)-induced spermatogonial cell differentiation (Tong et al. 2012).

In pachytene spermatocytes, AGO4 is localized to sites of asynapsis and the transcriptionally silenced X and Y chromosome subdomain (sex body; Modzelewski et al. 2012), thus adding another level of complexity to the relationship between miRNAs and spermatogonial cell sex chromosomes. The morphology of the sex body in *Ago4*-null male mice was abnormal, leading to influx of RNA Polymerase II and failure in MSCI, indicated by expression of many sex-linked mRNA transcripts. The expression of many miRNAs was also reduced in *Ago4*-null spermatocytes, a reduction that was not consistent among different miRNA families, indicating some degree of miRNA specificity for AGO4. Interestingly, the expression of all miRNAs encoded on X chromosomes was significantly decreased (Modzelewski et al. 2012).

Of the AGO family of proteins, AGO4 and AGO3 are highly expressed in the testes whereas AGO1 and AGO2 are more uniformly expressed in meiotic and non-meiotic tissues (Modzelewski et al. 2012). Some of the functions of mammalian AGO proteins in miRNAs' silencing overlap, implying a possible compensation among them (Su et al. 2009). Thus, it is not surprising that the expression of AGO3 was upregulated in testes of *Ago4*-null mice (Modzelewski et al. 2012). To date, AGO2 is the only member of the AGO family of proteins reported to possess a slicer catalytic activity, in both mouse (Cifuentes et al. 2010) and human (Meister et al. 2004). Furthermore, though all four AGO proteins can contribute to repression of siRNAs targets, AGO2 is more effective in siRNA-mediated gene silencing (Wu et al. 2008). In contrast, the contribution of AGO2 to repression of miRNAs targets is similar to that of the other three AGO proteins (Ma et al. 2014). Spermatogenesis was not hampered in cKO mice in which *Ago2* was deleted from germ cells, using *TNAP* promoter to drive the expression of Cre recombinase (Hayashi et al. 2008). However, as explained above, redundancy of AGO1–4 can account for the lack of phenotype in *Ago2* cKO mice.

12.4 Specific miRNAs and Their Contribution to Testes Functionality

Several miRNAs have been implicated to regulate spermatogonial stem cells' (SSCs) self-renewal; among them are miR-21, miR-20, and miR-106a, which are preferentially expressed in SSCs than in testicular somatic cells (He et al. 2013; Niu et al. 2011). Inhibition of miR-21 increased the number of germ cells undergoing apoptosis (Niu et al. 2011), whereas overexpression of miR-20 or miR-106a increased the proliferation rate of SSCs (He et al. 2013).

Adult mice SSCs undergo a series of mitotic divisions, producing undifferentiated spermatogonia cells (de Rooij and Russell 2000). RA regulates initiation of spermatogonia cell differentiation (Snyder et al. 2010) as well as entry into meiosis (Barrios et al. 2010). Exposure of spermatogonia to RA causes alteration in the expression of miRNAs, including the miR-17/92 and miR-106a/363 clusters (Tong et al. 2012), miR-221/222 cluster (Yang et al. 2013a), and miR-146 (Huszar and Payne 2013)—all are downregulated by RA and necessary for maintaining spermatogonia cells in an undifferentiated state. In contrast, RA increases the expression of members of the let-7 family miRNAs in spermatogonia cells, through suppression of LIN28 (Tong et al. 2011).

Overexpression of miR-184 promoted spermatogonia cell proliferation and caused a decrease in the expression of Nuclear receptor corepressor 2 (NCOR2) by targeting its 3′UTR region. miR-184 is also detected through the stages of primary and secondary spermatocytes and till the stage of round spermatids (Wu et al. 2011). The miR-449 cluster is upregulated in the testes upon meiotic initiation, as well as in testes of adult mice. In situ hybridization indicated that this cluster is expressed specifically in spermatocytes and spermatids, implying a possible role of the miR-449 cluster in initiation of meiosis and spermiogenesis (Bao et al. 2012), though mice lacking the miR-449 cluster exhibit normal spermatogenesis and fertility. This conflict could be explained by a functional redundancy of miR-449 cluster and miR-34b/c, which shares a cohort of target genes with the miR-449 cluster and possesses a similar expression pattern during male germ cell development (Bao et al. 2012).

Ample data exist concerning the importance of specific miRNAs during late stages of spermiogenesis. Conversion of nucleosomal chromatin to compact non-nucleosomal chromatin in elongating spermatids requires time-restricted regulation of the expression of "chromatin packing proteins," including protamines (PRM1, PRM2; Miller et al. 2010) and transition proteins (TNP1, TNP2; Adham et al. 2001; Meistrich et al. 2003). Spermatozoa of Heat shock transcription factor 2 (HSF2)-deficient mice have an increased rate of mild head abnormalities and altered levels of PRM1, PRM2, and TNP2 (Åkerfelt et al. 2008). The expression of HSF2 in the testes is regulated post-transcriptionally by miR-18 that belongs to the miR-17/92 cluster. Whereas miR-18 is expressed predominantly in spermatocytes, HSF2 is highly expressed in spermatogonia, decreases in spermatocytes, and reappears in round and elongating spermatids, while the level of miR-18 decreases

(Björk et al. 2010). Thus, miR-18 may have a role in regulating the time-restricted expression of "chromatin packing proteins" through HSF2. Other miRNAs are known to regulate the expression of "chromatin packing proteins": miR-469, which binds to the coding region of TNP2 and PRM2 mRNAs and inhibits translation (Dai et al. 2011), and miR-122a which binds to the 3′UTR region of TNP2, causing cleavage of the mRNA (Yu et al. 2005).

12.5 miRNAs in the Ovary

The general involvement of miRNAs in the ovary was studied using mice with cKO in key proteins that participate in the biogenesis of miRNAs, either in granulosa (follicular) cells or in oocytes. In the following section, we detail data obtained from these KO models and evaluate the comprehensive contribution of miRNAs to ovarian functionality. The role of specific miRNAs in the ovary will be discussed later.

12.5.1 miRNAs in Granulosa Cells

The possible role of miRNAs in the ovary was demonstrated by cKO of *Dicer1* from granulosa cells using the *Anti-Müllerian hormone receptor type 2* promoter to drive the expression of Cre recombinase (Jamin et al. 2002). A study by Nagaraja et al. (2008) demonstrated that folliculogenesis was not hampered in ovaries of adult *Dicer1* cKO mice, with the exception of increased follicular atresia. Upon controlled super-ovulation, *Dicer1* cKO mice yielded half the number of oocytes ovulated by control mice and histological examination of the ovaries showed oocytes trapped within luteinized follicles. When mated with wild-type males, fewer fertilized oocytes progressed to the two-cell stage, suggesting that the expression of *Dicer1* in granulosa cells affects also oocyte quality indirectly (Nagaraja et al. 2008). These findings were supported by other studies that showed increased follicular atresia (Lei et al. 2010) and lower ovulation rate (Hong et al. 2008) in *Dicer1* cKO mice. However, because DICER does not participate only in the biogenesis of miRNAs but also in the biogenesis of siRNAs, it cannot be determined whether miRNAs are solely responsible for this phenotype. Study by Otsuka et al. (2008) indicated that global reduction of *Dicer1* expression by a hypomorphic mutation causes infertility due to defects in ovarian angiogenesis and corpus luteum (CL) insufficiency. It is likely that these effects are miRNA dependent because in vivo overexpression of miR-17-5p and let-7b allowed partial vascular recovery in the CL of these *Dicer1* hypomorphic mice. Other ovarian functions such as folliculogenesis, oocyte maturation, and ovulation were not hampered. However, the investigators stated that it cannot be excluded that miRNAs regulate other ovarian functions, but the development and function of the CL appear to be more sensitive to the reduced expression of DICER (Otsuka et al. 2008).

12.5.2 miRNAs in Oocytes

To explore the role of miRNAs in oocytes, several models have been employed (see hereinafter). In cKO mouse model in which *Dicer1* was deleted, using the *zona pellucida glycoprotein 3* (*Zp3*) promoter to drive the expression of Cre recombinase, oocyte growth and development were not hampered (Murchison et al. 2007). However, the majority of *Dicer1* cKO oocytes did not complete the first meiotic division, either in vivo or in vitro, and were arrested at the first metaphase (MI). Oocytes that completed the first meiotic division and extruded the first polar body (PBI) frequently exhibited multiple spindles with misaligned chromosomes, a phenotype that was also observed in *Dicer1* cKO MI oocytes (Murchison et al. 2007). Microarrays revealed differences in the expression levels of mRNA transcripts from *Dicer1* cKO oocytes and control oocytes matured in vitro. A strong correlation was found between the mRNA transcripts whose level was increased in *Dicer1* cKO oocytes to mRNAs that are selectively degraded during meiotic maturation of wild-type oocytes (Murchison et al. 2007; Su et al. 2007). Furthermore, several putative miRNA target sites were identified within the upregulated mRNA transcripts, though many were not related to any known mouse miRNAs (Murchison et al. 2007). This was further supported by another study that showed an association between miRNA binding sites and specific transcript isoforms during the transition of oocytes from germinal vesicle (GV) stage to metaphase of the second meiotic division (MII) and degradation of maternal mRNA (Salisbury et al. 2009).

A similar method (*ZP3* promoter-mediated expression of Cre recombinase) was used to generate *Ago2* cKO mouse oocytes (Kaneda et al. 2009). These oocytes developed normally and matured in vivo. However, when ovulated, they exhibited abnormal spindles and their chromosomes were misaligned, similar to the phenotype of *Dicer1* cKO oocytes. Moreover, their mRNA expression profile was radically different than that of WT oocytes, concomitant with a decrease in the expression of most miRNAs, by more than 80 % (Kaneda et al. 2009). Like DICER, AGO2 is also common to the miRNA and siRNA biogenesis pathways. Hence, the phenotype of *Dicer1*- or *Ago2*-deficient oocytes can also be attributed to differences in the expression of siRNAs.

To distinguish between the role of miRNAs and siRNAs in mouse oocytes, Suh et al. (2010) used the *ZP3* promoter–Cre recombinase system to generate cKO mice with deleted *Dgcr8*, which is specific to the biogenesis of canonical miRNAs. Surprisingly, though the levels of their miRNAs were decreased, the oocytes matured normally and produced healthy-appearing offspring when fertilized in vitro with wild-type sperm. The deletion of *Dgcr8* had a negative impact on fecundity because *Dgcr8*-deficient mothers produced fewer offspring than control mothers. The investigators did not detect prominent changes in the mRNA expression profile between *Dgcr8* cKO oocytes and control oocytes (Suh et al. 2010), unlike the case in *Dicer1* cKO and *Ago2* cKO oocytes (Kaneda et al. 2009; Murchison et al. 2007). The lack of a profound phenotype of *Dgcr8* cKO oocytes may be partially explained by the presence of noncanonical miRNAs, specifically those that are DGCR8 independent. One of these noncanonical pathways, the mirtrons (short RNA duplexes mapped to short hairpin introns), is DICER dependent and DGCR8 independent (reviewed by Yang and Lai 2011), meaning that lack of mirtrons in *Dicer1* cKO oocytes and its presence in the *Dgcr8* cKO oocytes may be held responsible for the differences between the phenotypes of the *Dicer1* and *Dgcr8* cKO oocytes. To exclude this possibility, the contribution of mirtrons to the changes observed in mRNA transcripts of *Dicer1* KO was evaluated (Suh et al. 2010). However, no association was found between two examined mirtrons and mRNA transcripts that are upregulated in *Dicer1* KO oocytes. In contrast, siRNAs were mapped to many of the upregulated transcripts (Suh et al. 2010).

These findings imply that the phenotype of *Dicer1* cKO oocytes is a result of decreased expression of siRNAs rather than of miRNAs. However, several aspects need to be considered. Deletion of *Dgcr8* by *ZP3* promoter that drives the expression of Cre recombinase occurs only when the oocytes start growing (Epifano et al. 1995), leaving a window for synthesis of miRNAs that are necessary for their further growth and maturation. Moreover, several novel subgroups of noncanonical miRNAs that require the activity of DICER but not that of DGCR8 were discovered recently (Fig. 12.1). One biogenesis pathway shares similarity with the mirtron pathway, though distinguished from them as these noncanonical miRNAs are transcribed from long noncoding introns or independent genes, rather than short introns (Chong et al. 2010). Other pathways include miRNAs derived from small-nuclear RNAs (snoRNAs), endogenous short-hairpin RNAs (shRNAs), and tRNAs (reviewed by Miyoshi et al. 2010). These noncanonical miRNAs may contribute to the differences between the phenotype of the *Dicer1* and *Dgcr8* cKO oocytes.

It seems that when evaluating the activity of oocytes miRNAs, it is of great importance to distinguish between growing and fully grown GV oocytes. Ma et al. (2010) used the luciferase reporter assay to examine the activity of two endogenous miRNAs, let-7a and miR-30c, which are abundant in the oocyte (Tang et al. 2007). Several reporter constructs were used, each containing 2–4 bulged miRNA binding sites or one perfect complementary site (Ma et al. 2010). In growing oocytes, both let-7a and miR-30c suppressed the expression of the reporter luciferase, when under the regulation of either the bulged sites or the perfect complementary site. In fully grown oocytes, only miR-30c inhibited the expression of both bulged and perfect site reporters, whereas let-7a suppressed only the expression of the perfect site reporter (Ma et al. 2010). These findings imply that miRNA activity is slightly suppressed in fully grown GV oocytes, compared to growing oocytes, though miRNAs remain functional in fully grown GV oocytes.

Growing and fully grown GV oocytes differ in the lack of P-bodies within fully grown oocytes. P-bodies are involved in RNA-mediated silencing and translational inhibition. Liu et al. (2005) used reporter mRNAs that contain binding sites for endogenous and exogenous miRNAs and demonstrated that localization of mRNAs to P-bodies is miRNA dependent. It was further shown that the AGO family of proteins, including AGO2, are localized to P-bodies (Liu et al. 2005). Flemr et al. (2010) showed that P-bodies disperse as the oocytes grow, concomitantly with the formation of mRNA-containing aggregates in the subcortical zone of fully grown oocytes. These mRNA-containing aggregates do not contain AGO2 protein and disappear in mature MII oocytes (Flemr et al. 2010). However, Swetloff et al. (2009) detected P-body-like foci in fully grown mouse oocytes and their presence was sensitive to depletion of *Dicer1*, implying a possible occurrence of miRNA-mediated gene silencing in the P-body-like foci of fully grown GV oocytes. Like the abovementioned mRNA-containing aggregates, these P-body-like foci also disappeared in mature oocytes. Taken together, alterations in the localization of mRNAs, miRNAs, and RNA-binding proteins may partially explain the differences in the activity of miRNAs found between growing and fully grown mouse oocytes.

The activity of miRNAs is affected also by editing of adenosine to inosine (A–I), which can cause a decreased activity, retargeting of miRNAs, or degradation of pre-miRNA (reviewed by Nishikura 2006). García-López et al. (2013) examined the editing profile of several miRNAs that are expressed in mouse oocytes and early embryos and are edited in other cell types. The majority of these miRNAs were unedited, with only few mature edited forms expressed in oocytes and zygotes. On the contrary, edited miRNAs are relatively highly expressed in two-cell embryos and remain so in blastocysts (García-López et al. 2013). The susceptibility of other miRNAs expressed in fully grown oocytes to A–I editing needs to be explored in order to evaluate the possible role of A–I editing in regulating miRNA activity in oocytes.

Adenylation is also involved in the regulation of miRNA activity and stability, though the frequency of miRNA adenylation is usually very low (reviewed by Kim et al. 2010b). Lee et al. (2014) found that maternal miRNAs are adenylated in mouse MII oocytes, reaching a total of ~30 % and dropping down to 5 % in 8–16 cell embryos. However, the rate of adenylated miRNAs in growing and fully grown GV oocytes is yet to be determined.

12.6 Specific miRNAs and Their Contribution to Ovarian Functionality

Lately, a progress has been made in identifying the role of specific miRNAs and their target genes in the ovary, specifically in granulosa cells and oocytes. Herein, we discuss data regarding the involvement of these miRNAs in distinct processes occurring throughout folliculogenesis, oocyte maturation, and ovulation.

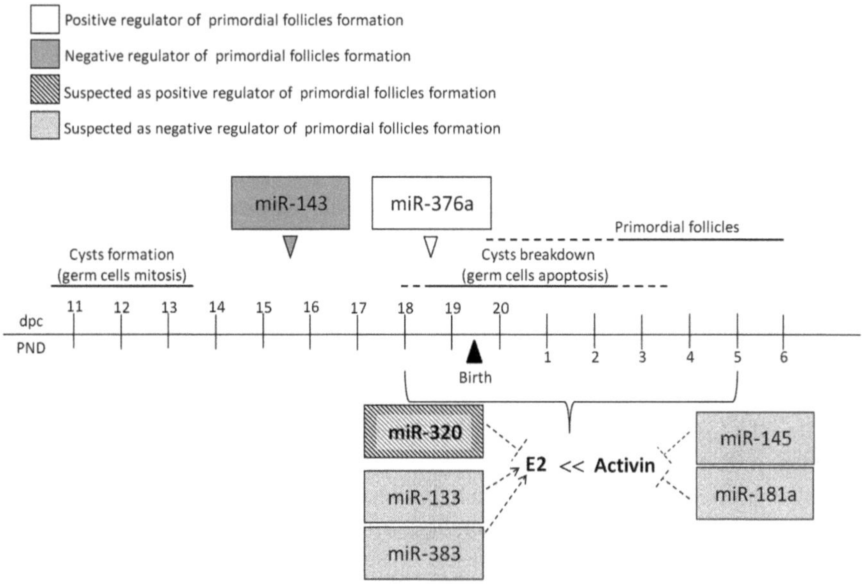

Fig. 12.2 MiRNA regulation of primordial follicle formation in the mouse. A scheme illustrating events that occur during early stages of mouse ovarian development is presented. Germline cysts are formed between 10.5 and 13.5 *dpc*, accompanied by multiple mitotic divisions. Around 13.5 *dpc*, first oocytes enter first meiosis and arrest at the diplotene stage between 17.5 *dpc* and 5 *dpp*. During this time (17.5 *dpc*–5 *dpp*), breakdown of germline cysts occurs followed by primordial follicle assembly. Overexpression of miR-143 in 15.5 *dpc* ovaries inhibits the formation of primordial follicles by suppressing proliferation of pre-granulosa cells. Ovaries of 18.5 *dpc* mice transfected with miR-376a have an increased number of primordial follicles. Several miRNAs modulate the secretion of E2, an inhibitor of primordial follicle formation (miR-320, miR-133, miR-383), or the activity of activin, an inducer of primordial follicle formation (miR-145, miR-181a)

12.6.1 Assembly of Primordial Follicles

Mouse primordial germ cells migrate to the genital ridge and undergo several mitotic divisions, forming clusters of oogonia, or what is termed cysts. At 13.5 *dpc*, oogonia enter the first meiosis and from then on are referred to as oocytes (reviewed by Pepling 2012). Oocytes will further progress through meiosis and arrest at the diplotene stage of the first prophase. This arrest takes place between 17.5 *dpc* and 5 *dpp* when the last oocytes reach the diplotene stage (reviewed by Pepling 2012). During this period approximately two-thirds of mouse oocytes will die, accompanied by cysts breakdown. The surviving oocytes become enclosed within a single layer of flattened pre-granulosa cells, forming the primordial follicles (Pepling and Spradling 2001; Fig. 12.2).

It is believed that apoptosis is the major mechanism involved in the loss of oocytes during cyst breakdown and primordial follicle assembly (reviewed by Tiwari et al. 2015). Xu et al. (2011) showed that proliferating cell nuclear antigen

(PCNA) is differentially expressed in oocytes during the critical time of cyst breakdown and primordial follicle assembly. The expression of PCNA decreases from 13.5 to 18.5 *dpc* and increases from 18.5 *dpc* to 5 *dpp*. RNA interference blocks the increase of PCNA expression in cultured 18.5 *dpc* mouse ovaries, thus dramatically reducing oocytes' apoptosis and increasing primordial follicles assembly (Xu et al. 2011). In another study conducted by the same group, miR-376a was found to downregulate PCNA expression and its expression was reciprocally correlated with that of PCNA mRNA in fetal mouse ovaries. Furthermore, cultured 18.5 *dpc* ovaries transfected with miR-376a exhibited reduced apoptosis of oocytes and an increase in the number of primordial follicles; this effect was mediated by miR-376a-inhibition of PCNA (Zhang et al. 2014a). These findings imply that by inhibiting oocytes' apoptosis miR-376a acts as a positive regulator of primordial follicles assembly.

The effect of miRNAs on primordial follicles assembly is not confined to the oocytes. miR-143 is predominantly expressed in pre-granulosa cells of prenatal mouse ovaries, as indicated by in situ hybridization; its expression increases during the time of primordial follicle assembly, from 15.5 *dpc* to 1 *dpp*, and remains elevated till 4 *dpp* (Zhang et al. 2013a). Overexpression of miR-143 in 15.5 *dpc* ovaries inhibits the formation of primordial follicles by suppressing proliferation of pre-granulosa cells. Accordingly, inhibition of miR-143 in ovaries of the same age resulted in an increase in the number of primordial follicles as well as in proliferation of pre-granulosa cells, thus implying that miR-143 acts as a negative regulator of primordial follicle formation (Zhang et al. 2013a).

To date, there is limited evidence supporting the involvement of specific miRNAs in primordial follicle assembly through regulation of gene expression in mouse pre-granulosa cells. However, several miRNAs have been implicated as participants in the regulation of estradiol (E2) production in granulosa cells, increasing E2 level by targeting Forkhead L2 (Foxl2; Dai et al. 2013) or RNA-binding motif, single-stranded interacting protein 1 (RBMS1; Yin et al. 2012), and decreasing E2 level by targeting the E2F1 and steroidogenic factor 1 (SF1) transcription factors (Yin et al. 2014). Because estrogen negatively regulates cyst breakdown and assembly of primordial follicles, it is possible that these miRNAs contribute to the development of a relatively low estrogenic environment, permissive of primordial follicle assembly (Kezele and Skinner 2003; Chen et al. 2007, 2009).

Unlike estrogen, activin A, a member of the TGF-β superfamily, causes an increase in the number of primordial follicles when administered to neonatal mice during the period of cyst breakdown and primordial follicle assembly (Bristol-Gould et al. 2006). The inverse effect of these two signaling pathways is further supported by the presence of multi-oocytic follicles (MOF) in ovaries of neonatal estrogen-treated mice (Kipp et al. 2007) and of mice models in which activin expression or its downstream signaling was suppressed (Bristol-Gould et al. 2005; Pangas et al. 2007). Furthermore, ovaries of neonatal mice treated with diethylstilbestrol (DES) or E2 had lower level of activin subunits and an attenuated activin signaling, as determined by the level of phosphorylated Smad-

2 (Kipp et al. 2007), implying that the estrogen and activin signaling pathways act in a reciprocal manner and are linked together as well. Interestingly, both miR-145 and miR-181a inhibit activin function in granulosa cells by targeting activin receptor IB (ActRIB) and IIA (ActRIIA), respectively, thereby suppressing proliferation of granulosa cells (Yan et al. 2012; Zhang et al. 2013b). Hence, miRNAs might have a role in maintaining a delicate balance between estrogen and activin, thus enabling cyst breakdown and primordial follicle assembly in a time-restricted manner.

12.6.2 Primordial to Primary Follicle Transition

Primordial follicles remain dormant for a period of time that varies between follicles. In the mouse, activation of few primordial follicles to start growing begins shortly after the formation of all primordial follicles. Most of the primordial follicles will be gradually recruited to the growing pool throughout postnatal life (Fortune et al. 2000; reviewed by Hirshfield 1991).

The involvement of miRNAs in the process of primordial follicle activation is largely obscure. Yang et al. (2013b) showed that 24 miRNAs are differentially expressed in mouse ovaries during the developmental period of primordial follicles into primary follicles. Whereas some miRNAs were downregulated, others were upregulated, implying they might have a role in maintaining the dormancy of primordial follicles or in initiating their development. Gene ontology analysis revealed many signaling pathways that are modulated by these differentially expressed miRNAs (Yang et al. 2013b); many of these signaling pathways are involved in initiation of primordial follicle development (Skinner 2005). The upregulated miR-145 is predominantly expressed in granulosa cells of primary follicles and is undetectable in pre-granulosa cells of primordial follicles (Yang et al. 2013b). Inhibition of miR-145 in whole ovaries during the time of primordial to primary transition causes a significant decrease in the number of primordial follicles, coincident with an increase in the number of primary follicles, with no effect on the total number of follicles (Yang et al. 2013b). Accelerated primordial follicle activation caused by anti-miR-145 is dependent on TGF-β receptor type 2, a direct target of miR-145 (Yang et al. 2013b), that belongs to the TGF-β superfamily, known to be involved in the development of primordial follicles (Trombly et al. 2009).

The possible involvement of other differentially expressed miRNAs during the primordial to primary transition is of great interest. One of these miRNAs, miR-125a-3p, is upregulated in primary follicles compared to primordial follicles (Yang et al. 2013b). We recently found that miR-125a-3p post-transcriptionally regulates the expression of Fyn, a Src-family kinase (SFK), in granulosa cells (Grossman et al. 2015). SFKs can either synergize or operate apart and can also compensate one another (Bock and Herz 2003; Zhang et al. 2014b). Of the SFKs, Src is essential for activation of primordial follicle by regulating the PI3K-PKC-

ERK1/2 pathway (Du et al. 2012). Along with the increase in miR-125a-3p during the time of primordial follicles activation, it may point at a possible involvement of miR-125a-3p in this process.

Though yet to be established, other miRNAs are suspected to be involved in activation of primordial follicles due to their proven ability to regulate key factors in this process. One of these miRNAs is miR-133b, which downregulates the expression of *Foxl2* in granulosa cells of growing follicles (Dai et al. 2013). The fundamental role of *Foxl2* in differentiation of granulosa cells is demonstrated by the block of the transition from squamous pre-granulosa cells to cuboidal granulosa cells in the absence of FOXL2 (Schmidt et al. 2004). Hence, as a negative regulator of *Foxl2*, miR-133b is expected to remain low during activation of primordial follicles. Currently there is no data regarding the expression pattern of miR-133b during these early stages of follicular development.

12.6.3 Follicular Growth

At early stages of folliculogenesis, granulosa cells of single-layered primary follicles undergo mitotic divisions to form multilayered secondary follicles surrounded with a layer of theca cells. As the small secondary follicles grow to become large pre-antral follicles, proliferation of granulosa and theca cells continues and is coordinated with oocyte growth. The early folliculogenesis stages are modulated primarily by intraovarian signaling molecules, acting in both autocrine and paracrine manners, including activins (reviewed by Oktem and Urman 2010).

Expression of activin βA and βB subunits as well as activin receptors (ActRIB, ActRIIA, and ActRIIB) was detected already in somatic cells and oocytes of early follicles from postnatal mouse ovaries (Bristol-Gould et al. 2006). Several studies have demonstrated that activins play an important role in the proliferation and differentiation of granulosa cells (Guzel et al. 2014; Smitz et al. 1998). Interestingly, miR-145 was found to inhibit granulosa cell proliferation; this effect is dependent upon the ability of miR-145 to inhibit the expression of ActRIB. miR-145 also targets and inhibits the expression of cyclin D2, whereas activin A exerts an opposite effect on the expression of cyclin D2 and causes a decrease in the expression of miR-145 (Yan et al. 2012). These data point at miR-145 as a dominant regulator of granulosa cell proliferation and demonstrate a feedback loop between miR-145 and the activin pathway.

Activin A also causes a decrease in the level of miR-181a in mouse primary granulosa cells (Zhang et al. 2013b). The expression of miR-181a is high in primary follicles and decreases dramatically in pre-antral and antral follicles, concomitant with an increase in the level of ActRIIA and granulosa cell proliferation. Furthermore, miR-181a inhibits proliferation of granulosa cells, by decreasing the expression of ActRIIA through direct binding to ActRIIA 3′UTR (Zhang et al. 2013b).

Activins are also well known for their ability to induce expression of FSH receptor (Nakamura et al. 1993) and potentiate FSH effect in granulosa cells by increasing the level of cAMP (El-Hefnawy and Zeleznik 2001). Though the exact stage in which follicles become FSH dependent is controversial, it is well established that antrum formation and growth of early antral to preovulatory antral follicles are dependent upon FSH and its interaction with intraovarian factors (Kumar et al. 1997; Sánchez et al. 2010). Many miRNAs are under the regulation of FSH, in both mouse and rat granulosa cells (Yao et al. 2009, 2010b), implying that miRNAs might participate in FSH signaling in granulosa cells. Furthermore, a specific miRNA, expressed in human granulosa cells and predicted to regulate ActRIIB and Smad-2, aligns with the intronic region of the FSH-receptor gene (Velthut-Meikas et al. 2013), adding another level of complexity to the involvement of miRNAs in modulating FSH signaling.

FSH induces aromatization of androgen to estrogen and expression of LH receptor in granulosa cells (Hillier 1994). In turn, estrogen induces proliferation of granulosa cells and facilitates the activity of FSH. Normal cyclic levels of estrogen and estrogen receptors are important for proper folliculogenesis as well as ovulation (reviewed by Drummond and Findlay 1999). Several miRNAs have been demonstrated to regulate secretion of estrogen from granulosa cells by inhibiting the expression of steroidogenesis key members. Recently, a feedback loop between two miRNAs that modulate E2 production differentially was demonstrated in mouse granulosa cells. Transactivation of miR-383 by SF1 causes an increase in E2 secretion (Yin et al. 2012). Overexpression of miR-383 promotes the transcription of pri-miR-320, causing an increase in the expression of mature miR-320, which in turn inhibits the expression of SF1, causing a decrease in E2 secretion and granulosa cell proliferation (Yin et al. 2012, 2014). Both miRNAs are downregulated by TGF-β1 and FSH, or its analog, PMSG (Yao et al. 2010a). On the contrary, TGF-β1 upregulates the expression of miR-224, which facilitates secretion of E2 from granulosa cells (Yao et al. 2010a). The expression of miR-224 is unaffected by overexpression of miR-383 (Yin et al. 2014), implying that miRNAs are involved, via different pathways, in regulation of E2 production in granulosa cells.

Granulosa cell function is also regulated by androgens, already from early stages of folliculogenesis. Both testosterone and dihydrotestosterone have a prosurvival effect on mouse primary granulosa cells. Sen et al. demonstrated that this effect of testosterone and dihydrotestosterone is mediated by an increase in the levels of expression of miR-125b, followed by a decrease in the expression of pro-apoptotic proteins (Sen et al. 2014). Overall miRNAs appear to have a prominent role in mediating granulosa cells function, proliferation, and survival in response to intraovarian and extraovarian factors.

12.6.4 Oocyte Maturation and Early Embryogenesis

Starting at embryonic life and all throughout oocytes' growth, they remain arrested at the prophase of the first meiotic division and are characterized by a prominent nucleus, the GV. LH surge induces oocytes within selected antral follicles to resume the first meiotic division and undergo oocyte maturation, manifested by chromosome condensation, GV breakdown (GVBD), spindle formation and migration toward the oocyte cortex, segregation of homologous chromosomes, extrusion of the first polar body (PBI), and an arrest at metaphase of the second meiotic division (MII). Ovulated MII oocytes resume their second meiotic division upon fertilization. The process of oocyte maturation is highly coordinated both in time and space and involves many signaling pathways (reviewed by Brunet and Maro 2005; Solc et al. 2010).

It is well established that fully grown GV mouse oocytes are trancriptionally quiescent. Because transcription activity begins only at late zygote stage, oocytes rely on pre-synthesized mRNA transcripts for the processes of maturation, fertilization, and early embryogenesis (Bouniol-baly et al. 1999; Zeng and Schultz 2005). Thus, posttranscriptional regulation of existing transcripts is of great importance. Several mechanisms of posttranscriptional regulation are indicated in mouse oocytes (reviewed by Kang and Han 2011); the one operated by maternal miRNAs is the most controversial mechanism, mainly, as discussed above, due to differences in the phenotypes of *Dicer1* and *Dgcr8* cKO oocytes (Suh et al. 2010). Though Suh et al. (2010) found that miRNAs are not essential for oocyte maturation and early embryo development, several studies suggest that miRNAs play a role in their regulation (elaborated hereinafter). miR-335-5p is highly expressed in GV oocytes throughout maturation and in MII oocytes, and a decrease in its expression is detected shortly after fertilization. Microinjection of miR-355-5p mimic or miR-355-5p inhibitor into GV oocytes caused high incidence of GVBD or MI-arrested oocytes that exhibit spindle abnormalities. Many of the MII oocytes that completed maturation had a two-cell-like morphology, indicating failure in asymmetric cell division. When injected into zygotes, both miR-355-5p mimic and miR-355-5p inhibitor had no visible effect on embryo development till the blastocyst stage. miR-355-5p inhibits the expression of dishevelled-associated activator of morphogenesis 1 (Daam1; Cui et al. 2013), which modulates actin microfilaments (Liu et al. 2008), implying that the effect of miR-335-5p on oocyte maturation is mediated by regulation of the actin cytoskeleton.

The expression of the noncanonical miR-320 (Kim and Choi 2012), which is detected in human follicular fluids, is correlated with the quality of day 3 embryos (Feng et al. 2015). Knockdown of miR-320 in mouse MII oocytes causes a decrease in the number of two-cell embryos and blastocysts, as well as an aberration in the expression of components of the WNT signaling pathway (Feng et al. 2015). Though the WNT signaling pathways were studied in oocytes mainly in regard to early embryonic development, the canonical WNT signaling pathway is active in mouse oocytes from the stage of secondary follicle; its activity increases as the follicle grows, concomitantly with an increase in oocyte diameter, but it is undetectable in mature MII oocytes (Usongo et al. 2012). Along with the possible role of WNT in modulating the actin cytoskeleton (Akiyama and Kawasaki 2006) and its cross talk with the PI3K/AKT signaling pathway (reviewed by Boyer et al. 2010), it is of great interest to explore the involvement of miR-320 in oocyte maturation and emphasize its role as a regulator of WNT signaling.

The role of miRNAs in oocytes should not be limited to a direct effect on maternal miRNAs, but should rather include indirect effects mediated via granulosa cell miRNAs. Kim et al. (2013) showed that four miRNAs (let-7b, let-7c, miR-27a, and miR-322) are differentially expressed in mouse granulosa cells derived from in vitro cultured, ovulation-induced antral follicles that contain either MII or MI oocytes (Kim et al. 2013). Maturation rate of oocytes in in vitro cultured antral follicles after transfection of granulosa cells with miR-27a mimic is decreased, but ovulation and fertilization rates are not affected. Transfection with miR-27a inhibitor had an inverse effect, causing an increase in the rate of oocyte maturation. Similar effects were obtained upon transfection with let-7c and miR-322 inhibitors, whereas their mimic sequences had no effect on all examined criteria (Kim et al. 2013). These findings imply that miRNAs expressed within granulosa cells are involved in specific regulation of oocyte maturation.

12.6.5 Ovulation and Formation of the Corpus Luteum

When exposed to ovulatory dose of LH/hCG, both mural and cumulus granulosa cells undergo massive changes in gene expression pattern, emphasizing the importance of transcription and posttranscriptional modifications during the peri-ovulatory period (reviewed by Stocco et al. 2007). In mouse ovaries, miRNAs regulate several processes occurring during the peri-ovulatory period, by targeting key proteins necessary for cumulus expansion, follicular wall rupture, luteinization, and cell survival.

During their expansion, cumulus cells synthesize and secrete PTX3, which then localizes to the matrix where it regulates the organization of hyaluronan (Salustri et al. 2004). The expression of PTX3 is downregulated directly by miR-224. Administration of hCG to equine chorionic-gonadotropin (eCG)-primed mice inhibits the expression of miR-224 in mural granulosa cells and cumulus–oocyte complexes (COCs), whereas the expression of PTX3 increases. Overexpression of miR-224, in in vitro cultured COCs, decreases the rate of cumulus expansion in response to EGF (Yao et al. 2014). Furthermore, when miR-224 is overexpressed in vivo, less COCs are recovered from the oviducts (Yao et al. 2014). As previously mentioned, miR-224 also facilitates proliferation of granulosa cells and E2 production (Yao et al. 2010a), demonstrating a distinct role played by the same miRNA in the follicle, depending on the stage of follicular development.

Several miRNAs are under the regulation of LH/hCG in mural granulosa cells; among them are miR-132 and miR-212, both driven from a single pri-miRNA and upregulated upon in vivo administration of hCG (Fiedler et al. 2008). In vitro studies revealed that both miR-132 and miR-212 regulate the expression of the C-terminal binding protein, CTBP1 (Fiedler et al. 2008), hence holding a great potential in mediating LH/hCG-induced changes in global genes transcription.

Another miRNA, miR-21, is also upregulated by hCG. Knockdown of miR-21 in in vitro cultured primary granulosa cells causes an increase in the level of cleaved caspase-3 and in cell apoptosis (Carletti et al. 2010). Furthermore, knockdown of miR-21 in ovaries overrides the anti-apoptotic effect of hCG triggering, causing increased apoptosis of granulosa cells and impaired ovulation (Carletti et al. 2010).

Contrary to miRNAs that are upregulated by hCG and transduce its effect in granulosa cells (Carletti et al. 2010; Fiedler et al. 2008), we have recently found that the expression of miR-125a-3p is transiently downregulated in mural granulosa cells exposed to hCG (Grossman et al. 2015). This miRNA directly downregulates the expression of Fyn kinase, thus inhibiting hCG-induced Fyn upregulation and granulosa cell migration. Overexpression of miR-125a-3p in ovaries of PMSG-hCG-treated mice decreased ovulation rate, similar to the effect of Fyn si-RNA, inferring that the decrease in miR-125a-3p in response to hCG, accompanied by increased Fyn level, supports proper ovulation. The ratio between miR-125a-3p and Fyn is higher in ovaries of anovulatory mice, treated prenatally with dihydrotestosterone, than in ovaries of control mice, thus implying that dysregulation of the expression of miR-125a-3p may contribute to anovulatory pathologies (Grossman et al. 2015).

Aside from its anti-migratory properties, miR-125a-3p has a pro-apoptotic effect in several cell lines (Jiang et al. 2013; Ninio-Many et al. 2014; Yin et al. 2015). It is well established that apoptosis must be inhibited to allow the formation and function of the corpus luteum (CL; reviewed by Stocco et al. 2007). We, therefore, examined whether the expression of miR-125a-3p in mouse luteinized granulosa cells remains low after in vivo administration of hCG. Our results present that it decreased 4 h after hCG administration, returned to control value 4 h later (Grossman et al. 2015), decreased again 16 h after hCG administration, and remained low for another 16 h (Fig. 12.3). We also found that overexpression of miR-125a-3p in immortalized human granulosa cell line (SVOG; Lie et al. 1996) causes a dose-dependent decreased secretion of amphiregulin (AREG; Fig. 12.4), a prosurvival factor of human CL (Ben-Ami et al. 2009). These findings imply that, in addition to its role in ovulation, miR-125a-3p might participate in the regulation of CL formation and maintenance.

Not surprisingly, the effect of miRNAs in ovulation is not confined to locally acting miRNAs but is also at the pituitary level. Hasuwa et al. (2013) showed that female mice lacking miR-200b and miR-429 (both share the same seed sequence) are anovulatory. These miRNAs are highly expressed in the pituitary, where they regulate the expression of the ZEB1 transcription factor. Elimination of miR-200b and miR-429 results in increased expression of ZEB1, followed by inhibition of the β subunit of LH and hence low concentration of circulating LH and a failure to release the LH surge (Hasuwa et al. 2013).

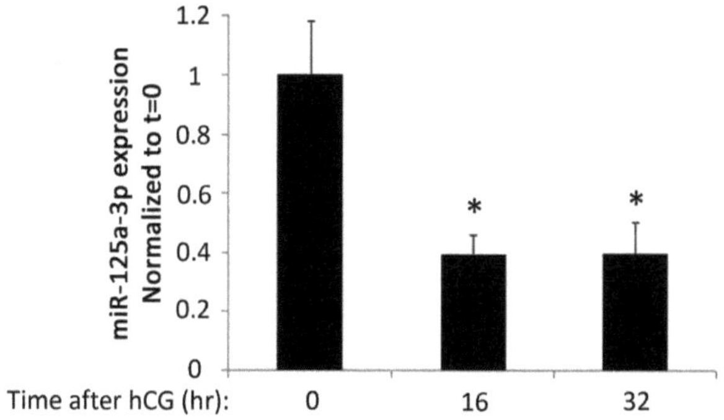

Fig. 12.3 MiR-125a-3p expression in luteinized mouse granulosa cells. Seven-weeks-old mice were administered with 5 IU PMSG and either killed 48 h later (control group; 0) or injected with 7 IU hCG and killed 16 or 32 h later. Ovaries were excised, and mural granulosa cells were isolated and subjected to quantitative PCR analysis. The expression of miR-125a-3p in granulosa cells of each mouse was normalized to the level of U6 snRNA expression; the average ratio for each treatment group is presented. Data were analyzed by one-way ANOVA. Bars are mean ± SEM ($n = 5$). $^*P < 0.05$

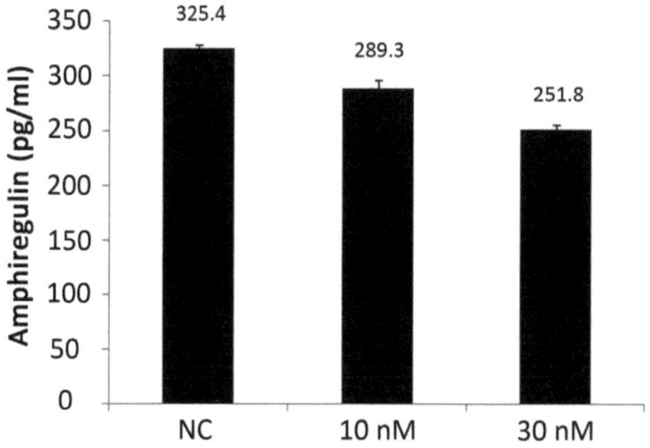

Fig. 12.4 Secretion of amphiregulin from miR-125a-3p overexpressing cells. SVOG cells were transfected with 10 nM scramble miR (negative control; NC), or with two concentrations of miR-125a-3p mimic (10 or 30 nM), using lipofectamine 2000. Cells were incubated for 12 h, culture media were refreshed, and cells were incubated for additional 24 h. Conditioned media were collected, centrifuged at 1200 rpm for 5 min and subjected to ELISA assay for the detection of amphiregulin. Experiment was repeated twice. *Bars* are mean (indicated on *top*) ± SEM

12.7 Conclusion

Gonad development is comprised of a series of events occurring at the gonadal level and regulated by a panel of hormones secreted from the hypothalamic–pituitary–gonadal axis. miRNAs have prominent roles in gonad development, exerting their effect either directly through silencing the expression of key proteins in gonadal somatic or germ cells or indirectly acting at the hypothalamus–pituitary level. Taking into account that one miRNA can regulate the expression of many genes depending on cell type, state of differentiation, and environment, the task of fully understanding the role of miRNAs in gonad development should be challenging.

Grant Support This work was supported by a grant from the Israel Science Foundation (ISF 470/14 to R.S.).

References

Abdelfattah AM, Choi MY (2015) Update on non-canonical microRNAs. Biomol Concepts 5:275–287

Adham IM, Nayernia K, Burkhardt-Gottges E et al (2001) Teratozoospermia in mice lacking the transition protein 2 (Tnp2). Mol Hum Reprod 7:513–520

Åkerfelt M, Henriksson E, Laiho A et al (2008) Promoter ChIP-chip analysis in mouse testis reveals Y chromosome occupancy by HSF2. Proc Natl Acad Sci U S A 105:11224–11229

Akiyama T, Kawasaki Y (2006) Wnt signalling and the actin cytoskeleton. Oncogene 25: 7538–7544

Bao J, Li D, Wang L et al (2012) MicroRNA-449 and microRNA-34b/c function redundantly in murine testes by targeting E2F transcription factor-retinoblastoma protein (E2F-pRb) pathway. J Biol Chem 287:21686–21698

Barrionuevo F, Bagheri-Fam S, Klattig J et al (2006) Homozygous inactivation of Sox9 causes complete XY sex reversal in mice. Biol Reprod 74:195–201

Barrios F, Filipponi D, Pellegrini M et al (2010) Opposing effects of retinoic acid and FGF9 on Nanos2 expression and meiotic entry of mouse germ cells. J Cell Sci 123:871–880

Bellvé AR, Cavicchia JC, Millette CF et al (1977) Spermatogenic cells of the prepuberal mouse. Isolation and morphological characterization. J Cell Biol 74:68–85

Ben-Ami I, Armon L, Freimann S et al (2009) EGF-like growth factors as LH mediators in the human corpus luteum. Hum Reprod 24:176–184

Björk JK, Sandqvist A, Elsing AN et al (2010) miR-18, a member of Oncomir-1, targets heat shock transcription factor 2 in spermatogenesis. Development 137:3177–3184

Bock HH, Herz J (2003) Reelin activates SRC family tyrosine kinases in neurons. Curr Biol 13: 18–26

Bouniol-Baly C, Hamraoui L, Guibert J et al (1999) Differential transcriptional activity associated with chromatin configuration in fully grown mouse germinal vesicle oocytes. Biol Reprod 587: 580–587

Boyer A, Goff AK, Boerboom D (2010) WNT signaling in ovarian follicle biology and tumorigenesis. Trends Endocrinol Metab 21:25–32

Bristol-Gould SK, Hutten CG, Sturgis C et al (2005) The development of a mouse model of ovarian endosalpingiosis. Endocrinology 146:5228–5236

Bristol-Gould SK, Kreeger PK, Selkirk CG et al (2006) Postnatal regulation of germ cells by activin: the establishment of the initial follicle pool. Dev Biol 298:132–148

Brunet S, Maro B (2005) Cytoskeleton and cell cycle control during meiotic maturation of the mouse oocyte: integrating time and space. Reproduction 130:801–811

Buchold GM, Coarfa C, Kim J et al (2010) Analysis of microRNA expression in the prepubertal testis. PLoS One 5:e15317

Carletti MZ, Fiedler SD, Christenson LK (2010) MicroRNA 21 blocks apoptosis in mouse periovulatory granulosa cells. Biol Reprod 83:286–295

Carthew RW, Sontheimer EJ (2009) Origins and mechanisms of miRNAs and siRNAs. Cell 136:642–655

Chang Y-F, Lee-Chang JS, Imam JS et al (2012) Interaction between microRNAs and actin-associated protein Arpc5 regulates translational suppression during male germ cell differentiation. Proc Natl Acad Sci U S A 109:5750–5755
Cheloufi S, Dos Santos CO, Chong MMW, Hannon GJ (2010) A dicer-independent miRNA biogenesis pathway that requires Ago catalysis. Nature 465:584–589
Chen Y, Jefferson WN, Newbold RR et al (2007) Estradiol, progesterone, and genistein inhibit oocyte nest breakdown and primordial follicle assembly in the neonatal mouse ovary in vitro and in vivo. Endocrinology 148:3580–3590
Chen Y, Breen K, Pepling ME (2009) Estrogen can signal through multiple pathways to regulate oocyte cyst breakdown and primordial follicle assembly in the neonatal mouse ovary. J Endocrinol 202:407–417
Chong MMW, Zhang G, Cheloufi S et al (2010) Canonical and alternate functions of the microRNA biogenesis machinery. Genes Dev 24:1951–1960
Cifuentes D, Xue H, Taylor DW et al (2010) A novel miRNA processing pathway independent of Dicer requires Argonaute2 catalytic activity. Science 328:1694–1698
Cui XS, Sun SC, Kang YK, Kim NH (2013) Involvement of microRNA-335-5p in cytoskeleton dynamics in mouse oocytes. Reprod Fertil Dev 25:691–699
Dai L, Tsai-Morris C-H, Sato H et al (2011) Testis-specific miRNA-469 up-regulated in gonadotropin-regulated testicular RNA helicase (GRTH/DDX25)-null mice silences transition protein 2 and protamine 2 messages at sites within coding region: implications of its role in germ cell development. J Biol Chem 286:44306–44318
Dai A, Sun H, Fang T et al (2013) MicroRNA-133b stimulates ovarian estradiol synthesis by targeting Foxl2. FEBS Lett 587:2474–2482
de Rooij DG, Russell LD (2000) All you wanted to know about spermatogonia but were afraid to ask. J Androl 21:776–798
Drummond AE, Findlay JK (1999) The role of estrogen in folliculogenesis. Mol Cell Endocrinol 151:57–64
Du X-Y, Huang J, Xu L-Q et al (2012) The proto-oncogene c-src is involved in primordial follicle activation through the PI3K, PKC and MAPK signaling pathways. Reprod Biol Endocrinol 10:58
El-Hefnawy T, Zeleznik AJ (2001) Synergism between FSH and activin in the regulation of proliferating cell nuclear antigen (PCNA) and cyclin D2 expression in rat granulosa cells. Endocrinology 142:4357–4362
Ender C, Krek A, Friedländer MR et al (2008) A Human snoRNA with MicroRNA-Like Functions. Mol Cell 32:519–528
Epifano O, Liang LF, Familari M et al (1995) Coordinate expression of the three zona pellucida genes during mouse oogenesis. Development 121:1947–1956
Ewen KA, Koopman P (2010) Mouse germ cell development: from specification to sex determination. Mol Cell Endocrinol 323:76–93
Feng R, Sang Q, Zhu Y et al (2015) MiRNA-320 in the human follicular fluid is associated with embryo quality in vivo and affects mouse embryonic development in vitro. Sci Rep 5:8689
Fiedler SD, Carletti MZ, Hong X, Christenson LK (2008) Hormonal regulation of MicroRNA expression in periovulatory mouse mural granulosa cells. Biol Reprod 79:1030–1037
Flemr M, Ma J, Schultz RM, Svoboda P (2010) P-body loss is concomitant with formation of a messenger RNA storage domain in mouse oocytes. Biol Reprod 82:1008–1017
Fortune JE, Cushman RA, Wahl CM, Kito S (2000) The primordial to primary follicle transition. Mol Cell Endocrinol 163:53–60
Gallardo T, Shirley L, John GB, Castrillon DH (2007) Generation of a germ cell-specific mouse transgenic Cre line, Vasa-Cre. Genesis 417:413–417
García-López J, Hourcade JDD, Del Mazo J (2013) Reprogramming of microRNAs by adenosine-to-inosine editing and the selective elimination of edited microRNA precursors in mouse oocytes and preimplantation embryos. Nucleic Acids Res 41:5483–5493
Greenlee AR, Shiao M-S, Snyder E et al (2012) Deregulated sex chromosome gene expression with male germ cell-specific loss of Dicer1. PLoS One 7:e46359

Gregory RI, Chendrimada TP, Cooch N, Shiekhattar R (2005) Human RISC couples microRNA biogenesis and posttranscriptional gene silencing. Cell 123:631–640

Grossman H, Chuderland D, Ninio-Many L et al (2015) A novel regulatory pathway in granulosa cells, the LH/human chorionic gonadotropin-microRNA-125a-3p-Fyn pathway, is required for ovulation. FASEB J 29(8):3206–3216

Guzel Y, Nur Şahin G, Sekeroglu M, Deniz A (2014) Recombinant activin A enhances the growth and survival of isolated preantral follicles cultured three-dimensionally in extracellular basement matrix protein (matrigel) under serum-free conditions. Gynecol Endocrinol 30:388–391

Ha M, Kim VN (2014) Regulation of microRNA biogenesis. Nat Rev Mol Cell Biol 15:509–524

Hasuwa H, Ueda J, Ikawa M, Okabe M (2013) MiR-200b and miR-429 function in mouse ovulation and are essential for female fertility. Science 341:71–74

Havens MA, Reich AA, Duelli DM, Hastings ML (2012) Biogenesis of mammalian microRNAs by a non-canonical processing pathway. Nucleic Acids Res 40:4626–4640

Hayashi K, Chuva de Sousa Lopes SM, Kaneda M et al (2008) MicroRNA biogenesis is required for mouse primordial germ cell development and spermatogenesis. PLoS One 3:e1738

He Z, Jiang J, Kokkinaki M et al (2013) MiRNA-20 and MiRNA-106a Regulate Spermatogonial Stem Cell Renewal at the Post-transcriptional Level via Targeting STAT3 and Ccnd1. Stem Cells 31:1–22

Hillier SG (1994) Current concepts of the roles of follicle stimulating hormone and luteinizing hormone in folliculogenesis. Hum Reprod 9:188–191

Hirshfield AN (1991) Development of follicles in the mammalian ovary. Int Rev Cytol 124: 43–101

Hong X, Luense LJ, McGinnis LK et al (2008) Dicer1 is essential for female fertility and normal development of the female reproductive system. Endocrinology 149:6207–6212

Huszar JM, Payne CJ (2013) MicroRNA 146 (Mir146) modulates spermatogonial differentiation by retinoic acid in mice. Biol Reprod 88:15

Hutvagner G, Simard MJ (2008) Argonaute proteins: key players in RNA silencing. Nat Rev Mol Cell Biol 9:22–32

Inui M, Martello G, Piccolo S (2010) MicroRNA control of signal transduction. Nat Rev Mol Cell Biol 11:252–263

Jamin SP, Arango NA, Mishina Y et al (2002) Requirement of Bmpr1a for Müllerian duct regression during male sexual development. Nat Genet 32:408–410

Jiang L, Chang J, Zhang Q et al (2013) MicroRNA hsa-miR-125a-3p activates p53 and induces apoptosis in lung cancer cells. Cancer Invest 31:538–544

Kaneda M, Tang F, O'Carroll D et al (2009) Essential role for Argonaute2 protein in mouse oogenesis. Epigenetics Chromatin 2:9

Kang MK, Han SJ (2011) Post-transcriptional and post-translational regulation during mouse oocyte maturation. BMB Rep 44:147–157

Kezele P, Skinner MK (2003) Regulation of ovarian primordial follicle assembly and development by estrogen and progesterone: endocrine model of follicle assembly. Endocrinology 144: 3329–3337

Kim B-M, Choi MY (2012) Non-canonical microRNAs miR-320 and miR-702 promote proliferation in Dgcr8-deficient embryonic stem cells. Biochem Biophys Res Commun 462:183–189

Kim G, Georg INA, Scherthan H et al (2010a) Dicer is required for Sertoli cell function and survival. Int J Dev Biol 875:867–875

Kim Y-K, Heo I, Kim VN (2010b) Modifications of small RNAs and their associated proteins. Cell 143:703–709

Kim YJ, Ku S-Y, Kim YY et al (2013) MicroRNAs transfected into granulosa cells may regulate oocyte meiotic competence during in vitro maturation of mouse follicles. Hum Reprod 28(11):3050–3061

Kipp JL, Kilen SM, Bristol-Gould S et al (2007) Neonatal exposure to estrogens suppresses activin expression and signaling in the mouse ovary. Endocrinology 148:1968–1976

Kobayashi A, Chang H, Chaboissier M-C et al (2005) Sox9 in testis determination. Ann N Y Acad Sci 1061:9–17

Korhonen HM, Meikar O, Yadav RP et al (2011) Dicer is required for haploid male germ cell differentiation in mice. PLoS One 6:e24821

Kotaja N (2014) MicroRNAs and spermatogenesis. Fertil Steril 101:1552–1562

Kumar TR, Wang Y, Lu N, Matzuk MM (1997) Follicle stimulating hormone is required for ovarian follicle maturation but not male fertility. Nat Genet 15:201–204

Lee M, Choi Y, Kim K et al (2014) Adenylation of maternally inherited microRNAs by Wispy. Mol Cell 56:696–707

Lei L, Jin S, Gonzalez G et al (2010) The regulatory role of Dicer in folliculogenesis in mice. Mol Cell Endocrinol 315:63–73

Lie BL, Leung E, Leung PC, Auersperg N (1996) Long-term growth and steroidogenic potential of human granulosa-lutein cells immortalized with SV40 large T antigen. Mol Cell Endocrinol 120:169–176

Liu J, Valencia-Sanchez MA, Hannon GJ, Parker R (2005) MicroRNA-dependent localization of targeted mRNAs to mammalian P-bodies. Nat Cell Biol 7:719–723

Liu W, Sato A, Khadka D et al (2008) Mechanism of activation of the Formin protein Daam1. Proc Natl Acad Sci U S A 105:210–215

Liu D, Li L, Fu H et al (2012) Biochemical and biophysical research communications inactivation of Dicer1 has a severe cumulative impact on the formation of mature germ cells in mouse testes. Biochem Biophys Res Commun 422:114–120

Lomelí H, Ramos-Mejia V, Nagy A (2000) Targeted insertion of Cre recombinase into the TNAP gene: excision in primordial germ cells. Genesis 26:8–10

Ma J, Flemr M, Stein P et al (2010) MicroRNA activity is suppressed in mouse oocytes. Curr Biol 20:265–270

Ma H, Zhang J, Wu H (2014) Designing Ago2-specific siRNA/shRNA to avoid competition with endogenous miRNAs. Mol Ther Nucleic Acids 3:e176

Maatouk DM, Loveland KL, McManus MT et al (2008) Dicer1 is required for differentiation of the mouse male germline. Biol Reprod 79:696–703

Meister G, Landthaler M, Patkaniowska A et al (2004) Human Argonaute2 mediates RNA cleavage targeted by miRNAs and siRNAs. Mol Cell 15:185–197

Meistrich ML, Mohapatra B, Shirley CR, Zhao M (2003) Roles of transition nuclear proteins in spermiogenesis. Chromosoma 111:483–488

Miller D, Brinkworth M, Iles D (2010) Paternal DNA packaging in spermatozoa: more than the sum of its parts? DNA, histones, protamines and epigenetics. Reproduction 139:287–301

Mishima T, Takizawa T, Luo SS et al (2008) MicroRNA (miRNA) cloning analysis reveals sex differences in miRNA expression profiles between adult mouse testis and ovary. Reproduction 136:811–822

Miyoshi K, Miyoshi T, Siomi H (2010) Many ways to generate microRNA-like small RNAs: non-canonical pathways for microRNA production. Mol Genet Genomics 284:95–103

Modzelewski AJ, Holmes RJ, Hilz S et al (2012) AGO4 regulates entry into meiosis and influences silencing of sex chromosomes in the male mouse germ line. Dev Cell 23:251–264

Modzelewski AJ, Hilz S, Crate EA et al (2015) Dgcr8 and Dicer are essential for sex chromosome integrity during meiosis in males. J Cell Sci 128:2314–2327

Mogilyansky E, Rigoutsos I (2013) The miR-17/92 cluster: a comprehensive update on its genomics, genetics, functions and increasingly important and numerous roles in health and disease. Cell Death Differ 20:1603–1614

Murchison EP, Stein P, Xuan Z et al (2007) Critical roles for Dicer in the female germline. Genes Dev 21:682–693

Nagaraja AK, Andreu-Vieyra C, Franco HL et al (2008) Deletion of Dicer in somatic cells of the female reproductive tract causes sterility. Mol Endocrinol 22:2336–2352

Nakamura M, Minegishi T, Hasegawa Y et al (1993) Effect of an activin A on follicle-stimulating hormone (FSH) receptor messenger ribonucleic acid levels and FSH receptor expressions in cultured rat granulosa cells. Endocrinology 133:538–544

Nguyen TA, Jo MH, Choi Y-G et al (2015) Functional anatomy of the human microprocessor. Cell 161:1374–1387

Ninio-Many L, Grossman H, Levi M et al (2014) MicroRNA miR-125a-3p modulates molecular pathway of motility and migration in prostate cancer cells. Oncoscience 1:250–261
Nishikura K (2006) Editor meets silencer: crosstalk between RNA editing and RNA interference. Nat Rev Mol Cell Biol 7:919–931
Niu Z, Goodyear SM, Rao S et al (2011) MicroRNA-21 regulates the self-renewal of mouse spermatogonial stem cells. Proc Natl Acad Sci U S A 108:12740–12745
O'Gorman S, Dagenais NA, Qian M, Marchuk Y (1997) Protamine-Cre recombinase transgenes efficiently recombine target sequences in the male germ line of mice, but not in embryonic stem cells. Proc Natl Acad Sci U S A 94:14602–14607
Oktem O, Urman B (2010) Understanding follicle growth in vivo. Hum Reprod 25:2944–2954
Otsuka M, Zheng M, Hayashi M et al (2008) Impaired microRNA processing causes corpus luteum insufficiency and infertility in mice. J Clin Invest 118:1944–1954
Pangas SA, Jorgez CJ, Tran M et al (2007) Intraovarian activins are required for female fertility. Mol Endocrinol 21:2458–2471
Panneerdoss S, Chang Y, Buddavarapu KC et al (2012) Androgen-responsive microRNAs in mouse Sertoli cells. PLoS One 7:e41146
Papaioannou MD, Pitetti J, Ro S et al (2009) Sertoli cell Dicer is essential for spermatogenesis in mice. Dev Biol 326:250–259
Papaioannou MD, Lagarrigue M, Vejnar CE et al (2011) Loss of dicer in Sertoli cells has a major impact on the testicular proteome of mice. Mol Cell Proteomics 10:1–14
Pepling ME (2012) Follicular assembly: mechanisms of action. Reproduction 143:139–149
Pepling ME, Spradling AC (2001) Mouse ovarian germ cell cysts undergo programmed breakdown to form primordial follicles. Dev Biol 234:339–351
Quick-Cleveland J, Jacob JP, Weitz SH et al (2014) The DGCR8 RNA-binding heme domain recognizes primary microRNAs by clamping the hairpin. Cell Rep 7:1994–2005
Rakoczy J, Fernandez-Valverde SL, Glazov EA et al (2013) MicroRNAs-140-5p/140-3p modulate Leydig cell numbers in the developing mouse testis. Biol Reprod 88:143
Real FM, Sekido R, Lupiáñez DG et al (2013) A microRNA (mmu-miR-124) prevents Sox9 expression in developing mouse ovarian cells. Biol Reprod 89:78
Ro S, Park C, Sanders KM et al (2007) Cloning and expression profiling of testis-expressed microRNAs. Dev Biol 311:592–602
Romero Y, Meikar O, Papaioannou MD et al (2011) Dicer1 depletion in male germ cells leads to infertility due to cumulative meiotic and spermiogenic defects. PLoS One 6:e25241
Ruby JG, Jan CH, Bartel DP (2007) Intronic microRNA precursors that bypass Drosha processing. Nature 448:83–86
Salisbury J, Hutchison KW, Wigglesworth K et al (2009) Probe-level analysis of expression microarrays characterizes isoform-specific degradation during mouse oocyte maturation. PLoS One 4:1–11
Salustri A, Garlanda C, Hirsch E et al (2004) PTX3 plays a key role in the organization of the cumulus oophorus extracellular matrix and in in vivo fertilization. Development 131: 1577–1586
Sánchez F, Adriaenssens T, Romero S, Smitz J (2010) Different follicle-stimulating hormone exposure regimens during antral follicle growth alter gene expression in the cumulus-oocyte complex in mice. Biol Reprod 83:514–524
Schmidt D, Ovitt CE, Anlag K et al (2004) The murine winged-helix transcription factor Foxl2 is required for granulosa cell differentiation and ovary maintenance. Development 131:933–942
Schwenk F, Baron U, Rajewsky K (1995) A cre-transgenic mouse strain for the ubiquitous deletion of IoxP-flanked gene segments including deletion in germ cells. Nucleic Acids Res 23: 5080–5081
Sen A, Prizant H, Light A et al (2014) Androgens regulate ovarian follicular development by increasing follicle stimulating hormone receptor and microRNA-125b expression. Proc Natl Acad Sci U S A 111:3008–3013
Skinner MK (2005) Regulation of primordial follicle assembly and development. Hum Reprod Update 11:461–471

Smitz J, Cortvrindt R, Hu Y, Vanderstichele H (1998) Effects of recombinant activin A on in vitro culture of mouse preantral follicles. Mol Reprod Dev 50:294–304

Snyder EM, Small C, Griswold MD (2010) Retinoic acid availability drives the asynchronous initiation of spermatogonial differentiation in the mouse. Biol Reprod 83: 783–790

Solc P, Schultz RM, Motlik J (2010) Prophase I arrest and progression to metaphase I in mouse oocytes: comparison of resumption of meiosis and recovery from G2-arrest in somatic cells. Mol Hum Reprod 16:654–664

Song R, Ro S, Michaels JD et al (2009) Many X-linked microRNAs escape meiotic sex chromosome inactivation. Nat Genet 41:488–493

Sree S, Radhakrishnan K, Indu S, Kumar PG (2014) Dramatic changes in 67 miRNAs during initiation of first wave of spermatogenesis in Mus musculus testis: global regulatory insights generated by miRNA-mRNA network analysis. Biol Reprod 91:69

Stefani G, Slack FJ (2008) Small non-coding RNAs in animal development. Nat Rev Mol Cell Biol 9:219–230

Stocco C, Telleria C, Gibori G (2007) The molecular control of corpus luteum formation, function, and regression. Endocr Rev 28:117–149

Su YQ, Sugiura K, Woo Y et al (2007) Selective degradation of transcripts during meiotic maturation of mouse oocytes. Dev Biol 302:104–117

Su H, Trombly MI, Chen J, Wang X (2009) Essential and overlapping functions for mammalian Argonautes in microRNA silencing. Genes Dev 23:304–317

Suh N, Baehner L, Moltzahn F et al (2010) MicroRNA function is globally suppressed in mouse oocytes and early embryos. Curr Biol 20:271–277

Swetloff A, Conne B, Huarte J et al (2009) Dcp1-bodies in mouse oocytes. Mol Biol Cell 20: 4951–4961

Tan T, Zhang Y, Ji W, Zheng P (2014) miRNA signature in mouse spermatogonial stem cells revealed by high-throughput sequencing. Biomed Res Int 2014:154251

Tang F, Kaneda M, O'Carroll D et al (2007) Maternal microRNAs are essential for mouse zygotic development. Genes Dev 21:644–648

Tiwari M, Prasad S, Tripathi A et al (2015) Apoptosis in mammalian oocytes: a review. Apoptosis 20:1019–1025

Tong M-H, Mitchell D, Evanoff R, Griswold MD (2011) Expression of Mirlet7 family microRNAs in response to retinoic acid-induced spermatogonial differentiation in mice. Biol Reprod 85:189–197

Tong M-H, Mitchell DA, McGowan SD et al (2012) Two miRNA clusters, Mir-17-92 (Mirc1) and Mir-106b-25 (Mirc3), are involved in the regulation of spermatogonial differentiation in mice. Biol Reprod 86:72

Trombly DJ, Woodruff TK, Mayo KE (2009) Roles for transforming growth factor beta superfamily proteins in early folliculogenesis. Semin Reprod Med 27:14–23

Ungewitter EK, Yao HHC (2012) How to make a gonad: cellular mechanisms governing formation of the testes and ovaries. Sex Dev 7:7–20

Usongo M, Rizk A, Farookhi R (2012) β-Catenin/Tcf signaling in murine oocytes identifies nonovulatory follicles. Reproduction 144:669–676

Velthut-Meikas A, Simm J, Tuuri T et al (2013) Research resource: small RNA-seq of human granulosa cells reveals miRNAs in FSHR and aromatase genes. Mol Endocrinol 27:1128–1141

Vidal VP, Chaboissier MC, de Rooij DG, Schedl A (2001) Sox9 induces testis development in XX transgenic mice. Nat Genet 28:216–217

Wainwright EN, Jorgensen JS, Kim Y et al (2013) SOX9 regulates microRNA miR-202-5p/3p expression during mouse testis differentiation. Biol Reprod 89:34

Wu L, Fan J, Belasco JG (2008) Importance of translation and nonnucleolytic ago proteins for on-target RNA interference. Curr Biol 18:1327–1332

Wu J, Bao J, Wang L et al (2011) MicroRNA-184 downregulates nuclear receptor corepressor 2 in mouse spermatogenesis. BMC Dev Biol 11:64

Wu Q, Song R, Ortogero N et al (2012) The RNase III enzyme DROSHA is essential for microRNA production and spermatogenesis. J Biol Chem 287:25173–25190

Xu B, Hua J, Zhang Y et al (2011) Proliferating cell nuclear antigen (PCNA) regulates primordial follicle assembly by promoting apoptosis of oocytes in fetal and neonatal mouse ovaries. PLoS One 6:e16046

Yan N, Lu Y, Sun H et al (2007) A microarray for microRNA profiling in mouse testis tissues. Reproduction 134:73–79

Yan G, Zhang L, Fang T et al (2012) MicroRNA-145 suppresses mouse granulosa cell proliferation by targeting activin receptor IB. FEBS Lett 586:3263–3270

Yang J-S, Lai EC (2011) Alternative miRNA biogenesis pathways and the interpretation of core miRNA pathway mutants. Mol Cell 43:892–903

Yang Q-E, Racicot KE, Kaucher AV et al (2013a) MicroRNAs 221 and 222 regulate the undifferentiated state in mammalian male germ cells. Development 140:280–290

Yang S, Wang S, Luo A et al (2013b) Expression patterns and regulatory functions of microRNAs during the initiation of primordial follicle development in the neonatal mouse ovary. Biol Reprod 89:126

Yao N, Lu C-L, Zhao J-J et al (2009) A network of miRNAs expressed in the ovary are regulated by FSH. Front Biosci 14:3239–3245

Yao G, Yin M, Lian J et al (2010a) MicroRNA-224 is involved in transforming growth factor-beta-mediated mouse granulosa cell proliferation and granulosa cell function by targeting Smad4. Mol Endocrinol 24:540–551

Yao N, Yang B-Q, Liu Y et al (2010b) Follicle-stimulating hormone regulation of microRNA expression on progesterone production in cultured rat granulosa cells. Endocrine 38:158–166

Yao G, Liang M, Liang N et al (2014) MicroRNA-224 is involved in the regulation of mouse cumulus expansion by targeting Ptx3. Mol Cell Endocrinol 382:244–253

Yin M, Lü M, Yao G et al (2012) Transactivation of microRNA-383 by steroidogenic factor-1 promotes estradiol release from mouse ovarian granulosa cells by targeting RBMS1. Mol Endocrinol 26:1129–1143

Yin M, Wang X, Yao G et al (2014) Transactivation of microRNA-320 by microRNA-383 regulates granulosa cell functions by targeting E2F1 and SF-1 proteins. J Biol Chem 289: 18239–18257

Yin F, Zhang JN, Wang SW et al (2015) MiR-125a-3p regulates glioma apoptosis and invasion by regulating Nrg1. PLoS One 10:e0116759

Yu Z, Raabe T, Hecht NB (2005) MicroRNA Mirn122a reduces expression of the posttranscriptionally regulated germ cell transition protein 2 (Tnp2) messenger RNA (mRNA) by mRNA cleavage. Biol Reprod 73:427–433

Zeng F, Schultz RM (2005) RNA transcript profiling during zygotic gene activation in the preimplantation mouse embryo. Dev Biol 283:40–57

Zhang J, Ji X, Zhou D et al (2013a) miR-143 is critical for the formation of primordial follicles in mice. Front Biosci 1:588–597

Zhang Q, Sun H, Jiang Y et al (2013b) MicroRNA-181a suppresses mouse granulosa cell proliferation by targeting activin receptor IIA. PLoS One 8:e59667

Zhang H, Jiang X, Zhang Y et al (2014a) MicroRNA 376a regulates follicle assembly by targeting Pcna in fetal and neonatal mouse ovaries. Reproduction 148:43–54

Zhang X, Simerly C, Hartnett C et al (2014b) Src-family tyrosine kinase activities are essential for differentiation of human embryonic stem cells. Stem Cell Res 13:379–389

Zimmermann C, Romero Y, Warnefors M et al (2014) Germ cell-specific targeting of DICER or DGCR8 reveals a novel role for endo-siRNAs in the progression of mammalian spermatogenesis and male fertility. PLoS One 9:e107023

Chapter 13
The Battle of the Sexes: Human Sex Development and Its Disorders

Anna Biason-Lauber

Abstract The process of sexual differentiation is central for reproduction of almost all metazoan and therefore for maintenance of practically all multicellular organisms. In sex development we can distinguish two different processes: First, sex determination is the developmental decision that directs the undifferentiated embryo into a sexually dimorphic individual. In mammals, sex determination equals gonadal development. The second process known as sex differentiation takes place once the sex determination decision has been made through factors produced by the gonads that determine the development of the phenotypic sex. Most of the knowledge on the factors involved in sexual development came from animal models and from studies of cases in whom the genetic or the gonadal sex does not match the phenotypical sex, i.e., patients affected by disorders of sex development (DSD). Generally speaking, factors influencing sex determination are transcriptional regulators, whereas factors important for sex differentiation are secreted hormones and their receptors. This review focuses on the factors involved in gonadal determination, and whenever possible, references on the "prismatic" clinical cases are given.

Abbreviations

AMH	Anti-Müllerian hormone
BPES	Blepharophimosis–ptosis–epicanthus inversus syndrome
CD	Campomelic dysplasia
CGD	Complete gonadal dysgenesis
CHD	Congenital heart defect
CTNNB1	Catenin (cadherin-associated protein) beta 1

A. Biason-Lauber, M.D. (✉)
Department of Medicine, University of Fribourg, Chemin du Musée 5, 1700 Fribourg, Switzerland
e-mail: anna.lauber@unifr.ch

R.P. Piprek (ed.), *Molecular Mechanisms of Cell Differentiation in Gonad Development*, Results and Problems in Cell Differentiation 58,
DOI 10.1007/978-3-319-31973-5_13

CYP17A1	Cytochrome P450, family 17, subfamily A, polypeptide 1
CYP19A1	Cytochrome P450, family 19, subfamily A, polypeptide 1
DAX1	DSS, adrenal hypoplasia critical region, on chromosome X, gene 1
DHH	Desert hedgehog
DMRT1 2, 3	Doublesex and mab-3 related, transcription factor 1, 2, 3
dpc	Days *post coitum*
DSD	Disorders of sex development
DSS	Dosage-sensitive sex reversal
Ems	Empty spiracles
EMX2	Empty spiracles homeobox 2
ESR1	Estrogen receptor 1
FGF9	Fibroblast growth factor 9
FGFR1, 2	Fibroblast growth factor receptors 1 and 2
FOG2	Friend of GATA protein 2
FOXL2	Forkhead box L2
FSH	Follicle-stimulating hormone
FST	Follistatin
FZD1	Frizzled 1
GATA4	GATA-binding protein 4
GNRHR	Gonadotropin-releasing hormone receptor
GOF	Gain of function
Gpr49	G-protein coupled receptor 49
GSK3-β	Glycogen synthase kinase 3 β
HMG	High mobility group
HSD3B2	3 Beta-hydroxysteroid dehydrogenase/delta 5-4-isomerase type 2
IHH	Indian hedgehog
INHA	Inhibin α-subunit
INSL3	Insulin-like factor 3
JNK	c-Jun NH2-terminal kinase
LHX9	LIM homeobox 9
LOF	Loss of function
LRP5, 6	Low-density lipoprotein receptor-related proteins 5 and 6
M33	Mouse polycomb group member M33
MAP3K1	Mitogen-activated protein kinase kinase kinase 1
MAP3K4	Mitogen-activated protein kinase kinase kinase 4
MAPK	Mitogen-activated protein kinase
MIS	Müllerian-inhibiting substance
MMTV	Mouse mammary tumor virus
NR0B1	Nuclear receptor subfamily 0 group B, member 1/DAX1
NR5A1	Nuclear receptor subfamily 5 group A member 1/SF1
Ods	Odsex
PcG	Polycomb group
PGD	Partial gonadal dysgenesis
PKC	Protein kinase C

POF	Premature ovarian failure
PRC 1, 2	Polycomb repressive complexes 1 and 2
PTCH1	Patched 1
RSPO1	R-spondin 1
SERKAL	Sex reversal with dysgenesis of kidneys, adrenals, and lungs
SF1	Steroidogenic factor-1
SHH	Sonic hedgehog
SOX9	SRY-related HMG box 9
SRY	Sex-determining region on the Y chromosome
STAR	Steroidogenic acute regulatory protein
TESCO	TES core sequence
WAGR	Wilms' tumor, aniridia, genitourinary, anomalies, and mental retardation
WNT4	Wingless-type MMTV integration site family, member 4
WT1	Wilms' tumor suppressor 1

13.1 Physiology of Human Sex Development

In sex development we can distinguish two different processes: sex determination, which is the developmental decision that directs the undifferentiated embryo into a sexually dimorphic individual, and sex differentiation, which takes place once the sex determination decision has been made through factors produced by the gonads that determine the development of the phenotypic sex. At the beginning of gestation (first and second weeks), embryos of the two sexes differ only in terms of their karyotypes. The turning point in mammalian sex development is the determination of the gonads. In humans, starting at third week, specific genes lead to the differentiation of the gonads that in turn produce hormones inducing anatomical and psychological differences, driving behavioral differences that are ultimately influenced by the social environment. At gestational weeks 6–7, the paramesonephric ducts (Müllerian ducts) develop next to the mesonephric ducts (Wolffian ducts). If testes develop and secrete testosterone, the mesonephric ducts increase in size and differentiate into epididymis, vas deferens, and prostate. A glycoprotein secreted from the Sertoli cells known as anti-Müllerian hormone (AMH) or Müllerian-inhibiting substance (MIS) results in Müllerian duct regression. If testes do not develop, the mesonephric ducts do not grow and eventually degenerate, whereas the paramesonephric ducts proliferate and the fallopian tubes, uterus, and upper third of the vagina develop (Fig. 13.1). In mammals, including humans, the differentiation of the gonads is the turning point of this whole process. The classical textbook theory says that in the presence of the sex-determining region on the Y chromosome (*SRY*), the default female pathway of sex determination will be inhibited and therefore testes will be formed. In the XX individual due to the absence of SRY, no inhibition of the default program will take place and ovaries will develop. Ovarian-determination factors might help the process of

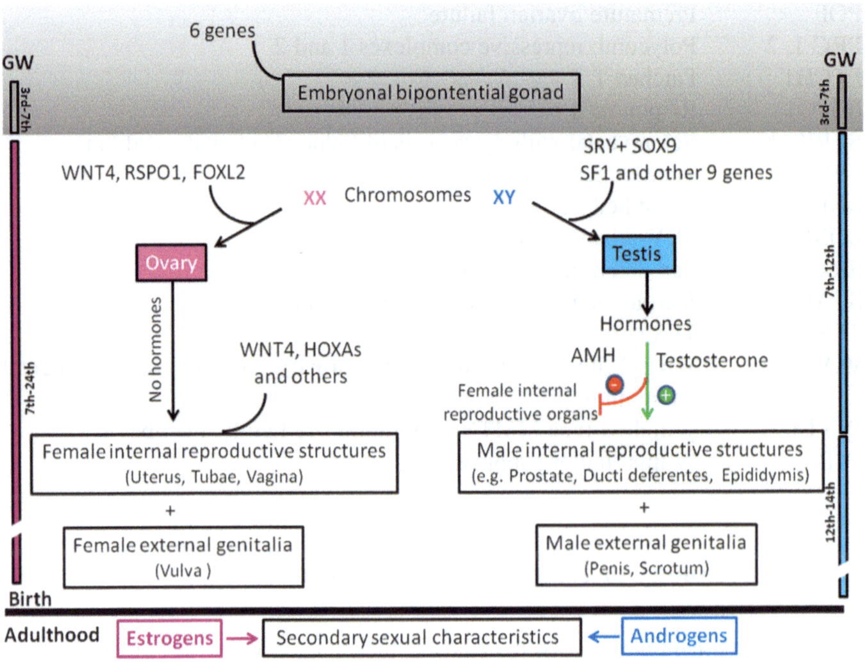

Fig. 13.1 Physiology of sex development. Simplified view of sex development in human life from intrauterine development to adulthood. *GW* gestational week. The main players supporting cell differentiation and the downstream sexual differentiation of other gonadal cell types are indicated. *SRY* sex-determining region Y, *SOX9* SRY-box 9, *WNT4* wingless-type MMMTV integration site family member 4, *RSPO1* R-spondin 1. In the fetal period, in contrast to the testis, the ovary is hormonally inactive. However and again in contrast to the testis, the entire oocyte patrimony is set during intrauterine life

differentiation. However, these factors are yet not fully determined. The second model called the Z-factor theory was proposed to explain the cases where XX individuals develop testes in the absence of SRY. According to this theory, the XX gonad expresses a factor that has both anti-testis and pro-ovary function. SRY in XY individuals acts as an inhibitor of the Z-factor to lift the block on the male pathway. In this case, the bipotential gonad will differentiate into a testis (McElreavey et al. 1993). It seems that SRY acts on a single gene, SRY-box 9 (*SOX9*), the expression of which is then rapidly reinforced by positive regulatory loops. SOX9 then drives Sertoli cell formation and, therefore, testis differentiation. If SRY is absent and fails to act in time or *SOX9* is otherwise silenced, the follicle cell, chiefly pre-granulosa cells, develops and ovary ensues, with β-catenin being one of the crucial components driving this process. In simple terms testis formation requires *SOX9* expression to be high (on), while ovary development needs *SOX9* to be low (off) (Fig. 13.2).

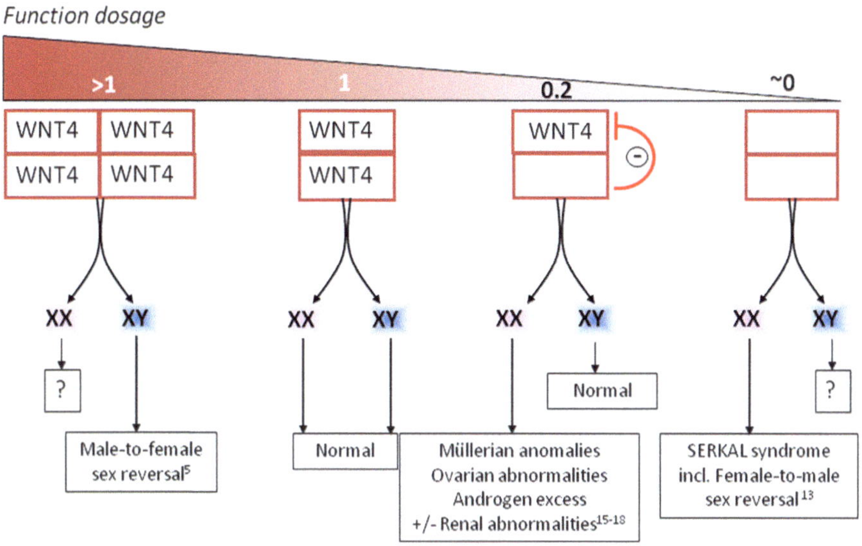

Fig. 13.2 Hypothetical relationship between WNT4 dosage and manifestation of disease in human. Normal function (and dosage) is set as 1.0. Heterozygote mutations have a dominant negative effect on the wild-type protein (*red line*) and lower the activity by 20 % instead of the expected 50 %

In the fetal period, in contrast to the testis, the ovary is hormonally inactive. However and again in contrast to the testis, the entire oocyte patrimony is set during intrauterine life (Fig. 13.1).

Although factors involved in male sexual development have been well studied, the pathways regulating female sexual differentiation remain incompletely defined. Until recently, no genes had been identified to play a similar role in ovarian development as was shown for *SRY* or *SOX9* gene in testicular development. Animal models and studies of patients with disorders/differences of sex development (DSD) allowed the discovery of factors actively involved in ovarian determination. WNT4 (wingless-type MMTV integration site family, member 4), RSPO1 (R-spondin 1), and FOXL2 (forkhead transcription factor 2) are some of a few factors with a demonstrated function in the ovarian-determination pathway. They suppress male sexual differentiation mainly by restraining *SOX9* expression and promote female development by sustaining Müllerian ducts differentiation and oocyte health (see below). Most of our knowledge on sex development derives from mouse models and the study of patients with disorders/differences of sex development (DSD, Table 13.1).

Table 13.1 Genes involved in the development of the bipotential gonad, testis development, and ovarian development in mice and humans

Gene	Protein function	Gonad phenotype in mouse models	Human syndrome/phenotype (DSD related)			
Genes involved in the development of the bipotential gonad						
Emx2	TXNF	*Emx2*$^{-/-}$ lacking kidneys, ureters, gonads, and genital tracts	6,XY DSD, a single kidney, intellectual disability			
Lhx9	TXNF	*Lhx9*$^{-/-}$ fail to develop bipotential gonad	[a]			
Nr5a1	Nuclear receptor/ TXNF	*Nr5a1*$^{-/-}$ fail to develop bipotential gonad	Embryonic testicular regression syndrome; XY gonadal dysgenesis			
			XX premature ovarian failure (POF)			
Wt1	TXNF	*Wt1*$^{-/-}$ mice fail to develop bi-potential gonad	Denys–Drash, WAGR, and Fraser syndromes			
Genes involved in testis development						
Dhh	Signaling molecule	*Dhh*$^{-/-}$ disruption of testis cord formation due to abnormal peritubular tissue	LOF: XY partial or complete gonadal dysgenesis			
Dmrt1	TXNF	*Dmrt1*$^{-/-}$ impaired testis development (P2) and loss of Sertoli and germ cells	Hemizygosity: XY gonadal dysgenesis			
			Deletion of 9p24 (including *DMRT1*): XY gonadal dysgenesis (varying degrees)			
Fgf9	Signaling molecule	*Fgf9*$^{-/-}$ XY male-to-female sex reversal and impaired development of Sertoli cells	[a]			
Fog2	Cofactor of Gata4	*Fog2*$^{-/-}$ reduced Sry levels, XY male-to-female sex reversal	Translocation including *FOG2*: XY hypergonadotropic hypogonadism combined with congenital heart defects (CHD)			

Gata4	TXNF	*Gata4*$^{-/-}$ embryo lethality (E7–E9.5)	LOF: XY ambiguous genitalia or reduced phallus length			
		Gata4^{ki} severe anomalies of testes	Deletion downstream of *GATA4* (also including *NEIL2*): XY complete gonadal dysgenesis combined with adrenal hypoplasia congenital (CAH)			
Map3k1	Kinase	[b](Dependent on genetic background)	XY partial or complete gonadal dysgenesis			
Map3k4	Kinase	XY male-to-female sex reversal	[a]			
Nr0b1	Nuclear receptor	XY impaired testis cord formation and spermatogenesis (dependant on genetic background)	LOF/deletions: congenital adrenal hypoplasia (CAH)			
			Duplications: XY gonadal dysgenesis with disorganized testis cords and hypogonadotropic hypogonadism			
Sox3	TXNF	[b]	Duplications (including *SOX3*) and deletion upstream of *SOX3*: XX testicular DSD			
Sox8	TXNF	*Sox8*$^{-/-}$ reduced fertility	[a]			
		Sox8$^{-/-}$, *Sox9*$^{-/-}$ XY variable degree of male-to-female sex reversal				

(continued)

Table 13.1 (continued)

Gene	Protein function	Gonad phenotype in mouse models	Human syndrome/phenotype (DSD related)			
Sox9	TXNF	*Sox9*$^{-/-}$ XY male-to-female sex reversal	LOF: XY gonadal dysgenesis combined with campomelic dysplasia (CD)			
		Ods (deletion upstream of *Sox9*) XX female-to-male sex reversal	GOF: XX female-to-male sex reversal (GOF)			
		Conditional knockout: XY ovarian development	Duplications including *SOX9*: XX testicular DSD			
		XX testicular development	Translocation upstream of SOX9: XY ovotesticular DSD			
			Duplication/triplication upstream of *SOX9*: XX testicular DSD			
			Deletion upstream of SOX9: XY ovarian DSD with acampomelic campomelic dysplasia (ACD), gonadal dysgenesis, female, or ambiguous external genitalia			
Sox10	TXNF	[b]	Duplication encompassing *SOX10* (includes also other genes): XX masculinized or incompletely feminized			
Sry	TXNF	*Sry*: XY male-to-female sex reversal	LOF: XY ovarian DSD			
		Sry translocation: XX female-to-male sex reversal	GOF/translocation: XX testicular DSD			

CBX2	TXNF	Testes: **0 %** Ovaries: **50 %** Testis +1x ovary (*not* ovotestis): **25 %** Undefined: **25 %** *Internal genitalia* Female (bipartite uterus): **75 %** Female + male (monopartite uterus + deferens): **25 %** Male (deferentes): **0 %** *External genitalia* Male: **16.6 %** Female: **33.3 %** Intersex: **50 %** *Fertility* All animals are *sterile*	–LOF XY: ovary-like gonads with oocytes –Female external genitalia –Müllerian structures present	Normal	*Gonads*: Testes: **0 %** Ovaries: **50 %** Testis +1x ovary (*not* ovotestis): **25 %** Undefined: **25 %** *Internal genitalia* Female (bipartite uterus): **75 %** Female + male (monopartite uterus + deferens): **25 %** Male (deferentes): **0 %** *External genitalia* Male: **16.6 %** Female: **33.3 %** Intersex: **50 %** *Fertility* All animals are *sterile*	*Gonads* Testes: **0 %** Ovaries (smaller): **87.5 %** Testis + 1x ovary: **0 %** Ovary/1x null: **12.5 %** *Internal genitalia* Female (bipartite uterus): **87.5 %** Female + male (monopartite uterus + deferens): **12.5 %** Male (deferentes): **0 %** *External genitalia* Male: 0 % Female: **75 %** Intersex: **25 %** *Fertility* All animals are *sterile*
Genes involved in ovary development						
Ctnnb1/β-catenin	TXNF	Conditional knockout in Sertoli cells	[a]			
		XX similar to $Wnt4^{-/-}$ and $Rspo1^{-/-}$				
Foxl2	TXNF	$Foxl2^{-/-}$ premature ovarian failure	BPES and premature ovarian failure (POF)			
		$Foxl2^{-/-}$, $Wnt4^{-/-}$ XX female-to-male sex reversal				

(continued)

Table 13.1 (continued)

Gene	Protein function	Gonad phenotype in mouse models	Human syndrome/phenotype (DSD related)			
Fst	Inhibitor of activin	XX partial female-to-male sex reversal, coelomic vessel formation	[a]			
Rspo1	Signaling molecule	XX partial female-to-male sex reversal, similar to *Wnt4*$^{-/-}$ and conditional *Ctnnb1* knockout	XX testicular and ovotesticular DSD			
			Duplication of 1p (including *WNT4* and *RSPO1*): XY gonadal dysgenesis (GOF)			
Wnt4	Signaling molecule	XX Müllerian duct agenesis, testosterone synthesis, and coelomic vessel formation	Duplication of 1p (including *WNT4* and *RSPO1*): XY gonadal dysgenesis (GOF) (male-to-female sex reversal)			
			LOF: XX Müllerian duct agenesis, testosterone synthesis, and coelomic vessel formation			

TXNF transcription factor
[a]No mutations described so far
[b]No gonadal phenotype

13.2 Disorders/Differences of Sex Development

DSD are congenital conditions in which development of chromosomal, gonadal, or anatomical sex is atypical. DSD covers a wide spectrum of different phenotypes, with hypospadias being the most common defect with an average of 1 in 250–350 male births. In addition, 1 in 4500 babies worldwide is born with ambiguous genitalia, and DSD account for 7.5 % of all birth defects (Groth et al. 2013).

Furthermore, DSD phenotypes are often associated with other syndromes, such as Mayer–Rokitansky–Küster–Hauser syndrome, Smith–Lemli–Opitz syndrome, or genito-palato-cardiac syndrome (Porter 2008; Sultan et al. 2009). Since the discovery of the sex-determining region on the Y chromosome (*SRY*) gene in 1990 (Sinclair et al. 1990), there have been considerable advances in understanding the genetic factors involved in gonad differentiation. Nevertheless, it has been estimated that a molecular diagnosis is made in only approximately 20 % of DSD cases (Groth et al. 2013) and that up to 50 % of 46,XY DSD patients cannot be provided with an accurate diagnosis. Furthermore, for approximately 80 % of 46, XY complete gonadal dysgenesis patients and about 20 % of 46,XX testicular DSD patients, the causative mutation remains unknown (Groth et al. 2013). Unpublished data from several research groups suggest the percentage of patients without a molecular diagnosis has decreased to approximately 70 % for 46,XY gonadal dysgenesis and 10 % for 46,XX testicular DSD, as more and more mutations are identified in known DSD genes (e.g., *SF1/NR5A1*). DSD represent a major pediatric concern, due to the difficulty of clinical management of these complex conditions and their common sequelae of gonad cancer and infertility. The cause of these DSD conditions is most often the breakdown of the complex network of gene regulation and gene expression, essential for proper development of testes or ovaries in the embryo.

Despite advances in understanding the genetic basis of human sexual determination and differentiation, in most subjects with DSD, the underlying genetic cause is unknown. Whereas pathogenic mutations have been identified in DSD patients, the phenotype can be highly variable, even within families, suggesting that other factors, including genetic variants, are influencing the expression of the phenotype (Bashamboo et al. 2010b). Unexplained DSD cases may be due to mutations in novel gonad-determining genes or genomic alterations affecting regulatory regions that lead to atypical gene expression. The identification of new genes involved in sex determination is therefore a goal of the utmost importance to the understanding of DSD and will allow for more accurate diagnosis and clinical management of these complex conditions (Baxter and Vilain 2013).

Given the importance of gonadal determination for human sex development, this chapter will focus on factors involved in gonadal development and the effect of their mutations in patients. Factors regulating sex differentiation, such as steroidogenic enzymes or end-organ receptors, will not be discussed.

13.3 Genes Required for the Development of the Bipotential Gonad

Mammalian gonads arise in both sexes from a bilateral, bipotential gonad (also called genital ridge), an organ that has the potential to develop either as an ovary or as a testis, depending on differentially expressed genes (Capel 1998, 2000; Swain and Lovell-Badge 1999). In mouse the bipotential gonads are first visible at 9.5 days *post coitum* (*dpc*), 1 day before the onset of *Sry* expression, which is the critical gene for initiating testis development in XY individuals. A number of genes have been shown to be required for the development of the bipotential gonad (Fig. 13.3; see also Table 13.1 for a summary of mouse and human phenotypes of genes described in this section).

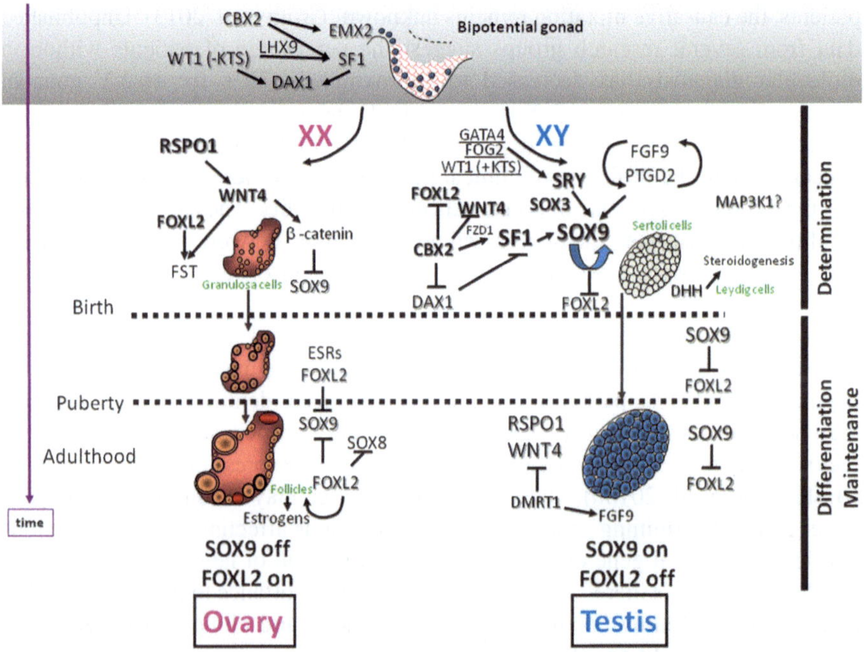

Fig. 13.3 Simplified model of sex development cascade. In early stages of determination, SRY in XY subjects initiates SOX9 upregulation, which is then maintained by SOX9 itself, SF1, prostaglandin D2 (PTGD2), and fibroblast growth factor 9 (FGF9). In the XX gonad WNT4/RSPO1 stabilize β-catenin that in turn suppresses SOX9. After birth, FOXL2 with the help of the estrogen receptors (ERs) maintains the transcriptional repression of SOX9, which is necessary to preserve ovarian function throughout life. Factors that when mutated are involved in DSD are shaded. Derived from Biason-Lauber (2012)

13.3.1 *Emx2*

The homeobox gene *Emx2* (empty spiracles homeobox 2) is a mouse homologue of the Drosophila head gap gene empty spiracles (*ems*) and is essential for the development of dorsal telencephalon in mice (Yoshida et al. 1997). In addition, *Emx2* is expressed in the epithelial components of the developing urogenital system. The kidneys, ureters, gonads, and genital tracts are completely absent in $Emx2^{-/-}$ mutant mice, whereas the adrenal glands and bladder develop normally. Recently, Piard et al. (2014) diagnosed a male patient who presented with 46,XY DSD, a single kidney, intellectual disability, and the smallest microdeletion including *EMX2* reported to date. *EMX2* haploinsufficiency in human is likely to explain the masculinization defect observed, similar to what has been described in mice.

13.3.2 *Lhx9*

The LIM homeobox 9 (*Lhx9*) is a member of the LIM homeobox gene family. In $Lhx9^{-/-}$ mice, germ cells migrate normally, but somatic cells of the genital ridge fail to proliferate and a discrete gonad fails to form. As a result of the absence of testosterone and anti-Müllerian hormone, genetically male mice develop as phenotypic females (Birk et al. 2000; Luo et al. 1994). The expression of steroidogenic factor-1 (*Sf1* also known as nuclear receptor subfamily 5 group A member 1 (*Nr5a1*), a nuclear receptor essential for gonadogenesis, is reduced to minimal levels in the LHX9-deficient genital ridge, indicating that *Lhx9* may lie upstream of *Nr5a1* in a developmental cascade in mouse (Birk et al. 2000). Furthermore, an in vitro biochemical analysis showed that LHX9 has the potential to bind and transactivate the *Nr5a1* promoter in conjunction with WT1 (Wilms' tumor suppressor 1) (Wilhelm and Englert 2002). However, mutation analysis of *LHX9* in a range of human DSD patients (including bilateral gonadal agenesis) did not reveal any mutations (Ottolenghi et al. 2001).

For **SF1/NR5A1**, **CBX2**, **WT1**, and **DAX1**, see below.

13.4 Male Sex Development in Humans: Health and Disease

13.4.1 *SRY*

The discovery of *Sry*, the mammalian Y-chromosomal sex-determining gene, promised to answer many questions relating to the genetics and developmental biology of sex determination (Gubbay et al. 1990; Sinclair et al. 1990). The simple transgenesis experiment of adding *Sry* function to an XX mouse to induce male phenotype confirmed the master switch role of this gene (Koopman et al. 1991). But since then, little else about this gene has proven to be simple or straightforward. The fact that *Sry* resides on the Y chromosome makes it vulnerable to degradation. As a result, it is minimally conserved and shows some functional flaws, surprising indeed for a gene on which survival and propagation of mammalian species so much depend on.

In the mouse embryo, *Sry* exhibits a tightly controlled and limited spatio-temporal profile of expression in the precursors of Sertoli cells of the XY gonad (Albrecht and Eicher 2001; Sekido et al. 2004; Wilhelm et al. 2005). Early studies revealed that *Sry* is first expressed around 10.5 *dpc*, shortly after the emergence of the genital ridges, reaches peak levels of expression at 11.5 *dpc*, and is extinguished shortly after 12.5 *dpc* in mouse (Hacker et al. 1995; Jeske et al. 1995; Koopman et al. 1990). SRY clearly throws a molecular switch to engage a male-specific cascade of molecular events, but continued expression of *Sry* is not required for these events to unfold.

Despite almost 20 years of work since the discovery of SRY, little is known about its biochemical function. The HMG domain, a DNA-binding and DNA-bending motif, plays a central role, being the only region conserved between species and the site of almost all clinical mutations causing XY gonadal dysgenesis. It seems that SRY acts on a single gene, *Sox9*, the expression of which is then rapidly reinforced by positive regulatory loops. SOX9 then drives Sertoli cell formation and, therefore, testis differentiation. If SRY is absent or fails to act in time, *Sox9* is silenced and the development of the follicle cell and ovary ensues, with β-catenin being one of the crucial components driving this process.

SRY and human disease Mutations in *SRY* are the cause of 46,XY DSD due to pure gonadal dysgenesis, originally described as XY, female with gonadal dysgenesis (Jager et al. 1992).

13.4.2 SOX9

Several lines of evidence indicate that *Sox9* is the best candidate for a direct SRY target gene (Sekido et al. 2004; Sekido and Lovell-Badge 2008; Wilhelm et al. 2005). First, *Sox9* expression is strongly upregulated soon after the expression of *Sry* begins, whereas it is downregulated in the ovary. Second, cell-fate mapping experiments show that SRY-positive cells exclusively become SOX9-positive Sertoli cells. Third, heterozygous mutations in SOX9 are responsible for the human skeletal malformation syndrome, campomelic dysplasia, in which most XY patients have male-to-female sex reversal. Similarly, targeted *Sox9* ablation in mice also leads to ovary development in XY embryos (Barrionuevo et al. 2006; Chaboissier et al. 2004).

However, the expression of *Sox9* in the genital ridges is not entirely dependent on SRY, and the normal gonadal *Sox9* transcriptional regulation consists of an SRY-independent initiation, an SRY-dependent upregulation, and SRY-independent maintenance in adult life.

SF1 is a good candidate for initiating or sensitizing *Sox9* expression because early sex-independent expression of *Sox9* is abolished in *Sf1* null mutant gonads (Lovell-Badge 2002; Sekido et al. 2004).

SF1 sensitizes *Sox9*, initiating a low level of expression in the genital ridge of both sexes at 10.5 *dpc*. In the male, SF1 (probably with other factors such as WT1−KTS) also activates *Sry* expression. *Sox9* expression is upregulated by the action of SRY together with SF1, whereas it is downregulated in the female. This downregulation is unlikely to be passive, implying the presence of one or more currently unknown repressors. After the transient expression of *Sry* has ceased, high levels of SOX9 are maintained by its direct autoregulation and via FGF9 signaling.

Regulatory elements

A recent breakthrough in the field of mammalian sex determination was the finding of a conserved testis-specific enhancer of *Sox9* in mice (mTES) that has revealed a co-transcriptional network of SRY, SF1, and SOX9 involved in the direct initiation, upregulation, and maintenance of *Sox9* expression in the mouse XY gonad (Sekido and Lovell-Badge 2008). A human homologous 3.2 kb enhancer element 12 kb upstream of the SOX9 coding region, termed human *TES* (*hTES*), has been identified as responsive to human SRY, SF1, and SOX9, thereby upregulating *SOX9* transcription in a testis-specific manner (Fig. 13.2) (Knower et al. 2011). Binding sites for SF1 and SRY/SOX9 are spread throughout the *hTES*, but the 1.4 kb core sequence *TESCO* is sufficient for enhancer activity. Indeed, overexpression of *Dax1* reduces *Sox9 mTES* activity in developing XY *Dax1* Tg gonads in vivo (Ludbrook et al. 2012) and inhibits the synergic activation of *TESCO* by SF1 and SRY and by SF1 and SOX9 in NT2/D1 cells, giving a feasible explanation of the dosage-sensitive sex-reversal (DSS) phenotype in 46,XY DSD patients with *DAX1* duplications (Ludbrook et al. 2012).

Moreover, it has been demonstrated that FOXL2 and ESR1 (estrogen receptor 1) synergize in repressing *TESCO* in mice granulosa cells, therefore maintaining the ovarian phenotype (Hyon et al. 2015). Also, ectopic activation of the WNT canonical pathway through stabilization of β-catenin leads to a male-to-female sex-reversal phenotype in XY mice inhibiting DNA binding of SF1 to *TESCO*, thus preventing *Sox9* transcription (Bernard et al. 2008).

Simultaneous mutation of SRY- and SF1 binding sites within *TESCO* abolished testis-specific expression in transgenic mice (Sekido and Lovell-Badge 2008).

SOX9 and human disease

Limbs abnormalities. As mentioned, mutations in *SOX9* lead to 46,XY DSD and skeletal abnormalities known as campomelic dysplasia (CD). Kurth et al. (2009) suggested that the duplications involved putative regulatory elements of *SOX9* and may induce *SOX9* misexpression and/or overexpression at specific time points during development, resulting in abnormal digit and nail development. In mouse embryo, *Sox9* was strongly expressed in the distal mesenchymal condensations that develop into terminal phalanges (Cook syndrome).

DSD and SOX9 regulatory elements Several point mutations and small insertions have been identified in patients with CD and 46,XY DSD (Cameron et al. 1996; Foster et al. 1994; Ninomiya et al. 2000; Pop et al. 2005; Wagner et al. 1994).

In recent years, copy-number variations (CNV) of a region upstream of SOX9 have been associated with 46,XX DSD or 46,XY DSD. Gains in copy number have been discovered in four males from two families with 46,XX *SRY*-negative testicular DSD and in three unrelated SRY-negative males with 46,XX ovotesticular DSD. In contrast, genomic deletion has been reported in two familial patients and three unrelated patients with 46,XY DSD (Uhlenhaut and Treier 2006). An overlapping 78 kb sex-determining region, named *RevSex*, located ~517–595 kb upstream of the *SOX9* gene was detected (Kusaka et al. 2010). Benko et al. (2011) proposed that the *RevSex* region encompasses gonad-specific *SOX9* transcriptional enhancer(s) and that copy-number variations of that interval result in genomic imbalances sufficient to activate or silence *SOX9* gonadal expression in a tissue-specific manner, switch sex determination, or result in an isolated DSD.

However, Lybæk et al. (2014) demonstrated that the *RevSex* duplication is associated with chromatin changes, predicting a better accessibility of the SRY-responsive TESCO enhancer region. According to this model, the chromatin environment in which *RevSex* is found would thus be either permissive for enhancer activity in males or repressive in females: *RevSex* duplication facilitates *SOX9* activation in the primordial gonads by inducing a conformational change to the *SOX9* transcription in the genital ridge, establishing a self-stimulatory SOX9 feedback loop. This may be a cause of testicular differentiation of primordial gonads despite the absence of SRY.

46,XX patients In a family with 46,XX testicular disorder of sex development in which three adult males (two brothers and a paternal uncle) were determined to be female according to karyotype (46,XX) and were negative for the *SRY* gene, a heterozygous 178 kb duplication 600 kb upstream of *SOX9* was identified (Cox et al. 2011). The duplication was arranged in tandem in wild-type orientation, and the joining points of the duplicated segments were uncorrupted. All affected family members carried the duplication in heterozygosity as did the proband's healthy, fertile 46,XY father. Affected individuals were infertile with azoospermia. In two men, the testes had been removed and prostheses placed during their 20s because of testicular pain secondary to testosterone replacement. Histologic exams showed the presence of Leydig and Sertoli cells, severely diminished and atrophied seminiferous tubules, and no spermatogenesis.

Similarly, in two 46,XX *SRY*-negative Italian brothers, who were phenotypically normal males but had hypotrophic testes and azoospermia, Vetro and collaborators identified heterozygosity for a 96 kb triplication located 500 kb upstream of the *SOX9* gene (that was not present in their two fertile sisters and mother (Vetro et al. 2011). The authors noted that, like the previously reported duplication (Cox et al. 2011), the triplication did not seem to have any effect on the XY background.

By performing CNV analysis in a cohort of 19 cases of *SRY*-negative 46,XX testicular or ovotesticular DSD, Kim et al. (2015) identified three unrelated individuals with heterozygous duplications upstream of the *SOX9* gene, one of which was shown to be paternally inherited. The three duplications and previously reported *SOX9* upstream duplication/triplication cases shared a common 68 kb duplicated region, located 516–584 kb upstream of *SOX9*, which they designated XY sex-reversal region XXSR for "XX sex-reversal region," noting that it was largely identical with the 78 kb *RevSex* region. The authors also defined a distinct 32.5 kb XY sex-reversal region (XYSR) upstream of the *SOX9* gene, based on 46, XY patients with deletions. The authors stated that the XYSR and XXSR intervals do not overlap, being separated by 23 kb, and proposed that each harbors a differently acting gonad-specific regulatory element.

Recently, Hyon et al. (2015) further defined the area by characterizing three phenotypically normal patients presenting with azoospermia and 46,XX testicular DSD. Two brothers carried a 83.8 kb duplication located ~600 kb upstream of *SOX9* that overlapped with the previously reported rearrangements. This duplication refines the minimal region associated with 46,XX *SRY*-negative DSD to a 40.7–41.9 kb element located ~600 kb upstream of *SOX9* (Hyon et al. 2015).

46,XY individuals In the same study, again Kim et al. (2015) analyzed 100 patients with *SRY*-positive 46,XY nonsyndromic partial or complete gonadal dysgenesis and were able to identify four unrelated individuals with heterozygous deletions upstream of the *SOX9* gene, including a patient from the family originally reported by German et al. (1978) and a patient from the family studied by Mann et al. (1983). Both of the latter deletions segregated with disease in the respective families. Together, the four deletions defined a 32.5 kb interval, which designated XYSR for "XY sex-reversal region," noting that it overlapped with previously described *SOX9* upstream deletions but not with the *RevSex* region. The authors also identified a distinct 68 kb XX sex-reversal region (XXSR) upstream of the *SOX9* gene, based on 46,XX patients with duplications, that was largely identical to the *RevSex* region. Kim et al. (2015) stated that the XYSR and XXSR intervals do not overlap, being separated by 23 kb, and proposed that each harbors a differently acting gonad-specific regulatory element. Testing of XYSR subfragments in cell transfection and transgenic experiments revealed a 1.9 kb SRY-responsive subfragment, designated F8, that drives expression specifically in Sertoli-like cells and contains consensus binding sites for SRY and WT1.

Hypertrichosis, congenital generalized, with or without gingival hyperplasia Microdeletions or microduplications upstream of *SOX9* have been also linked to congenital generalized hypertrichosis with or without gingival hyperplasia (DeStefano et al. 2014; Fantauzzo et al. 2012).

13.4.3 SF1/NR5A1

Human SF1/NR5A1 is a 461-amino-acid protein that shares structural homology with other members of the nuclear receptor superfamily (www.nursa.org). Critical functional domains of this protein include an amino-terminal 2 zinc finger DNA-binding domain (DBD), an accessory DNA-binding region, a hinge region, and a ligand-binding domain (LBD) that forms an AF-2 structure.

The expression pattern of *SF1* is consistent with its central role in regulating adrenal development, gonad determination and differentiation, and in the hypothalamic–pituitary control of reproduction and metabolism.

In mouse, *Sf1* is expressed in the early adrenogonadal primordium from 9 *dpc* and, thereafter, in the developing adrenal gland and gonad (Ikeda et al. 1994; Nef et al. 2005; Val et al. 2007). In humans, *SF1* expression has been shown in the developing adrenal gland and bipotential gonad at 32–33 days postconception (Hanley et al. 1999; Ramayya et al. 1997). Following testis determination (from around 42 days onward in humans), *SF1* expression is consistently maintained in the somatic cells of the early testis, where it may play a crucial role together with SRY in supporting *SOX9* expression (Sekido and Lovell-Badge 2008). In Sertoli cells, SF1 activates expression of *AMH* from around 7 weeks' gestation, which leads to the regression of Müllerian structures in the developing male fetus. In Leydig cells, SF1 activates the expression of steroidogenic enzyme systems from 8th week of gestation, which results in androgenization of the external genitalia. In females, persistent expression of *SF1* has been reported in early ovarian development in humans (Hanley et al. 1999), whereas *Sf1* expression may decline in mouse. However, SF1 is detectable in somatic cells (granulosa and theca cells) of the adult bovine ovary (Murayama et al. 2008). The exact role SF1 plays in the ovary is unclear, as the related nuclear receptor LRH1 (NR5A2) may also have a role in regulating aromatization and estrogen synthesis.

In addition to its expression in the adrenal gland and gonad, *SF1* is also expressed in the developing ventromedial hypothalamus (VMH), pituitary gonadotropes, and spleen (Kurrasch et al. 2007; Luo et al. 1994; Shinoda et al. 1995).

Deletion of the gene (*Nr5a1*) encoding SF1 in XY mice results in impaired adrenal development, complete testicular dysgenesis with Müllerian structures, and female external genitalia.

These studies in mice prompted the search for the human counterpart. Initial studies to identify *SF1/NR5A1* mutations in humans focused on patients with primary adrenal failure, 46,XY gonadal dysgenesis, and Müllerian structures persisted. The first prismatic case was identified by Achermann et al. (1999) in a patient with such phenotype.

Mutation in the human *SF1/NR5A1* gene has a phenotypic spectrum that ranges from complete testicular dysgenesis with Müllerian structures, through individuals with mild clitoromegaly or genital ambiguity, to severe penoscrotal hypospadias or even anorchia (Lin and Achermann 2008).

Adrenocortical insufficiency We described a female patient with adrenal insufficiency and no apparent defect in maturation, despite a heterozygous mutation in the *NR5A1* gene. Thus, *NR5A1* has a crucial role in adrenal gland formation in both sexes (Biason-Lauber and Schoenle 2000) but not essential for ovarian determination.

Adrenocortical tumors Using comparative genomic hybridization, Figueiredo et al. (2005) detected a consistent gain of chromosome 9q or a portion of it in eight of nine cases of pediatric adrenocortical tumors (ACTs) and amplification of 9q34 in the majority of these cases. They also examined if the *SF1* gene, which is located in this chromosomal region and plays an important role in the development and function of the adrenal cortex, is amplified in these ACT cases. They detected increased copy number of the *SF1* gene in all eight cases with 9q gain, suggesting an association between an increased copy number of the *SF1* gene and adrenocortical tumorigenesis.

Premature ovarian failure Although not essential for ovarian determination, SF1 appears to be necessary for maintenance of ovarian function. Sequencing of the *NR5A1* gene in four families with histories of both 46,XY disorders of sex development and 46,XX primary ovarian insufficiency and in 25 subjects with sporadic ovarian insufficiency (Lourenco et al. 2009) identified mutations in patients with premature ovarian failure (POF) as well as in patients with 46,XY disorders. None of the affected subjects had clinical signs of adrenal insufficiency. In-frame deletions and frameshift and missense mutations were detected. Functional studies indicated that these mutations substantially impaired NR5A1 transactivational activity.

Spermatogenic failure Bashamboo et al. (2010a) analyzed the candidate gene *NR5A1* in 315 men with idiopathic spermatogenic failure and identified heterozygous missense mutations in seven of them. This form of the disorder is designated spermatogenic failure-8 (SPGF8). None of the mutations were found in more than 2100 control samples, and analysis of the entire coding region of *NR5A1* in 370 fertile men (father of at least two children) or 359 normospermic men revealed no rare allelic variants.

46,XY DSD and spleen development abnormalities Zangen et al. (2014) identified a recessive mutation in SF1/NR5A1 in a child with both severe 46,XY DSD and asplenia. Interestingly, they demonstrated that SF1 is required for spleen development in humans via transactivation of *TLX1* and that mutations that only impair steroidogenesis, without altering the SF1/SRY transactivation of *SOX9*, can lead to 46,XY DSD (Zangen et al. 2014).

13.4.4 DAX1/NR0B1

DAX1 encodes a member of the orphan nuclear hormone receptor (NHR) family of transcriptional regulators. The C-terminal domain shows homology to the ligand-binding domain (LBD) of related NHRs, although a ligand remains unknown. The N-terminal domain represents a novel domain comprising 3.5 repeats of a 65–67-amino-acid motif containing two putative zinc fingers and possibly defining a nucleic acid-binding domain. DAX1 has a transcriptional silencing domain which has been proposed to interact directly with corepressors to mediate repression (Laudet et al. 1992). At least two corepressors, Alien and NCoR, interact in vitro with the DAX1 C-terminus (Iyer and McCabe 2004; Laudet 1997), one of which (Alien) is expressed in the testis (Ritchie et al. 1990).

In cultured cells, DAX1 represses SF1-mediated transcriptional activation of steroidogenic genes encoding enzymes, such as Cyp11A, Cyp17, and Cyp19 essential for the synthesis of androgens and estrogens; it interacts with the AF2 domain of ligand-bound estrogen receptors, ERα and ERβ, and represses ER-mediated activation of target reporter genes. It is probable that DAX1 and SF1 interact during gonadogenesis in a similar manner to that seen during steroidogenesis. Together, these studies suggest a common function for DAX1 as a repressor of SF1 transactivation in both steroidal and gonadal tissues (Lalli and Sassone-Corsi, 2003, and references therein).

DAX1 and human disease Inactivating mutations in *DAX1* (*NR0B1*) can lead to adrenal insufficiency with glucocorticoid and mineralocorticoid deficiency (McCabe 2001). These mutations are responsible for many, but not all, patients with the cytomegalic form of adrenal hypoplasia congenita (AHC). These mutations also cause hypogonadotropic hypogonadism (HH) which is consistent with the expression of *DAX1* in the hypothalamus and pituitary.

Duplications of the region of the X chromosome containing *DAX1* cause dosage-sensitive sex reversal (Bardoni et al. 1994; McCabe 2001). Transgenic XY mice with additional copies of the mouse *Dax1* ortholog expressed at high levels do not have male-to-female sex reversal but do show delayed testicular development (Swain et al. 1998).

13.4.5 DMRT1

Raymond et al. (1998) isolated the male sexual regulatory gene *mab3* (Shen and Hodgkin 1988) from the nematode *Caenorhabditis elegans* and found that it is related to the *Drosophila melanogaster* sexual regulatory gene “doublesex” (*dsx*) (Burtis and Baker 1989). Both genes encode proteins with a DNA-binding motif that Raymond et al. (1998) referred to as the “DM domain.” The same authors identified a human gene, which they designated *DMT1* that encodes a protein with a DM domain and found that *DMT1* is expressed only in testis.

Matson et al. (2011) found that *Dmrt1*$^{-/-}$ mouse gonads showed reprogramming of Sertoli cells into granulosa cells after birth. Testis in which deletion of *Dmrt1* was targeted to Sertoli cells appeared normal at birth, with normal expression of *Sox9*. However, *Dmrt1*-knockout Sertoli cells initiated expression of *Foxl2* (see below), which is normally enriched in ovary, and by postnatal day 28, few *Sox9*-expressing cells remained, and most intratubular cells strongly expressed *Foxl2*. Similar changes were found about 1 month after *Dmrt1* deletion in adult male mice. Loss of *Dmrt1* caused loss of Sertoli cell morphology and gain of a theca-like morphology associated with feminization in mRNA, enzyme expression, and response to gonadotropins. From this prospective, *Dmrt1* can be considered as a testis-maintaining factor analog to what *Foxl2* is for ovarian function.

Muroya et al. (2000) reported clinical and molecular findings in five 46,XY and one 46,XX patients with distal 9p monosomy. Some of the XY and the XX subjects had female external genitalia, one case showed ambiguous external genitalia, and another exhibited male external genitalia with left cryptorchidism and right intrascrotal testis. Gonadal explorations at gonadectomy revealed the presence of various morphologies, from streak gonad and right agonadism to hypoplastic testes. The first point mutation in *DMRT1* associated with 46,XY DSD allowed Murphy et al. (2015) to demonstrate that affecting DNA recognition of DMRT1 is associated with sex developmental defects in humans and an intersex phenotype in flies and indicated an evolutionary ancient molecular interaction underlying much of metazoan sexual development.

13.4.6 WT1

Wilms' tumor suppressor 1 (*WT1*) encodes a zinc finger DNA-binding protein, which is primarily expressed in embryonic mesodermal tissues such as the urogenital ridge, gonads, and mesonephros (Armstrong et al. 1993). *WT1* mutations in humans were first identified in patients with Wilms' tumor, a form of kidney cancer occurring primarily in children (Haber et al. 1990). This tumor is also seen as part of WAGR syndrome, which includes other clinical features such as aniridia, genitourinary anomalies, and mental retardation. This gonadal phenotype is recapitulated in other disorders associated with mutations in *WT1*, such as Denys–Drash syndrome (including gonadal abnormalities and renal failure) (Lee et al. 2011; Pelletier et al. 1991b) and Frasier syndrome (46,XY gonadal dysgenesis together with glomerulopathy) (Barbaux et al. 1997). The Frasier patients have been found to carry mutations that result in the loss of the WT1 +KTS isoform. This isoform has an additional three amino acids (K, lysine; T, threonine; and S, serine), which are located between the third and the fourth zinc finger of WT1. The two different isoforms –KTS and +KTS have been shown to play distinct roles during embryogenesis. The –KTS isoform has been shown to bind to the SRY promoter region, leading to the transactivation of *SRY* (Hossain and Saunders 2001). Furthermore, the –KTS isoform of Wt1 was shown to bind sequences within the *Nr5a1* promoter (Wilhelm and Englert 2002). Conversely, the +KTS isoform seems to play a role in the regulation of the *SRY* transcript. Knocking out this +KTS isoform of Wt1 specifically results in reduced *Sry* levels (Hammes et al. 2001), presumably mediated by its RNA-binding affinity. There is also evidence that this isoform may function synergistically with *Nr5a1* to increase Amh expression (Arango et al. 1999), which is essential for inhibiting the development of the female Müllerian structures. Mice carrying a mutated *Wt1* gene fail to develop kidneys and gonads, and heart defects often lead to embryo lethality (Kreidberg et al. 1993).

Mutations in the WT1 gene have been identified in patients with renal disease, such as Wilms' tumor and isolated diffuse mesangial sclerosis (IDMS), and complex diseases like WAGR syndrome, Frasier syndrome, Denys–Drash syndrome (DDS), and Meacham syndrome. The association of aniridia, hemihypertrophy, and other congenital anomalies with Wilms' tumor was first described by Miller et al. (1964). The syndrome subsequently became known as the WAGR syndrome (Wilms' tumor, aniridia, genitourinary anomalies, and mental retardation syndrome). Aniridia is due to the PAX6 gene, whereas the other features are due to WT1 gene defects (Breslow et al. 2000). Frasier syndrome is a rare disorder defined by 46,XY DSD and progressive glomerulopathy (Frasier et al. 1964). Patients present with normal female external genitalia, streak gonads, and XY karyotype and frequently develop gonadoblastoma. Wilms' tumor is not a feature of the syndrome. Frasier syndrome is caused by a mutation in the donor splice site in intron 9 of WT1, with the predicted loss of the +KTS isoform (Barbaux et al. 1997). In contrast with Frasier syndrome, most individuals with Denys–Drash syndrome (Denys et al. 1967; Drash et al. 1970) have ambiguous genitalia or a female phenotype, an XY karyotype, and dysgenetic gonads. Renal symptoms are characterized by diffuse mesangial sclerosis, usually before the age of 1 year, and the patients frequently develop Wilms' tumor. Pelletier identified mutations of various types in WT1 as the molecular cause of DDS (Pelletier et al. 1991a). Meacham described two 46,XY individuals with a novel constellation of genital, cardiac, and pulmonary malformations. The genital abnormalities consisted of a true double vagina, retention of Müllerian structures, and undervirilization of the external genitalia (Meacham et al. 1991). Although phenotype and sometimes phenotype overlap with that of DDS patients (Suri et al. 2007), the Meacham syndrome patients do not have renal disease.

13.4.7 CBX2

CBX2/M33 is a member of the Polycomb group (PcG) proteins, highly conserved regulatory factors initially discovered in Drosophila.

PcG genes are best known for their role in maintaining silent expression states of Hox genes during development. They act by regulating chromatin structure and chromosome architecture at their target loci.

Targeted ablation of the polycomb M33, the mouse homolog human CBX2, causes 46, XY sex reversal in these animals. Apart from their sterility, 50 % of M33 knock-out Sry positive mice are phenotypically females (ovaries with follicles, uterus and normal external genitalia), placing M33 upstream of Sry in the murine sex development cascade (Katoh-Fukui et al. 1998). Similarly, our group made the discovery of a loss-of-function double heterozygote mutation state in a 46, XY girl with ovaries at histology, normal uterus and external female genitalia, accidentally diagnosed because of a discrepancy between prenatal karyotype and phenotype at birth. Functional studies demonstrated that the mutated CBX2 does not properly bind to and does not adequately regulate the expression of target genes essential for sex development such as SF1/NR5A1. Our data identified CBX2 as essential for normal human male gonadal development, suggested that it lies upstream of SRY in the human sex development cascade and identified a novel autosomal recessive cause of DSD. From a more mechanistic point of view, we demonstrated that CBX2 might have a role as transactivator distinct from its known function as chromatin-modifier (Biason-Lauber et al. 2009). We demonstrated that CBX2 upregulates male-related genes e.g. SOX9, SOX3, FGF2, INSL3 as well as SF1 and SRY. Also, CBX2 negatively regulates female-related genes e.g. FZD1, PBX1and FOXL2. Mutated CBX2 lost the ability to regulate the selected target genes, confirming that our observations are CBX2-specific (Eid et al. 2015). Taken together, our findings represent the first comprehensive genome-wide study of CBX2 binding coupled with the expression profiling it further extends the role of CBX2 in human sexual differentiation, and it also unravels the complexity and context-dependence of CBX2-modulated responses. In fact, it appears that CBX2 on one hand acts in the testis by stimulating the expression of male specific genes, e.g. SOX9, and simultaneously suppresses the female pathway by negatively regulating FOXL2 and the WNT4 pathway. The mechanisms underlying this bifunctional role are speculation, but are most likely due to differences in the composition of regulatory complexes (Volkel et al. 2012) and confirm that, as nicely put by Sekido and Lovell-Badge "sex determination is a story of opposing forces and crucial alliances.[. . .]. It is a matter of timing (and expression level) that determines which pathway wins" (Sekido and Lovell-Badge 2009). Moreover, our study shed more light on novel and potential pathological mechanisms that could be involved in DSD: further characterization of how these new targets fit into an expanding CBX2-regulated network should reveal how CBX2 activation and suppression can impact our understanding of DSD pathogenesis and ultimately DSD diagnosis and management.

13.4.8 Other Players

13.4.8.1 GATA4 and FOG2

The evolutionarily conserved GATA family of tissue- and organ-specific vertebrate transcription factors (GATA-binding proteins 1–6; GATA1 to GATA6) consists of two zinc finger domains. The two zinc fingers have been shown to be necessary for DNA recognition and binding (C-terminal zinc finger), the stabilization of the protein–DNA interaction (N-terminal zinc finger), and the protein–protein interactions of the GATA family members with other transcription cofactors (Evans 1988; Molkentin 2000). All family members recognize the consensus target sequence (T/A)GATA(A/G). Of these six family members, two genes, *Gata2* and *Gata4*, are expressed in the fetal mouse gonads. *Gata2* expression is detected between 10.5 and 15.5 *dpc* in XX gonads and the mesonephroi of XX and XY gonads, but expression is missing in XX gonads lacking germ cells.

At 13.5 *dpc* the expression is restricted to germ cells of XX gonads, suggesting a role for GATA2 in ovarian germ cell development (Siggers et al. 2002). GATA4 is the only GATA family member that is expressed in somatic cells (and not germ cells) in the bipotential gonad. At 11.5 *dpc Gata4* is expressed in the somatic cells of the bipotential gonad of both sexes but becomes sexually dimorphic at 13.5 *dpc*, with its expression being upregulated in XY Sertoli cells and downregulated in interstitial cells. By contrast, *Gata4* expression is downregulated in all cell types in XX gonads. *Gata4* expression is maintained in Sertoli cells postnatally and regained in adult ovaries (predominantly in granulosa cells) (Heikinheimo et al. 1997; Viger et al. 2008). GATA4 cooperatively interacts with several proteins, including NR5A1 and FOG2 (Friend of GATA protein 2 or zinc finger protein, multitype 2; encoded by the *Zfpm2* gene), to regulate the expression of several - sex-determining genes. Among those are *Sry*, *Sox9*, and *Amh*. Those complexes regulate key steroidogenic genes such as *Star* (encoding steroidogenic acute regulatory protein), *Cyp19a1* (encoding aromatase), *Inha* (encoding inhibin a-subunit), and *Hsd3b2* (encoding 3 beta-hydroxysteroid dehydrogenase/delta 5-4-isomerase type 2) (Miyamoto et al. 2008; Nishida et al. 2008; Viger et al. 2008).

Gata4$^{-/-}$ mice die at around 7.0–9.5 *dpc* due to abnormalities in ventral morphogenesis and heart tube formation (Kuo et al. 1997; Molkentin et al. 1997) so analysis of gonadal differentiation is not possible. The role of GATA4 in gonad development has been highlighted by the use of *Gata4*ki mice that have a p.V217G mutation in the N-terminal zinc finger domain, which disrupts the protein–protein interaction of GATA4 with its cofactor FOG2 (Bouma et al. 2007). These mice have severe anomalies of testis development (Crispino et al. 2001; Tevosian et al. 2002).

In humans, mutations in *GATA4* have been associated with congenital heart defects (CHD), whereas other organs were described as being normal in all cases (Garg et al. 2003; Hirayama-Yamada et al. 2005; Nemer et al. 2006; Tomita-Mitchell et al. 2007). A recent study describes a heterozygous loss-of-function mutation in *GATA4* (p.G221R) in a family with three 46,XY DSD cases (with either ambiguous genitalia or reduced phallus length) and two 46,XX females with CHD. Functional studies showed failure of the mutated protein to bind and activate the *AHM* promoter and to bind to FOG2 (Lourenco et al. 2011). A 35 kb deletion downstream of *GATA4* was identified by White et al. in a patient with 46,XY complete gonadal dysgenesis and adrenal hypoplasia congenita. The authors argue that the deletion might affect a regulatory region of *GATA4*, which could in turn explain the lack of cardiac malformations in this patient (White et al. 2011).

In vitro, FOG2 represses GATA4-dependent transactivation of *AMH* in primary Sertoli cell cultures (Tremblay et al. 2001). Whether FOG2 acts as a transcriptional repressor or activator in the context of gonad development in vivo, in mouse as well as in human, still has to be elucidated, but a role in early gonadogenesis in mouse has been determined using *Fog2* null mice. These mice die at midgestation (around 14.5 *dpc*) from a cardiac defect and exhibit a block of gonad development (Crispino et al. 2001; Tevosian et al. 2002). Finelli et al. (2007) identified a translocation including *FOG2* in a male patient with hypergonadotropic hypogonadism, supporting a role for FOG2 in mammalian sex determination. More recently, mutations in *FOG2* have been found in 46,XY DSD patients with testicular dysgenesis (Bashamboo et al. 2014). As mechanism of disease the authors indicated an impaired ability of the mutant FOG2 proteins to interact with GATA4.

13.4.8.2 SOX8 and SOX10: The Other SoxE Genes

The *SOX* gene family consists of 20 members, which all encode for transcription factors. These genes have been divided into ten subgroups, A to J, according to their sequence homologies (Bowles et al. 2000). Besides *SRY*, three other members of this gene family are expressed in the XY gonad (Cory et al. 2007; Polanco et al. 2010; Schepers et al. 2003). These members are *SOX8*, *SOX9*, and *SOX10*, which together form the subgroup E within this gene family.

SOX9 was first shown to be associated with testis development in 1994, when mutations in and around *SOX9* were identified in human DSD patients (see above).

Initially, *Sox8* was shown to be specifically expressed in the mouse testis cords at 13.5 *dpc*. This and its ability to induce *Amh* expression in vitro suggested a role of this member of the *SOX* gene family member in male sex determination, testicular differentiation, or germ cell development (Schepers et al. 2003). However, the first studies of *Sox8*-ablated mice failed to result in abnormal sexual development and resulted only in idiopathic weight loss and reduced bone density phenotypes (Sock et al. 2001). Closer analysis of *Sox8*$^{-/-}$ mice by O'Bryan et al. (2008) showed that *Sox8*$^{-/-}$ males rarely produced litters, while *Sox8*$^{+/-}$ males and *Sox8*$^{-/-}$ females appeared reproductively normal. This study shows an essential role for *Sox8* in the maintenance of male fertility beyond the first wave of spermatogenesis. Loss of *Sox8* resulted in progressive degeneration of the seminiferous epithelium through perturbed physical interactions between Sertoli cells and the developing germ cells. Double knockout studies of *Sox8* and *Sox9* showed that *Sox8* reinforces *Sox9* in testis differentiation. Double knockout testis showed mainly complete absence of testis cord formation (tissue-specific *Sox9*$^{flox/flox}$ testis showed a variable degree testis cord formation and abnormal coelomic vessel formation) (Chaboissier et al. 2004). These results, together with the above mentioned in vitro studies regarding the ability of SOX8 to bind and synergistically activate *Amh* expression alongside SF1, support the idea of redundancy for *Sox8* and *Sox9* in testis differentiation. It has been postulated that the interchangeable roles of *Sox8* and *Sox9* are due to their shared ancestry and high sequence, as well as structural homology (Koopman 2005).

In *Sox10*-null mice, no testis phenotype has been described. Overexpression studies in XX gonads showed that *Sox10* is sufficient to induce testis differentiation (Polanco et al. 2010), suggesting that, although it might not be necessary for testis formation, *Sox10* still can function as a testis-determining gene. Human 46,XX testicular DSD patients who are masculinized or incompletely feminized and who have a duplication of the region encompassing *SOX10*, among a number of other genes, have been described (Aleck et al. 1999; Nicholl et al. 1994; Seeherunvong et al. 2004). The possibility that other genes within this duplicated region are responsible for this phenotype, either solely or in combination with *SOX10*, cannot be excluded. However, *SOX10* is a strong candidate for the causative gene within this region. It is interesting to note that *Sox9* is upregulated in *Sox10*-transgenic XX gonads. It would be worthwhile investigating whether overexpression of *Sox8* and or *Sox10* would be sufficient to rescue the *Sox9* knockout phenotype.

13.4.8.3 SOX3

SOX3 is another member of the *SOX* gene family of transcription factors. SOX3 shows near sequence identity to SRY, so it has been proposed that the Y-linked *Sry* gene evolved from the X-linked *Sox3* gene (Foster et al. 1992). This process occurred as the mammalian sex chromosomes evolved and differentiated. *Sox3* is not expressed in the developing gonads in either sex, and loss-of-function mutations in *Sox3* do not affect sex determination in humans or mice. However, a transgenic mouse line overexpressing *Sox3* showed ectopic expression of this gene in the bipotential gonad. XX embryos showed female-to-male sex reversal, and further analyses suggest that SOX3 induced testis differentiation by upregulation of *Sox9* expression in these animals through a similar mechanism to *Sry* (Sutton et al. 2011). The same study identified three 46,XX testicular DSD patients with genomic rearrangements: two duplications including *SOX3* and one deletion upstream of *SOX3*, in the putative *SOX3* regulatory region. It was hypothesized that these rearrangements caused ectopic expression of *SOX3* in the embryonic gonad, where it was able to substitute for the absence of *SRY* and drive testis development. Together, these findings suggest that *SOX3* and *SRY* are functionally interchangeable in testis determination and lend further support to the hypothesis that *SRY* evolved from *SOX3*.

13.4.8.4 DHH

Desert hedgehog (DHH) is a member of the hedgehog family of signaling molecules, which also includes sonic hedgehog (SHH) and Indian hedgehog (IHH) (Ingham 1998). Of the three family members, *Dhh* is the only one that is expressed in somatic cell population of the developing XY mouse gonad from 11.5 *dpc* and continues later in Sertoli cells. No expression can be detected in XX ovaries at any stage (Beverdam and Koopman 2006; Yao and Capel 2002). DHH binds to its receptor patched 1 (PTCH1), which is expressed shortly after DHH. PTCH1 is bound to the membrane of Leydig and peritubular myoid cells, and its expression is upregulated via DHH (Yao and Capel 2002). *Dhh* null mice show disrupted testis cord formation due to abnormal peritubular tissue (Clark et al. 2000). DHH seems to be necessary for the upregulation of *Sf1* in Leydig cells (Yao and Capel 2002). Several mutations of *DHH* have been described in patients affected by 46,XY partial or complete gonadal dysgenesis. The first case was described by Umehara et al. (2000). The patient was affected by 46,XY partial gonadal dysgenesis and minifascicular neuropathy. A homozygous missense mutation at the initiation codon in exon 1 (c.T2C; p.M1T) was identified in the patient. The father carried the same mutation in a heterozygous state, showing that the phenotype displays a recessive mode of inheritance. Further studies identified a homozygous substitution in exon 2 (c.T485C; p.L162P) in one patient with 46,XY complete gonadal dysgenesis (CGD) and a homozygous frameshift deletion in exon 3 (c.1086delG), which results in a premature stop codon four codons after the deletion in two patients with 46,XY CGD (Canto et al. 2004). The later mutation has also been identified in two patients with 46,XY partial gonadal dysgenesis (PGD) but in a heterozygous state (Canto et al. 2005). Recently, Das et al. (2011) described two novel heterozygous mutations (c.271_273delGAG; p.D60del and c.57_60dupAGCC resulting in premature translational termination) in *DHH* in two patients affected by 46,XY CGD.

13.4.8.5 FGF9

Fgf9 gene encodes fibroblast growth factor 9 (FGF9), one of a number of growth factors that play a role in various developmental processes such as cell proliferation, cell survival, cell migration, and cell differentiation. *Fgf9* is expressed in the bipotential gonad immediately after the expression of *Sry*. Mice null for *Fgf9* show male-to-female sex reversal along with impaired development of Sertoli cells. However, this is only evident on some genetic backgrounds, but not on others (Colvin et al. 2001; Schmahl et al. 2004). In the absence of *Fgf9*, *Sox9* expression is not maintained, Sertoli cells fail to differentiate, testis development is aborted and the resulting somatic cells express genes characteristic for ovarian development and the female pathway. There is data supporting the concept that SRY and Nr5a1 (SF1) initiate a positive feedback loop by upregulating *Sox9*, which in turn upregulates

Fgf9, further increasing *Sox9* expression (Kim et al. 2006). FGF9 mediates its molecular function through the membrane-bound receptor FGFR2 (fibroblast growth factor receptor 2). FGFR2 is integrated in the plasma membrane of progenitor Sertoli cells and is critical for Sertoli cell proliferation and differentiation in the developing testis. Although defects in FGFR2 are responsible for Apert syndrome, no mutations have yet been reported in human DSD patients (Kim et al. 2007). Due to its crucial role in Sertoli cell proliferation and differentiation, mutations in this gene could account for some unexplained cases of DSD.

13.4.8.6 MAP Kinases

MAPK signaling in gonad development has been recently demonstrated by the identification of a mutation in *Map3k4* in mouse, resulting in XY individuals with male-to-female sex reversal (Piard et al. 2014). More importantly, a related *MAP3K4* gene, *MAP3K1*, has been identified in humans as a novel gene causing 46,XY gonadal dysgenesis (Loke et al. 2014; Muzio et al. 2005). In fact, *Map3k1* gain-of-function mutations leads to decreased expression of *Sox9*, *Fgf9*, and *Fgfr2*, thus increasing expression of β-catenin and *Foxl2* in mice. Hence, gain-of-function *Map3k1* mutations mimic the ovarian-determining pathway, overriding the testis-determining signal from an expressed wild-type *Sry* gene (Ostrer 2014).

13.5 Female Sex Development in Humans: Health and Disease

13.5.1 WNT4

WNT4 is a member of the WNT family of secreted molecules that function in a paracrine manner to affect a number of developmental changes. WNT proteins bind to members of the Frizzled (FZ) family of cell-surface receptors and to the single-pass transmembrane protein LDL-receptor-related proteins 5 and 6 (LRP5 and LRP6) (He et al. 2004). The binding of WNT to FZ leads to activation of the phosphoprotein Dishevelled (DSH or DVL). The activation and membrane recruitment of DSH probably recruits AXIN and the destruction complex (including APC, Casein Kinase 1 and Glycogen Synthase Kinase 3) to the plasma membrane, where AXIN directly binds to the cytoplasmic tail of LRP5/6. AXIN is then degraded, which decreases β-catenin degradation. The activation of DSH also leads to the inhibition of Glycogen Synthase Kinase 3 (GSK3), which further reduces phosphorylation and degradation of beta-catenin with consequent beta-catenin-dependent induction of Wnt-responsive genes (Nusse 2003). Wnt proteins can also signal via a beta-catenin independent non-canonical pathway involving protein kinase C

(PKC) and c-jun NH2-terminal kinase (JNK) (Biason-Lauber and Chaboissier 2015).

In mice, Wnt4 and Rspo1, as well as the other elements of the canonical WNT signaling are initially expressed in gonads of both sexes and before sex determination, both genes are required for XY and XX gonadal development by stimulating proliferation of the coelomic cells that provide the precursors of the supporting cells (Chassot et al. 2012). Starting from 11.5 dpc Wnt4 is up-regulated in the ovary and Rspo1, Axin2 and Lef1 one day later (12.5 dpc) while other factors such as Lgr5/Gpr49 becomes female-specific at 13.5 dpc (Fig. 13.2) (Jameson et al. 2012b; Nef et al. 2005; Parma et al. 2006).

In XY gonads, Sry expression promotes the differentiation of these progenitors to Sertoli cells, mainly via fibroblast growth factor 9 (FGF9) and WT1 that contribute to the down-regulation of WNT signaling (Chang et al. 2008; Jameson et al. 2012a). Sertoli cell differentiation triggers the migration of mesonephric cells through the gonads. They assemble the coelomic vessel and the lateral vascularization under the regulation of the transforming growth factorActivin B and its momomer inhibin B (INHBB), which is in turn repressed by WNT signaling in XX gonads (Yao et al. 2006). Accordingly, one feature of the Wnt4, Rspo1 or Ctnnb1 loss-of-function mutant mice is the presence of coelomic vessel in XX gonads (Chassot et al. 2008b; Jeays-Ward et al. 2003; Liu et al. 2009; Tomizuka et al. 2008).

Also, WNT signaling prevents ectopic testosterone production in XX embryonic gonads (Heikkila et al. 2002, 2005; Tomizuka et al. 2008) probably due to a role of WNT signaling in the separation of the adrenal-gonadal lineages in the early developing gonads (Heikkila et al. 2002; Jeays-Ward et al. 2003; Liu et al. 2009).

In XX *Wnt4*$^{-/-}$ or *Rspo1*$^{-/-}$ gonads, sex reversal of the supporting cells begins around 16.5 *dpc* (Chassot et al. 2008b; Vainio et al. 1999). The precursors of the granulosa cells are mitotically arrested but precociously differentiate as granulosa cells from 15.5 *dpc*. Then they resume proliferation before they transdifferentiate into Sertoli cells (Maatouk et al. 2013). Depletion of germ cells by busulfan treatments from 11.5 *dpc* prevents premature differentiation of granulosa cells (AMH-positive cells) and subsequently transdifferentiation of these cells. This implies that WNT4 and RSPO1 initiate a signaling cascade required to prevent differentiation of the granulosa cell precursors via germ cell–somatic cell interactions. Ablation of *Wnt4* or *Rspo1* modifies the distribution of β-catenin building additional adherens junctions (Chassot et al. 2008b; Naillat et al. 2010). This increases likely the communication between the germ cells and the soma and might promote *Amh* expression in the somatic cells. What are these inducing factors? The answer cannot be found in adult ovaries since WNT4 is there implicated for *Amh* expression contrary to *Wnt4*$^{-/-}$ embryonic gonads (Prunskaite-Hyyrylainen et al. 2014). This implies that WNT4 ovarian function is different in fetal and adult stages. Thus the identity of the germ cell signals involved in granulosa cell differentiation in *Wnt4*$^{-/-}$ gonads is still an open question.

These signals induce either directly or indirectly Sertoli cell differentiation. Surprisingly, *Sry* and *Sox9* are dispensable for this process in the XX Rspo1$^{-/-}$

gonads (Lavery et al. 2012). Again assimilating this transdifferentiation with the antagonism between FOXL2 and DMRT1 requires for adult gonadal maintenance which might be a wrong approach (Matson et al. 2011; Uhlenhaut et al. 2009). Indeed *Dmrt1* is not expressed in XX *Wnt4*$^{-/-}$ gonads at the time of transdifferentiation (Maatouk et al. 2013). This suggests that DMRT1 is not involved in this sex-reversal model. Nevertheless this clearly shows that other key genes implicated in Sertoli cell differentiation are still unknown. And the story is far to be completed.

WNT4 and human disease The first report of disease caused by WNT4 abnormalities in humans, was the presence of more copies of WNT4 (due to duplication of chromosome 1p31-p35), in a patient with ambiguous genitalia, severe hypospadia, dysgenetic gonads and remnants of both Müllerian and Wolffian ducts, i.e., a 46, XY DSD (Jordan et al. 2001). Reversely, when both copies of the gene are inactive a intrauterine lethal clinical entity called the SERKAL syndrome results (Mandel et al. 2008). The syndrome was described in three 46, XX fetuses and is characterized by 46, XX DSD with ambiguous genitalia, gonadal morphology ranging from ovotestis to normal testis, renal agenesis, adrenal hypoplasia, lung and heart anomalies. Between these two extremes one might expect to find patients with intermediate defects of sex development. Searches for clinically relevant WNT4 mutations sometimes in large cohorts of these patients were unsuccessful (Domenice et al. 2004) until we described our first case. We described a woman with no uterine and fallopian tubes who had unilateral renal agenesis and clinical signs of androgen excess. Her phenotype resembles that of patients with the Mayer-Rokitansky-Küster-Hauser syndrome and is also strikingly similar to that of Wnt4-knockout female mice. This constellation prompted us to search for and eventually find mutations in the WNT4 gene in this patient. In this first patient, the mutant protein (p. E226G) appeared to be trapped inside the cell (Biason-Lauber et al. 2004), because of defective post-translational lipid modification, necessary for WNT proper function (Nusse 2003). The subsequent identification of other women with a similar phenotype and mutations in WNT4 confirmed its role in ovarian and female reproductive tract development (Biason-Lauber et al. 2007; Philibert et al. 2008); (Philibert et al. 2011). These additional patients also helped to refine the phenotype of WNT4 deficiency in humans. In fact, it appears that the absence of uterus (and not other Müllerian abnormalities) and the androgen excess are the pathognomic signs of WNT4 defects, suggesting that this might be a clinical entity distinct from the classical Mayer-Rokitansky-Kuster-Hauser syndrome.

From a more biological point of view, as often in sex development, this it is a dosage-sensitive process, as Jordan and co-workers already predicted (Jordan et al. 2001). In fact, while too much WNT4 activity (duplication) induces feminization of the male (46, XY DSD), too little WNT4 activity (homozygous loss-of-function mutation) induces exactly the opposite, i.e., masculinization of the female (46, XX DSD). Since WNT4 inhibits the male development in the female and males do not need WNT4 for their sex development (Vainio et al. 1999), situations between these two extremes are characterized by different degrees of masculinization of the female (reviewed in Biason-Lauber 2012). In recent years, an association between genetic variations in WNT4 gene and endometriosis has been identified (Mafra et al. 2015) suggesting a role in endometrium development.

13.5.2 RSPO1

R-spondin 1 (RSPO1) is a secreted factor that can stabilize β-catenin as part of the canonical WNT signaling pathway and is expressed at high levels in mouse, as well as human gonads around the critical time of gonad development (mouse: Chassot et al. 2008a; Parma et al. 2006; Smith et al. 2008; human: Tomaselli et al. 2011). *Wnt4* and *Rspo1* loss-of-function mouse embryos share similar phenotypes for the somatic lineages, but they show some differences for the germ cell fate (Chassot et al. 2008b; Maatouk et al. 2013; Vainio et al. 1999). Primordial germ cells get competent to gametogenesis and sexual differentiation under the influence of the gonadal environment (Hu et al. 2015). Following somatic sex determination, the germ cells respond to sex-determining signals around 12.5–13.5 *dpc* and initiate meiosis in the ovary (Adams and McLaren 2002). Since the primordial germ cells are primed with a bias toward the prospermatogonia fate, the somatic environment is decisive for oogonial differentiation (Jameson et al. 2012b). RSPO1 is involved in this germ cell sexual decision (Chassot et al. 2011). The removal of *Rspo1* results in germ cell defects ranging from developmental delay highlighted by the maintenance of the pluripotent marker OCT4 to sex reversal with expression of the prospermatogonia-specific marker Nanos2 (Chassot et al. 2008b, 2011).

The next question is how RSPO1 and/or WNT signaling influences oogonial differentiation. This is still pending but some clues come from work from different labs. RSPO1, WNT4, and LGR5 promote meiosis entry by providing the somatic environment capable of supporting germ cell development (Chassot et al. 2011; Naillat et al. 2010; Rastetter et al. 2014). Unexpectedly, *Ctnnb1* ablation in the soma allows germ cells to enter meiosis (Liu et al. 2009, 2010; Manuylov et al. 2008) suggesting that RSPO1, WNT4, and LGR5 act on germ cell independently of somatic β-catenin. However, *Ctnnb1* was conditionally disrupted in somatic cells of the gonads using the *Sf1*:cre that is highly efficient from 11.5 *dpc* (Bingham et al. 2006). In the light of the recent results (Hu et al. 2015), the ablation of *Ctnnb1* might have occurred too late to affect the somatic components controlling germ cell fate.

Further, WNT signaling is directly activated in XX germ cells as evidenced by *Axin2* expression in these cells (Chassot et al. 2011; Jameson et al. 2012b). Thus, the regulation of meiosis entry might be due to activation of WNT signaling within the germ cells. In addition to *Wnt4*, other *Wnts* like *Wnt2b* or *Wnt9a* are expressed within the ovary (Jameson et al. 2012b; Nef et al. 2005). Whereas *Wnt2b* does not seem to have a dramatic role in germ cell differentiation (Tsukiyama and

Yamaguchi 2012), *Wnt9a* came to attract our attention. *Wnt9a* is more closely related to Drosophila *DWnt4* than to other members of the WNT family in mammals (Bergstein et al. 1997). *DWnt4* is expressed in the somatic escort cells and promotes the passage from proliferation to germ cell differentiation via expression of RNA-binding protein Piwi by *DCtnnb1* in Drosophila (Hamada-Kawaguchi et al. 2014). The Piwi-interacting RNA pathway plays important roles in silencing of repeat elements through repressive histone modifications (Sienski et al. 2012). In mice, the polycomb repressive complexes PRC1 and PRC2 induce histone modifications by ubiquitination or methylation and subsequently regulate gene expression (Simon and Kingston 2013). In XX germ cells, PRC1 controls directly the expression of *Stra8* that is required for meiosis entry (Baltus et al. 2006; Yokobayashi et al. 2013). Then PRC1 is important for chromosome synapsis and progression of the meiosis (Crichton et al. 2014). In addition the ATP-dependent chromatin-remodeling factor LSH (lymphoid-specific helicase) represses activation of transposable elements by DNA methylation and thus favors meiotic chromosome synapsis (De La Fuente et al. 2006). Consequently, mutations in PRC components or LSH result in severe loss of oocytes. Fetal oocyte attrition is linked to retrotransposon activity (Malki et al. 2014). Interestingly 90 % of oocytes are eliminated in the *Wnt4* mutants (Vainio et al. 1999), but a direct link between epigenetic modifications and WNT signaling remains to be clarified in XX mouse germ cells. The importance of polycomb proteins in gonad determination has been recently strengthened by the discovery of the role of CBX2 in DSD (see Sect. 13.5).

RSPO1 and human disease The essential role of RSPO1 in human ovarian development was demonstrated by studying individuals with palmoplantar hyperkeratosis with squamous cell carcinoma of skin and 46,XX DSD in whom Parma et al. demonstrated mutations in the *RSPO1* gene (Parma et al. 2006). The authors concluded that RSPO1 is produced and secreted by fibroblasts and regulates keratinocyte proliferation and differentiation. The presence of "functional" testes in the sex-reversed individuals was confirmed by the absence of Müllerian derivatives and by the masculinization of the internal and external genitalia, presumably induced by functioning Sertoli and Leydig cells, respectively. All sex-reversed individuals were sterile. Notably, the normal reproductive phenotype of 46,XY individuals suggested that normal *RSPO1* is not required for testis differentiation and function.

In a 46,XX *SRY*-negative woman with true hermaphroditism and palmoplantar keratoderma whose parents were first cousins, in whom no mutations were found in several other gene, including *WNT4*, homozygosity for a splice site mutation in the *RSPO1* gene was identified (Tomaselli et al. 2008). The authors stated that this was the first patient in whom XX true hermaphroditism was associated with a single gene alteration in the absence of *SRY*.

13.5.3 FOXL2

FOXL2 is a single-exon gene encoding a forkhead/winged helix (fkh) transcription factor and a nuclear protein (Crisponi et al. 2001). In vertebrates, FOXL2 is one of the earliest known markers of ovarian differentiation (Cocquet et al. 2002). Thus, it may play a role in the early stage of development of the ovarian somatic compartment. As it is still strongly expressed in postnatal and adult follicular cells, it is thought to play a role in follicle development and/or maintenance during fertile life.

Despite the importance of FOXL2 in ovarian development and maintenance, only a few transcriptional targets have been described so far (Pannetier et al. 2006; Pisarska et al. 2004). FOXL2 seems to stimulate the expression of the gonadotropin-releasing hormone (GnRH) receptor. *Foxl2* expression precedes glycoprotein hormone α-subunit (α-GSU, common subunit to FSH, LH, and TSH) expression in the pituitary suggesting that the α-GSU gene is a downstream target of FOXL2. Indeed, it has been shown that expression of *Foxl2* activates the expression of α-GSU in cellular and transgenic mice models, by a direct effect on the α-GSU promoter (Ellsworth et al. 2006). Some data suggest that target specificity could stem from the interactions of FOXL2 with still unknown cofactors expressed in a tissue and/or cell specific manner (Pannetier et al. 2005). FOXL2 has also been shown to interact directly with the promoter of the steroidogenesis acute response (StAR) gene, to induce a robust inhibition of its basal transcriptional activity (Pisarska et al. 2004). StAR is a marker of late differentiation of granulosa cells in preovulating follicles and catalyzes cholesterol translocation from the outer to the inner mitochondrial membrane, where it can subsequently be processed in its way to yield pregnenolone and eventually steroid hormones. This translocation of cholesterol is the rate-limiting step in steroidogenesis. The role of FOXL2 in steroidogenesis in the ovary was further strengthened by the fact that it can upregulate the expression of aromatase (*Cyp19*), the enzyme responsible for the transformation of androgens to estrogens in granulosa cells (Pannetier et al. 2006).

In addition, FOXL2 seems to play a role in the reactive oxygen species (ROS) detoxification pathways, as several of its members are upregulated by the overexpression of *Foxl2*, namely, PPARGC1A, immediate early response 3 (IER3), and the mitochondrial manganese superoxide dismutase (MnSOD) (Moumne et al. 2008a) (and references therein). It is known that an increased resistance to oxidative stress correlates with longevity (Henderson and Johnson 2001). Thus, FOXL2 could play a major role in the regulation of ovarian senescence since its mutation leads to a phenotype similar to an accelerated ovarian aging (see below). FOXL2 also appears to be implicated in the regulation of apoptosis, as it activates the transcription of several genes encoding factors involved in apoptotic processes (Batista et al. 2007). In addition, FOXL2 seems to regulate players of inflammation processes such as several chemokine ligands and especially PTGS2/COX2. The latter is one of the two isoforms of cyclooxygenases involved in the synthesis of prostaglandins and catalyzes the rate-limiting step of conversion of arachidonic acid into prostaglandin H2, which is subsequently converted to other prostaglandins by specific synthases (Smith and Dewitt 1996). The fact that FOXL2 strongly activates PTGS2 points to an important role for prostaglandins in ovarian function and leads to the claims that ovulation is an inflammatory-like process and suggests that FOXL2 might act very early during gonadal determination and all the way through the latest stages of follicular maturation and ovulation.

The crucial role of FOXL2 in the life of the ovary has been supported by the seminal work of Uhlenhaut et al. (2009), who by deleting *Foxl2* in adult mice ovary, reprogrammed granulosa and theca cells into Sertoli- and Leydig-like cells. This work elegantly shows that the mammalian ovarian phenotype has to be maintained throughout life, mainly by active repression of the male-promoting gene *Sox9*.

FOXL2 and human disease At first, mutations in FOXL2 were found to be related to to blepharophimosis–ptosis–epicanthus inversus syndrome (BPES), a genetic disease leading to complex eyelid malformation and other mild craniofacial abnormalities, but can also present itself with premature ovarian failure (BPES I) or not (BPES II) (Crisponi et al. 2001).

In addition, Shah et al. analyzed 4 adult-type ovarian granulosa-cell tumor (GCT) specimens for GCT-specific mutations and identified a somatic point mutation, (C134W), in the FOXL2 gene in all four specimens (Shah et al. 2009). Since the very same mutation was found in several other additional GCTs and thecomas, but not in sex cord/stromal tumors of other types, the authors concluded that mutant FOXL2 is a potential driver in the pathogenesis of adult-type GCTs.

In search for a molecular mechanism and a genotype-phenotype correlation, Moumne and co-authors found a correlation between the transcriptional activity of FOXL2 variants in that a FOXL2 mutant completely lacking transactivation activity was likely to lead to BPES with POF. Among the targets that failed to be transactivated by mutant FOXL2 in BPES and POF (BPES II) were cell-cycle and DNA repair targets, whereas mutations leading to isolated craniofacial defects (and normal ovarian function, BPES II) activate them correctly. This evidence supports the idea that FOXL2 plays a key role in granulosa cell homeostasis, the failure of which is central to ovarian ageing and tumorigenesis.

13.6 Conclusion

The idea that the female sexual development happens by default was born in the middle of the last century after Jost carried out his innovative experiments to study the bases of differentiation of the reproductive tract and was based on the fact that the female reproductive tract develops even in the absence of any gonad. The term default (passive) attributed to the female developmental pathway, therefore established itself also for female determination, even if originally it was intended only for the internal and external genital organs.

Recent advances demonstrated that ovarian development and maintenance are active processes. In fact, WNT4 synergizes with RSPO1 to stabilize β-catenin that, in turn, suppresses expression of the male-specific gene *Sox9* (Troy et al. 2003). As a consequence migration of mesonephric endothelial and steroidogenic cells, formation of male-specific coelomic blood vessels, and production of steroids are prevented (Biason-Lauber et al. 1997; Jeays-Ward et al. 2004) with an overall "anti-testis" effect. WNT4, RSPO1, and FOXL2 also work as "pro-ovary" factors, WNT4/RSPO1 being necessary for early somatic cell differentiation and for protection of the germ cells in mice (Biason-Lauber et al. 2000) and in humans (Cavelti-Weder et al. 2012), while FOXL2 being essential for follicle formation and identity (Chassot et al. 2008a). These factors work in concert but at different time points: during determination in utero, the process is lead by the WNT4/RSPO1/β-catenin pathway; after birth and throughout life, FOXL2 (and the estrogen receptors) are the major regulators. Thus, it appears that while male sex determination is regulated by a single pathway of SOX9 activation, the female ovarian development is controlled by at least two ways of SOX9 suppression (Fig. 13.1).

From a more clinical point of view, mutations in these factors have consequences not only for ovarian ontogeny and sex development, but their interactions are necessary throughout the lifetime of the female to prevent ovarian dysfunction, including infertility, premature ovarian insufficiency, and perhaps polycystic ovary syndrome.

Grant support This work was supported by Grant 320000-116636 from the Swiss National Science Foundation.

References

Adams IR, McLaren A (2002) Sexually dimorphic development of mouse primordial germ cells: switching from oogenesis to spermatogenesis. Development 129(5):1155–1164

Albrecht KH, Eicher EM (2001) Evidence that Sry is expressed in pre-Sertoli cells and Sertoli and granulosa cells have a common precursor. Dev Biol 240(1):92–107

Aleck KA, Argueso L, Stone J et al (1999) True hermaphroditism with partial duplication of chromosome 22 and without SRY. Am J Med Genet 85(1):2–4

Arango NA, Lovell-Badge R, Behringer RR (1999) Targeted mutagenesis of the endogenous mouse Mis gene promoter: in vivo definition of genetic pathways of vertebrate sexual development. Cell 99(4):409–419

Armstrong JF, Pritchard-Jones K, Bickmore WA et al (1993) The expression of the Wilms' tumour gene, WT1, in the developing mammalian embryo. Mech Dev 40(1–2):85–97

Baltus AE, Menke DB, Hu YC et al (2006) In germ cells of mouse embryonic ovaries, the decision to enter meiosis precedes premeiotic DNA replication. Nat Genet 38(12):1430–1434

Barbaux S, Niaudet P, Gubler MC et al (1997) Donor splice-site mutations in WT1 are responsible for Frasier syndrome. Nat Genet 17(4):467–470

Bardoni B, Zanaria E, Guioli S et al (1994) A dosage sensitive locus at chromosome Xp21 is involved in male to female sex reversal. Nat Genet 7(4):497–501

Barrionuevo F, Bagheri-Fam S, Klattig J et al (2006) Homozygous inactivation of Sox9 causes complete XY sex reversal in mice. Biol Reprod 74(1):195–201

Bashamboo A, Ferraz-de-Souza B, Lourenco D et al (2010a) Human male infertility associated with mutations in NR5A1 encoding steroidogenic factor 1. Am J Genet 87(4):505–512. doi:10.1016/j.ajhg.2010.09.009
Bashamboo A, Ledig S, Wieacker P et al (2010b) New technologies for the identification of novel genetic markers of disorders of sex development (DSD). Sex Dev 4(4–5):213–224. doi:10.1159/000314917
Bashamboo A, Brauner R, Bignon-Topalovic J et al (2014) Mutations in the FOG2/ZFPM2 gene are associated with anomalies of human testis determination. Hum Mol Genet 23(14):3657–3665. doi:10.1093/hmg/ddu074
Batista F, Vaiman D, Dausset J, Fellous M, Veitia RA (2007) Potential targets of FOXL2, a transcription factor involved in craniofacial and follicular development, identified by transcriptomics. Proc Natl Acad Sci U S A 104(9):3330–3335. doi:10.1073/pnas.0611326104, 0611326104 [pii]
Baxter RM, Vilain E (2013) Translational genetics for diagnosis of human disorders of sex development. Annu Rev Genomics Hum Genet 14:371–392. doi:10.1146/annurev-genom-091212-153417
Benko S, Gordon CT, Mallet D et al (2011) Disruption of a long distance regulatory region upstream of SOX9 in isolated disorders of sex development. J Med Genet 48(12):825–830. doi:10.1136/jmedgenet-2011-100255
Bergstein I, Eisenberg LM, Bhalerao J et al (1997) Isolation of two novel WNT genes, WNT14 and WNT15, one of which (WNT15) is closely linked to WNT3 on human chromosome 17q21. Genomics 46(3):450–458. doi:10.1006/geno.1997.5041
Bernard P, Fleming A, Lacombe A, Harley VR, Vilain E (2008) Wnt4 inhibits beta-catenin/TCF signalling by redirecting beta-catenin to the cell membrane. Biol Cell 100(3):167–177. doi:10.1042/BC20070072
Beverdam A, Koopman P (2006) Expression profiling of purified mouse gonadal somatic cells during the critical time window of sex determination reveals novel candidate genes for human sexual dysgenesis syndromes. Hum Mol Genet 15(3):417–431. doi:10.1093/hmg/ddi463
Biason-Lauber A (2010) Control of sex development. Best Pract Res Clin Endocrinol Metab 24(2):163–186. doi:10.1016/j.beem.2009.12.002
Biason-Lauber A (2012) WNT4, RSPO1, and FOXL2 in sex development. Semin Reprod Med 30(5):387–395. doi:10.1055/s-0032-1324722
Biason-Lauber A, Schoenle EJ (2000) Apparently normal ovarian differentiation in a prepubertal girl with transcriptionally inactive steroidogenic factor 1 (NR5A1/SF-1) and adrenocortical insufficiency. Am J Hum Genet 67(6):1563–1568. doi:10.1086/316893
Biason-Lauber A, Chaboissier MC (2015) Ovarian development and disease: the known and the unexpected. Semin Cell Dev Biol 45:59–67. doi:10.1016/j.semcdb.2015.10.021
Biason-Lauber A, Leiberman E, Zachmann M (1997) A single amino acid substitution in the putative redox partner-binding site of P450c17 as cause of isolated 17,20-lyase deficiency. J Clin Endocrinol Metab 82(11):3807–3812
Biason-Lauber A, Zachmann M, Schoenle EJ (2000) Effect of leptin on CYP17 enzymatic activities in human adrenal cells: new insight in the onset of adrenarche. Endocrinology 141(4):1446–1454
Biason-Lauber A, Konrad D, Navratil F, Schoenle EJ (2004) A WNT4 mutation associated with Mullerian-duct regression and virilization in a 46, XX woman. N Engl J Med 351(8):792–798. doi:10.1056/NEJMoa040533
Biason-Lauber A, De Filippo G, Konrad D, Scarano G, Nazzaro A, Schoenle EJ (2007) WNT4 deficiency—a clinical phenotype distinct from the classic Mayer-Rokitansky-Kuster-Hauser syndrome: a case report. Hum Reprod 22(1):224–229. doi:10.1093/humrep/del360
Biason-Lauber A, Konrad D, Meyer M, DeBeaufort C, Schoenle EJ (2009) Ovaries and female phenotype in a girl with 46, XY karyotype and mutations in the CBX2 gene. Am J Hum Genet 84(5):658–663. doi:10.1016/j.ajhg.2009.03.016

Bingham NC, Verma-Kurvari S, Parada LF, Parker KL (2006) Development of a steroidogenic factor 1/Cre transgenic mouse line. Genesis 44(9):419–424
Birk OS, Casiano DE, Wassif CA et al (2000) The LIM homeobox gene Lhx9 is essential for mouse gonad formation. Nature 403(6772):909–913. doi:10.1038/35002622
Bouma GJ, Washburn LL, Albrecht KH, Eicher EM (2007) Correct dosage of Fog2 and Gata4 transcription factors is critical for fetal testis development in mice. Proc Natl Acad Sci U S A 104(38):14994–14999. doi:10.1073/pnas.0701677104
Bowles J, Schepers G, Koopman P (2000) Phylogeny of the SOX family of developmental transcription factors based on sequence and structural indicators. Dev Biol 227(2):239–255. doi:10.1006/dbio.2000.9883
Bracken AP, Helin K (2009) Polycomb group proteins: navigators of lineage pathways led astray in cancer. Nat Rev Cancer 9(11):773–784. doi:10.1038/nrc2736
Breslow NE, Takashima JR, Ritchey ML, Strong LC, Green DM (2000) Renal failure in the Denys-Drash and Wilms' tumor-aniridia syndromes. Cancer Res 60(15):4030–4032
Burtis KC, Baker BS (1989) Drosophila doublesex gene controls somatic sexual differentiation by producing alternatively spliced mRNAs encoding related sex-specific polypeptides. Cell 56(6): 997–1010
Cameron FJ, Hageman RM, Cooke-Yarborough C et al (1996) A novel germ line mutation in SOX9 causes familial campomelic dysplasia and sex reversal. Hum Mol Genet 5(10): 1625–1630
Canto P, Soderlund D, Reyes E, Mendez JP (2004) Mutations in the desert hedgehog (DHH) gene in patients with 46, XY complete pure gonadal dysgenesis. J Clin Endocrinol Metab 89(9): 4480–4483. doi:10.1210/jc.2004-0863
Canto P, Vilchis F, Soderlund D, Reyes E, Mendez JP (2005) A heterozygous mutation in the desert hedgehog gene in patients with mixed gonadal dysgenesis. Mol Hum Reprod 11(11): 833–836. doi:10.1093/molehr/gah216
Capel B (1998) Sex in the 90s: SRY and the switch to the male pathway. Annu Rev Physiol 60: 497–523. doi:10.1146/annurev.physiol.60.1.497
Capel B (2000) The battle of the sexes. Mech Dev 92(1):89–103
Chaboissier MC, Kobayashi A, Vidal VI et al (2004) Functional analysis of Sox8 and Sox9 during sex determination in the mouse. Development 131(9):1891–1901. doi:10.1242/dev.01087
Chang H, Gao F, Guillou F, Taketo MM, Huff V, Behringer RR (2008) Wt1 negatively regulates {beta}-catenin signaling during testis development. Development 135(10):1875–1885
Chassot AA, Gregoire EP, Magliano M, Lavery R, Chaboissier MC (2008a) Genetics of ovarian differentiation: Rspo1, a major player. Sex Dev 2(4–5):219–227. doi:10.1159/000152038
Chassot AA, Ranc F, Gregoire EP et al (2008b) Activation of beta-catenin signaling by Rspo1 controls differentiation of the mammalian ovary. Hum Mol Genet 17(9):1264–1277
Chassot AA, Gregoire EP, Lavery R et al (2011) RSPO1/beta-catenin signaling pathway regulates oogonia differentiation and entry into meiosis in the mouse fetal ovary. PLoS One 6(10): e25641
Chassot AA, Bradford ST, Auguste A et al (2012) WNT4 and RSPO1 together are required for cell proliferation in the early mouse gonad. Development 139(23):4461–4472. doi:10.1242/dev. 078972
Clark AM, Garland KK, Russell LD (2000) Desert hedgehog (Dhh) gene is required in the mouse testis for formation of adult-type Leydig cells and normal development of peritubular cells and seminiferous tubules. Biol Reprod 63(6):1825–1838
Cocquet J, Pailhoux E, Jaubert F et al (2002) Evolution and expression of FOXL2. J Med Genet 39(12):916–921
Colvin JS, Green RP, Schmahl J, Capel B, Ornitz DM (2001) Male-to-female sex reversal in mice lacking fibroblast growth factor 9. Cell 104(6):875–889
Combes AN, Wilhelm D, Davidson T et al (2009) Endothelial cell migration directs testis cord formation. Dev Biol 326(1):112–120

Cory AT, Boyer A, Pilon N, Lussier JG, Silversides DW (2007) Presumptive pre-Sertoli cells express genes involved in cell proliferation and cell signalling during a critical window in early testis differentiation. Mol Reprod Dev 74(12):1491–1504. doi:10.1002/mrd.20722

Cox JJ, Willatt L, Homfray T, Woods CG (2011) A SOX9 duplication and familial 46, XX developmental testicular disorder. N Engl J Med 364(1):91–93. doi:10.1056/NEJMc1010311

Crichton JH, Playfoot CJ, Adams IR (2014) The role of chromatin modifications in progression through mouse meiotic prophase. J Genet Genomics 41(3):97–106. doi:10.1016/j.jgg.2014.01.003

Crispino JD, Lodish MB, Thurberg BL et al (2001) Proper coronary vascular development and heart morphogenesis depend on interaction of GATA-4 with FOG cofactors. Genes Dev 15(7):839–844. doi:10.1101/gad.875201

Crisponi L, Deiana M, Loi A et al (2001) The putative forkhead transcription factor FOXL2 is mutated in blepharophimosis/ptosis/epicanthus inversus syndrome. Nat Genet 27(2):159–166. doi:10.1038/84781

Das DK, Sanghavi D, Gawde H, Idicula-Thomas S, Vasudevan L (2011) Novel homozygous mutations in Desert hedgehog gene in patients with 46, XY complete gonadal dysgenesis and prediction of its structural and functional implications by computational methods. Eur J Med Genet 54(6):e529–e534. doi:10.1016/j.ejmg.2011.04.010

De La Fuente R, Baumann C, Fan T, Schmidtmann A, Dobrinski I, Muegge K (2006) Lsh is required for meiotic chromosome synapsis and retrotransposon silencing in female germ cells. Nat Cell Biol 8(12):1448–1454. doi:10.1038/ncb1513

Denys P, van den Berghe H, Tanghe W, Proesmans W (1967) Association d'un syndrome anatomo-pathologique de pseudohermaphrodisme masculin, d'une tumeur de Wilms, d'une nephropathie parenchymateuse et d'un mosaicisme XX/XY. Arch Fr Pediatr 24:729–739

DeStefano GM, Kurban M, Anyane-Yeboa K et al (2014) Mutations in the cholesterol transporter gene ABCA5 are associated with excessive hair overgrowth. PLoS Genet 10(5):e1004333. doi:10.1371/journal.pgen.1004333

Domenice S, Correa RV, Costa EM et al (2004) Mutations in the SRY, DAX1, SF1 and WNT4 genes in Brazilian sex-reversed patients. Braz J Med Biol Res 37(1):145–150. doi:10.1590/S0100-879X2004000100020 [pii]

Drash A, Sherman F, Hartmann WH, Blizzard RM (1970) A syndrome of pseudohermaphroditism, Wilms' tumor, hypertension, and degenerative renal disease. J Pediatr 76(4):585–593

Eid W, Opitz L, Biason-Lauber A (2015) Genome-wide identification of CBX2 targets: insights in the human sex development network. Mol Endocrinol 29(2):247–257. doi:10.1210/me.2014-1339

Ellsworth BS, Egashira N, Haller JL et al (2006) FOXL2 in the pituitary: molecular, genetic, and developmental analysis. Mol Endocrinol 20(11):2796–2805. doi:10.1210/me.2005-0303, me.2005-0303 [pii]

Evans RM (1988) The steroid and thyroid hormone receptor superfamily. Science 240(4854):889–895

Fantauzzo KA, Kurban M, Levy B, Christiano AM (2012) Trps1 and its target gene Sox9 regulate epithelial proliferation in the developing hair follicle and are associated with hypertrichosis. PLoS Genet 8(11):e1003002. doi:10.1371/journal.pgen.1003002

Figueiredo BC, Cavalli LR, Pianovski MA et al (2005) Amplification of the steroidogenic factor 1 gene in childhood adrenocortical tumors. J Clin Endocrinol Metab 90(2):615–619. doi:10.1210/jc.2004-0942

Finelli P, Pincelli AI, Russo S et al (2007) Disruption of friend of GATA 2 gene (FOG-2) by a de novo t(8;10) chromosomal translocation is associated with heart defects and gonadal dysgenesis. Clin Genet 71(3):195–204. doi:10.1111/j.1399-0004.2007.00752.x

Foster JW, Brennan FE, Hampikian GK et al (1992) Evolution of sex determination and the Y chromosome: SRY-related sequences in marsupials. Nature 359(6395):531–533. doi:10.1038/359531a0

Foster JW, Dominguez-Steglich MA, Guioli S et al (1994) Campomelic dysplasia and autosomal sex reversal caused by mutations in an SRY-related gene. Nature 372(6506):525–530. doi:10.1038/372525a0

Frasier SD, Bashore RA, Mosier HD (1964) Gonadoblastoma associated with pure gonadal dysgenesis in monozygous twins. J Pediatr 64:740–745

Garg V, Kathiriya IS, Barnes R et al (2003) GATA4 mutations cause human congenital heart defects and reveal an interaction with TBX5. Nature 424(6947):443–447. doi:10.1038/nature01827

German J, Simpson JL, Chaganti RS, Summitt RL, Reid LB, Merkatz IR (1978) Genetically determined sex-reversal in 46, XY humans. Science 202(4363):53–56

Glinka A, Dolde C, Kirsch N et al (2011) LGR4 and LGR5 are R-spondin receptors mediating Wnt/beta-catenin and Wnt/PCP signalling. EMBO Rep 12(10):1055–1061

Groth KA, Skakkebaek A, Host C, Gravholt CH, Bojesen A (2013) Clinical review: Klinefelter syndrome—a clinical update. J Clin Endocrinol Metab 98(1):20–30. doi:10.1210/jc.2012-2382

Gubbay J, Collignon J, Koopman P et al (1990) A gene mapping to the sex-determining region of the mouse Y chromosome is a member of a novel family of embryonically expressed genes. Nature 346(6281):245–250

Haber DA, Buckler AJ, Glaser T et al (1990) An internal deletion within an 11p13 zinc finger gene contributes to the development of Wilms' tumor. Cell 61(7):1257–1269

Hacker A, Capel B, Goodfellow P, Lovell-Badge R (1995) Expression of Sry, the mouse sex determining gene. Development 121(6):1603–1614

Hamada-Kawaguchi N, Nore BF, Kuwada Y, Smith CI, Yamamoto D (2014) Btk29A promotes Wnt4 signaling in the niche to terminate germ cell proliferation in Drosophila. Science 343(6168):294–297. doi:10.1126/science.1244512

Hammes A, Guo JK, Lutsch G et al (2001) Two splice variants of the Wilms' tumor 1 gene have distinct functions during sex determination and nephron formation. Cell 106(3):319–329

Hanley NA, Ball SG, Clement-Jones M et al (1999) Expression of steroidogenic factor 1 and Wilms' tumour 1 during early human gonadal development and sex determination. Mech Dev 87(1–2):175–180

He X, Semenov M, Tamai K, Zeng X (2004) LDL receptor-related proteins 5 and 6 in Wnt/beta-catenin signaling: arrows point the way. Development 131(8):1663–1677. doi:10.1242/dev.01117131/8/1663[pii]

Heikinheimo M, Ermolaeva M, Bielinska M et al (1997) Expression and hormonal regulation of transcription factors GATA-4 and GATA-6 in the mouse ovary. Endocrinology 138(8):3505–3514. doi:10.1210/endo.138.8.5350

Heikkila M, Peltoketo H, Leppaluoto J, Ilves M, Vuolteenaho O, Vainio S (2002) Wnt-4 deficiency alters mouse adrenal cortex function, reducing aldosterone production. Endocrinology 143(11):4358–4365. doi:10.1210/en.2002-220275

Heikkila M, Prunskaite R, Naillat F et al (2005) The partial female to male sex reversal in Wnt-4-deficient females involves induced expression of testosterone biosynthetic genes and testosterone production, and depends on androgen action. Endocrinology 146(9):4016–4023

Henderson ST, Johnson TE (2001) daf-16 integrates developmental and environmental inputs to mediate aging in the nematode Caenorhabditis elegans. Curr Biol 11(24):1975–1980, S0960-9822(01)00594-2 [pii]

Hirayama-Yamada K, Kamisago M, Akimoto K et al (2005) Phenotypes with GATA4 or NKX2.5 mutations in familial atrial septal defect. Am J Med Genet A 135(1):47–52. doi:10.1002/ajmg.a.30684

Hossain A, Saunders GF (2001) The human sex-determining gene SRY is a direct target of WT1. J Biol Chem 276(20):16817–16823. doi:10.1074/jbc.M009056200

Hu YC, Nicholls PK, Soh YQ et al (2015) Licensing of primordial germ cells for gametogenesis depends on genital ridge signaling. PLoS Genet 11(3):e1005019. doi:10.1371/journal.pgen.1005019

Hyon C, Chantot-Bastaraud S, Harbuz R et al (2015) Refining the regulatory region upstream of SOX9 associated with 46, XX testicular disorders of Sex Development (DSD). Am J Med Genet A 167A(8):1851–1858. doi:10.1002/ajmg.a.37101

Ikeda Y, Shen WH, Ingraham HA, Parker KL (1994) Developmental expression of mouse steroidogenic factor-1, an essential regulator of the steroid hydroxylases. Mol Endocrinol 8(5):654–662

Ingham PW (1998) Transducing Hedgehog: the story so far. EMBO J 17(13):3505–3511. doi:10.1093/emboj/17.13.3505

Iyer AK, McCabe ER (2004) Molecular mechanisms of DAX1 action. Mol Genet Metab 83(1–2): 60–73

Jager RJ, Harley VR, Pfeiffer RA, Goodfellow PN, Scherer G (1992) A familial mutation in the testis-determining gene SRY shared by both sexes. Hum Genet 90(4):350–355

Jameson SA, Lin YT, Capel B (2012a) Testis development requires the repression of Wnt4 by Fgf signaling. Dev Biol 370(1):24–32

Jameson SA, Natarajan A, Cool J et al (2012b) Temporal transcriptional profiling of somatic and germ cells reveals biased lineage priming of sexual fate in the fetal mouse gonad. PLoS Genet 8(3):e1002575

Jeays-Ward K, Hoyle C, Brennan J et al (2003) Endothelial and steroidogenic cell migration are regulated by WNT4 in the developing mammalian gonad. Development 130(16):3663–3670

Jeays-Ward K, Dandonneau M, Swain A (2004) Wnt4 is required for proper male as well as female sexual development. Dev Biol 276(2):431–440. doi:10.1016/j.ydbio.2004.08.049

Jeske YW, Bowles J, Greenfield A, Koopman P (1995) Expression of a linear Sry transcript in the mouse genital ridge. Nat Genet 10(4):480–482

Jordan BK, Mohammed M, Ching ST et al (2001) Up-regulation of WNT-4 signaling and dosage-sensitive sex reversal in humans. Am J Hum Genet 68(5):1102–1109. doi:10.1086/320125, S0002-9297(07)61219-1 [pii]

Katoh-Fukui Y, Tsuchiya R, Shiroishi T et al (1998) Male-to-female sex reversal in M33 mutant mice. Nature 393(6686):688–692. doi:10.1038/31482

Katoh-Fukui Y, Miyabayashi K, Komatsu T et al (2012) Cbx2, a polycomb group gene, is required for Sry gene expression in mice. Endocrinology 153(2):913–924. doi:10.1210/en.2011-1055

Kim Y, Kobayashi A, Sekido R et al (2006) Fgf9 and Wnt4 act as antagonistic signals to regulate mammalian sex determination. PLoS Biol 4(6):e187. doi:10.1371/journal.pbio.0040187

Kim Y, Bingham N, Sekido R, Parker KL, Lovell-Badge R, Capel B (2007) Fibroblast growth factor receptor 2 regulates proliferation and Sertoli differentiation during male sex determination. Proc Natl Acad Sci U S A 104(42):16558–16563. doi:10.1073/pnas.0702581104

Kim GJ, Sock E, Buchberger A et al (2015) Copy number variation of two separate regulatory regions upstream of SOX9 causes isolated 46, XY or 46, XX disorder of sex development. J Med Genet 52(4):240–247. doi:10.1136/jmedgenet-2014-102864

Knower KC, Kelly S, Ludbrook LM et al (2011) Failure of SOX9 regulation in 46XY disorders of sex development with SRY, SOX9 and SF1 mutations. PLoS One 6(3):e17751. doi:10.1371/journal.pone.0017751

Koopman P (2005) Sex determination: a tale of two Sox genes. Trends Genet 21(7):367–370. doi:10.1016/j.tig.2005.05.006

Koopman P, Munsterberg A, Capel B, Vivian N, Lovell-Badge R (1990) Expression of a candidate sex-determining gene during mouse testis differentiation. Nature 348(6300):450–452

Koopman P, Gubbay J, Vivian N, Goodfellow P, Lovell-Badge R (1991) Male development of chromosomally female mice transgenic for Sry. Nature 351(6322):117–121

Kreidberg JA, Sariola H, Loring JM et al (1993) WT-1 is required for early kidney development. Cell 74(4):679–691

Kuo CT, Morrisey EE, Anandappa R et al (1997) GATA4 transcription factor is required for ventral morphogenesis and heart tube formation. Genes Dev 11(8):1048–1060

Kurrasch DM, Cheung CC, Lee FY, Tran PV, Hata K, Ingraham HA (2007) The neonatal ventromedial hypothalamus transcriptome reveals novel markers with spatially distinct patterning. J Neurosci 27(50):13624–13634

Kusaka M, Katoh-Fukui Y, Ogawa H et al (2010) Abnormal epithelial cell polarity and ectopic epidermal growth factor receptor (EGFR) expression induced in Emx2 KO embryonic gonads. Endocrinology 151(12):5893–5904. doi:10.1210/en.2010-0915

Lalli E, Sassone-Corsi P (2003) DAX-1, an unusual orphan receptor at the crossroads of steroidogenic function and sexual differentiation. Mol Endocrinol 17(8):1445–1453. doi:10.1210/me.2003-0159

Laudet V (1997) Evolution of the nuclear receptor superfamily: early diversification from an ancestral orphan receptor. J Mol Endocrinol 19(3):207–226

Laudet V, Hanni C, Coll J, Catzeflis F, Stehelin D (1992) Evolution of the nuclear receptor gene superfamily. EMBO J 11(3):1003–1013

Lavery R, Chassot AA, Pauper E et al (2012) Testicular differentiation occurs in absence of R-spondin1 and Sox9 in mouse sex reversals. PLoS Genet 8(12):e1003170. doi:10.1371/journal.pgen.1003170

Lee DG, Han DH, Park KH, Baek M (2011) A novel WT1 gene mutation in a patient with Wilms' tumor and 46, XY gonadal dysgenesis. Eur J Pediatr 170(8):1079–1082. doi:10.1007/s00431-011-1439-0

Lin L, Achermann JC (2008) Steroidogenic factor-1 (SF-1, Ad4BP, NR5A1) and disorders of testis development. Sex Dev 2(4–5):200–209

Liu CF, Bingham N, Parker K, Yao HH (2009) Sex-specific roles of beta-catenin in mouse gonadal development. Hum Mol Genet 18(3):405–417

Liu CF, Parker K, Yao HH (2010) WNT4/beta-catenin pathway maintains female germ cell survival by inhibiting activin betaB in the mouse fetal ovary. PLoS One 5(4):e10382. doi:10.1371/journal.pone.0010382

Loke J, Pearlman A, Radi O et al (2014) Mutations in MAP3K1 tilt the balance from SOX9/FGF9 to WNT/beta-catenin signaling. Hum Mol Genet 23(4):1073–1083. doi:10.1093/hmg/ddt502

Lourenco D, Brauner R, Lin L et al (2009) Mutations in NR5A1 associated with ovarian insufficiency. N Engl J Med 360(12):1200–1210. doi:10.1056/NEJMoa0806228

Lourenco D, Brauner R, Rybczynska M, Nihoul-Fekete C, McElreavey K, Bashamboo A (2011) Loss-of-function mutation in GATA4 causes anomalies of human testicular development. Proc Natl Acad Sci U S A 108(4):1597–1602. doi:10.1073/pnas.1010257108

Lovell-Badge R (2002) Sex-determining genes in mice: building pathways. Wiley, Chichester, UK

Ludbrook LM, Bernard P, Bagheri-Fam S et al (2012) Excess DAX1 leads to XY ovotesticular disorder of sex development (DSD) in mice by inhibiting steroidogenic factor-1 (SF1) activation of the testis enhancer of SRY-box-9 (Sox9). Endocrinology 153(4):1948–1958. doi:10.1210/en.2011-1428

Luo X, Ikeda Y, Parker KL (1994) A cell-specific nuclear receptor is essential for adrenal and gonadal development and sexual differentiation. Cell 77(4):481–490

Maatouk DM, Mork L, Chassot AA, Chaboissier MC, Capel B (2013) Disruption of mitotic arrest precedes precocious differentiation and transdifferentiation of pregranulosa cells in the perinatal Wnt4 mutant ovary. Dev Biol 383(2):295–306. doi:10.1016/j.ydbio.2013.08.026

Mafra F, Catto M, Bianco B, Barbosa CP, Christofolini D (2015) Association of WNT4 polymorphisms with endometriosis in infertile patients. J Assist Reprod Genet. doi:10.1007/s10815-015-0523-1

Malki S, van der Heijden GW, O'Donnell KA, Martin SL, Bortvin A (2014) A role for retrotransposon LINE-1 in fetal oocyte attrition in mice. Dev Cell 29(5):521–533. doi:10.1016/j.devcel.2014.04.027

Mandel H, Shemer R, Borochowitz ZU et al (2008) SERKAL syndrome: an autosomal-recessive disorder caused by a loss-of-function mutation in WNT4. Am J Hum Genet 82(1):39–47. doi:10.1016/j.ajhg.2007.08.005

Mann JR, Corkery JJ, Fisher HJ et al (1983) The X linked recessive form of XY gonadal dysgenesis with a high incidence of gonadal germ cell tumours: clinical and genetic studies. J Med Genet 20(4):264–270

Manuylov NL, Smagulova FO, Leach L, Tevosian SG (2008) Ovarian development in mice requires the GATA4-FOG2 transcription complex. Development 135(22):3731–3743

Matson CK, Murphy MW, Sarver AL, Griswold MD, Bardwell VJ, Zarkower D (2011) DMRT1 prevents female reprogramming in the postnatal mammalian testis. Nature 476(7358): 101–104. doi:10.1038/nature10239

McCabe ERB (2001) Adrenal hypoplasias and aplasias, 8th edn. McGraw-Hill, New York

McElreavey K, Vilain E, Abbas N, Herskowitz I, Fellous M (1993) A regulatory cascade hypothesis for mammalian sex determination: SRY represses a negative regulator of male development. Proc Natl Acad Sci U S A 90(8):3368–3372

Meacham LR, Winn KJ, Culler FL, Parks JS (1991) Double vagina, cardiac, pulmonary, and other genital malformations with 46, XY karyotype. Am J Med Genet 41(4):478–481

Miller RW, Fraumeni JF Jr, Manning MD (1964) Association of Wilms's tumor with aniridia, hemihypertrophy and other congenital malformations. N Engl J Med 270:922–927

Miyamoto Y, Taniguchi H, Hamel F, Silversides DW, Viger RS (2008) A GATA4/WT1 cooperation regulates transcription of genes required for mammalian sex determination and differentiation. BMC Mol Biol 9:44. doi:10.1186/1471-2199-9-44

Molkentin JD (2000) The zinc finger-containing transcription factors GATA-4, -5, and -6. Ubiquitously expressed regulators of tissue-specific gene expression. J Biol Chem 275(50): 38949–38952. doi:10.1074/jbc.R000029200

Molkentin JD, Lin Q, Duncan SA, Olson EN (1997) Requirement of the transcription factor GATA4 for heart tube formation and ventral morphogenesis. Genes Dev 11(8):1061–1072

Moumne L, Fellous M, Veitia RA (2005) Deletions in the polyalanine-containing transcription factor FOXL2 lead to intranuclear aggregation. Hum Mol Genet 14(23):3557–3564. doi:10.1093/hmg/ddi383, ddi383 [pii]

Moumne L, Batista F, Benayoun BA et al (2008a) The mutations and potential targets of the forkhead transcription factor FOXL2. Mol Cell Endocrinol 282(1–2):2–11. doi:10.1016/j.mce.2007.11.006, S0303-7207(07)00428-5 [pii]

Moumne L, Dipietromaria A, Batista F et al (2008b) Differential aggregation and functional impairment induced by polyalanine expansions in FOXL2, a transcription factor involved in cranio-facial and ovarian development. Hum Mol Genet 17(7):1010–1019. doi:10.1093/hmg/ddm373, ddm373 [pii]

Murayama C, Miyazaki H, Miyamoto A, Shimizu T (2008) Involvement of Ad4BP/SF-1, DAX-1, and COUP-TFII transcription factor on steroid production and luteinization in ovarian theca cells. Mol Cell Biochem 314(1–2):51–58

Muzio L, Soria JM, Pannese M, Piccolo S, Mallamaci A (2005) A mutually stimulating loop involving emx2 and canonical wnt signalling specifically promotes expansion of occipital cortex and hippocampus. Cereb Cortex 15(12):2021–2028. doi:10.1093/cercor/bhi077

Naillat F, Prunskaite-Hyyrylainen R, Pietila I et al (2010) Wnt4/5a signalling coordinates cell adhesion and entry into meiosis during presumptive ovarian follicle development. Hum Mol Genet 19(8):1539–1550

Nef S, Schaad O, Stallings NR et al (2005) Gene expression during sex determination reveals a robust female genetic program at the onset of ovarian development. Dev Biol 287(2):361–377

Nemer G, Fadlalah F, Usta J et al (2006) A novel mutation in the GATA4 gene in patients with Tetralogy of Fallot. Hum Mutat 27(3):293–294. doi:10.1002/humu.9410

Nicholl RM, Grimsley L, Butler L et al (1994) Trisomy 22 and intersex. Arch Dis Child Fetal Neonatal Ed 71(1):F57–F58

Niehrs C (2012) The complex world of WNT receptor signalling. Nat Rev Mol Cell Biol 13(12): 767–779. doi:10.1038/nrm3470

Ninomiya S, Yokoyama Y, Teraoka M et al (2000) A novel mutation (296 del G) of the SOX90 gene in a patient with campomelic syndrome and sex reversal. Clin Genet 58(3):224–227

Nishida H, Miyagawa S, Vieux-Rochas M et al (2008) Positive regulation of steroidogenic acute regulatory protein gene expression through the interaction between Dlx and GATA-4 for testicular steroidogenesis. Endocrinology 149(5):2090–2097. doi:10.1210/en.2007-1265

Nusse R (2003) Wnts and Hedgehogs: lipid-modified proteins and similarities in signaling mechanisms at the cell surface. Development 130(22):5297–5305. doi:10.1242/dev.00821, 130/22/5297 [pii]

Ostrer H (2014) Disorders of sex development (DSDs): an update. J Clin Endocrinol Metab 99(5): 1503–1509. doi:10.1210/jc.2013-3690

Ottolenghi C, Moreira-Filho C, Mendonca BB et al (2001) Absence of mutations involving the LIM homeobox domain gene LHX9 in 46, XY gonadal agenesis and dysgenesis. J Clin Endocrinol Metab 86(6):2465–2469. doi:10.1210/jcem.86.6.7539

Pannetier M, Renault L, Jolivet G, Cotinot C, Pailhoux E (2005) Ovarian-specific expression of a new gene regulated by the goat PIS region and transcribed by a FOXL2 bidirectional promoter. Genomics 85(6):715–726. doi:10.1016/j.ygeno.2005.02.011, S0888-7543(05)00043-1 [pii]

Pannetier M, Fabre S, Batista F et al (2006) FOXL2 activates P450 aromatase gene transcription: towards a better characterization of the early steps of mammalian ovarian development. J Mol Endocrinol 36(3):399–413

Parma P, Radi O, Vidal V et al (2006) R-spondin1 is essential in sex determination, skin differentiation and malignancy. Nat Genet 38(11):1304–1309. doi:10.1038/ng1907

Pelletier J, Bruening W, Kashtan CE et al (1991a) Germline mutations in the Wilms' tumor suppressor gene are associated with abnormal urogenital development in Denys-Drash syndrome. Cell 67(2):437–447

Pelletier J, Bruening W, Li FP, Haber DA, Glaser T, Housman DE (1991b) WT1 mutations contribute to abnormal genital system development and hereditary Wilms' tumour. Nature 353(6343):431–434. doi:10.1038/353431a0

Philibert P, Biason-Lauber A, Rouzier R et al (2008) Identification and functional analysis of a new WNT4 gene mutation among 28 adolescent girls with primary amenorrhea and mullerian duct abnormalities: a French collaborative study. J Clin Endocrinol Metab 93(3):895–900. doi:10.1210/jc.2007-2023

Philibert P, Biason-Lauber A, Gueorguieva I et al (2011) Molecular analysis of WNT4 gene in four adolescent girls with mullerian duct abnormality and hyperandrogenism (atypical Mayer-Rokitansky-Kuster-Hauser syndrome). Fertil Steril 95(8):2683–2686. doi:10.1016/j.fertnstert.2011.01.152

Piard J, Mignot B, Arbez-Gindre F et al (2014) Severe sex differentiation disorder in a boy with a 3.8 Mb 10q25.3-q26.12 microdeletion encompassing EMX2. Am J Med Genet A 164A(10): 2618–2622. doi:10.1002/ajmg.a.36662

Pisarska MD, Bae J, Klein C, Hsueh AJ (2004) Forkhead l2 is expressed in the ovary and represses the promoter activity of the steroidogenic acute regulatory gene. Endocrinology 145(7): 3424–3433

Polanco JC, Wilhelm D, Davidson TL, Knight D, Koopman P (2010) Sox10 gain-of-function causes XX sex reversal in mice: implications for human 22q-linked disorders of sex development. Hum Mol Genet 19(3):506–516. doi:10.1093/hmg/ddp520

Pop R, Zaragoza MV, Gaudette M, Dohrmann U, Scherer G (2005) A homozygous nonsense mutation in SOX9 in the dominant disorder campomelic dysplasia: a case of mitotic gene conversion. Hum Genet 117(1):43–53. doi:10.1007/s00439-005-1295-y

Porter FD (2008) Smith-Lemli-Opitz syndrome: pathogenesis, diagnosis and management. Eur J Hum Genet 16:535–541

Prunskaite-Hyyrylainen R, Shan J, Railo A et al (2014) Wnt4, a pleiotropic signal for controlling cell polarity, basement membrane integrity, and antimullerian hormone expression during oocyte maturation in the female follicle. FASEB J 28(4):1568–1581. doi:10.1096/fj.13-233247

Ramayya MS, Zhou J, Kino T, Segars JH, Bondy CA, Chrousos GP (1997) Steroidogenic factor 1 messenger ribonucleic acid expression in steroidogenic and nonsteroidogenic human tissues: Northern blot and in situ hybridization studies. J Clin Endocrinol Metab 82(6):1799–1806

Rastetter RH, Bernard P, Palmer JS et al (2014) Marker genes identify three somatic cell types in the fetal mouse ovary. Dev Biol. doi:10.1016/j.ydbio.2014.08.013

Ritchie HH, Wang LH, Tsai S, O'Malley BW, Tsai MJ (1990) COUP-TF gene: a structure unique for the steroid/thyroid receptor superfamily. Nucleic Acids Res 18(23):6857–6862

Schepers G, Wilson M, Wilhelm D, Koopman P (2003) SOX8 is expressed during testis differentiation in mice and synergizes with SF1 to activate the Amh promoter in vitro. J Biol Chem 278(30):28101–28108. doi:10.1074/jbc.M304067200

Schmahl J, Kim Y, Colvin JS, Ornitz DM, Capel B (2004) Fgf9 induces proliferation and nuclear localization of FGFR2 in Sertoli precursors during male sex determination. Development 131(15):3627–3636. doi:10.1242/dev.01239

Seeherunvong T, Perera EM, Bao Y et al (2004) 46,XX sex reversal with partial duplication of chromosome arm 22q. Am J Med Genet A 127a(2):149–151. doi:10.1002/ajmg.a.20630

Sekido R, Lovell-Badge R (2008) Sex determination involves synergistic action of SRY and SF1 on a specific Sox9 enhancer. Nature 453(7197):930–934. doi:10.1038/nature06944

Sekido R, Lovell-Badge R (2009) Sex determination and SRY: down to a wink and a nudge? Trends Genet 25(1):19–29. doi:10.1016/j.tig.2008.10.008

Sekido R, Bar I, Narvaez V, Penny G, Lovell-Badge R (2004) SOX9 is up-regulated by the transient expression of SRY specifically in Sertoli cell precursors. Dev Biol 274(2):271–279

Shah SP, Kobel M, Senz J et al (2009) Mutation of FOXL2 in granulosa-cell tumors of the ovary. N Engl J Med 360(26):2719–2729. doi:10.1056/NEJMoa0902542, NEJMoa0902542 [pii]

Shen MM, Hodgkin J (1988) mab-3, a gene required for sex-specific yolk protein expression and a male-specific lineage in C. elegans. Cell 54(7):1019–1031

Shinoda K, Lei H, Yoshii H et al (1995) Developmental defects of the ventromedial hypothalamic nucleus and pituitary gonadotroph in the Ftz-F1 disrupted mice. Dev Dyn 204(1):22–29

Sienski G, Donertas D, Brennecke J (2012) Transcriptional silencing of transposons by Piwi and maelstrom and its impact on chromatin state and gene expression. Cell 151(5):964–980. doi:10.1016/j.cell.2012.10.040

Siggers P, Smith L, Greenfield A (2002) Sexually dimorphic expression of Gata-2 during mouse gonad development. Mech Dev 111(1–2):159–162

Simon JA, Kingston RE (2013) Occupying chromatin: Polycomb mechanisms for getting to genomic targets, stopping transcriptional traffic, and staying put. Mol Cell 49(5):808–824. doi:10.1016/j.molcel.2013.02.013

Sinclair AH, Berta P, Palmer MS et al (1990) A gene from the human sex-determining region encodes a protein with homology to a conserved DNA-binding motif. Nature 346(6281): 240–244. doi:10.1038/346240a0

Smith WL, Dewitt DL (1996) Prostaglandin endoperoxide H synthases-1 and -2. Adv Immunol 62: 167–215

Smith CA, Shoemaker CM, Roeszler KN, Queen J, Crews D, Sinclair AH (2008) Cloning and expression of R-Spondin1 in different vertebrates suggests a conserved role in ovarian development. BMC Dev Biol 8:72. doi:10.1186/1471-213x-8-72

Sock E, Schmidt K, Hermanns-Borgmeyer I, Bosl MR, Wegner M (2001) Idiopathic weight reduction in mice deficient in the high-mobility-group transcription factor Sox8. Mol Cell Biol 21(20):6951–6959. doi:10.1128/mcb.21.20.6951-6959.2001

Sparmann A, van Lohuizen M (2006) Polycomb silencers control cell fate, development and cancer. Nat Rev Cancer 6(11):846–856. doi:10.1038/nrc1991

Sultan C, Biason-Lauber A, Philibert P (2009) Mayer-Rokitansky-Kuster-Hauser syndrome: recent clinical and genetic findings. Gynecol Endocrinol 25(1):8–11. doi:10.1080/09513590802288291

Suri M, Kelehan P, O'Neill D et al (2007) WT1 mutations in Meacham syndrome suggest a coelomic mesothelial origin of the cardiac and diaphragmatic malformations. Am J Med Genet A 143A(19):2312–2320

Sutton E, Hughes J, White S et al (2011) Identification of SOX3 as an XX male sex reversal gene in mice and humans. J Clin Invest 121(1):328–341. doi:10.1172/jci42580

Swain A, Lovell-Badge R (1999) Mammalian sex determination: a molecular drama. Genes Dev 13(7):755–767

Swain A, Narvaez V, Burgoyne P, Camerino G, Lovell-Badge R (1998) Dax1 antagonizes Sry action in mammalian sex determination. Nature 391(6669):761–767

Tevosian SG, Albrecht KH, Crispino JD, Fujiwara Y, Eicher EM, Orkin SH (2002) Gonadal differentiation, sex determination and normal Sry expression in mice require direct interaction between transcription partners GATA4 and FOG2. Development 129(19):4627–4634

Tomaselli S, Megiorni F, De Bernardo C et al (2008) Syndromic true hermaphroditism due to an R-spondin1 (RSPO1) homozygous mutation. Hum Mutat 29(2):220–226. doi:10.1002/humu.20665

Tomaselli S, Megiorni F, Lin L et al (2011) Human RSPO1/R-spondin1 is expressed during early ovary development and augments beta-catenin signaling. PLoS One 6(1):e16366. doi:10.1371/journal.pone.0016366

Tomita-Mitchell A, Maslen CL, Morris CD, Garg V, Goldmuntz E (2007) GATA4 sequence variants in patients with congenital heart disease. J Med Genet 44(12):779–783. doi:10.1136/jmg.2007.052183

Tomizuka K, Horikoshi K, Kitada R et al (2008) R-spondin1 plays an essential role in ovarian development through positively regulating Wnt-4 signaling. Hum Mol Genet 17(9):1278–1291

Tremblay JJ, Robert NM, Viger RS (2001) Modulation of endogenous GATA-4 activity reveals its dual contribution to Mullerian inhibiting substance gene transcription in Sertoli cells. Mol Endocrinol 15(9):1636–1650. doi:10.1210/mend.15.9.0692

Troy PJ, Daftary GS, Bagot CN, Taylor HS (2003) Transcriptional repression of peri-implantation EMX2 expression in mammalian reproduction by HOXA10. Mol Cell Biol 23(1):1–13

Tsukiyama T, Yamaguchi TP (2012) Mice lacking Wnt2b are viable and display a postnatal olfactory bulb phenotype. Neurosci Lett 512(1):48–52. doi:10.1016/j.neulet.2012.01.062

Uhlenhaut NH, Treier M (2006) Foxl2 function in ovarian development. Mol Genet Metab 88(3):225–234. doi:10.1016/j.ymgme.2006.03.005

Uhlenhaut NH, Jakob S, Anlag K et al (2009) Somatic sex reprogramming of adult ovaries to testes by FOXL2 ablation. Cell 139(6):1130–1142. doi:10.1016/j.cell.2009.11.021, S0092-8674(09)01433-0 [pii]

Umehara F, Tate G, Itoh K et al (2000) A novel mutation of desert hedgehog in a patient with 46, XY partial gonadal dysgenesis accompanied by minifascicular neuropathy. Am J Hum Genet 67(5):1302–1305. doi:10.1016/s0002-9297(07)62958-9

Vainio S, Heikkila M, Kispert A, Chin N, McMahon AP (1999) Female development in mammals is regulated by Wnt-4 signalling. Nature 397(6718):405–409. doi:10.1038/17068

Val P, Martinez-Barbera JP, Swain A (2007) Adrenal development is initiated by Cited2 and Wt1 through modulation of Sf-1 dosage. Development 134(12):2349–2358

Vetro A, Ciccone R, Giorda R et al (2011) XX males SRY negative: a confirmed cause of infertility. J Med Genet 48(10):710–712. doi:10.1136/jmedgenet-2011-100036

Viger RS, Guittot SM, Anttonen M, Wilson DB, Heikinheimo M (2008) Role of the GATA family of transcription factors in endocrine development, function, and disease. Mol Endocrinol 22(4):781–798. doi:10.1210/me.2007-0513

Volkel P, Le Faou P, Vandamme J, Pira D, Angrand PO (2012) A human Polycomb isoform lacking the Pc box does not participate to PRC1 complexes but forms protein assemblies and represses transcription. Epigenetics 7(5):482–491. doi:10.4161/epi.19741

Wagner T, Wirth J, Meyer J et al (1994) Autosomal sex reversal and campomelic dysplasia are caused by mutations in and around the SRY-related gene SOX9. Cell 79(6):1111–1120

White S, Ohnesorg T, Notini A et al (2011) Copy number variation in patients with disorders of sex development due to 46, XY gonadal dysgenesis. PLoS One 6(3):e17793. doi:10.1371/journal.pone.0017793

Wilhelm D, Englert C (2002) The Wilms tumor suppressor WT1 regulates early gonad development by activation of Sf1. Genes Dev 16(14):1839–1851. doi:10.1101/gad.220102

Wilhelm D, Martinson F, Bradford S et al (2005) Sertoli cell differentiation is induced both cell-autonomously and through prostaglandin signaling during mammalian sex determination. Dev Biol 287(1):111–124

Yao HH, Capel B (2002) Disruption of testis cords by cyclopamine or forskolin reveals independent cellular pathways in testis organogenesis. Dev Biol 246(2):356–365. doi:10.1006/dbio.2002.0663

Yao HH, Aardema J, Holthusen K (2006) Sexually dimorphic regulation of inhibin beta B in establishing gonadal vasculature in mice. Biol Reprod 74(5):978–983

Yokobayashi S, Liang CY, Kohler H et al (2013) PRC1 coordinates timing of sexual differentiation of female primordial germ cells. Nature 495(7440):236–240. doi:10.1038/nature11918

Yoshida M, Suda Y, Matsuo I et al (1997) Emx1 and Emx2 functions in development of dorsal telencephalon. Development 124(1):101–111

Zangen D, Kaufman Y, Banne E et al (2014) Testicular differentiation factor SF-1 is required for human spleen development. J Clin Invest 124(5):2071–2075. doi:10.1172/JCI73186

Chapter 14
Methods for the Study of Gonadal Development

Rafal P. Piprek

Abstract Current knowledge on gonadal development and sex determination is the product of many decades of research involving a variety of scientific methods from different biological disciplines such as histology, genetics, biochemistry, and molecular biology. The earliest embryological investigations, followed by the invention of microscopy and staining methods, were based on histological examinations. The most robust development of histological staining techniques occurred in the second half of the nineteenth century and resulted in structural descriptions of gonadogenesis. These first studies on gonadal development were conducted on domesticated animals; however, currently the mouse is the most extensively studied species. The next key point in the study of gonadogenesis was the advancement of methods allowing for the in vitro culture of fetal gonads. For instance, this led to the description of the origin of cell lines forming the gonads. Protein detection using antibodies and immunolabeling methods and the use of reporter genes were also invaluable for developmental studies, enabling the visualization of the formation of gonadal structure. Recently, genetic and molecular biology techniques, especially gene expression analysis, have revolutionized studies on gonadogenesis and have provided insight into the molecular mechanisms that govern this process. The successive invention of new methods is reflected in the progress of research on gonadal development.

14.1 An Outline of the History of Gonadal Development Studies

For centuries investigators have tried to elucidate the processes responsible for the creation of organ structure. Although Hippocrates (460–370 BC) and Aristotle (384–322 BC) made the first descriptions of fetal development, the question of how the differences between males and females arise had remained unanswered

R.P. Piprek (✉)
Department of Comparative Anatomy, Institute of Zoology, Jagiellonian University, Gronostajowa 9, 30-387 Kraków, Poland
e-mail: rafal.piprek@uj.edu.pl

R.P. Piprek (ed.), *Molecular Mechanisms of Cell Differentiation in Gonad Development*, Results and Problems in Cell Differentiation 58,
DOI 10.1007/978-3-319-31973-5_14

(Needham 1959). Later in the modern period, it became evident that the gonads are the first structures in the body to undergo sexual differentiation and they are primarily responsible for orchestrating the development of sex-specific features. It was interesting to study the mechanisms leading to the appearance of the earliest sex-specific traits, i.e., the differences between the testes and ovaries—organs that arise from common sexually undifferentiated anlage. Figure 14.1 summarizes main events and discoveries in the field of studies on the gonadal development.

The earliest observations of gonadal development were made during the first histological studies of mammalian fetuses. These early studies were possible after the invention of microscopy and staining techniques. The first staining and microscopic observations were made in the seventeenth century by Leeuwenhoek who used dyes isolated from plants (madder, saffron, indigo). In 1856, Perkin invented the first synthetic dye (mauveine later known as toluidine blue). In 1858, von Gerlach stained cerebellum with carmine isolated from cochineal insects. In 1876, Waldeyer used hematoxylin as a histological stain. In 1876, Wissowsky used hematoxylin and eosin as the first complex histological stain. This shows that in the second half of the nineteenth century, there was significant development of histological methods, which resulted in acceleration of embryological discoveries. Many early descriptions of gonadogenesis and meiosis come from this period. The earliest stages of gonadogenesis and genital ridge formation were described in rabbit (Coert 1898; Egli 1876; Janošik 1885, 1887a, b, 1890a, b, 1894), swine (Janošik 1885, 1887a, b, 1890a, b, 1894), sheep (Janošik 1885, 1887a, b, 1890a, b, 1894), cat (Coert 1898), and humans (Nagel 1889a, b). Sexual differentiation was also described early on in rabbit (Coert 1898; Janošik 1885), swine (Allen 1904), sheep (Janošik 1885, 1887a, b, 1890a, b, 1894), and humans (Nagel 1889a, b). Felix (1911) and Fuss (1911) were the first to describe the extragonadal origin of PGCs in human embryos. Diagnostic features useful for sex recognition of a developing gonad had been defined. The fetal testis had been described as an organ covered by a thick connective tissue capsule under superficial epithelium and filled with cords inside. The fetal ovary had been identified by structural division into the cortex and medulla, the division which according to current knowledge seems oversimplified and misleading. Today, the majority of studies on gonadal development focus on the laboratory mouse (*Mus musculus*). Despite the differences between human and mouse, the latter species became a powerful research tool owing to easy genetic modification, a sequenced genome, small size, quick reproduction, and easy breeding. The mouse is also a convenient model in the field of developmental biology. The availability of many species-specific probes, antibodies, and markers makes research on mice fast, very efficient, and cost-effective. In 1902, Cuénot conducted the first genetic study on mice. In 1909, Castle and Little developed the first "lab mouse"—inbred strain DBA (Castle and Little 1910). After World War I, Strong bred C3H and CBA mice in 1909, and Little in 1921 developed the C57BL strain (Crow 2002). The C57BL/6 (B6) mouse is the first mammal whose complete genome sequence was published (Chinwalla et al. 2002; Waterston et al. 2002). Most studies concerning murine gonadogenesis were conducted on the C57BL/6 strain, a few on the CD-1 strain (Bendel-Stenzel et al. 2000; Best et al. 2008;

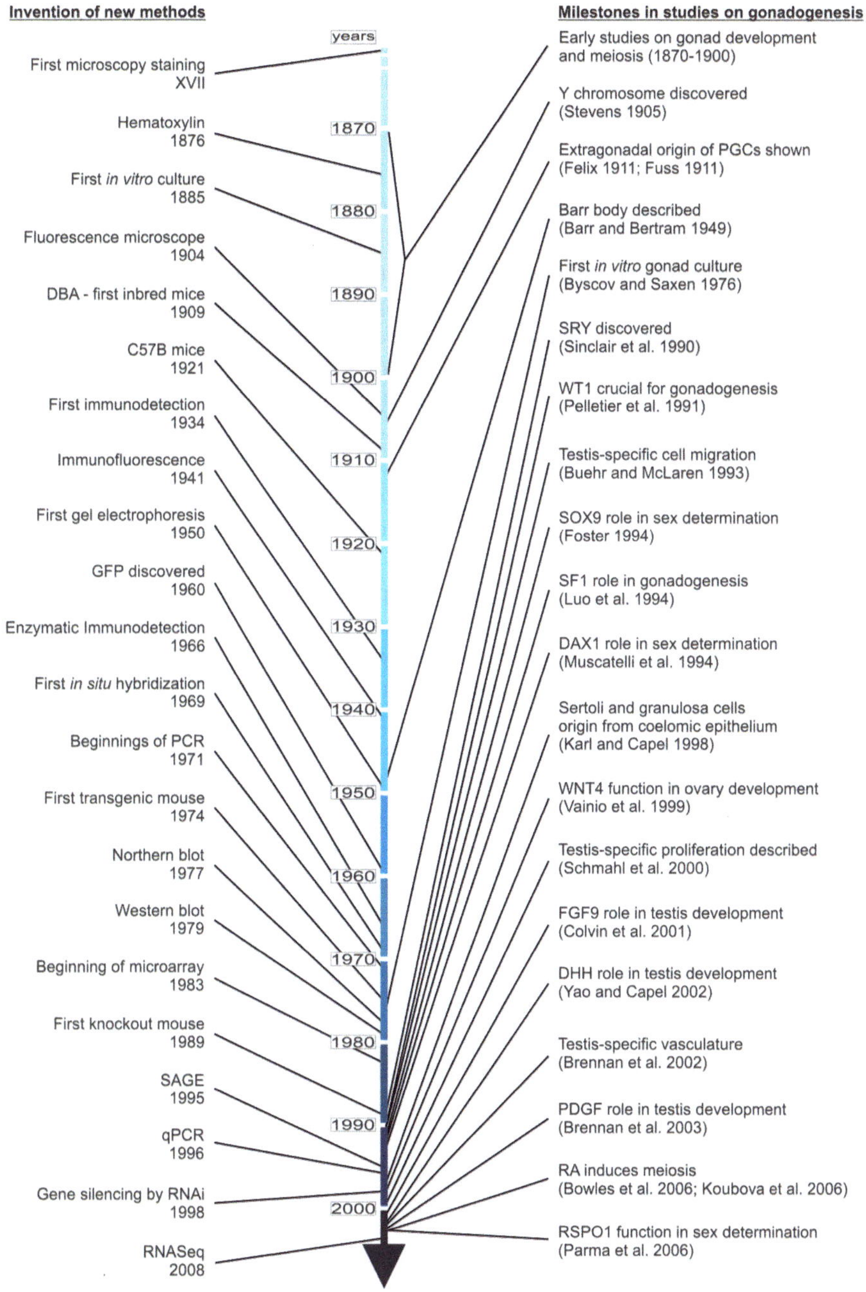

Fig. 14.1 A timeline depicting invention of new research methods that appeared crucial for studies on the gonadal development, and the main discoveries in the field of gonadogenesis studies

Bradford et al. 2009; Brennan et al. 2002; Combes et al. 2009a, b; Cool et al. 2008; DeFalco et al. 2011, 2014, 2015; Di Carlo et al. 2000; Karl and Capel 1998; Kim et al. 2006; Koubova et al. 2006; Martineau et al. 1997; Merchant-Larios and Moreno-Mendoza 1998; Munger et al. 2013; Schmahl and Capel 2003; Schmahl et al. 2000), and only exceptionally on FVB (Anderson et al. 1999; Bendel-Stenzel et al. 2000; Bishop et al. 2000; Bullejos and Koopman 2005), CBA (Buehr et al. 1993; Willerton et al. 2004), MF1 (Best et al. 2008), 129S1/SvImJ (Fleming et al. 2012), AKR/J (Bouma et al. 2007; Bullejos and Koopman 2005; Bouma et al. 2005), DBA/2J (Bouma et al. 2007), and Balb/C (Salas-Cortés et al. 1999) strains. The C57BL/6 mouse strain deserves special attention due to its exceptional sensitivity to disturbances in early gonadogenesis and is thus an efficacious tool for identifying genes involved in sex determination (Bouma et al. 2007). Valuable results are obtained when Y chromosomes, derived from other mouse strains, are placed on a different genetic background, e.g., the Y chromosome from *Mus domesticus poschiavinus* (Y^{POS}) carried on a C57BL/6 background causes ovarian development in XY but not on DBA/2J. This sensitivity is even stronger when the Y chromosome from AKR/J mice (Y^{AKR}) is on a C57BL/6 background (Bouma et al. 2007; Bullejos and Koopman 2005).

In the twentieth century, the rapid development of new methods, especially in the field of cytology, genetics, and molecular biology, propelled embryological studies and led to at least two crucial discoveries in research on sex determination. First, the Y chromosome was discovered in 1905 independently by American geneticist Nettie Maria Stevens and American zoologist and geneticist Edmund B. Wilson (Wilson 1905; Morgan 1912). Using cytogenetic methods and microscopy, they noticed that males of mealworms produce two kinds of sperm differing in the presence of a large or small variant of one of the chromosomes. It was deduced that the sex depends on the presence of the small (Y) chromosome. A second significant breakthrough in this discipline was the discovery of the *SRY* (the sex-determining region on the Y chromosome) gene in humans (Sinclair et al. 1990). This discovery was possible due to genetic analysis of four male patients with an XX karyotype; SRY was found in a translocated fragment of the Y chromosome. At that time, the *Sry* gene was described in mice (Gubbay et al. 1990). The role of *Sry* in sex determination, as a primary switch, was proved by XX male mice with an introduced *Sry* gene (Koopman et al. 1991).

14.2 Sex Markers

The *Sry* gene automatically became a suitable sex marker; however, before its discovery other markers were used to define the sex of an individual. Molecular sex markers were always eligible for studying the genes and processes involved in sex determination and development of the gonads due to difficulties or even incapability in distinguishing between male and female fetuses using structural features. Molecular sex markers can identify or predict the sex of an individual even before the gonads appear. However, before the widespread availability and use of sex-linked genes, the sex of fetuses was identified by the presence of Barr bodies (X chromatin bodies). Sex chromatin was discovered by Canadian physicians Murray L. Barr and Ewart G. Bertram in 1949 (Barr and Bertram 1949). By applying histological staining to neurons, they noticed an intranuclear intrusion present in about half of the studied animals. In gonadal development studies requiring the identification of the sex of an individual, the amnion membranes were dissociated and cell suspension was air dried on slides, stained with orcein or Giemsa dye, and Barr bodies were detected under a light microscope (Capel and Batchvarov 2008; Karl and Capel 1998; Palmer and Burgoyne 1991; Yao et al. 2004). Another sexing method involved the detection of activity of the X-linked enzyme glucose-6-phosphate dehydrogenase (G6PDH), used to identify the genetic sex in mice (Williams 1986).

Some studies on gonadogenesis rely on easy sexing based on the presence/ absence of testis cords. These elongated structures begin forming in the murine testes at 11.8 *dpc* (stage of 21 tail somites); however, 12.25 *dpc* is the earliest moment in which the sex can be recognized by the presence/absence of the testis cords under a dissecting microscope (Nel-Themaat et al. 2009; Wilhelm et al. 2005). A lack of cords indicates ovarian differentiation.

The aforementioned methods have been replaced by more accurate molecular techniques. Nowadays, the most extensively used method to define the sex of an individual is genotyping relying on sex-specific gene identification by PCR (polymerase chain reaction). This procedure involves tissue collection (e.g., a piece of earlobe or tail ideally from 10- to 21-day-old mice), DNA isolation, and simplex (one pair of primers for one template) or multiplex (two or more primer pairs) PCR. A series of genes are used as sex markers. First, Y-linked sequences, such as *Sry* (Chaboissier et al. 2004; Greenlee et al. 1998; Kunieda et al. 1992; Lambert et al. 2000; Lavrovsky et al. 1998; McClive and Sinclair 2001) or *Zfy* (Greenlee et al. 1998; Kunieda et al. 1992; Nagamine et al. 1990; Nef et al. 2005), can be amplified and detected indicating the male genetic sex. Since *Sry* is present only on the Y chromosome and has no equivalent on the X, and the *Zfy* gene has a homolog on the X chromosome (*Zfx*) with equal size, amplification of X-linked or autosomal genes is needed as a control (McFarlane et al. 2013; Palmer et al. 1990). The X-linked gene *Dxnds3* is an example of such an internal control in multiplex PCR (Greenlee et al. 1998; Kunieda et al. 1992). There are currently only three simplex PCR methods used for sex genotyping in mice. Sex identification is possible through the amplification of homologous genes *Uba1* and *Ube1y1* linked to the X and Y chromosomes, respectively, using a set of primers designed to amplify a deleted region in the *Ube1y1* gene, which results in two products of amplification in males and one large product in females with a small size difference (19 bp) between two amplicons (Best et al. 2008; Chuma and Nakatsuji 2001; McFarlane et al. 2013). Another method is based on amplification of X-linked *Kdm5c* (also known as *Smcx, Jarid1c*) and its Y-linked equivalent *Kdm5d* (*Smcy, Jarid1d*) with a 29 bp size difference between amplicons (Barrionuevo et al. 2006; Bullejos and Koopman 2005; Clapcote and Roder 2005; Jameson et al. 2012; Mroz et al. 1999;

Munger et al. 2009). A third example of simplex PCR sex genotyping involves a size difference between the Y-linked *Sly* gene and its X-linked equivalent *Xlr* that results from a 405-bp deletion in *Sly* (McFarlane et al. 2013). The primers were designed upstream and downstream of the deleted region giving two different size products in each sex. The last method is the most accurate due to size differences large enough to be easily detected during electrophoresis.

Previously, an alternative method of identification of *Sry* fragments via Southern blot was used to define the sex (Fechner et al. 1993; Gubbay et al. 1990; Tanoue et al. 1992). Identification of H-Y antigen encoded by *Kdm5d* gene was also used as a male-specific marker (Wolf 1998).

14.3 Staging and Timing of Early Gonadogenesis in Mouse and Other Vertebrates

Precise fetal age determination is essential for embryological studies especially if accurate spatiotemporal pattern of gene expression or description of structural changes in developing organs is required. Precise staging is crucial for studying the embryology of animals with a rapid rate of development such as rodents, in which some genes are expressed in a narrow time window.

In mice, it is necessary to use timed matings with E0.5 (or 0.5 *dpc*, *dies post coitum*) representing noon on the day a mating plug was detected. Usually mouse pairs are set up at ~5:00 pm; the presence of vaginal plugs is checked the next morning and noon is counted as 0.5 *dpc*. More accurate staging relies on counting the tail somites (ts). The first tail somite is counted as the first one posterior to the junction of the hindlimbs. In mice, early genital ridges begin to be visible around 10.5 *dpc* which corresponds to the presence of 8 ts (±2 ts); and onward 11.5 *dpc* corresponds to 18 ts (±2 ts), and 12.5 *dpc* corresponds to 30 ts (±3 ts) (Hacker et al. 1995). To obtain the key stages for the study of sexual differentiation of murine gonads, e.g., stages 15–17 ts, mice are sacrificed at 7:00 am on the 11th *dpc*, 18–20 ts at noon on the 11th day, and 28–30 ts at noon on the 12th day. Additionally, tables of Theiler stages (TS; from TS01 to TS28) are useful in determining the embryonic age, especially in fetuses older than 13.5 *dpc* staged by forelimb and hindlimb morphology (Theiler 1989). Mouse gonads start developing at TS16 (i.e., 9.5–10.75 *dpc*, corresponding to 30–34 total somite number), sex determination takes place at TS17–TS18 (i.e., 35–44 total somite number), and testis cords start forming at TS19 (i.e., 45–47 total somite number) and become discernible at TS20 (i.e., 48–51 total somite number).

Various standardized systems of staging are used for different vertebrates. For example, Carnegie stages are used to determine the human fetal age. These stages were applied in studies on human gonadogenesis (Hanley et al. 2000) but can be used for other vertebrates. Carnegie stages range from C01 to C23 covering the first 56 days of human gestation and later stages are described using the term of gestation. Sexual differentiation of gonads in humans occurs at C18–23. For chicken embryonic development, Hamburger and Hamilton stages (1951) are used and range from HH01 (laying) to HH46 (hatching). Undifferentiated gonads of chickens appear at HH25 (embryonic day E4.5), while sexual differentiation of gonads occurs at HH29 (E6) (Morais da Silva et al. 1996). Staging according to Greenbaum (2002) was used in a study of gonadal development in the red-eared slider (*Trachemys scripta*) (Smith et al. 2008b). In this reptile, the temperature-sensitive period begins at Greenbaum stage 14, and sexual differentiation of gonads begins at stages 18/19. In most anuran amphibians, embryonic and larval development is defined according to the Gosner system (from stage 1 to 46 at completion of metamorphosis) (1960). Undifferentiated gonads appear at about Gosner stage 26; however, the time of sexual differentiation varies among anuran species (Piprek 2013). The tadpoles of the model anuran species the African clawed frog (*Xenopus laevis*) and the Western clawed frog (*Silurana tropicalis*) differ in morphology from most anuran tadpoles and therefore another staging system is used to describe their development, i.e., Nieuwkoop and Faber stages (1967). This system ranges from NF1 (egg) to NF66 (completed metamorphosis). Undifferentiated gonads appear at NF49 and sexual differentiation occurs at about NF52/53 (Piprek 2013). In the model zebrafish (*Danio rerio*), development is staged according to somite number or time after fertilization. In this species, primordial germ cells accumulate at the sites of gonad formation at 2 weeks *pf* (post-fertilization), at 4 weeks *pf* the first ovarian development features are present, and at 7 weeks *pf* the first signs of testicular differentiation are visible (Kimmel et al. 1995; Maack and Segner 2003).

14.4 Histology: Light and Electron Microscopy

As previously mentioned, the earliest studies of gonadal development were based on structural analysis using histological techniques. Today, the histological studies have been replaced by molecular studies. Nevertheless, histological techniques are indispensable for instance if the effects of a gene mutation or knockout need to be assessed. Despite a wide range of staining methods, hematoxylin and eosin staining (H&E, HE) on paraffin sections (usually 3–8 μm thick) is mostly used. Sample preparation for histological analysis begins with tissue dissection and fixation. Fetal gonads are usually fixed in 4 % paraformaldehyde (PFA in PBS at 4 °C overnight) (Bagheri-Fam et al. 2008; Chang et al. 2008; Hummitzsch et al. 2013; Moniot et al. 2009; Nef et al. 2003; Nel-Themaat et al. 2009) or in Bouin's solution (room temp. or 4 °C for 1–2 days) (Britt et al. 2002; Chassot et al. 2008; El Jamil et al. 2008; Gassei et al. 2008; Hiramatsu et al. 2009; Koubova et al. 2006; Maatouk et al. 2008; MacLean et al. 2009; McLaren and Southee 1997; Qing et al. 2008; Tang et al. 2008). Other fixation methods are rarely used, e.g., Serra fixation (Barrionuevo et al. 2006, 2009). Bouin-fixed samples are dehydrated and embedded in paraffin; however, PFA-fixed samples can be both frozen or dehydrated and embedded in paraffin. Bouin's solution provides strong fixation of gonadal tissue ensuring maintenance of structure; poorer resolution is usually obtained in PFA fixation. PFA-fixed frozen sections are the most suitable for immunolocalization. Histological staining on cryosections and dyes other than H&E are rarely used. Trichrome staining (e.g., Masson's trichrome; Britt et al. 2002) provides more structural detail showing three different colors in the tissue (differently stained nuclei, cytoplasm, and extracellular matrix). Histological images analyzed via specialist software can also be used for 3D reconstruction of internal organ structure. This method was applied in the reconstruction of the structure of testis cords within developing mouse testes (Nel-Themaat et al. 2009) and ovarian follicles in *Xenopus* (Bilinski et al. 2010).

Ultrastructure of gonads can be studied using transmission electron microscopy (TEM) techniques (Best et al. 2008; Britt et al. 2002; El Jamil et al. 2008; Hummitzsch et al. 2013; Merchant-Larios and Moreno-Mendoza 1998; Ottolenghi et al. 2005; Tanimura and Iwasawa 1991). In the majority of these studies, tissues were fixed in Karnovsky's fixative (PFA, glutaraldehyde, cacodylate buffer), then postfixed in osmium tetroxide, dehydrated, and embedded in resin (Ito and Karnowsky 1968). Such samples can be cut for semithin (0.5–1 μm) or ultrathin sections (50–70 nm). Semithin sections are placed on microscope slides, stained with methylene blue and azur II and viewed under a light microscope. Ultrathin sections are placed on copper (sometimes nickel for immunolocalization) grids, contrasted with uranyl acetate and lead citrate, and viewed with TEM. Other histological staining techniques and even immunolocalization can also be used for both semithin (on slides) and ultrathin (on grids) sections; however, chemical removal of resin is necessary.

Scanning electron microscopy (SEM) has been used to study gonadal development and especially the external structure of gonads (Piprek et al. 2014; Wylie et al. 1976). In this procedure, samples are usually fixed in Karnovsky's fixative, postfixed in osmium tetroxide, dehydrated, dried using liquid carbon dioxide, sputter-coated with gold, and viewed using SEM (Ito and Karnowsky 1968).

14.5 Molecule Identification: Cytochemistry, Immunolocalization

Cyto- and histochemistry is the science of localization of chemical compounds within cells or tissues using chemical reactions or specific antibodies. Among cytochemical techniques, PAS staining, alkaline phosphatase detection, and especially immunohistochemistry are widely used in gonadal development studies. The PAS (periodic acid-Schiff reaction) method detects polysaccharides such as glycogen, glycoproteins, glycolipids, and mucins in tissues. Primordial germ cells (PGCs) have abundant cytoplasmic glycogen and thus the PAS method can be used to identify these cells (Chassot et al. 2008; De Felici et al. 2004; Gassei et al. 2008; Hiramatsu et al. 2009).

Alkaline phosphatase (AP, ALP) is an enzyme removing phosphate groups from many types of molecules and acts most efficiently in an alkaline environment. This enzyme is especially widespread on the surfaces of stem cells including PGCs and thus has been used to identify migrating PGCs prior to gonad development (Chiquoine 1954; De Felici and Dolci 1989; Di Carlo et al. 2000; Hanley et al. 2000; Maitland and Ullmann 1993).

Specific recognition and detection of proteins via antibodies became an extensively used method in embryological research and revolutionized this discipline. In direct immunolabeling, an antibody is covalently labeled with a marker enabling detection; in indirect immunolabeling, a primary antibody binds an antigen and a secondary antibody conjugated with a marker binds the primary antibody. Secondary antibodies can be labeled with (1) an enzyme that produces a colored compound (usually alkaline phosphatase with BCIP/NBT substrate, or horseradish peroxidase—HRP with DAB as a substrate) detected in bright-field light microscopy, (2) fluorochrome (usually derivatives of fluorescein or rhodamine) detectable in fluorescence microscopy, and (3) colloid gold for transmission electron microscopy (TEM), scanning electron microscopy (SEM), or light microscopy (LM). Several compounds such as biotin, avidin, and streptavidin are used to increase the number of bound secondary antibodies or enzymes and thus to intensify signal. Double immunolabeling is possible due to the application of two primary antibodies (differing in species origin) against two different proteins. This technique enables the detection of potential co-localization of two proteins. Interestingly, the first use of antibodies for detection was reported in 1934 (Marrack 1934), the first fluorescent detection using antibodies took place in 1941 (Coons et al. 1941), and the first enzymatic labeling was described in 1966 (Avrameas and Uriel 1966).

Immunolabeling can be classified as (1) immunocytochemistry (ICC)—immunolabeling of cells usually after *in vitro* culture, (2) immunohistochemistry (IHC)—detection in tissues on paraffin or cryosections or in whole organs (whole mount, WM-IHC), (3) immunofluorescence (IF)—localization via fluorochrome using fluorescence (also confocal) microscopy, and (4) immunogold staining (IGS)—detection of colloid gold under an electron or light microscope.

There are two types of immunolabeling in respect to sequence of labeling and embedding: pre-embedding and post-embedding. In the pre-embedding method, immunolabeling is done prior to embedding and sectioning (i.e., fixation, immunolabeling, embedding in paraffin, resin, or freezing, cutting, imaging). In the post-embedding method, the tissue after fixation is first embedded in paraffin, resin, or frozen, then cut, and immunolabeling is performed on sections (i.e., fixation, embedding, cutting, immunolabeling, imaging).

There is a plethora of gonadal development studies in which immunolabeling was used, usually IHC or IF (Abramyan et al. 2009; Bagheri-Fam et al. 2008; Barrionuevo et al. 2006, 2009; Bendel-Stenzel et al. 2000; Best et al. 2008; Beverdam and Koopman 2006; Bogani et al. 2009; Bradford et al. 2009; Britt et al. 2002; Chang et al. 2008; Chassot et al. 2008; Chen et al. 2012; Childs et al. 2011; DeFalco et al. 2015; DiNapoli and Capel 2007; Dumond et al. 2011; Dupont et al. 2000; Fleming et al. 2012; Fröjdman et al. 1992; Hanley et al. 2000; Hiramatsu et al. 2009; Hummitzsch et al. 2013; Jameson et al. 2012; Karl and Capel 1998; Nef et al. 2005; Li et al. 2012; Li and Kim 2004; Liu et al. 2009, 2015; Maatouk et al. 2008; MacLean et al. 2009; Matzuk et al. 1995; Molyneaux et al. 2003; Moniot et al. 2009; Morais da Silva et al. 1996; Mork et al. 2014; Nicol and Yao 2015; Ohe et al. 2002; Oshima et al. 2005; Ross et al. 2009; Salas-Cortés et al. 1999; Schmahl et al. 2004; Smith et al. 2008a, b; Soyal et al. 2000; Tevosian et al. 2002; Wilhelm et al. 2005; Willerton et al. 2004; Yao et al. 2004). For IHC, in most cases tissues are fixed with 4 % PFA or Bouin's solution, dehydrated, and embedded in paraffin. Such strong fixation entails later antigen retrieval which is usually done in citrate buffer at ~95 °C (HIER, heat-induced epitope retrieval). Samples for IF are usually fixed in 4 % PFA, frozen, and later cryosectioned; in the case of cryosections, antigen retrieval is unnecessary. Cryosections and deparaffinized sections after antigen retrieval, both on slides, are usually blocked with 6 % BSA in PBS (instead of BSA, Casein or serum from species matching the secondary antibody can be used), which is followed by incubation with primary antibodies (usually overnight at 4 °C), and then after rinsing (PBS with detergent), incubation with secondary antibodies is done (usually for 1–2 h at room temp.); after rinsing, color development (for IHC) and counterstain (for IHC and IF) is carried out.

In the pre-embedding version of IHC, also known as **whole mount immunohistochemistry** (WM-IHC or WIHC), the whole fetal gonads or urogenital ridges (gonads attached to mesonephroi) are fixed usually in 4 % PFA (or in a methanol: DMSO mixture), permeabilized in a solution of detergent to allow large molecules to penetrate, and then immersed in blocking buffer followed by incubation with primary and then secondary antibodies (Albrecht and Eicher 2001; Barske and Capel 2010; Bendel-Stenzel et al. 2000; Bogani et al. 2009; Bouma et al. 2005, 2007; Brennan et al. 2002; Bullejos and Koopman 2005; Combes et al. 2009a, b, 2011; Coveney et al. 2008; Cool et al. 2008; DeFalco et al. 2014, 2015; Karl and Capel 1998; Maatouk et al. 2008; Martineau et al. 1997; Nicol and Yao 2015; Schmahl et al. 2000; Schmahl and Capel 2003; Tang et al. 2008; Tevosian et al. 2002; Warr et al. 2009; Wilhelm et al. 2007). After immunolabeling, the gonads can be embedded and sectioned.

Immunogold staining (IGS) permits the detection of a protein using electron microscopy (TEM or SEM). Colloid gold or nanogold conjugated to secondary antibody are used in this technique. Both the pre-embedding and post-embedding versions of the IGS method were presented in detail by Bilinski et al. (2010). In most commonly used post-embedding IGS methods, the tissue is gently fixed in PFA with or without glutaraldehyde (strong fixation and osmium tetroxide postfixation should be omitted in order not to disrupt immunoreactivity), dehydrated, embedded in hydrophilic resin, and cut on formvar-carbon-coated nickel grids; the grids with ultrathin sections are applied onto drops of the following solutions: blocking solution (BSA), primary antibodies, rinsing, secondary antibodies, rinsing, and contrast staining with uranyl acetate and lead citrate. The subcellular localization of a protein of interest can be investigated with this method (Best et al. 2008).

The most important processes involved in development include cell proliferation and apoptosis. Both can be studied using appropriate antibodies. Cell divisions can be detected via immunolocalization of proliferation markers such as nuclear protein PCNA (Bogani et al. 2009), phosphohistone-H3 (pHH3) (Bogani et al. 2009;

Manuylov et al. 2007), and nuclear antigen Ki-67 (Dupont et al. 2000). However, the most popular technique used to identify proliferating cells is immunolocalization of bromodeoxyuridine (BrdU) after injection of this compound. In studies on cell proliferation in developing gonads of the mouse, BrdU is injected intraperitoneally (IP) to a pregnant female (in amount 10–50 mg/kg) and is incorporated into DNA during replication (phase S) in female and fetal tissues. Females are sacrificed 2–5 h after injection. Fetal gonads are fixed and immunolocalization with anti-BrdU is performed on paraffin or frozen sections (DeFalco et al. 2015; Dupont et al. 2000; Nef et al. 2003; Schmahl et al. 2000). BrdU pulse chase is a special version of this assay, which can determine when precursors for specific cell types were dividing (Schmahl et al. 2000). There is also a series of descriptions of proliferation studies in developing gonads in in vitro culture (DeFalco et al. 2011; Hiramatsu et al. 2009; Martineau et al. 1997; Schmahl et al. 2004; Schmahl and Capel 2003). In such assays, BrdU is dissolved in culture medium, and after 1–3 h, organs are rinsed and cultured in medium without BrdU for 9–12 h and then fixed and immunolabeled.

Administration of proliferation inhibitors enables to test if proliferation plays a role in a specific developmental process. Proliferation inhibitors such as 5-fluorouracil or methotrexate (MTX) were administered intraperitoneally to pregnant mice to study the role of proliferation in gonad development (Schmahl and Capel 2003). Alternatively, Affi-gel beads soaked in aphidicolin, a substance that inhibits the cell cycle at the S phase, demonstrated the role of proliferation in testis development (Schmahl and Capel 2003).

The contribution of apoptosis in development is typically studied using the TUNEL assay (Li and Kim 2004; Manuylov et al. 2007; Matzuk et al. 1995). In the TUNEL (terminal deoxynucleotidyl transferase dUTP nick end labeling) method, the fragmentation of DNA, common for cell death, is detected. Terminal deoxynucleotidyl transferase TdT binds 3′ nicks in DNA and catalyzes the addition of dUTPs labeled with a marker. This visualizes apoptotic cells on frozen or paraffin sections using chromogen or fluorochrome. Additionally, immunolocalization of an apoptosis marker—caspase3—is useful in cell death studies (Tang et al. 2008).

14.6 Identification of mRNA: In Situ Hybridization

Protein identification is especially useful for cell-specific marker identification and structure viewing; however, it is less accurate if gene expression is studied because a protein is present even after gene expression has ceased and moreover a protein can be detected in places other than its site of expression (e.g., in secreted proteins). Therefore, identification of specific mRNA is a more useful and accurate method in gene expression analyses.

The most commonly used method to detect gene expression in situ (in tissue) is **RNA in situ hybridization (ISH)**. The first report of detection of DNA:RNA hybridization was given by Gall and Pardue (1969). This method is frequently used in gonadal development studies in both the post-embedding version (ISH) (Abramyan et al. 2009; Best et al. 2008; Bradford et al. 2009; Chassot et al. 2008; El Jamil et al. 2008; Hanley et al. 2000; Kim et al. 2006; Maatouk et al. 2008; Martineau et al. 1997; Matzouk et al. 1995; Moniot et al. 2009; Morais da Silva et al. 1996; Nef et al. 2003; Oulad-Abdelghani et al. 1996; Parma et al. 2006; Schmahl et al. 2004; Sekido et al. 2004; Wilhelm et al. 2005) and pre-embedding or whole mount-ISH (WM-ISH or WISH) (Barrionuevo et al. 2006; Bishop et al. 2000; Bogani et al. 2009; Combes et al. 2009a,2011; Coveney et al. 2008; Jameson et al. 2012; Koubova et al. 2006; Manuylov et al. 2007; Matsuyama et al. 2005; Nef et al. 2005; Rolland et al. 2011; Smith et al. 2008a, b; Stebler et al. 2004). The core idea in the ISH technique is the hybridization of a labeled complementary probe to mRNA allowing the identification of a specific gene transcript under a microscope. A probe consists of complementary DNA (cDNA) or more commonly complementary RNA (cRNA, riboprobe). This analysis can be done on paraffin sections, frozen sections on slides for bright-field light or fluorescence microscopy, or even on grids for electron microscopy (TEM) (Kloc et al. 2001). This method involves probe design (250–1500 nt), probe in vitro synthesis, and probe labeling usually with digoxigenin (DIG) or biotin. DIG is a steroid with high antigenicity recognized by antibodies that can be conjugated with fluorochrome or enzyme (AP or HRP). Alternatively, probes can be biotinylated, i.e., conjugated to biotin which is bound by avidin or streptavidin conjugated to an enzyme. This method is termed CIHS (chromogenic in situ hybridization) in the case of enzymatic labeling and the appearance of color product indicating the presence of the mRNA of interest. In the case of fluorescence labeling, ISH is termed FISH (fluorescence in situ hybridization). In ISH, tissues are usually fixed in 4 % PFA, rinsed, frozen, or dehydrated and embedded in paraffin. Sections on slides undergo hybridization with probes overnight at 65 °C and samples are viewed under a microscope after labeling with antibodies. In WM-ISH, the whole organ is fixed in 4 % PFA, rinsed, hybridized in buffers, and after labeling the sample is viewed in toto under a dissection microscope or dehydrated, embedded, and sectioned and viewed on slides under light or fluorescence microscopes. In situ hybridization can also be used to localize a DNA fragment on chromosomes, e.g., in sex determination studies (Bishop et al. 2000; Parma et al. 2006; Yoshimoto et al. 2008).

14.7 Reporter Genes

One of the most advanced methods currently applied for visualization and specific cell type identification is the utilization of reporter genes. This method involves the use of transgenic animals. A reporter gene gives an easily identifiable product. The most commonly used reporter genes are *LacZ* and *Gfp*. These genes enable identifying the expression of a gene of interest and can also be used to trace specific cell lines. A reporter gene is introduced into a genome and is expressed under the control of a given promoter of interest.

LacZ encodes β-galactosidase (β-gal), an enzyme converting X-gal substrate to a blue product. *LacZ* with a promoter of the gene of interest is introduced into a genome. After *LacZ* expression, the β-gal protein is present in cells in which the gene of interest is normally expressed. Organs of the transgenic animal are dissected and fixed in PFA, and X-gal staining can be conducted on slides or in the whole organ (whole mount technique). This method is widely used in studies of gonadogenesis to visualize cells of a special type or to investigate the spatiotemporal pattern of gene expression (Anderson et al. 1999; Brennan et al. 2002; Capel et al. 1999; Liu et al. 2015; Maatouk et al. 2013; Manuylov et al. 2007; Martineau et al. 1997; McLaren and Southee 1997; Merchant-Larios and Moreno-Mendoza 1998; Miyamoto et al. 2008; Tang et al. 2008). This method was also applied in an assessment of the action of β-catenin in sex determination (Chassot et al. 2008).

Fluorescent proteins such as GFP (green fluorescent protein) act in a different manner. A signal is detectable after excitation (e.g., ultraviolet for GFP). The *Gfp* gene was isolated from the jellyfish *Aequorea victoria* in 1960 (Tsuji 2010). Several mutations were introduced in order to modify and improve GFP fluorescence and thus GFP derivatives were obtained, e.g., EGFP (enhanced GFP) and YFP (yellow fluorescent protein). Red fluorescent proteins include mCherry or tdTomato. In gonadogenesis studies, fluorescent reporter genes are used to identify and trace specific cells, to visualize gonadal structure, and to isolate cells of interest (Albrecht and Eicher 2001; Anderson et al. 1999; Best et al. 2008; Beverdam and Koopman 2006; Cool et al. 2008; DeFalco et al. 2011, 2014, 2015; Jameson et al. 2012; Maatouk et al. 2013; Matzuk et al. 1995; Molyneaux et al. 2001; Nef et al. 2005; Nel-Themaat et al. 2009; Qing et al. 2008; Rolland et al. 2011). For example GFP, YFP, mCherry, and tdTomato were used to detect cells expressing *Sf1*, *Sry*, *Sox9*, *Oct4*, or *αSma*, markers of germ cells, or gonadal somatic cells (Albrecht and Eicher 2001; Beverdam and Koopman 2006; Cool et al. 2008; DeFalco et al. 2014; Jameson et al. 2012; Matzuk et al. 1995; Molyneaux et al. 2001; Nef et al. 2005; Nel-Themaat et al. 2009; Qing et al. 2008; Rolland et al. 2011).

Luciferase may be used as a reporter gene because of light emission via the process of bioluminescence after oxidation of substrate (luciferin) by this enzyme. A luciferase reporter gene assay was used to trace the expression of *Sry* and *Amh* (Miyamoto et al. 2008; Wilhelm et al. 2007).

14.8 Cell Markers

The aforementioned methods such as IHC, ISH, and reporter genes can be used for the identification of a cell of interest due to the presence of cell markers. Each cell type has its own unique transcriptome and proteome, and thus, cell-specific markers, unique for a given cell type, can be identified. Such cell markers can be used for cell visualization in an organ, for tracing migrating cells, cell sorting, and isolation. It must be emphasized that cell markers revolutionized gonadal development studies and have been used to describe the origins and fates of cell lines during gonadogenesis.

A series of markers useful for identification of specific cell lines (e.g., Sertoli cells, follicular cells, fetal Leydig cells, endothelial cells, germ line cells, meiotic

Table 14.1 Cell markers used in the studies on the mouse gonadal development

	Cell type	Markers	References
Early genital ridge	Gonadal precursor cells	CK18, CK19, EMX2, GATA4, LHX9, SF1, WT1	Barrionuevo et al. (2006); Birk et al. (2000); Hu et al. (2013); Kusaka et al. (2010); Luo et al. (1994); Miyamoto et al. (1997, 2008)
	Pre-Sertoli cells	AMH, DHH, GDNF, N-Cadherin, NR5A1, PDGFa, Ptgds, SOX9, SPRR2, SRY, TUBB3 FGFR2 localized in nucleoplasm	Chassot et al. (2008); DeFalco et al. (2015); Fleming et al. (2012); Lee et al. (2009); Schmahl et al. (2004); Tang et al. (2008)
	Pre-follicular cells (= pre-granulosa cells)	ADAMTS19, BMP2, DAX1, FOLLISTATIN, FOXL2, LGR5, NR5A1, RSPO1, SPRR2, WNT4 FGFR2 localized in cytoplasm	Bouma et al. (2007); Chassot et al. (2008); Fleming et al. (2012); Lee et al. (2009); Liu et al. (2009); Rastetter et al. (2014); Schmahl et al. (2004)
	Mesonephros-derived migrating cells	PECAM1	Brennan et al. (2002); Smith et al. (2005)
	Vascular cells	PECAM1	Albrecht and Eicher (2001); Tang et al. (2008)
	Stromal/interstitial cells	ARX, MAFB, C-MAF, PECAM1, PDGFRα, PTCH1, VCAM1	Brennan et al. (2003); DeFalco et al. (2011); Kitamura et al. (2002); Yao et al. (2002)
Testis	Sertoli cells	ABP, AMH, GATA1, SOX9	Sharpe et al. (2003)
	Fetal Leydig cells	CYP11A1 = SCC, CYP17A1, HSD17B3, INSL3; 3βHSD	Chassot et al. (2008); Liu et al. (2009); Tang et al. (2008)
	Adult Leydig cells	CYP11A1, LHR, 3βHSD	Gassei et al. (2008); Teerds et al. (1999)
	Peritubular myoid cells (PMCs)	α-Actin, ACTA2, IGHMBP2	DeFalco et al. (2015); Gassei et al. (2008); Willerton et al. (2004)
Ovary	Follicular cells (=granulosa)	AMH, CYP19a1, FSHR	Hatzirodos et al. (2015); Rastetter et al. (2014)
	Theca interna	CYP17a1, INSL3, LHR	Hatzirodos et al. (2015)

(continued)

Table 14.1 (continued)

	Cell type	Markers	References
Germ cells	Germ cells in fetal gonads	E-Cadherin, CD31, DDX4, GCNA1, MRE11, MVH, OCT4, PECAM1, VASA, TRA98	Albrecht and Eicher (2001); Barrionuevo et al. (2009); Chang et al. (2008); Chassot et al. (2008); Gassei et al. (2008); Liu et al. (2009); Tang et al. (2008); Wilhelm et al. (2009)
	Oogonia	C-Kit, NANOG, OCT4, SSEA4	Byskov et al. (2011) and references therein
	Spermatogonia stem cells (SSCs)	GFRA1, OCT4, PLZF, PAX7	Reviewed in Chap. 10, Table 10.1
	Differentiated spermatogonial cells	C-Kit, ZBTB16	DeFalco et al. (2015)
	Meiotic cells	DMC1, FIGα, SCP3, SYN/COR, γH2AX	Barrionuevo et al. (2006); Yao et al. (2004)

cells) in developing gonads are summarized in Table 14.1. The most crucial cell markers include GATA4, LHX9, SF1, and WT1 for coelomic epithelium-derived gonadal precursor cells; SRY, SOX9, and AMH for pre-Sertoli cells; 3βHSD for fetal Leydig cells; OCT4 for germ cells; PECAM1 for germ and endothelial cells; and SYN/COR for meiotic cells.

14.9 Structure Visualization Using Cell Markers and Tracers

Labeled cells can be visualized continuously under in vitro conditions. This valuable method traces migrating cells within developing gonads in a **time-lapse** course (Coveney et al. 2008; DeFalco et al. 2011; Molyneaux et al. 2001; Nel-Themaat et al. 2009, 2010). Coveney et al. (2008) studied the dynamics of testis morphogenesis using an in vitro culture of fetal gonads of transgenic mice with expression of GFP in endothelial cells. In this study, *Cre* recombinase was expressed under a control of *Tie2* or *Flk1* promoters (markers of endothelium) and cleaved a stop codon upstream of *GFP* gene, which led to GFP presence in the endothelial cells.

An assembly of cell labeling images by specialist software enables the **three-dimensional modeling** of the internal structure of the developing gonads. This stereological analysis technique was especially useful in describing the process of testis cord formation. Barrionuevo et al. (2009) conducted 3D visualization of developing testes via immunolabeling for P450scc (a marker of Leydig cells); DeFalco et al. (2015) used immunolocalization of several cell markers to obtain a 3D model of the developing testis (Sertoli cells—TUBB3, PMCs—ACTA2, macrophages—MHCII, spermatogonia—ZBTB16, blood vessels and basement membranes—collagen IV). Nel-Themaat et al. (2009) used transgenic mice with expression of EGFP under control of the *Sox9* gene to visualize Sertoli cells and to present a 3D model of testiculogenesis.

14.10 In Vitro Culture Methods

In vitro organ culture is one of the most commonly used techniques to study the mechanisms of gonadogenesis in mice. In this method, developing gonads are isolated from fetus, placed in medium, and cultured for up to several days. This enables (1) observation of development in a time-lapse course, (2) studying cell migration between the gonad reassembled with the mesonephros (migration assay in coculture), and (3) analysis of the molecular mechanisms governing gonadogenesis via culturing gonads in medium supplemented with growth factors, their inhibitors, or antibodies blocking given proteins.

In vitro culture techniques were originally developed for embryological trials. In vitro culture of chicken embryo organs by Roux in 1885 and frog embryonic cells by Harrison in 1907 were maintained ex vivo at first. Byskov and Saxen (1976) conducted early in vitro cultures of developing gonads. Today, several techniques of in vitro culture of developing gonads are used, among which culture on agar blocks is the most popular. Dissected fetal mouse gonads are always cultured under conditions standard for mammal cells, i.e., at 37 °C and 5 % CO_2/95 % air. In most studies, developing gonads are cultured in DMEM medium.

In an early trial, developing gonads were maintained in vitro "glued" to a nucleopore filter using 1 % agar in Eagle's balanced salt solution (EBSS) supplemented with 10 % fetal calf serum (FCS) (Byskov and Saxen 1976). In subsequent studies, gonads were cultured at an air-medium interface on agar blocks (Evans et al. 1982; McLaren and Buehr 1990). The first recombinant cultures (a gonad reassembled with the mesonephros) were used by Buehr et al. (1993) and showed migration of mesonephric cell into developing testes (Combes et al. 2009b; Cool et al. 2008; Martineau et al. 1997). Buehr et al. (1993) also provided the original description of fetal gonad culture on agar blocks, currently the most common method for maintaining gonads in vitro (Barrios et al. 2010; Best et al. 2008; Bogani et al. 2009; Capel and Batchvarov 2008; Combes et al. 2009b; Coveney et al. 2008; DeFalco et al. 2011; Di Carlo et al. 2000; Hiramatsu et al. 2009; Karl and Capel 1998; Kim et al. 2006; Koubova et al. 2006; Martineau et al. 1997; McLaren and Southee 1997; Nishino et al. 2001; Wilhelm et al. 2005; Tang et al. 2008). Alternatively, a droplet method is used to maintain whole fetal gonads under in vitro conditions (Cupp et al. 2003; DeFalco et al. 2014; Maatouk et al. 2008). An organ is cultured in a droplet on the lid of Petri dish while the dish is floating on water in a larger Petri dish. Fetal gonads were also cultured on Millicell filters (Cupp et al. 2003; MacLean et al. 2009; Nel-Themaat et al. 2009). A slice method of in vitro culture was used to study the regulation of PGC migration from the hindgut to the genital ridges (Bendel-Stenzel et al. 2000; Molyneaux et al. 2001, 2003).

The fetal gonads are usually cultured for 48 h on 1.5 % agar blocks in DMEM medium supplemented with 10 % FCS or FBS and ampicillin. Exceptionally, F12/DME medium supplemented with 10 % FBS is used (Liu et al. 2015; Nishino et al. 2001). This method of fetal gonad culture on agar blocks was also used for studying fetal *T. scripta* gonads (Mork et al. 2014). In this case, the L15 medium was supplemented with 10 % FBS. However, in vitro culture of fetal chicken gonads was done on an isopore filter in DMEM medium supplemented with FCS and chicken serum (Smith et al. 2008a).

14.11 In Vitro Studies of Cell Signaling

In vitro gonad culture was also effective in determining the function of signaling molecules in the regulation of gonad development. The method is based on culturing dissected gonads in medium supplemented with exogenous growth factors, their inhibitors, or antibodies binding and thus blocking the action of particular proteins. This technique was used to study the role of FGF9 in testis development. Gonads were cultured in medium supplemented with FGF9 (25, 50, or 100 ng/ml) (Barrios et al. 2010; Hiramatsu et al. 2009; Kim et al. 2006; Wilhelm et al. 2005) or in the vicinity of agarose beads soaked with FGF9 growth factor for up to 2 days (Bogani et al. 2009; DeFalco et al. 2011). The function of neurotrophin 3 in gonadogenesis was studied by implementing a blocker of its receptor (AG879) added to medium (Cupp et al. 2003). In a study on neurotrophic tyrosine receptor kinases, the medium was supplemented with neurotrophin 3, antibodies binding and blocking neurotrophin 3, antibodies binding receptor NTRK3, inhibitors of tyrosine kinase, NGF, or antibodies binding NGF (Gassei et al. 2008). γ-secretase inhibitor (DAPT) added to medium inhibited Notch signaling and elucidated the function of this pathway in progenitor Leydig cells (Tang et al. 2008). The role of hedgehog (HH) signaling was examined via addition of cyclopamine (inhibitor of HH pathway) to medium (Liu et al. 2015). Sex determination was also investigated by the addition of prostaglandin D2 to medium (Wilhelm et al. 2005). Interestingly, the role of blood vessels in gonad morphogenesis was tested via blocking of vasculature formation with an inhibitor of VEGF receptor tyrosine kinase (VEGFR TKI II) (DeFalco et al. 2014). Moreover, the influence of blood vessels on gonadogenesis was studied by addition of anti-VE-cadherin antibodies to medium which disrupted vasculature formation (Combes et al. 2009b). WNT signaling was investigated by adding LiCl to medium, which increased the function of β-catenin and WNT pathways (Maatouk et al. 2008). The method of signaling modulation in vitro was also useful for studying the induction of meiosis by retinoic acid. Medium was supplemented with all-trans retinoic acid (RA), agonists of retinoic acid receptors (RAR), and ketoconazole that inhibits the Cyp26 enzyme from degrading RA (Bowles et al. 2006; Koubova et al. 2006; Li and Kim 2004; Piprek et al. 2013).

14.12 Cell Tracing and Migration Assay

Cell tracing is a method revealing cell migration within developing organs. This technique also provides evidence for the origin of different cells or cell types. In gonadal development studies, cell tracing involves cell labeling with a marker (usually fluorescent dye applied onto the surface of the gonad) (DeFalco et al. 2011; Karl and Capel 1998; Li and Kim 2004; Yao et al. 2004). The fluorescent dye penetrates the superficial cells of the developing gonad and then labeled cells carry the marker while migrating inward into the developing organ. Identification of labeled cells after in vitro culture of the whole organ shows the ingression of superficial cells. This method demonstrated the origin of Sertoli and granulosa cells from coelomic epithelium (Karl and Capel 1998). Karl and Capel (1998) incorporated fluorescent CM-DiI dye to superficial gonadal cells using ionophoresis. A single superficial cell was touched by an electrode with the applied dye; the current was pulsed until a single cell was positive for CM-DiI. In the same study, an alternative method of surface cell labeling was used; the dye was pipetted onto the surface of the developing gonad, and after washing, the gonads were cultured in vitro for 30–48 h. The labeled cells were localized under a microscope. DeFalco et al. (2011) labeled superficial cells of developing mouse gonads via pipetting MitoTracker Orange dye onto the gonads dissected from mouse fetuses; after 45 min of incubation with dye, gonads were washed and cultured in vitro for 24–48 h and then viewed. Two studies used a mixture of two fluorescent dyes (MitoTracker Red and rhodamine derivative, 5-carboxytetramethylrhodamine, succinimidyl ester) applied onto the surface of the developing gonad of *T. scripta* and *X. laevis* (Piprek 2013; Yao et al. 2004). After incubation, gonads were washed and cultured in vitro for 48 h, then fixed, and processed for immunohistochemistry and imaging.

Different cell tracing methods were used to study cell migration from the fetal mesonephros to developing gonads (**migration assay**). In this technique, in vitro coculture of the fetal gonad assembled with the mesonephros (usually derived from another individual) is conducted, with expression of reporter genes enabling cell identification or with labeled cells. The first experiment showing cell migration from the mesonephros to the fetal gonads was carried out by Buehr et al. (1993). They detached developing mouse gonads from the adjacent mesonephroi and cultured isolated gonads and mesonephroi in vitro separately, or the organs were reassembled and grafted into the fragments of limb or heart, or cultured while separated by a membrane. The migration of cells from mesonephroi into developing testes was observed in reassembly of wild-type embryonic testes with transgenic mesonephroi from murine embryos carrying a nuclear marker (~1000 copies of the β-globin gene). Also other techniques, such as in situ hybridization and immunolocalization, permit identification of mesonephros-derived cells in the testes. Li and Kim (2004) tested mesonephric cell migration to developing rat gonads via in vitro culture of reassembled embryonic gonads with mesonephros labeled with fluorescent cell-permeable dye. The mesonephros, isolated from a fetus, was cultured in medium with dye (5(6)-carboxyfluorescein diacetate, succinimidyl ester, CFDA-SE) for 20 min; after washing, the stained mesonephros was reassembled with a fetal gonad and cultured in vitro for 3 days, and then the organs were fixed and viewed using fluorescent microscopy. Capel et al. (1999) and Martineau et al. (1997) studied cell migration from the mesonephros to the developing testes using in vitro culture of mesonephros from ROSA26 transgenic mice showing constitutive expression of β-gal reassembled with fetal wild-type mouse gonads. After 42–48 h of in vitro culture, the organs were fixed, stained for β-gal, embedded, and viewed for detection of (mesonephros-derived) blue cells in the testes. Brennan et al. (2002) used coculture of wild-type gonads with mesonephroi from mice with endothelial cells marked with β-gal expressed under the control of endothelial markers (*Flt1* or *Tie2*).

Most recent studies on cell migration to developing gonads reassemble fetal wild-type mouse gonads with mesonephros derived from transgenic mice with constitutional expression of fluorescent protein GFP (Bogani et al. 2009; Brennan et al. 2002; Combes et al. 2009a; Cool et al. 2008; Cui et al. 2004; DeFalco et al. 2011; Hiramatsu et al. 2009; Nishino et al. 2001; Yao et al. 2004). After 24–64 h of in vitro culture, the organs are fixed, processed for immunolabeling, and viewed to detect green signal in the gonads. Yao et al. (2004) studied gonads isolated from the slider *T. scripta* reassembled with mouse mesonephros or mouse gonads and mesonephros in various sandwich combinations.

14.13 Separation of Cell Lines

The gonad consists of cells of many lines, which differ in origin, structure, function, and gene expression pattern. The first analyses of gene expression were based on RNA isolated from whole developing gonads and thus genes expressed in many different cell lines were studied. Nowadays, the use of cell markers facilitates the isolation of a single cell line from the heterogeneous mass of tissues forming the gonad. Moreover, even a single cell can be removed from the whole organ or from a section on a microscopic slide (LMD, laser microdissection) and its transcriptome can be studied. For instance, genes expressed in epithelial cells covering developing gonads were studied after laser dissection of frozen sections (Kusaka et al. 2010).

Isolation of a specific cell line from an organ was propelled by the application of transgenic animals with the expression of a reporter gene in a cell type of interest. This method requires isolation of the gonads, their enzymatic digestion, and cell sorting (e.g., fluorescence-activated cell sorting—FACS or cell isolation with magnetic beads). Tissue digestion is usually conducted by treatment with one or more enzymes at 37 °C followed by mechanical dissociation (pipetting, filtering) and cell resuspension in medium or PBS. Trypsin/EDTA, collagenase, DNase, and dispase are enzymes used for dissociation of developing gonads (Beverdam and Koopman 2006; Childs et al. 2011; DeFalco et al. 2014, 2015; Jameson et al. 2012; McLaren and Southee 1997; Molyneaux et al. 2003; Munger et al. 2013; Nef et al. 2005; Nishino et al. 2001; Qing et al. 2008; Rolland et al. 2011; Smith et al. 2008a; Wilhelm et al. 2005).

FACS is a specialized type of flow cytometry that can separate a heterogeneous mixture of cells of various lines into two or more components based upon specific light emission of the cells. For example, in this method a fetal gonad of a transgenic mouse with expression of GFP under control of the *Sox9* gene (pre-Sertoli cell marker) can be dissected and digested; the mixture of cells can be sorted with FACS and two pools of cells (GFP^+ pre-Sertoli cells and the rest—GFP^-) are obtained. Such isolated cells (of the same type) can be used for protein or RNA isolation and thus for gene expression analysis. FACS was used to separate cell lines from developing gonads in several studies (Anderson et al. 1999; Bendel-Stenzel et al. 2000; Beverdam and Koopman 2006; DeFalco et al. 2014, 2015; Jameson et al. 2012; Nef et al. 2005; Nishino et al. 2001; Rolland et al. 2011). In an excellent study, Jameson et al. (2012) presented a method to separate the main cell lines from developing mouse gonads. They used several transgenic mouse strains to specifically label cell lines; *EGFP* under control of the *Sry* promoter (*Sry:EGFP*) resulted in a green signal in XY but also XX supporting cells and enabled the separation of these cells (Albrecht and Eicher 2001); *Sox9:EGFP* can be used to isolate XY supporting cells, *Mafb:EGFP* and *α-Sma:EYFP* to isolate stromal/interstitial cells, *Flk1:mCherry* to isolate endothelial cells, and *Oct4:EGFP* to isolate germ cells. As an alternative method, cell purification via magnetic beads was used to isolate primordial germ cells (PGCs) (Pellegrini et al. 2008; Pesce and De Felici 1995; Qing et al. 2008). In this technique, anti-SSEA1 (a marker of PGCs and other stem cells) primary antibodies are added to a mix of cells after organ digestion, and then magnetic beads coated with secondary antibodies are introduced; the beads facilitate isolation of PGCs. Pellegrini et al. (2008) used CD117 magnetic microbeads to isolate Kit-positive spermatogonia.

Previously, several methods allowing gonadal cell isolation and separation were developed. Separation of somatic and germ cells isolated from fetal or newborn gonads is based on differences in their adhesion to a culture dish (O and Baker 1978). Somatic cells have a tendency to adhere to a culture dish, whereas the germ cells do not; thus, during rinsing the germ cells can be removed from the culture. A method for Sertoli cell isolation via a sequence of gradual enzymatic treatment and centrifugation of tissue has been described (Chang et al. 2008; Chapin et al. 1987; Gassei et al. 2008; Hadley et al. 1985; Mackay et al. 1999; Willerton et al. 2004). Enzymes used to digest gonadal tissue include collagenase, DNase, trypsin, and hyaluronidase. Tissue dissociation via enzyme and EDTA treatment was used for germ cell isolation from gonads (Barrios et al. 2010; De Felici and McLaren 1982; Di Carlo et al. 2000; Pellegrini et al. 2008).

14.14 Gene Expression Analysis

Contemporary analyses of gene expression take advantage of a series of techniques that provide a wealth of knowledge about an organ or cell of interest. From early development, the gonad is composed of many cells with different gene expression patterns. The large number of genes expressed in a given cell complicates gene expression analysis. For instance, more than 15,000 genes are expressed in human testis (Ramsköld et al. 2009). Some gene expression techniques allow an analysis of only one or a few genes (low-plex analyses such as the use of reporter genes, Northern blot, ISH, RT-PCR, qPCR, RNase protection assay); however, other techniques enable simultaneous analyses of the expression of many or all genes in a given organ, tissue, or even a single cell (high-plex analyses such as SAGE, DNA microarray, RNA-Seq). Most of these methods have been crucial for studying the mechanisms driving gonadal development.

In this chapter, several techniques of gene expression analysis were already mentioned, such as in situ hybridization (ISH) and the use of reporter genes. Immunolocalization of proteins also provides data on spatiotemporal gene expression. These three methods are based on gene expression product analysis in situ, in a single cell, tissue, or whole organ. Another series of techniques of gene expression analysis entails RNA isolation from the whole organ, tissue, or cell and is conducted ex situ, "in a test tube." In studies on gonadal development, the gonads are dissected and immediately immersed in a solution (usually RNAlater) permeating the tissue and thereby stabilizing and protecting RNA due to the inactivation of ubiquitous RNases. Later the gonads are homogenized and RNA is isolated using column-based techniques or rarely using oligo(dT)-conjugated magnetic beads (Anderson et al. 1999). The most commonly used solution for RNA isolation is TRIzol (Best et al. 2008; Cupp et al. 2003; Dumond et al. 2011; Hiramatsu et al. 2009; Hummitzsch et al. 2013; Koubova et al. 2006; Maatouk et al. 2008; Molyneaux et al. 2003; Munger et al. 2013; Nef et al. 2003; Schmahl et al. 2004; Smith et al. 2008a; Tang et al. 2008). The RNA isolation method via TRIzol is based on guanidinium thiocyanate–phenol–chloroform extraction. The amount of mRNA in fetal gonads is very low and sometimes pooling of samples from several individuals is necessary. The dissection of a gonad and its separation from adjoining tissues (mesonephros) is especially difficult at early stages. In this case, RNA isolation is facilitated by removing the gonads along with adjoining tissues and initially immersing them in RNA later solution in which the gonads harden and become easy to separate from the mesonephros. In the meantime, the individuals can be genotyped and then pooled according to sex and stages. After tissue homogenization followed by column-based RNA isolation, RNA suspended in RNase-free water can be stored frozen. Total RNA is obtained after isolation (including all types of RNA of which only 1–5 % constitutes mRNA representing the expression of protein-coding genes). In order to analyze protein-coding genes, mRNA can be isolated from total RNA by hybridizing the mRNA poly(A) tail to oligo(d)T connected to a carrier (e.g., magnetic beads) (Anderson et al. 1999; Parma et al. 2006).

14.14.1 Northern Blot

Among ex situ, low-plex methods (i.e., analysis of one or a few transcripts after RNA isolation from cells), the **Northern blot assay** (introduced in 1977) enables identification of individual RNA in a heterogeneous RNA sample separated by electrophoresis. In this method, RNA samples are separated by agarose gel electrophoresis and then transferred onto a nylon membrane through a vacuum or capillary blotting system. RNA is immobilized by linkage to the membrane using UV light or heat and a labeled probe is applied onto the membrane leading to the visualization of a transcript. A probe consists of cDNA or cRNA complementary to the sequence of interest and can be labeled with radioactive (^{32}P) or a chemiluminescent signal. The Northern blot assay was useful in studying the expression of genes in developing gonads (Gallardo et al. 2007; Higgy et al. 1995; Pask et al. 2000; Seligman and Page 1998).

14.14.2 RNase Protection Assay

The **RNase protection assay (RPA)** is another method used for the identification of individual RNA molecules accompanied by the degradation of other RNAs with enzymes. In RPA, isolated RNA is mixed with a labeled complementary probe (cRNA or cDNA) and hybridized to form double-stranded structures (cRNA: mRNA or cDNA:mRNA). Then RNases (e.g., RNase A and/or RNase T1) are added to the mixture. The enzymes digest only single-stranded molecules leaving double-stranded molecules intact. The RNA molecules from double-stranded hybrids are analyzed using gel electrophoresis. This method helped to describe the exact timing of *Sry* expression in the developing gonads of mice (Hacker et al. 1995) and also the expression of other genes (Hacker et al. 1995; Morais da Silva et al. 1996; Smith et al. 2009; Soyal et al. 2000).

PCR (polymerase chain reaction) (known since 1971) methods are also commonly used to analyze gene expression. Standard PCR and its implementation in genotyping were previously described in this chapter. However, gene expression analysis uses RT-PCR and qRT-PCR. The basis of all types of PCR is amplification of a given sequence using specific primers, a polymerase enzyme, a template (DNA or RNA), and reaction substrates. Amplification proceeds by repeated cycling of three steps: denaturation (at ~95 °C the DNA helix is split into single threads), annealing (at 45–65 °C primers bind to complementary sequences), and extension (at ~72 °C polymerase extends primers by adding nucleotides in a sequential manner). The products of amplification are detected at a subsequent stage.

14.14.3 PCR-Based Methods of Gene Expression Analysis

In **RT-PCR** (reverse transcription PCR), RNA as a template is converted into complementary cDNA using reverse transcriptase (RT) and resulting cDNA is amplified as in typical PCR. Gene expression data are collected by analysis of bands after agarose gel electrophoresis. RT-PCR is commonly used to study gene expression in gonadogenesis (Abramyan et al. 2009; Albrecht and Eicher 2001; Anderson et al. 1999; Barrionuevo et al. 2006; Barrios et al. 2010; Bendel-Stenzel et al. 2000; Best et al. 2008; Beverdam and Koopman 2006; Bradford et al. 2009; Chang et al. 2008; Chassot et al. 2008; Childs et al. 2011; Coveney et al. 2008; DeFalco et al. 2014; Dumond et al. 2011; Dupont et al. 2000; Hummitzsch et al. 2013; Jameson et al. 2012; Kocer et al. 2008; Koubova et al. 2006; Li et al. 2012; Molyneaux et al. 2003; Moniot et al. 2009; Nef et al. 2003, 2005; Nishino et al. 2001; Oshima et al. 2005; Oulad-Abdelghani et al. 1996; Parma et al. 2006; Rolland et al. 2011; Schmahl et al. 2004; Smith et al. 2008a, b; Tang et al. 2008; Wilhelm et al. 2005; Yoshimoto et al. 2008, 2010).

In **qRT-PCR** (quantitative reverse transcriptase PCR or real-time PCR, qPCR), the first stage consists of converting RNA into cDNA via PCR with reverse transcriptase; however, during the following steps the fluorescently labeled primers or probes are used, enabling monitoring of the amount of product during amplification. qRT-PCR is commonly used to study gene expression during gonadogenesis (Abramyan et al. 2009; Barrionuevo et al. 2006; Barske and Capel 2010; Beverdam and Koopman 2006; Bogani et al. 2009; Bouma et al. 2005, 2007; Bradford et al. 2009; Britt et al. 2002; Chang et al. 2008; Chassot et al. 2008; Chen et al. 2012; Childs et al. 2011; Combes et al. 2011; Cupp et al. 2003; DeFalco et al. 2014; Hiramatsu et al. 2009; Houmard et al. 2009; Koubova et al. 2006; Li et al. 2012; Liu et al. 2015; Maatouk et al. 2008; Manuylov et al. 2007; Molyneaux et al. 2003; Moniot et al. 2009; Munger et al. 2013; Nef et al. 2003, 2005; Nicol and Yao 2015; Parma et al. 2006; Rolland et al. 2011; Ross et al. 2009; Smith et al. 2008a, b, 2009; Tang et al. 2008; Tevosian et al. 2002; Wilhelm et al. 2005). In the majority of studies of gonad development, primers labeled with SYBR Green fluorochrome are used although TaqMan probes have also been applied (Bogani et al. 2009; Liu et al. 2015; Smith et al. 2008a, b). Multigene qRT-PCR is used to analyze numerous genes simultaneously (Bouma et al. 2004, 2005, 2007). The accuracy of results obtained by qRT-PCR strongly depends on accurate normalization using stably expressed genes (Guenin et al. 2009). In studies on gonad development, normalization is usually carried out with *Gapdh* (Chang et al. 2008; DeFalco et al. 2014; Dumond et al. 2011; Hiramatsu et al. 2009; Liu et al. 2015; Manuylov et al. 2007; Nef et al. 2003, 2005; Moniot et al. 2009; Munger et al. 2013), *Hprt1* (Albrecht and Eicher 2001; Bogani et al. 2009; Chassot et al. 2008; Coveney et al. 2008; Nef et al. 2003; Smith et al. 2008a,2009; Tang et al. 2008), 18S rRNA (Liu et al. 2015; Wilhelm et al. 2005), *Tbp* (Barrionuevo et al. 2006), or *Actb* (Houmard et al. 2009).

14.14.4 SAGE

SAGE (serial analysis of gene expression) developed in 1995 is a high-plex ex situ method, enabling analysis of the transcriptome and the expression of even thousands of genes in a single assay (Yamamoto et al. 2001). In the SAGE method, RNA is reverse transcribed using biotinylated primers, and then the biotin-labeled cDNAs are bound by streptavidin beads and cleaved by restriction enzymes (anchoring enzyme, AE). In result, partially digested cDNAs of different length remain on the beads; this material is divided into two portions (A and B) and exposed to one of two adapter oligonucleotides (A or B). After ligation, cDNAs with adapters are removed from the beads using tagging enzyme (TE), and in effect short tags of original cDNAs are detached from beads. Cleaved cDNA tags are repaired with DNA polymerase to produce blunt ends; tags from A and B samples are mixed and ditags form due to blunt ends. Ditags are then cleaved with AE and linked with other ditags to form a cDNA concatemer; concatemers are transformed into bacteria and thus amplified. Isolated concatemers are sequenced using high-throughput DNA sequencers; bioinformatics methods can quantify the recurrence of individual tags (Lee et al. 2010), yielding quantitative data. SAGE methods were used to study the transcriptome in cell lines in developing gonads (Chan et al. 2006; Lee et al. 2006; Wu et al. 2004).

14.14.5 Microarray

Another high-plex method is the **microarray** technique (developed in 1983) that is based on applying a sample onto a solid surface (usually a glass slide) with an array of probes (Trevino et al. 2007). A series of microarray techniques can be used for analysis of gene expression (DNA microarray), protein identification (protein microarray), tissue imaging (tissue array), cell biology (cellular microarray), etc. All subtypes of microarrays miniaturize and multiplex the analysis ensuring the study of multiple samples simultaneously. Gene expression analysis via this method has the capacity to cover thousands of transcripts at once on a single microarray (or gene chip) wherein each spot contains multiple identical strands of DNA, representing a single gene. In gonadogenesis studies, DNA microarrays have provided a large amount of data and have significantly contributed to knowledge on gene expression in developing gonads. DNA microarray techniques require RNA isolation and its conversion into cDNA through reverse transcription; cDNA is labeled with fluorochrome, then applied onto chips, and hybridized to probes (cDNA or oligonucleotides) previously attached to the chip; after rinsing, only labeled sequences that bound complementary probes remain on the chip and signals emitted by them are detected. In gonadogenesis studies, microarrays were used for gene expression analysis of RNA isolated from whole developing gonads and thus from many different cell lines (Combes et al. 2011; Coveney et al. 2008;

Gallardo et al. 2007; Garcia-Ortiz et al. 2009; Grimmond et al. 2000; Houmard et al. 2009; Liu et al. 2015; Munger et al. 2013; Nicol and Yao 2015; Ottolenghi et al. 2007; Rolland et al. 2011; Small et al. 2005), whereas other microarray studies were conducted on particular gonadal cell lines, e.g., on germ cells (Jameson et al. 2012; Molyneaux et al. 2003; Pan et al. 2005), pre-Sertoli and pre-granulosa cells (Beverdam and Koopman 2006; Bouma et al. 2007; Jameson et al. 2012; Nef et al. 2005), epithelial cells of gonadal primordium (Hummitzsch et al. 2013; Kusaka et al. 2010), and interstitial and endothelial cells (Jameson et al. 2012). The microarray technique was also used to identify noncoding RNAs in developing gonads (Chen et al. 2012).

14.14.6 RNA Sequencing

RNA sequencing (RNA-Seq) is one of the latest high-throughput methods used for gene expression analysis (from 2008). RNA-Seq involves next-generation sequencing (NGS) to reveal a snapshot of RNA presence and quantity. In this method, mRNA is isolated and converted to cDNA; sequencing adapters are added and then a cDNA library is prepared; sequencing of cDNA is performed using an NGS platform. RNA-Seq has been used in studies on sex determination and gonad development (Ayers et al. 2015; Gong et al. 2013).

14.15 Protein Analysis Methods

Proteins can be identified in situ, i.e., in a given organ, tissue, or cell, and examples of such methods (immunohistochemistry, IHC, or immunocytochemistry, ICC) were previously described in this chapter. However, there is also a wide range of ex situ methods based on protein analysis in vitro after their isolation from organs, tissues, or cells. In many trials, it is crucial to isolate proteins separately from the nucleus or the cytoplasm (Miyamoto et al. 2008; Niksic et al. 2004; Oulad-Abdelghani et al. 1996; Tremblay and Viger 2003).

14.15.1 Western Blot

There are thousands of different proteins in a single cell type; thus before analysis proteins are usually separated using the **Western blot** technique (used since 1979). In this method, proteins are separated via vertical electrophoresis in a polyacrylamide gel. An anionic detergent, SDS (sodium dodecyl sulfate), is added to a sample to linearize proteins and to impart an even distribution of charge per unit mass which results in proteins being fractioned in the electric field only by their size. Electrophoresis in a polyacrylamide gel in the presence of SDS is called SDS-PAGE. Analogically to nucleic acid electrophoresis, in Western blot smaller proteins migrate from the cathode (–) toward the anode (+) faster than larger proteins. Proteins separated via electrophoresis are transferred onto a membrane (usually nitrocellulose) using an electric current. The proteins of interest are identified using specific primary and secondary antibodies; the latter are usually linked to horseradish peroxidase (HRP) or alkaline phosphatase (AP). The most commonly used HRP enzyme cleaves a chemiluminescent substrate (e.g., luminol); the intensity of the luminescent signal is proportional to the amount of protein. Western blot analysis was useful in many studies on gonadal development (Bradford et al. 2009; Barrios et al. 2010; Chang et al. 2008; Dumond et al. 2011; Dupont et al. 2000; Kocer et al. 2008; Miyamoto et al. 2008; Oulad-Abdelghani et al. 1996; Paranko and Pelliniemi 1992; Parma et al. 2006; Salas-Cortés et al. 1999; Tremblay and Viger 2003).

14.15.2 Phosphorylation Analysis

Proteins are modified by, e.g., phosphorylation, glycosylation, SUMOylation, etc. Such posttranslational modifications change the function or determine the subcellular location of a protein. In gonadogenesis studies, phosphorylation of proteins was tested using antibodies specifically binding phosphorylated forms of a given protein and has revealed the presence and subcellular location of unphosphorylated vs. phosphorylated proteins. Phosphorylation of proteins can also be studied using recombinant proteins under in vivo and in vitro conditions (Tremblay and Viger 2003).

14.16 Methods for Studying Interactions Between Molecules

14.16.1 EMSA

EMSA (electrophoretic mobility shift assay) is a method used to study protein–DNA or protein–RNA interactions and to test if a protein is capable of binding a given sequence of DNA or RNA. In this technique, a mixture of protein–DNA or protein–RNA is electrophoretically separated on a polyacrylamide or agarose gel for a short period. Protein–DNA interactions (binding) result in an increase of molecular weight and lower velocity of migration in an electric current. Thus, a shift in bands indicates an interaction between a protein and given sequence of DNA (or RNA). Protein interactions with DNA were studied via EMSA to elucidate mechanisms of gonadogenesis (Bernard et al. 2008; Miyamoto et al. 2008; Wilhelm et al. 2007; Yoshimoto et al. 2010).

14.16.2 Immunoprecipitation

Immunoprecipitation (pull-down techniques) encompasses a group of similar techniques enabling the precipitation of a protein out of a solution using antibodies specifically binding to a particular protein.

Individual protein immunoprecipitation (**IP**) involves using specific antibody that binds a particular protein in a mixture of many different proteins (Kaboord and Perr 2008). Proteins are usually precipitated using agarose beads that are coated with protein A or G which show affinity to immunoglobins. In studies on gonad development, IP is typically used to isolate recombinant proteins (Barrios et al. 2010; Tremblay and Viger 2003).

Protein complex immunoprecipitation (**Co-IP**) is useful in identifying protein–protein interactions. In this technique, an antibody specifically binding to a particular protein that is thought to form a complex with other proteins is added to a mixture of different proteins (e.g., all proteins isolated from an organ). After antibody binding, the whole protein complex is pulled out of the mixture. The complex is disintegrated and proteins of the complex are separated via SDS-PAGE. Then the separated proteins are isolated from bands and identified using liquid chromatography-mass spectrometry (LC-MS). Co-immunoprecipitation has been valuable in studies on direct interactions between proteins involved in sex determination and other aspects of gonad development, such as interactions between factors including SOX9, GATA4, WT1, FOXL2, vinexin, and E-cadherin (Di Carlo et al. 2000; Matsuyama et al. 2005; Miyamoto et al. 2008; Uhlenhaut et al. 2009).

Chromatin immunoprecipitation (**ChIP**) can determine DNA binding sites in the genome for a protein of interest. Thus, this technique reveals DNA–protein interactions and is especially useful in studying transcription factors and histones that bind DNA. In this method, formaldehyde is used to cross-link proteins to DNA and then cells are lysed and DNA is fragmented via sonication. Specific antibodies are used to precipitate a particular protein in a complex with specific DNA sequences. Subsequently, the isolated DNA–protein complexes are heated to release cross-linking and purified DNA is then amplified and sequenced. Considering the great number of binding sites in a genome, DNA microarray techniques (ChIP-on-chip) can be useful in detecting all binding sites for a given protein. In studies on gonad development, ChIP has been useful in studying factors binding to promoters, e.g., SOX9 binding to the *Pdgs* promoter, GATA4 to the *Sry* promoter, and SRY and SOX9 binding to the *Cerbelin4* promoter (Bradford et al. 2009; Miyamoto et al. 2008; Wilhelm et al. 2007).

The RNA immunoprecipitation (**RIP**) method examines protein–RNA interactions. Cells are lysed and the protein of interest along with bound RNA is precipitated using specific antibodies. RNA is separated from protein, extracted, reverse transcribed, and analyzed using cDNA sequencing or RT-PCR. RNA–protein interactions were intensively studied with this method in sex determination of *Drosophila* (Vied et al. 2003).

14.17 Genetic Engineering of Cells Cultured In Vitro

14.17.1 Tagged Proteins (Fusion Proteins) and In Vitro Cell Culture

A pull-down assay can be facilitated by using tagged proteins. A protein tag is a sequence genetically grafted to the N- or C-terminus. The recombinant protein, recognized by a specific antibody due to the presence of a tag, is easy to detect or purify. Several tags can be used for different purposes. One of the most commonly applied protein tags is glutathione-S-transferase (GST), which was used to study such proteins as SRY or GATA4 (Bernard et al. 2008; Best et al. 2008; Salas-Cortés et al. 1999; Tremblay and Viger 2003). A protein of interest along with bound GST constitutes a fusion protein, purified via the pull-down assay owing to GST affinity to GSH (glutathione); usually GSH-coated beads are used. After pull-down, a tag can be removed by a proteolytic enzyme. Other protein tags used in studies on gonad development include Myc-tag (Bradford et al. 2009; Kim et al. 2006; Sekido et al. 2004), FLAG-tag (Bernard et al. 2008; Matsuyama et al. 2005), HA-tag (Bernard et al. 2008), His-tag (Miyamoto et al. 2008), and TRX-tag (Dumond et al. 2011). Such fusion proteins can be purified using antibodies specifically binding the tags. Fluorescent protein tags, such as GFP or YFP, can also be used to visualize a protein and to study its subcellular localization. As previously mentioned, GFP and YFP are used to study a protein in the organ or to visualize or isolate a given cell line; however, these tags can also be used to identify a recombinant protein in in vitro cell culture, e.g., *Cerbelin4* and *Sdmgl* in a study on mouse sex determination (Best et al. 2008; Bradford et al. 2009). Recombinant proteins may also be used to produce specific antibodies (Best et al. 2008; Matsuyama et al. 2005; Salas-Cortés et al. 1999).

Fusion proteins are synthesized in cells under in vitro conditions via cell transfection. The most commonly used cell lines in studies on sex determination and mechanisms of gonad development are HEK293 (Barrios et al. 2010; Bernard et al. 2008; Munger et al. 2013; Wilhelm et al. 2007), HeLa (Moniot et al. 2009; Ohe et al. 2002; Salas-Cortés et al. 1999; Wilhelm et al. 2005), COS7 (Li and Kim 2004; Ohe et al. 2002; Wilhelm et al. 2005), NT2/D1 (Bernard et al. 2008; Moniot et al. 2009; Ohe et al. 2002), NIH3T3 (Best et al. 2008), C3H10T1/2 (Matsuyama et al. 2005), MCF7 (Bradford et al. 2009), and Sertoli cells SK11 (Best et al. 2008). These cell lines are usually maintained in DMEM medium supplemented with 10 % FBS at 37 °C and 5 % CO_2. Lipofection is typically applied to transfect these cells with an exogenous genetic construct; here liposomes containing genetic material merge with the cell membrane. In some cases, fusion proteins were synthesized in *E. coli* (Best et al. 2008; Dumond et al. 2011; Wilhelm et al. 2007).

Cells cultured in vitro are sometimes maintained on a layer of cells that produce nutrients supporting cell growth. Such feeder cells include embryonic primary fibroblasts (EMFI) used for culture of mesonephros-derived migrating cells (Nishino et al. 2001) or the STO fibroblast cell line used for culture of PGCs (Di Carlo et al. 2000). Cells can also be cultured on a three-dimensional matrix scaffold, e.g., Matrigel. A 3D cell culture was used to culture primary Sertoli cells to reconstitute their 3D aggregations (Gassei et al. 2008).

14.17.2 RNA Interference

Cells cultured in vitro are useful in studying gene function via silencing of expression, i.e., genetic knockdown (KD). For example, RNA interference (RNAi) is used to silence gene expression. In studies on the role of genes in sex determination, shRNA (small hairpin RNA) was introduced via viral vectors or lipotransfection to cells cultured in vitro to silence the expression of a gene of interest. shRNA binds the RISC complex that recognizes mRNA due to the antisense sequence of shRNA and degrades mRNA leading to a decrease in gene expression. For instance, Munger et al. (2013) used lentiviruses to introduce shRNA to gonadal primary cells cultured in vitro. shRNA was designed to silence the expression of candidate genes presumably involved in sex determination in mice. Best et al. (2008) used shRNA introduced into Sertoli SK11 cells via lipotransfection in order to silence the expression of the *Sdmg1* gene. Smith et al. (2009) used an avian retroviral vector to introduce shRNA to chicken egg to silence *Dmrt1* gene expression, which resulted in feminization of ZZ individuals and proved the role of *Dmrt1* in avian sex determination (Smith et al. 2009).

14.17.3 Labeling of Cells Cultured In Vitro (Immunocytochemistry)

Proteins can be detected in cells cultured in vitro using antibodies. In immunocytochemical labeling (ICC), cells attached to microscope slides are usually fixed with 2–4 % PFA, permeabilized with detergent, blocked with BSA, incubated with primary and then secondary antibodies, and viewed (Barrios et al. 2010; Best et al. 2008; Gassei et al. 2008; Qing et al. 2008; Willerton et al. 2004). Alternatively, incubation of non-permeabilized cells with antibodies may precede fixation (Best et al. 2008).

14.18 Genetically Modified Mice

Genetic modification is a powerful tool for studies in the field of developmental biology. The first transgenic mouse was obtained in 1974 by Rudolf Jaenish via viral DNA insertion into an early-stage mouse embryo (Jaenisch and Mintz 1974). Since then, genetically modified mice have revolutionized studies on gonad development. The aforementioned mice with expression of GFP or other reporter genes are useful for studying gene expression and to trace the organogenesis process. A reporter gene such as *GFP* or *LacZ* can be expressed in all cells of an individual (Capel et al. 1999; Combes et al. 2009b; Martineau et al. 1997; Merchant-Larios and Moreno-Mendoza 1998; Nishino et al. 2001; Qing et al. 2008). Alternatively, a reporter gene can be expressed only in a cell line of interest; e.g., expression of *EGFP* under control of the *Oct4* promoter is useful to trace and even isolate primordial germ cells (Anderson et al. 1999; Maatouk et al. 2013; Molyneaux et al. 2001; Rolland et al. 2011). A cell line giving rise to Sertoli cells can be visualized and isolated employing strains with *EGFP* under control of *Sf1* (Beverdam and Koopman 2006; Nef et al. 2005), *Sry* (Albrecht and Eicher 2001; Kim et al. 2006), or *Sox9* (Nel-Themaat et al. 2009). Lists of numerous transgenic strains with labeled cell lines facilitating studies on gonad development have been published (Brennan et al. 2002; DeFalco et al. 2011, 2014, 2015; Jameson et al. 2012).

Transgenic mice also facilitate the functional annotation of a gene of interest. The best method to study gene function is to delete this gene from a genome or introduce a nonfunctional sequence of a gene (knockout, KO). There are two types of KO: constitutive and conditional. Ablation of gene expression in all cells of an individual determines a constitutive knockout and has been used to study the function of *Wnt4*, *Foxl2*, and *Fgf9* in gonadogenesis (Coveney et al. 2008; Kim et al. 2006; Matzuk et al. 1995; Ottolenghi et al. 2005). However, a disadvantage of this method is the lethality caused by deletion of some genes, precluding analysis due to the early death of mutant embryos. A conditional knockout can sidestep lethality after gene deletion. In the conditional knockout, gene expression is depleted only in a particular cell type. There is a series of methods of such tissue-specific knockout, among which the Cre-loxP system is commonly applied in embryological studies. In this method, Cre recombinase, expressed under the control of a particular promoter, cleaves and thus deletes a given gene that is flanked with *loxP* sequences. This method has been used to study mouse gonadal development (Bagheri-Fam et al. 2008; Barrionuevo et al. 2006, 2009; Bouma et al. 2005; Buscara et al. 2009; Chang et al. 2008; Chassot et al. 2008; Coveney et al. 2008; Liu et al. 2009, 2015; Maatouk et al. 2008, 2013; Moniot et al. 2009; Tang et al. 2008). A gene knockout in the precursors of Sertoli and granulosa cells is possible due to *Cre* expression under the control of promoters of steroidogenic factor 1—*Sf1* (Chassot et al. 2008; Liu et al. 2009; Maatouk et al. 2008; Tang et al. 2008) and cytokeratin 19—*Ck19* (Bagheri-Fam et al. 2008; Barrionuevo et al. 2006; Moniot et al. 2009); however, a knockout in differentiating Sertoli cells can be achieved with the aid of *Amh-Cre* (Barrionuevo et al. 2009; Chang et al. 2008; Moniot et al. 2009). The *Oct4-Cre* strain permits gene deletion in the germ cell line (Sabour et al. 2014). In these studies, genes such as *Sox9*, *Fgf9*, *Fgfr2*, *Wt1*, and *Ctnnb* were successfully deleted in a tissue-specific manner. A tissue-specific knockout is usually induced by tamoxifen injected into an individual. Cre recombinase is fused with estrogen receptor (ER), and after intraperitoneal injection of tamoxifen, this compound binds ER and activates Cre recombinase which deletes a specific gene. Tamoxifen-inducible recombination has improved investigations owing to the possibility of controlling the timing of gene deletion.

Alternatively, knowledge on gene function can be obtained by the introduction of additional copies of a gene into a genome leading to the overexpression of a gene (knockin, KI). For instance, the effects of overexpression of goat *Rspo1* introduced into mice deprived of *Rspo1* gene expression (*Rspo1*$^{-/-}$, *Rspo1*-KO) have been studied (Buscara et al. 2009). The KI method has also revealed the function of *Gata4* in gonad development after the insertion of a sequence encoding GATA4 with disrupted potentiality to bind the FOG2 cofactor (Tevosian et al. 2002). Insertion of reporter gene sequences into the genome is also an example of gene knockin.

There are two main methods to obtain genetically modified organisms. First, transgenic mice can be generated via injection of genetic material to a male pronucleus in the zygote (Albrecht and Eicher 2001; Anderson et al. 1999; Buscara et al. 2009). Second, a transgene can be introduced via injection or electroporation to ES cells (embryonic stem cells) that are then introduced into a blastocyst. The second method was used to generate Cre-loxP knockout mice in studies on gonad development (Chassot et al. 2008; Dupont et al. 2000; Maclean et al. 2009; Nel-Themaat et al. 2009; Ottolenghi et al. 2005).

14.19 Methods for Studies of Steroidogenesis

One of the most crucial processes during gonad development is differentiation of steroidogenic cells and the onset of sex steroid synthesis orchestrating the establishment of secondary sex features. The first method to study steroidogenesis is identification of steroidogenic enzymes via immunolabeling or their expression via RT-PCR, qRT-PCR, or in situ hybridization. The most important steroidogenic enzymes for studies on gonad development include cytochrome P450scc (CYP11a1), 3β-hydroxysteroid dehydrogenase (3β-HSD), 17β-hydroxysteroid dehydrogenase (17β-HSD), 3α-hydroxysteroid dehydrogenase (3α-HSD), and cytochrome P450 17A1 (CYP17a1), which mark steroidogenic cells such as Leydig and theca cells and aromatase (CYP19a1) expressed mainly in granulosa cells (Chassot et al. 2008; Payne et al. 1992; Tang et al. 2008). Immunolabeling or gene expression analysis, however, only suggests steroid hormone production; other methods are needed to prove sex steroid synthesis. The assessment of hormone synthesis can be done by investigation of enzymatic activity via histochemical methods or by assaying hormone levels in the blood.

The histochemical method for studying steroidogenic enzyme activity is usually carried out on frozen sections containing tissue gently fixed in 4 % PFA. A staining solution is applied onto the sample to conduct a chromogenic reaction (Baillie et al. 1965, 1966). The staining solution contains a substrate (e.g., dehydroepiandrosterone for 3β-HSD; testosterone for 17β-HSD; androsterone for 3α-HSD), nicotinamide adenine dinucleotide (NAD^+), and NBT/BCIP. Enzymatic activity gives rise to NADH and H^+, which leads to reduction of yellow NBT to blue insoluble diformazan. A blue signal indicates enzyme activity.

Steroid hormone concentration can be assessed via the RIA (radioimmunoassay) method (Habert and Picon 1982; Migrenne et al. 2012). Rosalyn Yalow, the inventor of this method, was awarded the Nobel Prize in 1977. RIA is a very sensitive technique measuring the concentration of hormones, e.g., in blood plasma. In this method, a known quantity of antigen (hormone) is labeled with the radioactive isotope of iodine attached to tyrosine. Radiolabeled antigen is mixed with a known quantity of antibody specifically binding the antigen. Then a tested sample with an unknown quantity of antigen is added to the mixture. The unlabeled antigens from the tested sample compete with the radiolabeled antigen for binding to antibodies. As the concentration of unlabeled antigen increases, it displaces radiolabeled antigen. This leads to a decrease of the ratio between antibody-bound radiolabeled antigen to free radiolabeled antigen. The radioactivity of unbound radiolabeled antigen in the supernatant is measured, indicating the quantity of hormone in the tested sample.

Another method allowing for the detection of steroid hormones or any other antigen is ELISA (enzyme-linked immunosorbent assay). In this method, a multi-well plate is used; each well is coated with capture antigen (indirect ELISA) or antibody that specifically binds a given antigen (sandwich ELISA) (Voller et al. 1978). A plate with antibody-coated wells is used to study the presence and concentration of antigen such as steroid hormone in blood plasma. Samples are applied to each well, and after incubation and rinsing, the enzyme-conjugated secondary antibody is added to trigger a signal, which permits the detection of the antigen. A plate is then analyzed in an imager, enabling quantitative measurement of the steroid. This method was used by Liu et al. (2015) as a steroid hormone multiplex immunoassay to study the concentrations of DHEA, estradiol, progesterone, and testosterone during the ovarian development.

14.20 Germ Cell Depletion

There is a series of studies based on gonadogenesis analysis after the removal of germ cells. Such studies elucidate the role of germ cells in gonadal development. Several methods are used to deplete germ cells in developing gonads. Usually chemical ablation of germ cells or mutated mice deprived of germ cells are used.

Chemical depletion of germ cells is typically acquired via busulfan treatment. Busulfan (butane-1,4-diyl dimethanesulfonate) is an alkylating compound that forms cross-links between the DNA bases A-G and G-G (Iwamoto et al. 2004). DNA cross-linking prevents DNA replication. Moreover, due to the presence of intrastrand cross-links, DNA cannot be repaired in the treated cells, leading to cell apoptosis. Originally it was shown that the intraperitoneal administration of busulfan in rat caused a lack of germ cells (Bollag 1953). Germ cell depletion and infertility is a serious drawback for busulfan use for human chemotherapy (López-Ibor and Schwartz 1986). Nevertheless, busulfan has been useful in studying the effects of germ cell depletion in many vertebrate species. To obtain sterile mouse embryos, pregnant females are intraperitoneally (IP) injected with a solution of busulfan (usually 20–50 mg/kg in 50 % DMSO) at 9.5, 10.5, or 11.5 *dpc* (Koubova et al. 2006; Liu et al. 2015; Maatouk et al. 2013; Nef et al. 2005). In chicken, busulfan treatment and germ cell depletion is done by administration of the solution onto the surface of the egg (DiNapoli and Capel 2007). In anuran amphibians, germ cell depletion can be achieved by the addition of busulfan dissolved in DMSO to water with living tadpoles to a final concentration of 0.12 mM (Piprek et al. 2012).

Another method of obtaining sterile gonads involves using mice with mutation. Here a viable allelic form W^v in the dominant white spotting (*W*) is used in a homozygous configuration (W^v/W^v) resulting in almost complete sterility (Russell 1979). Sterile individuals (W^v/W^v) are obtained by mating heterozygotes $W^v/^+$. The factor originally termed W is the c-Kit proto-oncogene and is a type of transmembrane receptor tyrosine-protein kinase. Signaling from the c-Kit receptor is essential for primordial germ cell (PGC) survival (De Miguel et al. 2002). W^v/W^v mutants have been used to study sterile gonads (Chen et al. 2012; Garcia-Ortiz et al. 2009; Koubova et al. 2006; Rolland et al. 2011).

Uhlenhaut et al. (2009) used transgenic mice to delete oocytes in postnatal ovaries. A strain with *Cre* recombinase expression under the *Gdf9* promoter was mated with the R26DTA strain. *Gdf9* is expressed in oocytes and thus Cre recombinase in these cells removes the stop codon triggering the expression of diphtheria toxin receptor (DTR) in oocytes. Eight-week-old females were injected with diphtheria toxin (DT) in PBS (50 μg/kg). DT via DTR receptor disrupts protein synthesis and thus causes cell death. Here diphtheria toxin receptor-mediated targeted cell ablation led to the total absence of oocytes in postnatal ovaries.

14.21 Conclusion

In this chapter, I showed how various techniques have been used to investigate the mechanisms controlling gonad development and how scientific methods have evolved through the decades. Figure 14.1 shows how new discoveries in this field correlate with inventions of new scientific methods and tools. Initially, exclusively histological methods were used to study gonad development and thus only structural changes in gonad development were described. Advances in cell and molecular biology drastically accelerated the rate of embryological studies. Some of the most powerful tools revealing the molecular mechanisms controlling development involve techniques of gene expression analysis, such as reverse transcription PCR, quantitative PCR, in situ hybridization, and DNA microarray. Additionally, transgenic animals have significantly contributed to our understanding of the genetic control of gonadogenesis and sex determination. Knockout mice have been especially informative concerning the functions of some genes in these processes. In vitro culture of gonads was valuable for studying the development of these organs. This method helped trace cell migration shedding light on their origins and contributions to gonad structures. Immunolabeling and reporter genes have been applied in investigations of changes taking place in gonad structure during development. These methods enabled not only visualization of gonad structure, but also facilitated the isolation of particular cell lines and thus also advanced the study of gene expression in a given cell type. The issues raised in this chapter reveal that although we accumulated a lot of information about the mechanisms orchestrating gonad development and sex determination, there are still many developmental mechanisms awaiting discovery. It is known that molecular mechanisms are complex and involve a large number of interactions between the genes driving development. Further studies of gonad development will undoubtedly lead to the discovery of new mechanisms controlling the development of an individual into a male or female.

References

Abramyan J, Feng CW, Koopman P (2009) Cloning and expression of candidate sexual development genes in the cane toad (*Bufo marinus*). Dev Dyn 238:2430–2441

Albrecht KH, Eicher EM (2001) Evidence that *Sry* is expressed in pre-Sertoli cells and Sertoli and granulosa cells have a common precursor. Dev Biol 240:92–107

Allen BM (1904) The embryonic development of the ovary and testis of the mammalia. Am J Anat 3:89–146

Anderson R, Fassler R, Georges-Labouesse E et al (1999) Mouse primordial germ cells lacking β1 integrins enter the germline but fail to migrate normally to the gonads. Development 126:1655–1664

Avrameas S, Uriel J (1966) Method of antigen and antibody labelling with enzymes and its immunodiffusion application. C R Acad Sci Hebd Seances Acad Sci D 262:2543–2545

Ayers KL, Lambeth LS, Davidson NM et al (2015) Identification of candidate gonadal sex differentiation genes in the chicken embryo using RNA-seq. BMC Genomics 16:704

Bagheri-Fam S, Sim H, Bernard P et al (2008) Loss of *Fgfr2* leads to partial XY sex reversal. Dev Biol 314:71–83

Baillie AH, Calman KC, Ferguson MM et al (1965) Histochemical demonstration od 20-β-hydroxysteroid dehydrogenase. J Endocrinol 32:337–339

Baillie AH, Calman KC, Ferguson MM et al (1966) Histochemical utilization of 3α-, 6β-, 11α-, 12α-, 16α-, 17α-, 21- and 24-hydroxysteroids. J Endocrinol 34:1–12

Barr ML, Bertram EG (1949) A morphological distinction between neurones of the male and female, and the behaviour of the nucleolar satellite during accelerated nucleoprotein synthesis. Nature 163:676–677

Barrionuevo F, Bagheri-Fam S, Klattig J et al (2006) Homozygous inactivation of *Sox9* causes complete XY sex reversal in mice. Biol Reprod 74:195–201

Barrionuevo F, Georg I, Scherthan H et al (2009) Testis cord differentiation after the sex determination stage is independent of Sox9 but fails in the combined absence of *Sox9* and *Sox8*. Dev Biol 15:301–312

Barrios F, Filipponi D, Pellegrini M et al (2010) Opposing effects of retinoic acid and *Fgf9* on *Nanos2* expression and meiotic entry of mouse germ cells. J Cell Sci 123:871–880

Barske LA, Capel B (2010) Estrogen represses SOX9 during sex determination in the red-eared slider turtle Trachemys scripta. Dev Biol 341:305–314

Bendel-Stenzel MR, Gomperts M, Anderson R et al (2000) The role of cadherins during primordial germ cell migration and early gonad formation in the mouse. Mech Dev 91:143–152

Bernard P, Sim H, Knower K et al (2008) Human SRY inhibits beta-catenin-mediated transcription. Int J Biochem Cell Biol 40:2889–2900

Best D, Sahlender DA, Walther N et al (2008) Sdmg1 is a conserved transmembrane protein associated with germ cell sex determination and germline-soma interactions in mice. Development 135:1415–1425

Beverdam A, Koopman P (2006) Expression profiling of purified mouse gonadal somatic cells during the critical time window of sex determination reveals novel candidate genes for human sexual dysgenesis syndromes. Hum Mol Genet 15:417–431

Bilinski SM, Jaglarz MK, Dougherty MT, Kloc M (2010) Electron microscopy, immunostaining, cytoskeleton visualization, in situ hybridization, and three-dimensional reconstruction of Xenopus oocytes. Methods 51:11–19

Birk OS, Casiano DE, Wassif CA et al (2000) The LIM homeobox gene Lhx9 is essential for mouse gonad formation. Nature 403:909–913

Bishop CE, Whitworth DJ, Qin Y et al (2000) A transgenic insertion upstream of sox9 is associated with dominant XX sex reversal in the mouse. Nat Genet 26:490–494

Bogani D, Siggers P, Brixey R (2009) Loss of mitogen-activated protein kinase kinase kinase 4 (MAP3K4) reveals a requirement for MAPK signalling in mouse sex determination. PLoS Biol 7(9):e1000196

Bollag W (1953) The effect of myleran on rat gonads. Experientia 9:268

Bouma GJ, Hart GT, Washburn LL et al (2004) Using real time RT-PCR analysis to determine multiple gene expression patterns during XX and XY mouse fetal gonad development. Gene Expr Patterns 5:141–149

Bouma GJ, Albrecht KH, Washburn LL et al (2005) Gonadal sex reversal in mutant Dax1 XY mice: a failure to upregulate Sox9 in pre-Sertoli cells. Development 132:3045–3054

Bouma GJ, Washburn LL, Albrecht KH et al (2007) Correct dosage of Fog2 and Gata4 transcription factors is critical for fetal testis development in mice. Proc Natl Acad Sci U S A 104: 14994–14999

Bowles J, Knight D, Smith C et al (2006) Retinoid signaling determines germ cell fate in mice. Science 312:596–600

Bradford ST, Hiramatsu R, Maddugoda MP et al (2009) The cerebellin 4 precursor gene is a direct target of SRY and SOX9 in mice. Biol Reprod 80:1178–1188

Brennan J, Karl J, Capel B (2002) Divergent vascular mechanisms downstream of Sry establish the arterial system in the XY gonad. Dev Biol 244:418–428

Brennan J, Tilmann C, Capel B (2003) Pdgfr-alpha mediates testis cord organization and fetal Leydig cell development in the XY gonad. Genes Dev 17:800–810

Britt KL, Kerr J, O'Donnell L et al (2002) Estrogen regulates development of the somatic cell phenotype in the eutherian ovary. FASEB J 16:1389–1397

Buehr M, Gu S, McLaren A (1993) Mesonephric contribution to testis differentiation in the fetal mouse. Development 117:273–281

Bullejos M, Koopman P (2005) Delayed *Sry* and *Sox9* expression in developing mouse gonads underlies B6-Y(DOM) sex reversal. Dev Biol 278:473–481

Buscara L, Montazer-Torbati F, Chadi S et al (2009) Goat RSPO1 over-expression rescues sex-reversal in Rspo1-knockout XX mice but does not perturb testis differentiation in XY or sex-reversed XX mice. Transgenic Res 18:649–654
Byskov AG, Saxen L (1976) Induction of meiosis in fetal mouse testis in vitro. Dev Biol 52: 193–200
Byskov AG, Høyer PE, Yding Andersen C et al (2011) No evidence for the presence of oogonia in the human ovary after their final clearance during the first two years of life. Hum Reprod 26: 2129–2139
Capel B, Batchvarov J (2008) Sex chromatin staining in amnion cells. CSH Protoc 3:11
Capel B, Albrecht KH, Washburn LL et al (1999) Migration of mesonephric cells into the mammalian gonad depends on Sry. Mech Dev 84:127–131
Castle WE, Little CC (1910) On a modified Mendelian ration among yellow mice. Science 32:868–870
Chaboissier MC, Kobayashi A, Vidal VI et al (2004) Functional analysis of *Sox8* and *Sox9* during sex determination in the mouse. Development 131:1891–1901
Chan WY, Lee TL, Wu SM et al (2006) Transcriptome analyses of male germ cells with serial analysis of gene expression (SAGE). Mol Cell Endocrinol 250:8–19
Chang H, Gao F, Guillou F et al (2008) Wt1 negatively regulates β-catenin signaling during testis development. Development 135:1875–1885
Chapin RE, Phelps JL, Miller BE et al (1987) Alkaline phosphatase histochemistry discriminates peritubular cells in primary rat testicular cell culture. J Androl 8:155–161
Chassot AA, Ranc F, Gregoire EP et al (2008) Activation of beta-catenin signaling by Rspo1 controls differentiation of the mammalian ovary. Hum Mol Genet 17:1264–1277
Chen H, Palmer JS, Thiagarajan RD et al (2012) Identification of novel markers of mouse fetal ovary development. PLoS One 7(7):e41683
Childs AJ, Cowan G, Kinnell HL et al (2011) Retinoic acid signalling and the control of meiotic entry in the human fetal gonad. PLoS One 6:e20249
Chinwalla AT, Cook LL, Delehaunty KD et al (2002) Initial sequencing and comparative analysis of the mouse genome. Nature 420:520–562
Chiquoine AD (1954) The identification, origin, and migration of the primordial germ cells in the mouse embryo. Anat Rec 118:135–146
Chuma S, Nakatsuji N (2001) Autonomous transition into meiosis of mouse fetal germ cells in vitro and its inhibition by gp130-mediated signaling. Dev Biol 229:468–479
Clapcote SJ, Roder JC (2005) Simplex PCR assay for sex determination in mice. Biotechniques 38:702–706
Coert HJ (1898) Over de Ontwikkeling en den Bouw van de Geschlachtsklier bij de Zoogdieren meer in het bijzonder van den Eierstok. Dissertation, Leiden University
Combes AN, Lesieur E, Harley VR et al (2009a) Three-dimensional visualization of testis cord morphogenesis, a novel tubulogenic mechanism in development. Dev Dyn 238:1033–1041
Combes AN, Wilhelm D, Davidson TL et al (2009b) Endothelial cell migration directs testis cord formation. Dev Biol 326:112–120
Combes AN, Bowles J, Feng CW et al (2011) Expression and functional analysis of Dkk1 during early gonadal development. Sex Dev 5:124–130
Cool J, Carmona FD, Szucsik JC, Capel B (2008) Peritubular myoid cells are not the migrating population required for testis cord formation in the XY gonad. Sex Dev 2:128–133
Coons AH, Creech HJ, Jones RN (1941) Immunological properties of an antibody containing a fluorescent group. Proc Soc Exp Biol Med 47:200–202
Coveney D, Cool J, Oliver T, Capel B (2008) Four-dimensional analysis of vascularization during primary development of an organ, the gonad. Proc Natl Acad Sci U S A 105:7212–7217
Crow JF (2002) C.C. Little, cancer and inbred mice. Genetics 161:1357–1361
Cuénot L (1902) La loi de Mendel et l'hérédité de la pigmentation chez les souris. Arch Zool Exp Gen 3:27–30

Cui S, Ross A, Stallings N et al (2004) Disrupted gonadogenesis and male-to-female sex reversal in Pod1 knockout mice. Development 131:4095–4105

Cupp AS, Uzumcu M, Skinner MK (2003) Chemotactic role of neurotropin 3 in the embryonic testis that facilitates male sex determination. Biol Reprod 68:2033–2037

DeFalco T, Takahashi S, Capel B (2011) Two distinct origins for Leydig cell progenitors in the fetal testis. Dev Biol 352:14–26

DeFalco T, Bhattacharya I, Williams AV et al (2014) Yolk-sac-derived macrophages regulate fetal testis vascularization and morphogenesis. Proc Natl Acad Sci U S A 111:E2384–E2393

DeFalco T, Potter S, Williams AV et al (2015) Macrophages contribute to the spermatogonial niche in the adult testis. Cell Rep 12:1107–1119

De Felici M, Dolci S (1989) In vitro adhesion of mouse fetal germ cells to extracellular matrix components. Cell Differ Dev 26:87–96

De Felici M, McLaren A (1982) Isolation of mouse primordial germ cells. Expl Cell Res 142: 476–482

De Felici M, Scaldaferri ML, Lobascio M (2004) In vitro approaches to the study of primordial germ cell lineage and proliferation. Hum Reprod Update 10:197–206

De Miguel M, Cheng L, Holland EC et al (2002) Dissection of the c-Kit signaling pathway in mouse primordial germ cells by retroviral-mediated gene transfer. Proc Natl Acad Sci U S A 99:10458–10463

Di Carlo AD, Travia G, De Felici M (2000) The meiotic specific synaptonemal complex protein SCP3 is expressed by female and male primordial germ cells of the mouse embryo. Int J Dev Biol 44:241–244

DiNapoli L, Capel B (2007) Germ cell depletion does not alter the morphogenesis of the fetal testis or ovary in the red-eared slider turtle (*Trachemys scripta*). J Exp Zool B Mol Dev Evol 308: 236–241

Dumond H, Al-Asaad I, Chesnel A et al (2011) Temporal and spatial SOX9 expression patterns in the course of gonad development of the caudate amphibian Pleurodeles waltl. J Exp Zool B Mol Dev Evol 316B:199–211

Dupont S, Krust A, Gansmuller A et al (2000) Effect of single and compound knockouts of estrogen receptors alpha (ERalpha) and beta (ERbeta) on mouse reproductive phenotypes. Development 127:4277–4291

Egli T (1876) Beitrage zur Anatomie der Entwickelungsgeschichte der Geschlechtsorgane. I. Zur Entwickelung des Urogenitalsystems beim Kaninchen. Dissertation, University of Zurich

El Jamil A, Kanhoush R, Magre S et al (2008) Sex-specific expression of SOX9 during gonadogenesis in the amphibian Xenopus tropicalis. Dev Dyn 237:2996–3005

Evans CW, Robb DI, Tuckett F, Challoner S (1982) Regulation of meiosis in the foetal mouse gonad. J Embryol Exp Morphol 68:59–67

Fechner PY, Marcantonio SM, Jaswaney V et al (1993) The role of the sex-determining region Y gene in the etiology of 46, XX maleness. J Clin Endocrinol Metab 76:690–695

Felix W (1911) Die Entwicklung der Harn- und Geschlechtsorgane. In: Keibel F, Mall FP (eds) Handbuch der Entwicklungageschichte des Menshen, vol 2. Hirzel, Leipzig, pp 732–795

Fleming A, Ghahramani N, Zhu MX et al (2012) Membrane β-catenin and adherens junctions in early gonadal patterning. Dev Dyn 241:1782–1798

Fröjdman K, Paranko J, Virtanen I, Pelliniemi LJ (1992) Intermediate filaments and epithelial differentiation of male rat embryonic gonad. Differentiation 50:113–123

Fuss A (1911) Uber extraregionare Geschlechtszellen bei einem menschlichen Embryo von 4 Wochen. Anat Am 39:407–409

Gall JG, Pardue ML (1969) Formation and detection of RNA-DNA hybrid molecules in cytological preparations. Proc Natl Acad Sci U S A 63:378–383

Gallardo T, Shirley L, John GB, Castrillon DH (2007) Generation of a germ cell-specific mouse transgenic Cre line, Vasa-Cre. Genesis 45:413–417

Garcia-Ortiz JA, Pelosi E, Omari S et al (2009) *Foxl2* functions in sex determination and histogenesis throughout mouse ovary development. BMC Dev Biol 9:36

Gassei K, Ehmcke J, Wood MA et al (2008) Immature rat seminiferous tubules reconstructed *in vitro* express markers of Sertoli cell maturation after xenografting into nude mouse hosts. Mol Hum Reprod 16:97–110
Gong W, Pan L, Lin Q et al (2013) Transcriptome profiling of the developing postnatal mouse testis using next-generation sequencing. Sci China Life Sci 56:1–12
Gosner KL (1960) A simplified table for staging anuran embryos and larvae with notes on identification. Herpetologica 16:183–190
Greenbaum E (2002) A standardized series of embryonic stages for the emydid turtle *Trachemys scripta*. Can J Zool 80:1350–1370
Greenlee AR, Krisher RL, Plotka ED (1998) Rapid sexing of murine preimplantation embryos using a nested, multiplex polymerase chain reaction (PCR). Mol Reprod Dev 49:261–267
Grimmond S, Van Hateren N, Siggers P et al (2000) Sexually dimorphic expression of protease nexin-1 and vanin-1 in the developing mouse gonad prior to overt differentiation suggests a role in mammalian sexual development. Hum Mol Genet 9:1553–1560
Gubbay J, Collignon J, Koopman P et al (1990) A gene mapping to the sex-determining region of the mouse Y chromosome is a member of a novel family of embryonically expressed genes. Nature 346:245–250
Guenin S, Mauriat M, Pelloux J et al (2009) Normalization of qRT-PCR data: the necessity of adopting a systematic, experimental conditions-specific, validation of references. J Exp Bot 60: 487–493
Habert R, Picon R (1982) Control of testicular steroidogenesis in foetal rat: effect of decapitation on testosterone and plasma luteinizing hormone-like activity. Acta Endocrinol 99:466–473
Hacker A, Capel B, Goodfellow P, Lovell-Badge R (1995) Expression of *Sry*, the mouse sex determining gene. Development 121:1603–1614
Hadley MA, Byers SW, Suarez-Quian CA et al (1985) Extracellular matrix regulates Sertoli cell differentiation, testicular cord formation, and germ cell development in vitro. J Cell Biol 101: 1511–1522
Hamburger V, Hamilton HL (1951) A series of normal stages in the development of the chick embryo. J Morphol 88:49–92
Hanley NA, Hagan DM, Clement-Jones M et al (2000) SRY, SOX9, and DAX1 expression patterns during human sex determination and gonadal development. Mech Dev 91:403–407
Hatzirodos N, Hummitzsch K, Irving-Rodgers HF, Rodgers RJ (2015) Transcriptome comparisons identify new cell markers for theca interna and granulosa cells from small and large antral ovarian follicles. PLoS ONE 10(3):e0119800
Higgy NA, Zackson SL, van der Hoorn FA (1995) Cell interactions in testis development: overexpression of c-mos in spermatocytes leads to increased germ cell proliferation. Dev Genet 16:190–200
Hiramatsu R, Matoba S, Kanai-Azuma M et al (2009) A critical time window of Sry action in gonadal sex determination in mice. Development 136:129–138
Houmard B, Small C, Yang L et al (2009) Global gene expression in the human fetal testis and ovary. Biol Reprod 81:438–443
Hu YC, Okumura LM, Page DC (2013) Gata4 is required for formation of the genital ridge in mice. PLoS Genet 9(7), e1003629
Hummitzsch K, Irving-Rodgers HF, Hatzirodos N et al (2013) A new model of development of the mammalian ovary and follicles. PLoS One 8(2):e55578
Ito S, Karnowsky MJ (1968) Formaldehyde glutaraldehyde fixatives containing trinitro compounds. J Cell Biol 36:168
Iwamoto T, Hiraku Y, Oikawa S et al (2004) DNA intrastrand cross-link at the 5'-GA-3' sequence formed by busulfan and its role in the cytotoxic effect. Cancer Sci 95:454–458
Jaenisch R, Mintz B (1974) Simian virus 40 DNA sequences in DNA of healthy adult mice derived from preimplantation blastocysts injected with viral DNA. Proc Natl Acad Sci U S A 71: 1250–1254

Jameson SA, Natarajan A, Cool J et al (2012) Temporal transcriptional profiling of somatic and germ cells reveals biased lineage priming of sexual fate in the fetal mouse gonad. PLoS Genet 8 (3):e1002575
Janošik J (1885) Histologish embryologische Untersuchungen über das Urogenitalsystem. Sitz-Ber Akad Wiss Wien Math Nat Kl 91:32–38
Janošik J (1887a) Zur Histologie des Ovariums. Sitz-Ber Akad Wiss Wien Math Nat Kl 96:45–48
Janošik J (1887b) Zwei junge menschliche Embryonen. Arch Mikr Anat 30:87–90
Janošik J (1890a) Berichtigung zu Nagels'Arbeit: Ueber die Entwickelung des Urogenitalsystems des Menschen. Ebenda 35:3–9
Janošik J (1890b) Bemerkungen über die Entwickelung des Genitalsystems. Sitz-Ber Akad Wiss Wien Math Nat Kl 99:46–54
Janošik J (1894) Ueber die Entwickelung der Vorniere und des Vornierenganges bei Säugern. Arch Mikr Anat 64:214–234
Kaboord B, Perr M (2008) Isolation of proteins and protein complexes by immunoprecipitation. Methods Mol Biol 424:349–364
Karl J, Capel B (1998) Sertoli cells of the mouse testis originate from the coelomic epithelium. Dev Biol 203:323–333
Kim Y, Kobayashi A, Sekido R (2006) Fgf9 and Wnt4 act as antagonistic signals to regulate mammalian sex determination. PLoS Biol 4:e187
Kimmel CB, Ballard WW, Kimmel SR et al (1995) Stages of embryonic development of the zebrafish. Dev Dyn 203:253–310
Kitamura K, Yanazawa M, Sugiyama N et al (2002) Mutation of ARX causes abnormal development of forebrain and testes in mice and X-linked lissencephaly with abnormal genitalia in humans. Nat Genet 32:359–369
Kloc M, Bilinski S, Chan AP, Etkin LD (2001) Mitochondrial ribosomal RNA in the germinal granules in Xenopus embryos revisited. Differentiation 67:80–83
Kocer A, Pinheiro I, Pannetier M et al (2008) R-spondin1 and FOXL2 act into two distinct cellular types during goat ovarian differentiation. BMC Dev Biol 8:36
Koopman P, Gubbay J, Vivian N et al (1991) Male development of chromosomally female mice transgenic for Sry. Nature 351:117–121
Koubova J, Menke DB, Zhou Q et al (2006) Retinoic acid regulates sex-specific timing of meiotic initiation in mice. Proc Natl Acad Sci U S A 103:2474–2479
Kunieda T, Xian M, Kobayashi E et al (1992) Sexing of mouse preimplantation embryos by detection of Y chromosome-specific sequences using polymerase chain reaction. Biol Reprod 46:692–697
Kusaka M, Katoh-Fukui Y, Ogawa H et al (2010) Abnormal epithelial cell polarity and ectopic epidermal growth factor receptor (EGFR) expression induced in Emx2 KO embryonic gonads. Endocrinology 151:5893–5904
Lambert JF, Benoit BO, Colvin GA et al (2000) Quick sex determination of mouse fetuses. J Neurosci Methods 95:127–132
Lavrovsky Y, Song CS, Chatterjee B, Roy AK (1998) A rapid and reliable PCR-based assay for gene transmission and sex determination in newborn transgenic mice. Transgenic Res 7:319–320
Lee TL, Alba D, Baxendale V et al (2006) Application of transcriptional and biological network analyses in mouse germ-cell transcriptomes. Genomics 88:18–33
Lee TL, Li Y, Alba D et al (2009) Developmental staging of male murine embryonic gonad by SAGE analysis. J Genet Genomics 36:215–227
Lee TL, Li Y, Cheung HH et al (2010) GonadSAGE: a comprehensive SAGE database for transcript discovery on male embryonic gonad development. Bioinformatics 26:585–586
Li H, Kim KH (2004) Retinoic acid inhibits rat XY gonad development by blocking mesonephric cell migration and decreasing the number of gonocytes. Biol Reprod 70:687–693
Li Y, Taketo T, Lau YF (2012) Isolation of fetal gonads from embryos of timed-pregnant mice for morphological and molecular studies. Methods Mol Biol 825:3–16

Liu CF, Bingham N, Parker K, Yao HH (2009) Sex-specific roles of beta-catenin in mouse gonadal development. Hum Mol Genet 18:405–417
Liu C, Peng J, Matzuk MM, Yao HH (2015) Lineage specification of ovarian theca cells requires multicellular interactions via oocyte and granulosa cells. Nat Commun 6:6934
López-Ibor B, Schwartz AD (1986) Gonadal failure following busulfan therapy in an adolescence girl. Am J Pediatr Hematol Oncol 8:85–87
Luo X, Ikeda Y, Parker KL (1994) A cell-specific nuclear receptor is essential for adrenal and gonadal development and sexual differentiation. Cell 77:481–490
Maack G, Segner H (2003) Morphological development of the gonads in zebrafish. J Fish Biol 62:895–906
Maatouk DM, DiNapoli L, Alvers A (2008) Stabilization of beta-catenin in XY gonads causes male-to-female sex-reversal. Hum Mol Genet 17:2949–2955
Maatouk DM, Mork L, Chassot AA et al (2013) Disruption of mitotic arrest precedes precocious differentiation and transdifferentiation of pregranulosa cells in the perinatal Wnt4 mutant ovary. Dev Biol 383:295–306
Mackay S, Booth SH, MacGowan A, Smith RA (1999) Ultrastructural studies demonstrate that epithelial polarity is established in cultured mouse pre-Sertoli cells by extracellular matrix components. J Electron Microsc 48:159–165
Maclean G, Dolle P, Petkovich M (2009) Genetic disruption of CYP26B1 severely affects development of neural crest derived head structures, but does not compromise hindbrain patterning. Dev Dyn 238:732–745
Maitland P, Ullmann SL (1993) Gonadal development in the opossum, *Monodelphis domestica*: the rete ovarii does not contribute to the steroidogenic tissues. J Anat 183:43–56
Manuylov NL, Fujiwara Y, Adameyko II et al (2007) The regulation of Sox9 gene expression by the GATA4/FOG2 transcriptional complex in dominant XX sex reversal mouse models. Dev Biol 307:356–367
Marrack J (1934) Nature of antibodies. Nature 133:292
Martineau J, Nordqvist K, Tilmann C et al (1997) Male specific cell migration into the developing gonad. Curr Biol 7:958–968
Matsuyama M, Mizusaki H, Shimono A et al (2005) A novel isoform of Vinexin, Vinexin gamma, regulates Sox9 gene expression through activation of MAPK cascade in mouse fetal gonad. Genes Cells 10:421–434
Matzuk MM, Finegold MJ, Mishina Y et al (1995) Synergistic effects of inhibins and müllerian-inhibiting substance on testicular tumorigenesis. Mol Endocrinol 9:1337–1345
McClive PJ, Sinclair AH (2001) Rapid DNA extraction and PCR-sexing of mouse embryos. Mol Reprod Dev 60:225–226
McFarlane L, Truong V, Palmer JS, Wilhelm D (2013) Novel PCR assay for determining the genetic sex of mice. Sex Dev 7:207–211
McLaren A, Buehr M (1990) Development of mouse germ cells in cultures of fetal gonads. Cell Diff Dev 31:185–195
McLaren A, Southee D (1997) Entry of embryonic germ cells into meiosis. Dev Biol 187:107–113
Merchant-Larios H, Moreno-Mendoza N (1998) Mesonephric stromal cells differentiate into Leydig cells in the mouse fetal testis. Exp Cell Res 244:230–238
Migrenne S, Moreau E, Pakarinen P et al (2012) Mouse testis development and function are differently regulated by follicle-stimulating hormone receptors signaling during fetal and prepubertal life. PLoS One 7(12):e53257
Miyamoto Y, Taniguchi H, Hamel F et al (2008) A GATA4/WT1 cooperation regulates transcription of genes required for mammalian sex determination and differentiation. BMC Mol Biol 9:44
Miyamoto N, Yoshida M, Kuratani S et al (1997) Defects of urogenital development in mice lacking Emx2. Development 124:1653–1664
Molyneaux KA, Stallock J, Schaible K, Wylie C (2001) Time-lapse analysis of living mouse germ cell migration. Dev Biol 240:488–498

Molyneaux K, Zinszner H, Kunwar P et al (2003) The chemokine SDF1/CXCL12 and its receptor CXCR4 regulate mouse germ cell migration and survival. Development 130:4279–4286
Moniot B, Declosmenil F, Barrionuevo F et al (2009) The PGD2 pathway, independently of FGF9, amplifies SOX9 activity in Sertoli cells during male sexual differentiation. Development 136:1813–1821
Morais da Silva S, Hacker A, Harley V et al (1996) Sox9 expression during gonadal development implies a conserved role for the gene in testis differentiation in mammals and birds. Nat Genet 14:62–68
Morgan TH (1912) The scientific work of Miss N. M. Stevens. Science 36:468–470
Mork L, Czerwinski M, Capel B (2014) Predetermination of sexual fate in a turtle with temperature-dependent sex determination. Dev Biol 386:264–271
Mroz K, Carrel L, Hunt PA (1999) Germ cell development in the XXY mouse: evidence that X chromosome reactivation is independent of sexual differentiation. Dev Biol 207:229–238
Munger SC, Aylor DL, Syed HA et al (2009) Elucidation of the transcription network governing mammalian sex determination by exploiting strain-specific susceptibility to sex reversal. Genes Dev 23:2521–2523
Munger SC, Natarajan A, Looger LL et al (2013) Fine time course expression analysis identifies cascades of activation and repression and maps a putative regulator of mammalian sex determination. PLoS Genet 9(7):e1003630
Nagamine CM, Chan K, Hake LE, Lau YF (1990) The two candidate testis-determining Y genes (Zfy-1 and Zfy-2) are differentially expressed in fetal and adult mouse tissues. Genes Dev 4:63–74
Nagel W (1889a) Ueber das Vorkommen von Primordialeiern auberhalb der Keimdrusenanlage beim Menschen. Anat Anz 4:12–19
Nagel W (1889b) Ueber die Entwickelung des Urogenitalsystmes des Menschen. Arch Mikr Anat 34:269–384
Needham J (1959) A history of embryology. Abelard-Schuman, New York
Nef S, Verma-Kurvari S, Merenmies J et al (2003) Testis determination requires insulin receptor family function in mice. Nature 426:291–295
Nef S, Schaab O, Stallings NR et al (2005) Gene expression during sex determination reveals a robust female genetic program at the onset of ovarian development. Dev Biol 287:361–377
Nel-Themaat L, Vadakkan TJ, Wang Y et al (2009) Morphometric analysis of testis cord formation in Sox9-EGFP mice. Dev Dyn 238:1100–1110
Nel-Themaat L, Gonzalez G, Akiyama H, Behringer RR (2010) Illuminating testis morphogenesis in the mouse. J Androl 31:5–10
Nicol B, Yao HH (2015) Gonadal Identity in the Absence of Pro-Testis Factor SOX9 and Pro-Ovary Factor Beta-Catenin in Mice. Biol Reprod 93:35
Nieuwkoop PD, Faber J (1967) Normal table of *Xenopus laevis* (Daudin): a systematical and chronological survey of the development from the fertilized egg till the end of metamorphosis, 2nd edn. North Holland, Amsterdam
Niksic M, Slight J, Sanford JR et al (2004) The Wilms' tumour protein (WT1) shuttles between nucleus and cytoplasm and is present in functional polysomes. Hum Mol Genet 13:463–471
Nishino K, Yamanouchi K, Naito K, Tojo H (2001) Characterization of mesonephric cells that migrate into the XY gonad during testis differentiation. Exp Cell Res 267:225–232
O WS, Baker TG (1978) Germinal and somatic cell interrelationships in gonadal sex differentiation. Ann Biol Anim Biochim Biophys 18:351–357
Ohe K, Lalli E, Sassone-Corsi P (2002) A direct role of SRY and SOX proteins in pre-mRNA splicing. Proc Natl Acad Sci U S A 99:1146–1151
Oshima Y, Hayashi T, Tokunaga S, Nakamura M (2005) Wnt4 expression in the differentiating gonad of the frog Rana rugosa. Zoolog Sci 22:689–693
Ottolenghi C, Omari S, Garcia-Ortiz JE et al (2005) Foxl2 is required for commitment to ovary differentiation. Hum Mol Genet 14:2053–2062

Ottolenghi C, Pelosi E, Tran J et al (2007) Loss of Wnt4 and Foxl2 leads to female-to-male sex reversal extending to germ cells. Hum Mol Genet 16:2795–2804
Oulad-Abdelghani M, Bouillet P, Décimo D et al (1996) Characterization of a premeiotic germ cell-specific cytoplasmic protein encoded by Stra8, a novel retinoic acid-responsive gene. J Cell Biol 135:469–477
Palmer SJ, Burgoyne PS (1991) The Mus musculus domesticus Tdy allele acts later than the Mus musculus musculus Tdy allele: a basis for XY sex-reversal in C57BL/6-YPOS mice. Development 113:709–714
Palmer MS, Berta P, Sinclair AH et al (1990) Comparison of human ZFY and ZFX transcripts. Proc Natl Acad Sci U S A 87:1681–1685
Pan H, O'Brien MJ, Wigglesworth K et al (2005) Transcript profiling during mouse oocyte development and the effect of gonadotropin priming and development in vitro. Dev Biol 286:493–506
Paranko J, Pelliniemi LJ (1992) Differentiation of smooth muscle cells in fetal rat testis and ovary: localization of alkaline phosphatase, smooth muscle myosin, F-actin, and desmin. Cell Tissue Res 268:521–530
Parma P, Radi O, Vidal V et al (2006) R-spondin1 is essential in sex determination, skin differentiation and malignancy. Nat Genet 38:1304–1309
Pask AJ, Harry JL, Renfree MB, Marshall Graves JA (2000) Absence of SOX3 in the developing marsupial gonad is not consistent with a conserved role in mammalian sex determination. Genesis 27:145–152
Payne DW, Packman JN, Adashi EY (1992) Follicle-stimulating hormone inhibits granulosa activity. J Biol Chem 267:13348–13355
Pellegrini M, Filipponi D, Gori M et al (2008) ATRA and KL promote differentiation toward the meiotic program of male germ cells. Cell Cycle 7:3878–3888
Pesce M, De Felici M (1995) Purification of mouse primordial germ cells by MiniMACS magnetic separation system. Dev Biol 170:722–725
Piprek RP (2013) Gonadogenesis in anuran amphibians: cellular and molecular mechanisms of sexual differentiation of gonads. Dissertation, Jagiellonian University, Krakow
Piprek RP, Pecio A, Kubiak JZ, Szymura JM (2012) Differential effects of busulfan on gonadal development in five divergent anuran species. Reprod Toxicol 34:393–401
Piprek RP, Pecio A, Laskowska-Kaszub K et al (2013) Retinoic acid homeostasis regulates meiotic entry in developing anuran gonads and in Bidder's organ through Raldh2 and Cyp26b1 proteins. Mech Dev 130:613–627
Piprek RP, Pecio A, Kloc M et al (2014) Evolutionary trend for metamery reduction and gonad shortening in Anurans revealed by comparison of gonad development. Int J Dev Biol 58:929–934
Qing T, Liu H, Wei W et al (2008) Mature oocytes derived from purified mouse fetal germ cells. Hum Reprod 23:54–61
Ramsköld D, Wang ET, Burge CB, Sandberg R (2009) An Abundance of ubiquitously expressed genes revealed by tissue transcriptome sequence data. PLoS Comput Biol 5(12):e1000590
Rastetter RH, Bernard P, Palmer JS et al (2014) Marker genes identify three somatic cell types in the fetal mouse ovary. Dev Biol 394:242–252
Rolland AD, Lehmann KP, Johnson KJ et al (2011) Uncovering gene regulatory networks during mouse fetal germ cell development. Biol Reprod 84:790–800
Ross DG, Bowles J, Hope M et al (2009) Profiles of gonadal gene expression in the developing bovine embryo. Sex Dev 3:273–283
Russell ES (1979) Hereditary anemias of the mouse: a review for geneticists. Adv Genet 20: 357–459
Sabour D, Xu X, Chung ACK et al (2014) Germ cell nuclear factor regulates gametogenesis in developing gonads. PLoS One 9(8):e103985
Salas-Cortés L, Jaubert F, Barbaux S et al (1999) The human SRY protein is present in fetal and adult Sertoli cells and germ cells. Int J Dev Biol 43:135–140

Schmahl J, Capel B (2003) Cell proliferation is necessary for the determination of male fate in the gonad. Dev Biol 258:264–276
Schmahl J, Eicher EM, Washburn LL, Capel B (2000) *Sry* induces cell proliferation in the mouse gonad. Development 127:65–73
Schmahl J, Kim Y, Colvin JS et al (2004) Fgf9 induces proliferation and nuclear localization of FGFR2 in Sertoli precursors during male sex determination. Development 131:3627–3636
Sekido R, Bar I, Narvaez V et al (2004) SOX9 is up-regulated by the transient expression of SRY specifically in Sertoli cell precursors. Dev Biol 274:271–279
Seligman J, Page DC (1998) The Dazh gene is expressed in male and female embryonic gonads before germ cell sex differentiation. Biochem Biophys Res Commun 245:878–882
Sharpe RM, McKinnell C, Kivlin C, Fisher JS (2003) Proliferation and functional maturation of Sertoli cells, and their relevance to disorders of testis function in adulthood. Reproduction 125: 769–784
Sinclair AH, Berta P, Palmer MS et al (1990) A gene from the human sex-determining region encodes a protein with homology to a conserved DNA-binding motif. Nature 346:240–244
Small CL, Shima JE, Uzumcu M et al (2005) Profiling gene expression during the differentiation and development of the murine embryonic gonad. Biol Reprod 72:492–501
Smith CA, McClive PJ, Hudson Q, Sinclair AH (2005) Male-specific cell migration into the developing gonad is a conserved process involving PDGF signalling. Dev Biol 284:337–350
Smith CA, Roeszler KN, Bowles J et al (2008a) Onset of meiosis in the chicken embryo; evidence of a role for retinoic acid. BMC Dev Biol 8:85
Smith CA, Shoemaker CM, Roeszler KN et al (2008b) Cloning and expression of R-Spondin1 in different vertebrates suggests a conserved role in ovarian development. BMC Dev Biol 8:72
Smith CA, Roeszler KN, Ohnesorg T et al (2009) The avian Z-linked gene DMRT1 is required for male sex determination in the chicken. Nature 461:267–271
Soyal SM, Amleh A, Dean J (2000) FIGalpha, a germ cell-specific transcription factor required for ovarian follicle formation. Development 127:4645–4653
Stebler J, Spieler D, Slanchev K et al (2004) Primordial germ cell migration in the chick and mouse embryo: the role of the chemokine SDF-1/CXCL12. Dev Biol 272:351–361
Tang H, Brennan J, Karl J et al (2008) Notch signaling maintains Leydig progenitor cells in the mouse testis. Development 135:3745–3753
Tanimura A, Iwasawa H (1991) Proliferative activity of somatic cells during gonadal development in the Japanese pond frog, *Rana nigromaculata*. J Exp Zool 259:365–370
Tanoue A, Nakamura T, Endo F et al (1992) Sex-determining region Y (SRY) in a patient with 46, XX true hermaphroditism. Jpn J Hum Genet 37:311–320
Teerds KJ, de Boer-Brouwer M, Dorrington JH et al (1999) Identification of markers for precursor and leydig cell differentiation in the adult rat testis following ethane dimethyl sulphonate administration. Biol Reprod 60:1437–1445
Tevosian SG, Albrecht KH, Crispino JD et al (2002) Gonadal differentiation, sex determination and normal Sry expression in mice require direct interaction between transcription partners GATA4 and FOG2. Development 129:4627–4634
Theiler K (1989) The house mouse. Development and normal stages from fertilization to 4 weeks of age, 2nd edn. Springer, Berlin
Tremblay JJ, Viger RS (2003) Transcription factor GATA-4 is activated by phosphorylation of serine 261 via the cAMP/PKA pathway in gonadal cells. J Biol Chem 278:22128–22135
Trevino V, Falciani F, Barrera-Saldana HA (2007) DNA microarrays: a powerful genomic tool for biomedical and clinical research. Mol Med 13:527–541
Tsuji FI (2010) Early history, discovery, and expression of *Aequorea* green fluorescent protein, with a note on an unfinished experiment. Microsc Res Tech 73:785–796
Uhlenhaut NH, Jakob S, Anlag K et al (2009) Somatic sex reprogramming of adult ovaries to testes by FOXL2 ablation. Cell 139:1130–1142

Vied C, Halachmi N, Salzberg A, Horabin JI (2003) Antizyme is a target of sex-lethal in the Drosophila germline and appears to act downstream of hedgehog to regulate sex-lethal and cyclin B. Dev Biol 253:214–229
Voller A, Bartlett A, Bidwell DE (1978) Enzyme immunoassays with special reference to ELISA techniques. J Clin Pathol 31:507–520
Warr N, Siggers P, Bogani D et al (2009) Sfrp1 and Sfrp2 are required for normal male sexual development in mice. Dev Biol 326:273–284
Waterston RH, Lindblad-Toh K, Birney E et al (2002) Initial sequencing and comparative analysis of the mouse genome. Nature 420:520–562
Wilhelm D, Martinson F, Bradford S et al (2005) Sertoli cell differentiation is induced both cell-autonomously and through prostaglandin signaling during mammalian sex determination. Dev Biol 287:111–124
Wilhelm D, Hiramatsu R, Mizusaki H et al (2007) SOX9 regulates prostaglandin D synthase gene transcription in vivo to ensure testis development. J Biol Chem 282:10553–10560
Wilhelm D, Washburn LL, Truong V et al (2009) Antagonism of the testis- and ovary-determining pathways during ovotestis development in mice. Mech Dev 126:324–336
Willerton L, Smith RA, Russell D, Mackay S (2004) Effects of FGF9 on embryonic Sertoli cell proliferation and testicular cord formation in the mouse. Int J Dev Biol 48:637–643
Williams TJ (1986) A technique for sexing mouse embryos by a visual colorimetric assay of the X-linked enzyme, glucose 6-phosphate dehydrogenase. Theriogenology 25:733–739
Wilson EB (1905) Studies on chromosomes. I. The behavior of the idiochromosomes in hemiptera. J Exp Zool 2:371–405
Wolf U (1998) The serologically detected H-Y antigen revisited. Cytogenet Cell Genet 80: 232–235
Wu SM, Baxendale V, Chen Y et al (2004) Analysis of mouse germ-cell transcriptome at different stages of spermatogenesis by SAGE: biological significance. Genomics 84:971–981
Wylie CC, Bancroft M, Heasman J (1976) The formation of the gonadal ridge in Xenopus laevis. II. A scanning electron microscope study. J Embryol Exp Morphol 35:139–148
Yamamoto M, Wakatsuki T, Hada A, Ryo A (2001) Use of serial analysis of gene expression (SAGE) technology. J Immunol Methods 250:45–66
Yao HH, DiNapoli L, Capel B (2004) Cellular mechanisms of sex determination in the red-eared slider turtle, Trachemys scripta. Mech Dev 121:1393–13401
Yao HH, Whoriskey W, Capel B (2002) Desert Hedgehog/Patched 1 signaling specifies fetal Leydig cell fate in testis organogenesis. Genes Dev 16:1433–1440
Yoshimoto S, Okada E, Umemoto H et al (2008) A W-linked DM-domain gene, DM-W, participates in primary ovary development in *Xenopus laevis*. Proc Natl Acad Sci U S A 105:2469–2474
Yoshimoto S, Ikeda N, Izutsu Y et al (2010) Opposite roles of DMRT1 and its W-linked paralogue, DM-W, in sexual dimorphism of *Xenopus laevis*: implications of a ZZ/ZW-type sex-determining system. Development 137:2519–2526

Index

R.P. Piprek (ed.), *Molecular Mechanisms of Cell Differentiation in Gonad Development*, Results and Problems in Cell Differentiation 58,
DOI 10.1007/978-3-319-31973-5

E

F

G

MIX
Papier aus verantwortungsvollen Quellen
Paper from responsible sources
FSC® C105338

If you have any concerns about our products,
you can contact us on
ProductSafety@springernature.com

In case Publisher is established outside the EU,
the EU authorized representative is:
Springer Nature Customer Service Center GmbH
Europaplatz 3, 69115 Heidelberg, Germany

Printed by Libri Plureos GmbH
in Hamburg, Germany